Jets and Winds in Pulsar Wind Nebulae, Gamma-Ray Bursts and Blazars

From left to right:

G. Pavlov
A. Bykov
R. Willingale
M. Falanga
G. Romero
M. Hoshino
M. Lemoine
S. Reynolds
G. Pelletier
S. Mereghetti
E. Waxman

M. Pohl
F. Tavecchio
B. Cerutti
S. Komissarov
F. Daigne
R. Buehler
E. Amato
B. Olmi (front)
A. MacFadyen (back)
A. Tchekhovskoy (sitting)
S. Markoff

M. Tavani
K. Asano
J. Arons
A. Lamberts
A. Rudy
O. Porth
M. Boettcher
Y. Yuan (front)
R. Blandford (back)
L. Sironi
T. Piran

Andrei Bykov • Elena Amato • Jonathan Arons •
Maurizio Falanga • Martin Lemoine • Luigi Stella •
Rudolf von Steiger
Editors

Jets and Winds in Pulsar Wind Nebulae, Gamma-Ray Bursts and Blazars

Chapters 2–10 were previously published in *Space Science Reviews* Volume 207, Issues 1–4, 2017

Springer

Editors
Andrei Bykov
High Energy Astrophysics
Ioffe Institute
St. Petersburg, Russia

Elena Amato
Osservatorio Astrofisico di Arcetri
INAF
Firenze, Italy

Jonathan Arons
Department of Astronomy
University of California, Berkeley
Berkeley, CA, USA

Maurizio Falanga
International Space Science Institute
Bern, Switzerland

Martin Lemoine
Institut d'Astrophysique de Paris
CNRS–UPMC
Paris, France

Luigi Stella
Osservatorio Astronomico
di Monteporzio Catone
INAF
Rome, Italy

Rudolf von Steiger
International Space Science Institute
Bern, Switzerland

ISSN 1385-7525 Space Sciences Series of ISSI
ISBN 978-94-024-1291-8

Library of Congress Control Number: 2018937681

Cover Image: Simulated ultraviolet (8 eV) image of a pulsar wind nebula with a bow shock similar to that observed around PSR J0437-4715 with the Hubble Space Telescope, overlaid upon the DSSII optical (2.6 eV) image of the region. For more details see the chapter "Pulsar wind nebulae with bow shocks: non-thermal radiation and cosmic ray leptons" in this book.
Credit: A.M. Bykov, A.E. Petrov, A.M. Krassilchtchikov (Ioffe Insitute)

Printed on acid-free paper

This Springer imprint is published by the registered company Springer Science+Business Media B.V. part of Springer Nature.
The registered company address is: Van Godewijckstraat 30, 3311 GX Dordrecht, The Netherlands

Contents

Editorial: Topical Collection on Jets and Winds in Pulsar Wind Nebulae, Gamma-Ray Bursts and Blazars

Physics of Extreme Energy Release

A. Bykov[1,2,3] · E. Amato[4,5] · J. Arons[6] · M. Falanga[3] · M. Lemoine[7] · L. Stella[8] · R. von Steiger[3]

1 Jets and Winds in Pulsar Wind Nebulae, Gamma-Ray Bursts and Blazars

The ever increasing sensitivity and improving accuracy of multi-wavelength observations have offered in recent decades an unprecedented insight into the physics of relativistic astrophysical sources, most notably, gamma-ray bursts, active galactic nuclei, microquasars, and pulsar wind nebulae. The superb angular resolution images of pulsar wind nebulae and extragalactic jets provided by the modern radio interferometers, the Hubble Space Telescope and Chandra X-ray observatory, as well as time resolved spectra obtained with modern gamma-ray telescopes are beautiful examples of these achievements. They allow us to reveal the intimate details of synchrotron nebulae and jets, which result from dissipation of highly relativistic magnetized winds originating in the central engines of such objects.

Relativistic objects in astrophysics present a diverse set of phenomena, cover vast ranges of time scales and spatial scales, but their descriptions rely on a similar backbone: produc-

✉ A. Bykov
byk@astro.ioffe.ru

1 Ioffe Institute, 194021 St. Petersburg, Russia

2 St. Petersburg Polytechnical University, St. Petersburg, Russia

3 International Space Science Institute, Bern, Switzerland

4 INAF Osservatorio Astrofisico di Arcetri, Largo E. Fermi 5, 50125 Firenze, Italy

5 Dipartimento di Fisica ed Astronomia, Universit' a degli Studi di Firenze, Via G. Sansone 1, 50019 Sesto Fiorentino, Italy

6 Astronomy Department and Theoretical Astrophysics Center, University of California, Berkeley, 501 Campbell Hall, Berkeley, CA 94720, USA

7 Institut d'Astrophysique de Paris, CNRS-UPMC, 98 bis boulevard Arago, 75014 Paris, France

8 INAF-Osservatorio Astronomico di Roma, via Frascati 33, 00078 Monte Porzio Catone, Roma, Italy

tion of a relativistic outflow in a central compact object and dissipation of the outflow at large radii accompanied by acceleration of particles up to very high energies and production of secondary non-thermal electromagnetic radiation. Studies in this field thus tackle a number of fundamental physical problems, such as: how the gravitational, rotational and electro-magnetic energies of the central engine are converted into the observed radiation, how magneto-hydrodynamic flows are launched and pushed to relativistic velocities, how the energy of a bulk flow is transferred to a minority of ultra-relativistic particles, and so on. Such astrophysical objects also harbour matter and radiation at extreme physical conditions in terms of electro-magnetic field strengths, matter densities, flow velocities, or particle energies, which lie well beyond the reach of any terrestrial laboratory, thus providing us with a unique opportunity to test the fundamental laws of physics.

This volume is opened with the review by G. Romero et al., who present a broad discussion of the physics of relativistic jets on large spatial scales, found in active galactic nuclei and in microquasars.

Studies of gamma-ray bursts (GRBs) have become a hot and rapidly advancing field of today's astrophysics as soon as their cosmological origin was established about twenty years ago. Multi-wavelength observations of GRBs and some other related transients, including ultra-long bursts and tidal disruption events, are discussed in detail by R. Willingale and P. Meszaros in a dedicated review. GRB production by relativistic jets that start in a highly opaque zone near the central source is discussed by A.M. Beloborodov and P. Meszaros, who provide a theoretical description of photospheric emission models of the prompt phase of GRBs. They discuss dissipative and radiative processes in the jet and the radiation transfer mechanisms, which shape the observed nonthermal emission spectrum released from the GRB photosphere.

Pulsars emit relativistic outflows in a variety of forms. Their magnetized outflows consist of some combination of highly relativistic $e^{\pm}$ pairs and, perhaps, ions. Nearby pulsar wind nebulae produced by the Crab, Vela and Geminga pulsars allowed us to perform detailed imaging observations of jets, torii and different types of fine emitting structures of synchrotron origin, being excellent laboratories for studies of the complex structure formation and particle acceleration in relativistic outflows. The physics of pulsar wind nebulae is discussed in a number of comprehensive reviews. The paper by B. Cerutti and A.M. Beloborodov presents a detailed review of the basic physical processes taking place in pulsar magnetospheres, at the base of the wind outflow, and of current efforts in the kinetic numerical simulation of these complex phenomena. It is followed by the review of O. Porth et al., who provide a detailed discussion of the morphology of pulsar wind nebulae and its numerical simulation in relativistic magneto-hydrodynamics. This discussion is nicely complemented by the review of S.P. Reynolds et al. on the observational properties of pulsar wind nebulae and magnetars in various wavelength domains. Then, the review paper by A.M. Bykov, E. Amato et al. offers an exhaustive discussion of the physics governing the so-called bow shock pulsar wind nebulae. Numerical modeling, radiative signatures and a possible origin of the fascinating X-ray structures observed in the Vela, Geminga, Guitar and the Lighthouse nebulae are discussed within the bow shock scenario. Pulsars and their wind nebulae are producing electron-positron outflows and therefore are considered as possible contributors to the enhanced cosmic ray positron fraction observed in PAMELA and AMS-02 space experiments—the issue also addressed in this review.

Fast energy release in the central engines producing relativistic outflows of different magnetization and particle acceleration capability is the key physical process at the core of the modeling of non-thermal spectra of astrophysical sources. Two reviews are therefore dedicated to the physics of fast energy release in relativistic astrophysical objects. The paper by

R. Blandford et al. discusses magnetoluminescence—an important concept of a rapid energy release in catastrophic reconnection events, a scenario which is strongly motivated by the recent discovery of rapid variability and flares of GeV range radiation in the Crab nebula and in some blazars. Relativistic collisionless shock waves, which form at the interface between relativistic outflows and the external medium, have long been thought to give rise to an efficient Fermi process of particle acceleration. The underlying physical principles together with current efforts at their numerical modeling employing state of the art particle-in-cell and Monte Carlo techniques, are discussed in some detail by G. Pelletier et al.

Overall, this volume provides a timely and comprehensive review of the main physical processes involved in formation and dissipation of relativistic astrophysical outflows and the multi-wavelength observational appearances of this objects. It will be particularly useful to postgraduate students and researchers active in various areas of astrophysics and space science.

This volume is based on an international workshop which took place at ISSI Bern from November 16 to November 20, 2015, gathering about forty leading experts in the physics of high-energy astrophysical objects. The Workshop was the first in the series of the three subsequent ISSI workshops devoted to physics and astrophysics of the astrophysical objects with extreme energy release. The next two ISSI workshops of the series will be devoted to supernovae and clusters of galaxies. The Editors would like to thank the participants of the workshop for their participation, and for their contributions to this volume. The scope, the scientific level and the pedagogical aspect of the present reviews testify of the high scientific level of the workshop and of its collaborative spirit. Special thanks are due to the reviewers, who helped improve the balance and the quality of the articles through their constructive reports. Obviously, this workshop could not have taken place without the strong support of the ISSI Bern directorate, science committee and staff. The ISSI staff have provided the warm atmosphere, generous hospitality and assistance. We would like to thank particularly Silvia Wenger, Jennifer Fankhauser, Andrea Fischer, Alexandra Lehmann and Saliba Saliba. Finally we thank the production staff of Space Science Reviews, as well as the production staff of Springer for their competence.

Space Sci Rev (2017) 207:5–61
DOI 10.1007/s11214-016-0328-2

Relativistic Jets in Active Galactic Nuclei and Microquasars

Gustavo E. Romero[1] · M. Boettcher[2] · S. Markoff[3] · F. Tavecchio[4]

Received: 3 August 2016 / Accepted: 22 December 2016 / Published online: 3 January 2017

Abstract Collimated outflows (jets) appear to be a ubiquitous phenomenon associated with the accretion of material onto a compact object. Despite this ubiquity, many fundamental physics aspects of jets are still poorly understood and constrained. These include the mechanism of launching and accelerating jets, the connection between these processes and the nature of the accretion flow, and the role of magnetic fields; the physics responsible for the collimation of jets over tens of thousands to even millions of gravitational radii of the central accreting object; the matter content of jets; the location of the region(s) accelerating particles to TeV (possibly even PeV and EeV) energies (as evidenced by γ-ray emission observed from many jet sources) and the physical processes responsible for this particle acceleration; the radiative processes giving rise to the observed multi-wavelength emission; and the topology of magnetic fields and their role in the jet collimation and particle acceleration processes. This chapter reviews the main knowns and unknowns in our current understanding of relativistic jets, in the context of the main model ingredients for Galactic and extragalactic jet sources. It discusses aspects specific to active Galactic nuclei (especially blazars) and microquasars, and then presents a comparative discussion of similarities and differences between them.

✉ M. Boettcher
Markus.Bottcher@nwu.ac.za

G.E. Romero
romero@iar-conicet.gov.ar

S. Markoff
s.b.markoff@uva.nl

F. Tavecchio
fabrizio.tavecchio@brera.inaf.it

1 Instituto Argentino de Radioastronomía (IAR), C.C. No. 5, 1894, Buenos Aires, Argentina

2 Centre for Space Research, North-West University, Private Bag X6001, Potchefstroom, 2520, South Africa

3 Anton Pannekoek Institute for Astronomy, University of Amsterdam, 1098XH Amsterdam, The Netherlands

4 INAF-Osservatorio Astronomico di Brera, Via Bianchi 47, 23807 Merate, Italy

Keywords Jets, outflows and bipolar flows · Jets and bursts · Galactic winds and fountains · Active and peculiar galaxies and related systems · X-ray binaries · Radiation mechanisms: non-thermal

1 Introduction

Relativistic jets are collimated outflows of plasma and fields produced by accreting compact objects. The bulk motion of the plasma is close to the speed of light, so relativistic effects play an important role in the physical processes in these jets. The jet phenomenon appears to be common wherever mass accretion onto a central object occurs, and they become relativistic in the case of accretion onto compact objects, such as neutron stars or black holes, both in Galactic systems (neutron star and black-hole X-ray/γ-ray binaries, hereafter termed generically XRBs—termed microquasars when they exhibit prominent radio jets) and in extragalactic systems (active galactic nuclei [AGN], gamma-ray bursts [GRBs] and recently Tidal Disruption Events [TDEs]). Relativistic jets are ionized and produce non-thermal radiation from radio to gamma-rays in some cases. Observationally, in most cases, the presence of non-thermal radio emission has been the primary signature of relativistic jets, distinguishing jet-dominated radio-loud AGN from their radio-quiet counterparts (with very weak or absent jets), and microquasars (XRB states with prominent radio jets) from radio-quiet XRB states. In this chapter, we will focus on the relativistic jets from AGN and microquasars, while the physics of GRB and TDE jets is discussed extensively in Chaps. 6–8, and jets from pulsars in Chap. 2.

The existence of jet-like structures was known as a peculiarity of some galaxies such as M87 since the early twentieth century (Curtis 1918). By the mid 1950s it was recognized that the extended radio emission in M87 was the result of synchrotron radiation (Burbidge 1956). The synchrotron hypothesis was soon applied to explain the extended radio lobes observed in some galaxies (Ginzburg 1961; Shklovskii 1961). After the discovery of quasars (Schmidt 1963) at large cosmological redshifts, it became clear that some of the radio sources were extremely distant and the size and total energy output of the radio emitting clouds was huge. The cooling time of the electrons responsible for the radiation is much shorter than the dynamical timescales, thus indicating a need for in-situ acceleration of relativistic particles along the jets and in the radio lobes. Fully non-thermal radio jets were soon detected in many radio galaxies using interferometric techniques. Structural time variations were also detected in these jets, and superluminal motions, originally predicted by Rees (1966) were found, first in the jet of the quasar 3C273, and later in many other sources (see Zensus and Pearson 1987, and references therein).

In our own galaxy, relativistic jets were associated with the emission line star SS433 (e.g., Abell and Margon 1979). The jets of this source, a compact object accreting from a companion star, precess with a 164-day period. Only much later, in the 1990s, other X-ray binary sources mimicking the behavior of extragalactic quasars were found (Mirabel et al. 1992) and superluminal motions observed within the Galaxy (Mirabel and Rodríguez 1994). XRBs with relativistic radio jets were dubbed "microquasars". About 15 such sources are currently classified as microquasars because of explicit measurements of relativistic proper motion, but $\sim$50 XRBs have by now been observed in the jet-dominated 'hard state' (Tetarenko et al. 2016, and see Sect. 4), and many more are suspected to exist in the Milky Way.

With the advent of orbiting X-ray observatories such as *Einstein*, *EXOSAT*, *ROSAT*, and more recently *RXTE*, *XMM-Newton*, *Chandra*, *Swift*, and *NuSTAR*, and the opening-up of the γ-ray sky especially with the Compton Gamma-Ray Observatory in the 1990s and more recently with *AGILE* and the *Fermi* Gamma-Ray Space Telescope, it became soon clear that both Galactic and extragalactic jets are high-energy sources. Their spectral energy distribution (SED) is strongly dominated by non-thermal emission from radio up to γ-ray frequencies, and in many cases it is dominated by the γ-ray component that extends up to GeV and even TeV energies (see, e.g., Maraschi and Tavecchio 2001; Abdo et al. 2010). The emission from these objects tends to be highly variable at all wavelengths, with often complex—sometimes correlated, but sometimes also uncorrelated among different wavelength regimes—flux and spectral variability patterns, that may change character even within the same source among different episodes of activity. The study of these complex radiative features of relativistic jets is the only way to gain insight into the physical conditions of the outflowing plasma and the mechanisms that result in the jet launching and subsequent particle acceleration. However, even after over three decades of intensive multi-wavelength observing campaigns (including observations from the radio to γ-rays) and remarkable progress in general relativistic magnetohydrodynamic (GRMHD) and particle-in-cell (PIC) simulations of plasma phenomena relevant to jet launching, the dynamics, and particle acceleration, many fundamental aspects of the jet phenomenon are still very poorly understood or even constrained. This chapter attempts to summarize the cutting edge, as well as elucidate potential avenues for future research, in order to gain insight into the answers to these open questions:

- What is the primary physical mechanism powering the acceleration and collimation of the jet? Is the primary energy source the rotational energy of the black hole (Blandford and Znajek 1977), or is power released in the accretion flow the primary driver (Blandford and Payne 1982)? In the latter case, one would expect a tight connection between the observational signatures of the accretion flow and the presence/absence of relativistic jets, as is, indeed, observed in many Galactic systems, but also evidence for a correlation between black-hole spin and jet power has been claimed (e.g., McClintock et al. 2014, see Sect. 4 for a discussion). In general, what is the role of magnetic fields in the acceleration and collimation of relativistic jets?
- To what extent can the accretion and jet processes involving supermassive black holes in AGN be considered as simply scaled-up versions of the microquasar phenomenon? While the source of the accreted material and thus the accretion-flow geometry is quite different, many aspects, such as mass-scaled short-term variability patterns and the characteristic two-component, non-thermal spectral energy distributions, appear to indicate a large degree of analogy between the two phenomena on different scales (e.g., Markoff et al. 2015).
- What is the strength and topology of magnetic fields along relativistic jets, and what role do they play in the collimation of jets out to tens of thousands of gravitational radii, as well as the acceleration of relativistic particles in-situ along the jets? What determines where particle acceleration begins, a region often modeled as a single zone for e.g., blazars (see Sect. 3), but also becoming a focus now for XRBs (see Sect. 4).
- What is the matter content of material flowing along the jets? Is it primarily an electron-positron pair plasma, or an ordinary electron-proton plasma (see, e.g., Sikora and Madejski 2000)? If protons are present in significant numbers, are they also subject to relativistic particle acceleration?
- What are the dominant radiation mechanisms leading to the observed two-component non-thermal emission seen from relativistic jet sources? Are these primarily leptonic processes, with protons (if present) not contributing to the radiative output, or are protons also accelerated to ultrarelativistic energies so that hadronic processes can make a significant contribution (see, e.g., Böttcher et al. 2013, for a comparative analysis of both

types of models in the case of blazars)? If the latter is the case, then AGN may be the elusive sources of ultra-high-energy cosmic rays (UHECRs, i.e., cosmic-rays with energies $E > 10^{19}$ eV). Also, the acceleration of UHECRs in AGN environments is likely to result in charged-pion production, leading to the emission of PeV neutrinos, as recently confidently detected by the IceCube neutrino observatory (Aartsen et al. 2014).

In this chapter, we will discuss the physics and phenomenology that is common to relativistic jets on different size scales, from astronomical units to kiloparsecs. We will focus on the jets of microquasars and blazars, with emphasis on their radiative output. We will first introduce the basic, common ingredients required for any model of relativistic jets of different sizes and velocities (Sect. 2). Against this background, we will offer an overview of the current knowledge about particle acceleration within the jets of blazars (Sect. 3) and microquasars (Sect. 4). The final section (5) provides a comparative discussion of the prevalent issues in blazar and microquasar jet physics, and an outlook towards future experiments that may shed light on these issues. The reader interested in extensive treatments of this subject may consult the recent monographs and collections by Beskin (2009), Belloni (2010), Böttcher (2010), Markoff (2010) and Romero and Vila (2014).

2 Basic Model Ingredients: Knowns and (Educated) Guesses

Wherever accretion of matter with angular momentum and magnetic fields into a gravitational well occurs, an accretion disk is formed, often accompanied by the expulsion of collimated outflows. When the gravitating object is compact (a weakly magnetized neutron star or a black hole), it is capable of launching relativistic outflows, although the exact mechanism is not yet clear. Models of relativistic jets were first developed for Active Galactic Nuclei at the end of the 1970s and early 1980s (e.g., Blandford and Königl 1979; Marscher 1980a,b; Marscher and Gear 1985) and later adapted to microquasars (by then not known with this name) by Band and Grindlay (1986), Hjellming and Johnston (1988), Levinson and Blandford (1996).

In these early models for the broadband emission from jets, it is assumed that only leptons (electrons and positrons) are accelerated, and thus contributing significantly to the radiation, with protons (if present) not achieving sufficiently high energy. Such models are termed *leptonic models*, and they have achieved an increasing degree of sophistication over the past several decades, both for Galactic (e.g., Markoff et al. 2001a, 2005; Georganopoulos et al. 2002; Kaufman Bernadó et al. 2002; Bosch-Ramon et al. 2006; Zdziarski et al. 2014) and extragalactic jets (e.g., Maraschi et al. 1992; Dermer and Schlickeiser 1993; Blandford and Levinson 1995; Levinson and Blandford 1995; Marcowith et al. 1995; Romero et al. 1995; Böttcher et al. 1997; Böttcher and Chiang 2002; Markoff et al. 2015).

Alternatively, if protons are accelerated to ultrarelativistic energies they may contribute significantly to the radiative output, and the corresponding class of models is termed *hadronic models* or *lepto-hadronic models*, to indicate that even in those models, relativistic leptons make a significant contribution to the SED, especially at low frequencies. For lepto-hadronic jet models for Galactic sources, see, e.g., Romero et al. (2003, 2005), Romero and Vila (2008), Vila and Romero (2010), Vila et al. (2012), Yan and Zhang (2015), Cerruti et al. (2015), Pepe et al. (2015) and for extragalactic sources, Mannheim (1993), Rachen and Biermann (1993), Mücke et al. (2003), Reynoso et al. (2011, 2012), Böttcher et al. (2013).

As indicated in Sect. 1, many aspects of the jet geometry, magnetization, mode of particle acceleration and the location of the dominant energy dissipation and particle acceleration zone are currently still unknown, leaving a significant amount of freedom in the choice of model parameters. The various models of relativistic jets differ in the characteristics of the injection of relativistic particles, the properties of the region in the jet where the radiation is produced, the particles and physical interactions involved in the production of the radiation, the external conditions, the treatment of the transport of particles in the radiative zone, the inclusion of dynamical effects, the boundary and external medium conditions, the radiation reprocessing in the source, and many other factors.

The emission of high-energy and very-high-energy γ-rays requires the acceleration of particles to GeV and even TeV energies, which is likely to occur in relatively small, localized regions along the jet. Due to their short cooling time scale, those particles are expected to emit high-energy radiation locally. The concept of a single, dominant, small emission region is further strengthened by the observed rapid variability which, due to the causality constraint, restricts the size of the emission region, in many cases, to the order of the gravitational radius of the central accreting object (e.g., Begelman et al. 2008). This gives credence to single-zone models, where the emitting region is considered to be homogeneous (usually assumed spherical but see, e.g. Giannios et al. 2009) and the physical conditions uniform. Such models have been popular because of their relative simplicity. However, the often observed lack of strict correlations between variability patterns in different wavelength regimes demonstrates that they may be applicable primarily only to short-term, isolated flaring events in which the emission is likely to be dominated by a single, small emission region. Nevertheless, single-zone models may be characterized by a limited number of parameters, many of which can be reasonably well constrained from observables (for details on such observational parameter constraints, see Sects. 3 and 4). Hence, they have resulted in significant insight into the physical conditions and physical processes in the dominant high-energy emission region.

The necessity of inhomogeneous (multi-zone) models has already been pointed out, e.g., by Ghisellini et al. (1985), but their implementation in a self-consistent way has been hampered by the large number of (poorly constrained) parameters required for such models. Only recently inhomogeneous models with an adequate treatment of particle transport and all radiative processes have been developed. However these models have still not been made self-consistent with the magnetohydrodynamics (MHD) governing the plasma properties and flow characteristics, and which may determine how structures that accelerate particles are created. Ultimately the number of free parameters will be reduced once the macro- and microphysics can be linked, but this type of work is still in its early stages.

In what follows we will discuss the most important physical processes that enter relativistic jet models (both leptonic and lepto-hadronic) for both Galactic and extragalactic sources in a generic model geometry as illustrated in Fig. 1. The discussion presented here rests upon the work by Markoff et al. (2005), Vila et al. (2012), Pepe et al. (2015), Diltz and Böttcher (2014), Reynoso et al. (2012), Polko et al. (2014), Diltz et al. (2015), Zhang et al. (2014, 2015).

2.1 Energetics and Geometry

A compact object (hereafter assumed to be a black hole) of mass $M_{\rm BH}$ accretes matter from its environment (in the case of a microquasars the accretion flow comes from a companion star). We write the accretion power $L_{\rm accr}$ in terms of the Eddington luminosity of the black hole as

$$L_{\rm accr} \equiv \dot{M}c^2 = q_{\rm accr}\, L_{\rm Edd} \approx 1.3 \times 10^{46} q_{\rm accr} \left(\frac{M_{\rm BH}}{10^8\, M_\odot}\right) \,{\rm erg\,s^{-1}}, \tag{1}$$

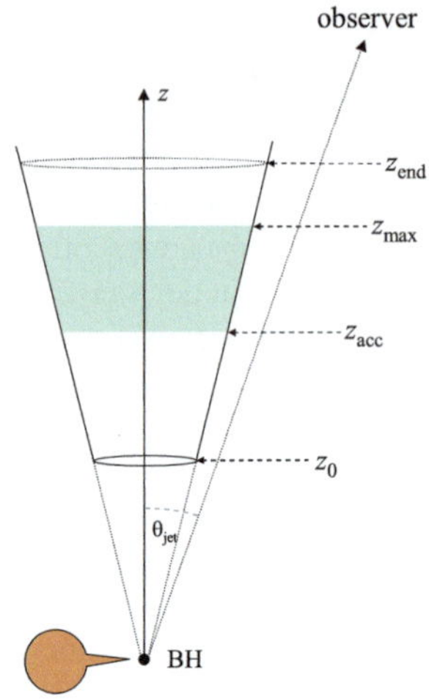

Fig. 1 Detail of a generic conical jet model, including the acceleration region (not to scale). Some relevant parameters are indicated

where $\dot{M}$ is the mass accretion rate, $M_\odot$ is the solar mass, and q_{accr} is a dimensionless parameter. The accretion disk extends from an inner radius R_{in} to an outer radius R_{out}. For particularly lower luminosity (sub-Eddington) sources, within R_{in} the plasma is thought to become radiatively inefficient, "inflating" to form a hot, optically thin corona. The nature and geometry of this corona is a matter of significant debate (see, e.g. Nowak et al. 2011), however the existence of a population of hot electrons is necessary in any scenario.

A fraction of the accreted matter is ejected into two symmetrical jets, each carrying a power

$$L_{jet} = \frac{1}{2} q_{jet} L_{accr}, \tag{2}$$

where $q_{jet} < 1$. Note however that this assumption implicitly assumes the Blandford and Payne (1982) mechanism, thus if enhancement from the black hole spin is relevant (Blandford and Znajek 1977) this linear relationship with the accretion disk luminosity may no longer hold.

The jets are launched perpendicularly to the plane of the inner accretion disk at a distance z_0 from the black hole. The scale z_0 is the characteristic scale of the coronal plasma that is thought to surround the black hole (Romero et al. 2010). The jet might be the result of the evacuation of the corona by magnetocentrifugal effects, a hypothesis supported by the fact that jets in XRBs are observed only in the hard state, when the accretion disk is truncated at a distance of the order of z_0. A spherical geometry for the corona seems to be the more appropriate on the basis of the intensity of the Compton reflection feature observed in some sources (e.g. Poutanen et al. 1997). It has been also suggested that the corona is essentially identical to the base of the jet (Markoff et al. 2005). In any case, the full formation of the collimated outflow takes place at a distance z_0, and from there on the jet can be treated as a distinct physical component of the system.

The jet axis makes an angle θ_{jet} with the line of sight to an observer on Earth. The exact geometry of the jet will be determined by the balance of internal and external pressure, and thus strongly depends on the plasma properties at launch, which are currently undetermined. There is evidence, e.g. in the case of M87, that the jet base is described by an approximately paraboloidal shape, transitioning to a conical structure (Asada and Nakamura 2012; Potter and Cotter 2013) consistent with predictions of a collimated, magnetically-dominated flow that is converted to a kinetically dominated flow. At large distances ($z \gg 10^5 r_g$, where $r_g = GM_{\rm BH}/c^2$ is the gravitational radius of the black hole), jets appear to be well collimated in a quasi-cylindrical structure. The simplest assumption, often adopted in the literature, is thus a constant jet radius over the section of the jet responsible for the majority of the radiative output. Alternatively, a conical geometry is often assumed, so that the jet cross-sectional radius grows as

$$r(z) = r_0\left(\frac{z}{z_0}\right). \tag{3}$$

If magnetic energy is the primary driver of the outflow (i.e., the plasma is ejected by some magnetocentrifugal mechanism or driven by Poynting flux), the magnetic energy density at the base of the jet must be enough to set the plasma in motion. This requires at least as much energy in the magnetic field as kinetic energy the jet will achieve. In this case, the value B_0 of the magnetic field at z_0 may be constrained by the condition of equipartition between the magnetic and the kinetic energy densities, $U_{\rm mag}(z_0) \geq U_{\rm kin}(z_0)$. Then, assuming magnetic-flux conservation, the magnetic energy density decreases as the jet expands. One may parameterize the dependence on z of the magnetic field strength as

$$B(z) = B_0\left(\frac{z_0}{z}\right)^m, \tag{4}$$

where the index $m \geq 1$ depends on the magnetic-field topology (e.g. Krolik 1999). In the case of a conical jet and a purely poloidal magnetic field (i.e., $B = B_z$), $m = 2$, whereas for a purely toroidal B-field (i.e., $B = B_\phi$), $m = 1$.

However, in the case of an MHD-driven outflow, such conservation is likely not valid. In a widely accepted model of jet acceleration (see, for example, Spruit 2010), a fraction of the magnetic energy is converted into kinetic energy of the plasma. The bulk Lorentz factor of the jet, $\Gamma_{\rm jet}$, then increases up to a maximal value. The behavior of $\Gamma_{\rm jet}$ with the distance to the black hole has been studied both analytically and numerically in the context of ideal MHD (e.g., Lyubarsky 2009; Tchekhovskoy et al. 2008, 2010; Polko et al. 2014).

In the context of this simple model we parameterize dissipation in the region $z_{\rm acc} \leq z \leq z_{\rm max}$, where some fraction of the jet power is transformed into kinetic energy of relativistic electrons (and protons in the case of lepto-hadronic models). Possible acceleration mechanisms include diffusive shock acceleration (DSA, see, e.g., Summerlin and Baring 2012) or, in particular for jets with high magnetization, fast magnetic reconnection in presence of turbulence (e.g., Lazarian et al. 2015). If DSA is the dominant acceleration mechanism, the value of $z_{\rm acc}$ is constrained by the fact that, for shock waves to develop, the magnetic energy density of the plasma must be in sub-equipartition with the kinetic energy density (Komissarov et al. 2007). The power transferred to the relativistic particles is parameterized as

$$L_{\rm rel} = q_{\rm rel} L_{\rm jet}, \tag{5}$$

where $q_{\rm rel} < 1$. The value of $L_{\rm rel}$ is the sum of the power injected in both protons and electrons

$$L_{\rm rel} = L_p + L_e \equiv (1 + a)\, L_e \tag{6}$$

where we have defined an energy partition parameter a such that $L_p = a\, L_e$. If $a > 1$ the jet is proton-dominated (i.e., hadronic or lepto-hadronic models), whereas for $a < 1$ most of the power is injected in relativistic electrons (i.e., leptonic models).

2.2 Microphysical Particle Treatment

The relativistic particles in the jet lose energy by radiation and also adiabatic losses in the case of an expanding jet. Several processes contribute to the radiative cooling, since particles can interact with the magnetic, radiation, and matter fields of the jet.

The most important interaction channel with the magnetic field of the jet is synchrotron radiation. For a charged particle of mass m, energy $E = \gamma mc^2$, and charge $q = Ze$, the synchrotron energy-loss rate is (e.g., Blumenthal and Gould 1970)

$$\left.\frac{d\gamma}{dt}\right|_{\text{synchr}} = -\frac{4}{3}\left(\frac{m_e}{m}\right)^3 \frac{c\sigma_T U_{\text{mag}}}{m_e c^2} Z^4 \gamma^2, \tag{7}$$

where c is the speed of light, m_e is the electron mass, σ_T is the Thomson cross-section, and U_{mag} is the magnetic energy density. The ratio $(m_e/m)^3$ in Eq. (7) makes synchrotron cooling particularly efficient for the lightest particles.

Relativistic electrons also cool through inverse Compton (IC) scattering off the jet and potentially also external radiation fields. Writing the characteristic photon energy of the target photon field as $\epsilon = h\nu/(m_e c^2)$, the Compton scattering process with electrons is elastic in the electron rest frame if $\epsilon\gamma \ll 1$, which denotes the Thomson regime. In this regime, the Compton cooling rate is identical to Eq. (7), except for replacing the energy density U_{rmmag} by the energy density of the target radiation field, U_{rad}. Due to the same suppression factor $(m_e/m)^3$, Compton scattering is very inefficient for particles heavier than electrons/positrons, and is therefore usually considered only for leptons. Only in cases where extreme energies for protons are possible, this channel must be taken into account also for heavier particles (e.g., Aharonian 2000, 2002). For the more general case, in which the Thomson condition is not fulfilled (i.e., including the Klein-Nishina regime with $\epsilon\gamma \gtrsim 1$), the energy-loss rate needs to be evaluated considering recoil in the rest frame, using the full Klein-Nishina cross section, e.g. following Blumenthal and Gould (1970); see Böttcher et al. (1997), Dermer and Menon (2009), Böttcher et al. (2012) for detailed calculations of the Compton loss rates specific to blazars.

An important and ubiquitous target photon field is the synchrotron radiation co-spatially produced within the jet. This is referred to as the synchrotron-self-Compton (SSC) process, which often dominates the X-ray (and sometimes also γ-ray) emission in relativistic jet models. In a homogeneous single zone of radius R in steady-state, the radiation energy density U_{rad} is related to the emissivity j_{synchr} (i.e., the energy production rate per unit volume in synchrotron radiation) by

$$U_{\text{synchr}}^{\text{homog}} = \frac{3}{4}\frac{R}{c} j \tag{8}$$

In more complex geometries, a convenient treatment lies in the "local approximation" of Ghisellini et al. (1985):

$$U_{\text{synchr}}^{\text{local}}(z) \approx j_{\text{synchr}}(z)\,\frac{r_{\text{jet}}(z)}{c}. \tag{9}$$

Additional target photon fields for Compton scattering can originate outside the jet and include the radiation field from the accretion disk and of the stellar companion, in the case of microquasars, as well as line-dominated radiation from the Broad Line Region, Narrow Line Region, and a dusty torus around the central accretion flow in the case of AGN (see Sects. 3 and 4 for more details).

The interaction of relativistic protons with radiation can create electron-positron pairs ("photopair" production via the Bethe-Heitler process)

$$p + \gamma \rightarrow p + e^- + e^+. \tag{10}$$

The photon threshold energy for this process is ~ 1 MeV in the rest frame of the proton. At higher energies, proton-photon collisions can also yield pions ("photomeson" production). The two main channels are

$$p + \gamma \rightarrow p + a\pi^0 + b(\pi^+ + \pi^-) \tag{11}$$

and

$$p + \gamma \rightarrow n + \pi^+ + a\pi^0 + b(\pi^+ + \pi^-), \tag{12}$$

where the integers a and b are the pion multiplicities. Both channels have approximately the same cross-section, and a photon threshold energy of ~ 145 MeV in the proton rest-frame, corresponding to a proton energy, in the laboratory frame, of

$$E_p^{\text{thr}} = \frac{m_p c^2 \, m_\pi c^2}{2\, E_{\text{ph}}} \left(1 + \frac{m_\pi}{2 m_p}\right) \sim 7 \times 10^{16}\, E_{\text{eV}}^{-1}\ \text{eV} \tag{13}$$

where E_{eV} is the characteristic photon energy of the target photon field in units of eV. The subsequent decay of charged pions injects leptons and neutrinos

$$\pi^+ \rightarrow \mu^+ + \nu_\mu, \quad \mu^+ \rightarrow e^+ + \nu_e + \overline{\nu}_\mu, \tag{14}$$

$$\pi^- \rightarrow \mu^- + \overline{\nu}_\mu, \quad \mu^- \rightarrow e^- + \overline{\nu}_e + \nu_\mu, \tag{15}$$

whereas neutral pions decay into two photons,

$$\pi^0 \rightarrow \gamma + \gamma. \tag{16}$$

For photomeson and photopair production, the proton energy-loss rate can be calculated as in Begelman et al. (1990). Hadronic/lepto-hadronic models involving photo-meson production require protons to be accelerated to energies of at least the threshold energy given in Equation (13). Confining protons of this energy to the small emission region sizes inferred from rapid variability, typically requires magnetic fields of $B \gtrsim 10$–100 G. In the presence of such strong magnetic fields, the electron-synchrotron radiation field is expected to be strongly dominant over any potential external radiation fields. Therefore, one typically only considers the synchrotron photons of primary electrons as the target photon field for photo-pair and photo-pion production. γ-rays are produced through the $p\gamma$ interaction channels primarily by synchrotron radiation of charged secondaries ($e^\pm$, $\mu^\pm$, $\pi^\pm$), and the decay of neutral pions.

High-energy protons can also interact with thermal particles in the jet plasma. Due to the low densities present in extragalactic jet sources, proton-proton interactions are usually negligible in AGN jets, but they can make a significant contribution to the proton loss rate and radiation output in the case of the much denser Galactic jet sources. If the energy of the relativistic proton is higher than the threshold value for π^0 production (≈ 1.22 GeV), the collision with a thermal proton can create pions

$$p + p \rightarrow p + p + a\pi^0 + b(\pi^+ + \pi^-), \tag{17}$$

$$p + p \rightarrow p + n + \pi^+ + a\pi^0 + b(\pi^+ + \pi^-). \tag{18}$$

As in proton-photon interactions, proton-proton collisions inject photons by means of the decay of neutral pions. Charged pions inject electron-positron pairs and neutrinos through the decay chains of Eqs. (14) and (15).

The energy-loss rate for a proton of energy E is given by (e.g., Begelman et al. 1990)

$$\left.\frac{dE}{dt}\right|_{pp} = -n_p \, c \, E \, \kappa_{pp}\sigma_{pp}, \tag{19}$$

where n_p is the number density of thermal protons in the co-moving frame,[1] σ_{pp} is the proton-proton cross-section, and $\kappa_{pp} \approx 0.5$ is the total inelasticity of the interaction. A convenient parametrization for σ_{pp} is presented in Kelner et al. (2006). Taking the assumptions about distribution of jet power above, we can calculate the value of n_p in the comoving frame (see also Bosch-Ramon et al. 2006):

$$n_p = (1 - q_{\rm rel}) \frac{L_{\rm jet}}{\Gamma_{\rm jet}^2 \pi r_{\rm jet}^2 v_{\rm jet} m_p c^2}, \tag{20}$$

where m_p is the proton mass.

Finally, in the case of an expanding jet, particles also lose energy due to adiabatic cooling. The corresponding adiabatic energy loss rate is given by

$$\left.\frac{dE}{dt}\right|_{\rm ad} (E, z) = -\frac{E}{3}\frac{dV}{dt}, \tag{21}$$

which obviously depends on the jet geometry (see, e.g., Bosch-Ramon et al. 2006).

The total energy-loss rate is simply the sum of the relevant loss rates for the particle species in consideration

$$\left.\frac{dE}{dt}\right|_{\rm tot} (E, z) = \sum_i \left.\frac{dE}{dt}\right|_i (E, z). \tag{22}$$

For particles whose gyro-radii are smaller than the acceleration/radiation region (this is typically always the case for electrons/positrons), the maximum energy of primary electrons and protons can be estimated by equating the total energy-loss rate to the acceleration rate

$$-\left.\frac{dE}{dt}\right|_{\rm tot} (E_{\rm max}) = \left.\frac{dE}{dt}\right|_{\rm acc} (E_{\rm max}). \tag{23}$$

However, for protons, the confinement condition (i.e., the gyro-radius equalling the size of the acceleration region) is sometimes the dominant factor limiting the maximum energy, rather than radiative cooling. The magnitude and energy dependence of the particle acceleration rate obviously depends on the dominant acceleration mechanisms. However, it is often convenient to parameterize diffusive shock acceleration processes through an acceleration

[1] Henceforth, we refer to as "comoving" or "jet frame" the reference frame fixed to the jet plasma, where the outflow bulk velocity is zero. The "observer frame" corresponds to the frame in which the jet bulk velocity is $v_{\rm jet}$.

efficiency parameter $\eta < 1$ so that the acceleration time scale is a multiple $\eta^{-1} > 1$ of the gyro-timescale. Consequently, the acceleration rate is then given by (e.g., Aharonian 2004).

$$\left.\frac{dE}{dt}\right|_{\rm acc} = \eta\, e\, c\, B(z). \tag{24}$$

Pions and muons can also interact before decaying. This effect becomes important when particles are very energetic and the physical conditions in the jet are such that the cooling time is shorter than the mean lifetime (Reynoso and Romero 2009). Pion and muon synchrotron radiation (and cooling) may be neglected if the maximum primary proton Lorentz factor and the magnetic field satisfy the conditions (Böttcher et al. 2013):

$$B\,\gamma_p \ll \begin{cases} 7.8 \times 10^{11}\ \mathrm{G} & \text{for pions} \\ 5.6 \times 10^{10}\ \mathrm{G} & \text{for muons} \end{cases} \tag{25}$$

The injection of secondary particles is not a consequence of diffusive shock acceleration, but a by-product of the interaction of primary protons and electrons. The maximum energy of pions, muons, and secondary electron-positron pairs is then not fixed by Eq. (23). It is instead determined by the characteristics of the particular process through which the different particles were injected, and by the maximum energy of the primary particles (which is constrained by Eq. (23)).

The injection of primary particles into the emission region may represent either the entrainment of external material by the flow, or a very rapid (e.g., first-order Fermi) acceleration process, on a time scale much shorter than all other relevant time scales. Such injection is thus usually treated as a phenomenological source term $Q(E, z)$ in the co-moving jet frame, which is often parameterized either as a δ function in particle energy,

$$Q(E, z)|_{\rm entrainment} = Q_0\, \delta\big(E - \Gamma_{\rm jet}\, mc^2\big) \tag{26}$$

in the case of entrainment by the relativistic flow of Lorentz factor $\Gamma_{\rm jet}$ (e.g., Pohl and Schlickeiser 2000), or as a power-law in energy multiplied by an exponential cutoff,

$$Q(E, z) = Q_0\, E^{-p} \exp\big[-E/E_{\rm max}(z)\big]\, f(z), \tag{27}$$

in the case of diffusive shock acceleration. In the latter case, the value of the spectral index is typically in the range $1.5 \lesssim p \lesssim 2.4$. The cutoff energy $E_{\rm max}$ depends on z, and is calculated according to Eq. (23). The function $f(z)$ describes the injection profile along the jet that defines the length of the acceleration region.

In most studies in the literature, the form of the injection function $Q(E)$—especially the spectral index p—is left as a free parameter, whose value may be used as an indication of the dominant particle acceleration mechanism. In the case of $p \sim 2$, non-relativistic shocks present themselves as a standard solution to achieve such a particle spectrum, while very hard spectra (down to $p \sim 1$) have been shown to be possible in the case of magnetic reconnection (e.g., Sironi and Spitkovsky 2014; Guo et al. 2016). Diffusive shock acceleration at relativistic parallel shocks produces particle spectra with $\alpha \sim 2.2$–2.3 (e.g., Kirk and Duffy 1999). Mildly relativistic, oblique shocks are capable of producing spectral indices in a wide range, from $p \sim 1$ to very steep spectra, depending primarily on the obliquity, i.e., the angle of the ordered component of the magnetic field with respect to the shock normal (Summerlin and Baring 2012). Thus, unfortunately, the diagnostic of determining p from model fitting of spectral energy distributions of jet sources, is not sufficient to pin down the dominant particle acceleration mechanism, one of the fundamental unknowns in our understanding of relativistic jets (but see some additional discussion in Sect. 5).

Finally, the value of the normalization constant Q_0 is obtained from the total power injected in each particle species

$$L_i = \int_V d\vec{r} \int_{E_{\min}}^{E_{\max}} dE\, E\, Q_i(E, z), \tag{28}$$

where the index i runs over protons and electrons. The interaction of relativistic protons with matter and radiation injects charged pions, muons, and secondary electron-positron pairs. Electron-positron pairs are also created through photon-photon annihilation. These processes define the specific expressions for $Q(E, z)$ for the respective secondary particles.

The sources of charged pions are proton-photon and proton-proton interactions. Suitable expressions for the corresponding pion injection functions were presented in Atoyan and Dermer (2003), Kelner et al. (2006), and Kelner and Aharonian (2008), while expressions for the consequent muon injection may be found in Lipari et al. (2007). Finally, electrons and positrons are injected through muon decay, photopion production, and $\gamma\gamma$ absorption. Convenient expressions for the electron/positron injection rate from muon decay are presented in Schlickeiser (2002). Expressions for the pair injection rate from photopair production may be found in Chodorowski et al. (1992) and Mücke et al. (2000). The rate of pair production from photon-photon annihilation

$$\gamma + \gamma \rightarrow e^+ + e^- \tag{29}$$

is derived in Böttcher and Schlickeiser (1997). Two-photon annihilation is also a photon sink. Its effect on the escape of photons from the jet is discussed below (Sect. 2.3).

The evolution of the particle distributions is governed by transport equations of the form

$$\begin{aligned} &\frac{\partial N(E,t,z)}{\partial t} + \frac{\partial(\Gamma_{\rm jet} v_{\rm jet} N(E,t,z))}{\partial z} + \frac{\partial(b(E,t,z)N(E,t,z))}{\partial E} \\ &\quad - \frac{\partial}{\partial E}\left[\frac{E^2}{(a+2)\,t_{\rm acc}}\frac{\partial N(E,t,z)}{\partial E}\right] + \frac{N(E,t,z)}{T_{\rm dec}(E)} + \frac{N(E,t,z)}{T_{\rm esc}(E)} = Q(E,t,z) \end{aligned} \tag{30}$$

The second term on the l.h.s. of Eq. (30) represents particle convection along the jet, the third term describes systematic energy losses/gains, where $b(E, z) = -dE/dt|_{\rm tot}$ encompasses all energy loss plus first-order acceleration terms. The fourth term on the l.h.s. represents diffusion in momentum space and describes second-order (stochastic) shock acceleration on a characteristic time scale $t_{\rm acc}$. Here, $a = v_s^2/v_A^2$ is the square of the ratio of shock speed to the Alfvén speed in the co-moving jet plasma. $T_{\rm dec}$ is the decay time scale. This term vanishes for stable particles. However, a corresponding term should be included for electrons and positrons to account for the loss of pairs due to pair annihilation. $T_{\rm esc}$ is the particle escape time scale.

This equation is appropriate for a very general case of a spatially extended, inhomogeneous jet, in which the physical parameters may vary in time and space along the jet. In time-independent (steady-state) models, the first term $\partial N/\partial t$ is set to zero. Homogeneous, single-zone models neglect the second (convection) term. In many treatments, momentum diffusion (and, thus, stochastic acceleration) is also neglected, thus leaving out the fourth term on the l.h.s. Note that Eq. (30) still assumes that the jet is homogeneous across the jet cross section (i.e., no radial dependence) and that the particle distributions are locally isotropic in the co-moving jet frame. Also, spatial diffusion has been neglected (see, e.g., Chen et al. 2015, 2016, for a blazar jet model that includes spatial diffusion).

2.3 Radiative Output

We now proceed to calculate the total radiative spectrum of all particle species produced by each of the interaction processes described in Sect. 2.2. In this exposition, we will restrict ourselves to quoting simple, δ-function approximations for the respective emissivities. Detailed expressions for the emissivities can be found, e.g., in Atoyan and Dermer (2003), Romero and Vila (2008), Reynoso and Romero (2009), Dermer and Menon (2009), Böttcher et al. (2012), and references therein.

The emissivities $j(\epsilon) = dE/(dV\, dt\, d\epsilon\, d\Omega)$ [erg cm^{-3}s^{-1}] at photon energy $h\nu = \epsilon\, m_e c^2$ are commonly first evaluated in the co-moving jet frame, and then transformed to the observer frame. The relevant transformations are governed by the Doppler factor

$$D = \left[\Gamma_{\rm jet}(1 - \beta_{\rm jet}\cos\theta_{\rm jet})\right]^{-1} \tag{31}$$

where $\theta_{\rm jet}$ is the jet viewing angle and $\beta_{\rm jet} = v_{\rm jet}/c$.

Relativistic Doppler effects then result in photon energies being blue/red shifted as $\epsilon_{\rm obs} = D\,\epsilon$, and variability time scales appear contracted as $\delta t_{\rm obs} = \delta t/D$. The relativistic boosting of radiative flux depends on the geometry (see, e.g., Lind and Blandford 1985). In the case of a continuous, extended jet, the flux will be enhanced by a factor of D^2 so that the observed flux can be calculated as

$$F_{\epsilon_{\rm obs},{\rm ext}} = \frac{1}{d_L^2}\int_{z_{\rm acc}}^{z_{\rm max}} r_{\rm jet}^2(z) D^2(z)\, j(\epsilon_{\rm obs}/D, z)\, dz, \tag{32}$$

where d_L is the luminosity distance to the source.

In the case of single emission zone with co-moving volume V, the flux transformation is determined by a factor D^3, i.e.,

$$F_{\epsilon_{\rm obs},{\rm zone}} = \frac{V\, j(\epsilon_{\rm obs}/D)\, D^3}{d_L^2} \tag{33}$$

where the additional factor of D originates from the volume integration (which is taken over z in the laboratory frame in Eq. (32)), taking into account Lorentz contraction and light-travel-time effects.

The dominant radiation mechanisms resulting from high-energy, accelerated particles are: synchrotron radiation (of primary electrons and protons as well as secondary pions, muons, and electrons/positrons), inverse Compton scattering (relevant only for electrons and positrons), and π^0 decay.

The synchrotron spectrum from an ensemble of charged particle of mass m, charge $q = Z\,e$, and energy $\gamma\, mc^2$ moving at random directions with respect to a magnetic field $B = B_{\rm G}$ G peaks at a frequency

$$\nu_c^{\rm sy} = \frac{3qB}{4\pi mc}\gamma^2 \equiv \nu_0\,\gamma^2 \approx 4.2\times 10^6\, B_{\rm G}\gamma^2\, Z\,\frac{m_e}{m}\ {\rm Hz} \tag{34}$$

Based on a simple δ-function approximation for the synchrotron emissivity of a single particle, the synchrotron spectrum of an ensemble of particles with energy distribution $n(\gamma)$ can be evaluated as[2]

$$j_\nu^{{\rm sy},\delta} = \frac{4}{9}c\left(\frac{q^2}{m\,c^2}\right)^2 U_{\rm mag}\,\nu_0^{-3/2}\,\nu^{1/2}\, n\left(\sqrt{\frac{\nu}{\nu_0}}\right). \tag{35}$$

[2]Note a typo in Eqs. (3.38) and (3.39) in Böttcher et al. (2012), where a factor c is missing

In the Thomson regime ($\epsilon\,\gamma \ll 1$), a target photon of energy ϵ will be scattered to higher energies by an isotropic ensemble of relativistic electrons of energy $\gamma\, m_e c^2$ to an average scattered photon energy $\epsilon_s \approx \gamma^2\,\epsilon$. The Compton emissivity resulting from Compton scattering of a target photon field with photon number distribution $n_{\rm ph}(\epsilon)$ by a relativistic electron distribution $n_e(\gamma)$ in the Thomson regime can be approximated as

$$j_\nu^{\rm Thomson,\delta}(\epsilon_s) \approx \frac{h\,c\,\sigma_T\,\epsilon_s^2}{8\pi} \int\limits_{\epsilon_s}^{\infty} d\gamma\,\frac{n_e(\gamma)}{\gamma^4} \int\limits_{\epsilon_s/(2\gamma^2)}^{\infty} d\epsilon\,\frac{n_{\rm ph}(\epsilon)}{\epsilon^2} \tag{36}$$

where a simple δ function approximation for the Compton cross section has been employed. More general expressions for the synchrotron and Compton emissivities, including Compton scattering in the Klein-Nishina regime, may be found, e.g., in Blumenthal and Gould (1970), Rybicki and Lightman (1979), Dermer and Menon (2009), Böttcher et al. (2012).

In the decay of a π^0 meson of energy $E_{\pi^0} = \gamma_{\pi^0} m_\pi c^2$, two photons with energy $E_\gamma \approx (1/2) E_{\pi^0}$ are produced, while the production of π^0 mesons depends on the (energy-dependent) $p\gamma$ cross section, branching ratio (i.e., fraction of $p\gamma$ pion production events resulting in the production of a π^0) and inelasticity (i.e., the fraction of the proton's energy transferred to the newly produced pion). A convenient approximation for the calculation of the resulting photon production spectrum can be found in Atoyan and Dermer (2003), Romero and Vila (2008).

The photon emission spectrum must be corrected for various absorption processes. At low (radio) frequencies, synchrotron-self absorption (SSA) will be dominant (see, e.g., Rybicki and Lightman 1979; Böttcher et al. 2012), while at high energies ($E \gg 1$ GeV) γ-ray absorption can be caused by photon-photon annihilation into electron-positron pairs (see Eq. (29)). The $\gamma\gamma$ opacity for a photon of energy ϵ_γ in a target photon field $n_{\rm ph}(\epsilon)$ is

$$\kappa_{\gamma\gamma}(\epsilon_\gamma) = \frac{1}{2} \int\limits_{-1}^{1} d\mu \int\limits_{\epsilon_{\rm th}}^{\infty} d\epsilon\,\sigma_{\gamma\gamma}(\epsilon_\gamma, \epsilon, \mu)\, n_{\rm ph}(\epsilon)\,(1-\mu) \tag{37}$$

where μ is the cosine of the collision angle, and $\sigma_{\gamma\gamma}$ is the annihilation cross-section (e.g., Gould and Schréder 1967), and $\epsilon_{\rm th}$ is the threshold energy,

$$\epsilon_{\rm th} = \frac{2}{\epsilon_\gamma\,(1-\mu)}. \tag{38}$$

The $\gamma\gamma$ optical depth $\tau_{\gamma\gamma}$ is then evaluated as the length integral over $\kappa_{\gamma\gamma}$ along the photon propagation direction. Equation (37) is simplified when the target photon distribution is isotropic. A convenient expression for $\tau_{\gamma\gamma}$ in this particular case is obtained in Gould and Schréder (1967).

If the absorbing photon field is located outside the emission region (as, e.g., in the case of VHE γ-ray absorption by the Extragalactic Background Light, see Sect. 3), the $\gamma\gamma$ opacity simply leads to a correction factor $exp(-\tau_{\gamma\gamma})$ with respect to the unabsorbed flux spectrum, $F_{\rm intr}(\epsilon)$. However, in case emission and absorption are taking place co-spatially, the solution to the radiative transfer equation becomes more involved. In the case of a homogeneous source, this solution is given by

$$F_{\rm obs}^{\rm int}(\epsilon_\gamma) = F_{\rm intr}(\epsilon_\gamma)\,\frac{1-e^{-\tau_{\gamma\gamma}(\epsilon_\gamma)}}{\tau_{\gamma\gamma}(\epsilon_\gamma)} \tag{39}$$

In the optically-thin limit ($\tau_{\gamma\gamma} \ll 1$), Eq. (39) simply reduces to $F_{\rm obs}^{\rm opt.\,thin} = F_{\rm intr}$, as expected, while in the optically-thick limit ($\tau_{\gamma\gamma} \gg 1$), it becomes $F_{\rm obs}^{\rm opt.\,thick} = F_{\rm intr}/\tau_{\gamma\gamma}$. If the target photon field has a power-law spectrum $n_{\rm ph}(\epsilon) \propto \epsilon^{-(\alpha_t+1)}$ (i.e., an energy index of α_t), then the opacity also has a power-law dependence on energy, $\tau_{\gamma\gamma}(\epsilon_\gamma) \propto \epsilon_\gamma^{\alpha_t}$. Consequently, internal $\gamma\gamma$ absorption leads to a spectral break by $\Delta\alpha_\gamma = \alpha_t$ around the photon energy where $\tau_{\gamma\gamma}(\epsilon_\gamma) = 1$. The presence or absence of such spectral breaks in the γ-ray spectra of relativistic jet sources can be used to constrain the bulk Lorentz factor of the emission region (e.g., Baring 1993; Dondi and Ghisellini 1995; Böttcher et al. 2012). As we will elaborate on in Sect. 3.3, $\gamma\gamma$ absorption in the immediate environment (especially, the Broad Line Region) of powerful blazars also provides constraints on the location of the γ-ray emission region in those objects.

In addition to photons, if hadronic charged-pion production processes are relevant, also electron and muon neutrinos at $\sim$ TeV–PeV energies are produced. The detection of astrophysical neutrinos and their unique identification with relativistic jet sources would therefore uniquely identify sites where protons are being accelerated to ultrarelativistic energies exceeding the pion production threshold (see Eq. (13)), possibly even reaching the energies of ultra-high-energy cosmic rays (UHECRs) with energies $E \gtrsim 10^{19}$ eV.

Convenient templates for the pion production spectra from relativistic $p\gamma$ and $p-p$ interactions in the case where pion and muon synchrotron losses may be neglected, can be found in Kelner and Aharonian (2008), Kelner et al. (2006), respectively. If pion and muon synchrotron emission can not be neglected, their energy losses prior to decay need to be taken into account. A formalism to evaluate the secondary-particle production rates in this more general case, has been developed in Hümmer et al. (2010).

2.4 Polarization

Relativistic jets are expected to produce polarized synchrotron radiation. Such radiation is observed from radio to X-rays. At higher energies, inverse Compton upscattering of the polarized synchrotron photons can yield polarized X/gamma-rays (e.g., Bonometto and Saggion 1973), and also anisotropic Compton scattering of unpolarized radiation fields by non-relativistic or mildly relativistic electrons can introduce polarization (e.g., Poutanen 1994; Krawczynski 2012). Evidence for polarization in high-energy radiation from relativistic jet sources has been found, e.g., in Cygnus X-1 (Laurent et al. 2011; Romero et al. 2014), and there are several developments for X-ray and γ-ray polarimeters currently on-going and proposed, such as the X-Ray Imaging Polarimeter Experiment (XIPE, Soffitta et al. 2013), or the Polarization Spectroscopic Telescope Array (PolSTAR, Krawczynski et al. 2016).

Both in the case of optically-thin synchrotron and synchrotron-self-Compton scattering, the preferred polarization direction indicates the predominant orientation of the magnetic field, as the Electric Vector Position Angle (EVPA) is perpendicular to the projection of the B-field onto the plane of the sky. A non-vanishing synchrotron polarization therefore clearly indicates an at least partially ordered magnetic field. If the B-field is perfectly ordered, then optically-thin synchrotron radiation by a non-thermal power-law particle distribution with index p yields a degree of polarization Π of

$$\Pi_{\rm sy} = \frac{p+1}{p+\frac{7}{3}} \tag{40}$$

corresponding to polarization degrees of $\Pi \sim 70$–75 % for typical spectral indices of $p \sim 2$–3 (e.g., Rybicki and Lightman 1979). The measured degree of optical polarization in relativistic jet sources is typically in the range $\Pi \lesssim 30\ \%$, and often significantly smaller than that. This indicates that the emission must originate from regions with partially disordered magnetic fields.

Especially in the case of blazars, it is known that the polarization properties are highly variable, both in degree of polarization and polarization angle (PA). Particularly noteworthy in this context are large swings in the PA (Romero et al. 1995), which sometimes correlate with multi-wavelength flares (Marscher et al. 2008, 2010; Abdo et al. 2010; Blinov et al. 2015, 2016). Such PA swings naturally suggest the existence of helical fields (e.g., Zhang et al. 2014, 2016). Models based on random fluctuations in a turbulent field have been also proposed (Marscher 2014).

Time-dependent multiwavelength polarization measurements diagnose the topology of the magnetic field in the emission region and its change during variability events. Such observations may therefore be a unique tool to identify the role of magnetic fields in the formation of jets and in the relativistic particle acceleration processes leading to the emission of high-energy radiation. The diagnostic power of (time-dependent) polarization signatures in blazars will be discussed in more detail in Sect. 3.

At low (radio) frequencies, the propagation of polarized emission through a magnetized medium leads to a rotation of the EVPA, known as Faraday Rotation (e.g. Pacholczyk 1963). This may occur even within the source, and thus lead to effective Faraday depolarization (e.g., Burn 1966), as the EVPAs of radiation produced in different locations along the line of sight within the source appear rotated by different amounts. The amount of Faraday rotation, $\Delta\psi$, i.e., the angle by which the EVPA is rotated, depends on the electron density n_e, the projection of the magnetic field $\overrightarrow{B}$ along the line of sight $\overrightarrow{s}$, and the radiation wavelength λ as

$$\Delta\psi \propto \lambda^2 \int n_e(s)\, \overrightarrow{B}(s) \cdot d\overrightarrow{s} \tag{41}$$

Due to the λ^2 scaling, this effect is generally only relevant at radio frequencies and is negligible at optical or higher frequencies.

3 Blazars

The class of peculiar extragalactic sources dubbed *blazars* represents a quite small, but remarkably interesting fraction of the entire population of AGN (e.g. Urry and Padovani 1995). While a formal definition is difficult to establish, the canonical features used in their classification include the presence of a compact radio core, with flat or even inverted spectrum, extreme variability (both in timescale and in amplitude) at all frequencies, and a high degree of optical and radio polarization. More recently, intense gamma-ray emission (although not ubiquitous, but in many cases dominating the total bolometric radiative power output) has also been used for blazar selection. A recent census counts about 3000 known blazars (Massaro et al. 2009).

Phenomenologically, blazars are further divided in two subgroups depending on special observational characteristics: 1) *BL Lacertae* objects are characterized by an extreme weakness (equivalent width < 5 Å) or even absence of emission lines in their optical spectra, a feature generally attributed to a low accretion rate onto the central supermassive black hole and an environment largely depleted of potential fuel for black-hole accretion, not exhibiting a prominent broad-line region; 2) if the optical spectra display broad emission lines typical of quasars (or QSOs = Quasi Stellar Objects), blazars will be instead classified as *Flat*

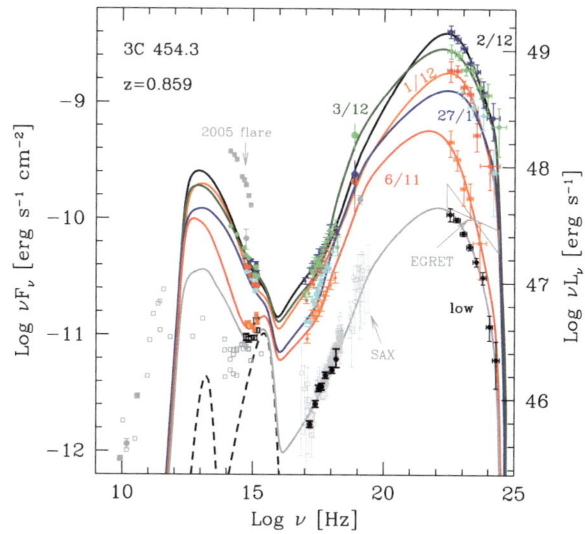

Fig. 2 Spectral energy distributions of the FSRQ 3C454.3 during different activity states in 2012 (*colored*) and a quiescent phase (*gray*). The data clearly show the typical "double humped" shape of the SED. *Curves* show the result of a one-zone leptonic emission model. The reader can appreciate the extreme variability, especially in the γ-ray band. From Bonnoli et al. (2011)

Spectrum Radio Quasars (FSRQ). FSRQs are generally more powerful than BL Lacs. The FSRQ population follows a positive cosmological evolution close to that of normal quasars, being quite rare in the local Universe and with a density peaking around $z \sim 2$ (e.g., Dunlop and Peacock 1990). The BL Lac population, instead, displays a quite different trend. In particular, the less powerful high-frequency peaked BL Lacs (HBL), follow a *negative* cosmological evolution (Rector et al. 2000; Ajello et al. 2014), i.e. they tend to be more numerous in the local Universe, with their density decreasing with redshift. The different evolutions could be related to a generic connection between FSRQs and BL Lacs, the latter representing the final stage of the FSRQ activity, after the fuel for black-hole accretion has been exhausted (Cavaliere and D'Elia 2002; Böttcher and Dermer 2002).

Relativistic beaming offers a simple explanation for the peculiarities displayed by blazars (Blandford and Rees 1978; Blandford and Königl 1979), in particular the inverted radio spectrum, the huge brightness temperature, the existence of superluminal moving features, the short variability timescale and the powerful γ-ray emission. Different methods agree to determine typical Doppler factors of localized emission regions in blazar jets in the range $D \simeq 10$–20 (but reaching in some cases values as large as 50), implying a similar bulk Lorentz factor for the plasma outflowing in the jet (Lister et al. 2013; Ghisellini and Tavecchio 2015, and references therein) and a viewing angle as small as few degrees. Blazars, therefore, represent radio sources in which the jet velocity points almost along the line of sight. In this geometry, the relativistically beamed non-thermal continuum produced within the jet easily outshines any other emission component from the active nucleus or the host galaxy. In the view of the *unification scheme*, BL Lacs and FSRQ are considered the beamed counterparts of the FRI (low power) and FRII (high power) radiogalaxies, respectively (e.g., Urry and Padovani 1995).

The remarkably smooth continuum emission of blazars, extending over the entire electromagnetic spectrum, from the radio band to γ-ray energies (see Fig. 2 for an example), is characterized in the $\nu F(\nu)$ representation (the so-called spectral energy distribution, SED) by a typical "double humped" shape (e.g., Sambruna et al. 1996; Fossati et al. 1998). While the characteristics of the low energy component were already well known and fairly well understood as synchrotron radiation of relativistic electrons in the 1980s (e.g., Marscher and Gear 1985), it was realized only after the advent of the γ-ray telescope EGRET (Energetic Gamma-Ray Experiment Telescope) onboard the *Compton Gamma-ray Observatory* in the 1990s (Fichtel et al. 1994) that many blazars exhibit significant γ-ray emission, often dominating the SEDs. Including the γ-ray output, the apparent bolometric luminosities of blazars reach, in some cases, $L_\gamma = 10^{49}$ erg s^{-1} (e.g., Bonnoli et al. 2011). These extreme luminosities, combined with the observed short (often intra-day) variability, provide one line of evidence for relativistic beaming in blazars (see, e.g., Schlickeiser 1996, for a review).

Improved γ-ray instruments in space (*Fermi* and *AGILE*) and the development of Cherenkov telescope facilities on the ground (H.E.S.S., MAGIC, VERITAS, and the future CTA), together with intense multifrequency coordination efforts provide increasingly complete coverage of the rapidly changing SEDs, which is instrumental in the understanding of the physical processes shaping the blazar phenomenon. The Cherenkov telescopes, sensitive to very high energy (VHE) γ-rays ($E > 50$ GeV), have revealed an important population of blazars (BL Lacs, in particular) characterized by an intense emission at TeV energies which, besides jet physics, can be exploited as a probe for cosmic backgrounds and intergalactic magnetic field.

3.1 Emission Models: The Standard Scenario

As clearly indicated by the high degree of radio and optical polarization, the low-energy spectral component of blazars is produced through the synchrotron mechanism, most likely by a population of highly relativistic electrons. The origin of the high-energy component is still debated. Two general scenarios for the production of high energy radiation are discussed (see also Sect. 2), namely *hadronic* or *leptonic* models (e.g., Böttcher et al. 2013).

3.1.1 Leptonic Models

The most popular class of models adopts the leptonic scenario, in which the high-energy component is the result of the inverse Compton scattering of soft photons by the same electrons responsible for the synchrotron emission. Different sources of soft photons can be considered; the emission will be dominated by inverse Compton scattering of the photon component with the largest energy density (as measured in the co-moving jet frame). Depending on the nature of soft photons dominating the IC process, the leptonic high-energy emission is comprised of SSC or *External Compton* (EC) radiation.

In the SSC model (e.g., Maraschi et al. 1992; Tavecchio et al. 1998) it is assumed that target photons are dominated by the synchrotron photons themselves. In this class of models synchrotron and IC emission are strongly coupled. This affords the possibility to derive robust constraints on the basic physical quantities of the jet from the observed SED (Tavecchio et al. 1998). In the EC model one assumes that the low energy radiation produced in the central regions of the AGN (directly emitted by the accretion disk or reprocessed by the gas in the Broad Line Region or by the parsec-scale molecular torus) dominates over the synchrotron radiation. However, even in this case, SSC emission is still expected to make a non-negligible contribution, especially at X-ray energies. A possibility (Dermer and Schlickeiser 1993) is thus to consider the direct UV emission from the accretion disk; however for sufficiently large distances from the disk, the de-beaming suffered by the radiation directly coming from the disk causes a strong depression of this contribution in the jet frame. The primary disk radiation is partly reflected/reprocessed by the gas of the BLR and beamed in the jet frame, providing a strong contribution to the emission (Sikora et al. 1994). Błażejowski et al. (2000) pointed out that IC scattering of the thermal near-infrared

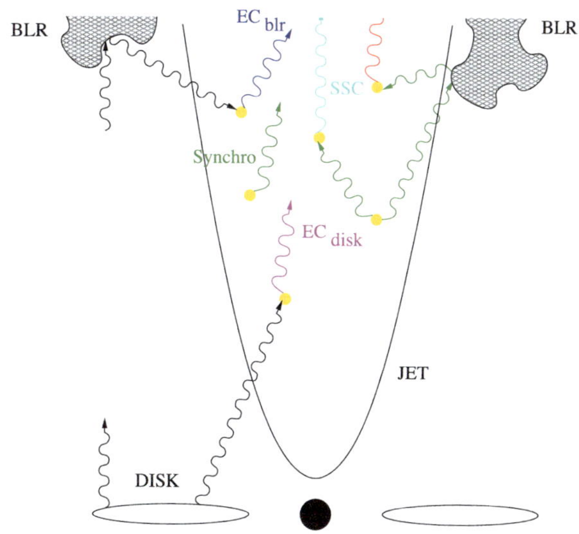

Fig. 3 Sketch illustrating the basic emission model for blazars. Relativistic electrons in the jet produce radiation through synchrotron emission and IC scattering. Soft target photons for the IC process are the synchrotron photons themselves (*SSC radiation*) and those produced in the external environment (*EC radiation*), namely produced in the disk, or reprocessed in the BLR. A third class of models (*the mirror model*) consider also the synchrotron photons "reflected" by the BLR. Further out also the IR radiation emitted by the molecular torus can be scattered

radiation ($T \simeq 10^3$ K) emitted by the dust of the parsec-scale torus expected to surround the central AGN region could provide the dominant contribution to the high energy emission, especially in the energy band 10 keV–100 MeV. This could be the case especially for FSRQ detected above few tens of GeV (see below), however for BL Lacs this component is less likely to be important as recent evidence suggests the dusty torus dwindles at lower accretion rates (Plotkin et al. 2012a).

Not only the optical/UV emission from the accretion disk, but also the beamed synchrotron emission from the jet can be "reflected" by the BLR and thus serve as target photon field for Compton scattering. The double change of frame (source $\rightarrow$ lab frame $\rightarrow$ source) leads to a great amplification of the energy density of the soft radiation (Ghisellini and Madau 1996; Tavani et al. 2015). However, as discussed by Böttcher and Dermer (1998) and Bednarek (1998), when the constraints posed by the travel time of the radiation are taken into account, the relative contribution of this mechanism to the total emission may be severely suppressed.

The question of the dominant target photon field is intimately linked to the location of the emission region along the jet, which will provide additional clues towards the dominant particle acceleration mechanism. This aspect will be discussed in more detail in Sect. 3.3

The general set-up of leptonic blazar emission models is sketched in Fig. 3. The jet flow is expected to become dissipative around 10^2–10^3 Schwarzschild radii (corresponding to about 0.01–0.1 parsec for a BH mass of $10^9 M_\odot$), likely through shocks or magnetic reconnection events which can accelerate particles to ultrarelativistic energies (sometimes this is defined as the "blazar zone"). Typical variability timescales of $\delta t_{\rm var} \sim$ few hours (although shorter timescales events have been occasionally observed), together with typical Doppler factors of $D \sim 10$–20, limit the size of the emission region to $R \lesssim \delta t_{\rm var}\, c\, D/(1+z) \sim 10^{15}$–$10^{16}$ cm.

SED modeling of a large number of blazars (e.g., Ghisellini et al. 2010) suggests that SSC and EC processes dominate the high energy emission of different types of blazars: in the case of FSRQ, which display bright thermal features (emission lines and the *blue bump*), γ-ray production is likely dominated by EC (often with a contribution from SSC in the X-ray band), while in high-frequency peaked BL Lac objects (characterized by very weak or even absent thermal components) the SSC process dominates. However, the situation is

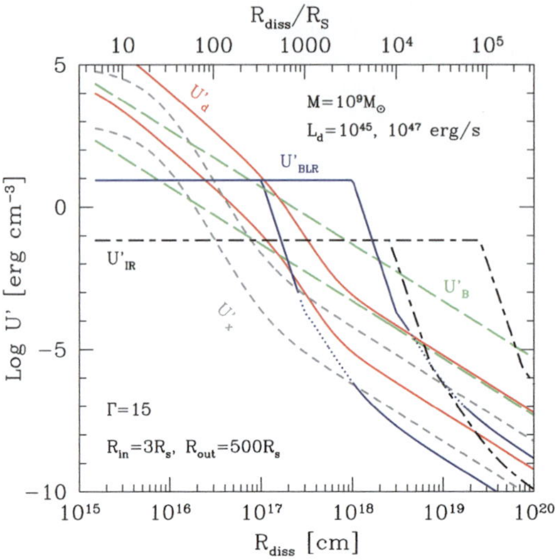

Fig. 4 Comparison of different energy densities as measured in the jet reference frame. The jet is assumed to have a bulk Lorentz factor $\Gamma = 15$ and the black hole mass is assumed to be $M = 10^9 M_\odot$. The two sets of *lines* correspond to two disk luminosities: $L_d = 10^{45}$ and 10^{47} erg s^{-1}. The magnetic energy density (*long dashed lines*) is calculated assuming that a fraction $\epsilon = 0.1$ of the jet power is carried as magnetic field. From Ghisellini and Tavecchio (2009)

less straightforward for intermediate (relatively powerful) BL Lac objects, in which the EC emission could play some role (Böttcher 2010). Moreover, for FSRQ, the actual external radiation field ruling the EC emission depends on the position of the emitting region along the jet. This is shown in Fig. 4, in which the energy densities of the various radiation fields (in the rest frame of the jet) are plotted as a function of the distance from the central engine. Within the radius marking the distance where the BLR clouds are located[3] the energy density is dominated by the constant BLR radiation field. Beyond the BLR radius, U_{BLR} falls rapidly with distance and the radiation from the torus starts to dominate. The great majority of FSRQs can be modeled assuming that the emitting region is at distances smaller than the BLR radius (e.g., Ghisellini et al. 2010, 2014), thus considering only EC with the BLR photons. However, for FSRQ detected at VHE, the emission likely occurs beyond this radius due to $\gamma\gamma$ absorption constraints (see Sect. 3.3).

3.1.2 Hadronic Models

Even in the case of the leptonic models described above, protons are expected to be present in the jet in order to ensure charge neutrality, unless a pure electron-positron pair plasma is assumed. However, in order to contribute significantly to the radiative output (through proton synchrotron and $p\gamma$ pion production), they need to be accelerated to energies $E_p \gg 10^{16}$ eV (see Sect. 2.2). In order for such acceleration to be possible within the constrained size of the blazar emission region, large magnetic fields of $B \gtrsim 10$ G are required.

If the acceleration of protons to the required energies does happen in blazar jets, the hadronic radiation processes discussed in Sects. 2.2 and 2.3 become relevant. Due to the high magnetic fields, the radiative output from leptons in the emission region will then be strongly dominated by synchrotron radiation, so that the high-energy X-ray through γ-ray emission is likely dominated by proton synchrotron radiation and synchrotron emission of secondaries resulting from photopion production. Such models are termed *hadronic* or *lepto-hadronic*

[3]For simplicity we are assuming that the clouds producing the broad lines are located at the same distance. In reality, different emission lines are likely produced in a stratified BLR, in which the ionization factor decreases with distance (Netzer 2008).

Fig. 5 Leptonic (*red*) and hadronic (*green*) fits to the simultaneous SED of the intermediate BL Lac object 3C66A. Different line styles indicate different radiation components: *solid* = total SED, including correction for EBL absorption; *dotted* = primary electron synchrotron; *red dashed* = EC (torus); *red dot-dashed* = SSC; *green dot-dashed* = proton synchrotron; *green dashed* = total proton synchrotron + cascade emission

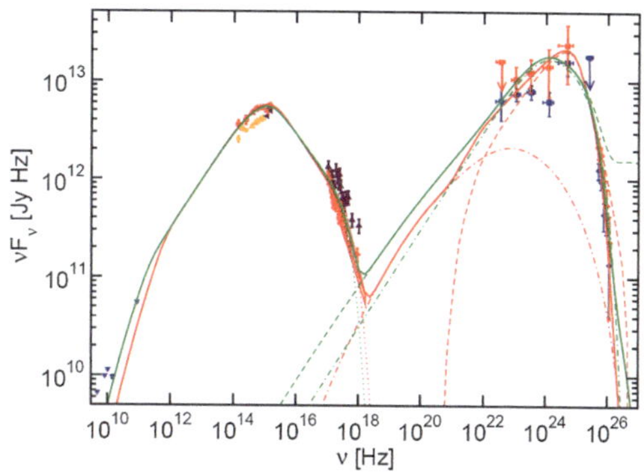

models (e.g., Mannheim 1993; Mannheim and Biermann 1992; Mücke and Protheroe 2001; Mücke et al. 2003; Reynoso et al. 2011; Petropoulou and Mastichiadis 2012; Böttcher et al. 2013; Petropoulou et al. 2015)

As the particle densities in blazar jets are very low ($n \lesssim 10^3$ cm^{-3}), pion production processes are strongly dominated by pγ interactions, in which case the secondary particles (pions, muons, electrons/positrons) are produced with large Lorentz factors of $\gamma \gtrsim 10^7$. π^0 decay photons and synchrotron photons from electrons/positrons at these energies emerges at $\gg 1$ TeV energies, where the emission region is highly opaque to $\gamma\gamma$ absorption, producing additional electron/positron pairs in a cascading process. Due to the extreme energies in protons and the relatively low radiative efficiency of protons compared to leptons, as well as the increased bulk mass to accelerate, hadronic jet models of blazars typically require several orders of magnitude larger jet powers than leptonic models for the same blazars. In some cases, the required jet powers even exceed the Eddington luminosity of the central supermassive black hole, thus requiring a revision of our current understanding of the formation of relativistic jets and related energy considerations (see, e.g., Sect. 2.1; Zdziarski and Böttcher 2015).

In terms of fitting SEDs of blazars, hadronic models have met with similar success as leptonic models (see, e.g., Fig. 5). Hence, SED modeling is generally not sufficient to distinguish leptonic from hadronic models and, thus, to constrain the jet energetics, and additional diagnostics need to be developed. One possibility lies in the different variability patterns expected in leptonic and hadronic models, due to the vastly different radiative cooling times of protons compared to electrons/positrons. Due to the complexity of hadronic models (including 8 different particle species + photons, with a rather complicated energy dependence of pγ pion production processes), the development of time-dependent hadronic models has only recently made significant progress (e.g., Mastichiadis et al. 2013; Diltz et al. 2015; Weidinger and Spanier 2015).

Another alternative to distinguish leptonic from hadronic models consists of the fact that only hadronic processes result in the production of high-energy neutrinos (e.g., Petropoulou et al. 2015). Thus, the identification of blazars with the sources of astrophysical TeV–PeV neutrinos, as now confidently detected by IceCube (Aartsen et al. 2014) would unambiguously prove the presence of ultrarelativistic protons in blazar jets. However, unfortunately, the angular resolution of IceCube (and other neutrino observatories) is currently too poor to provide such a unique identification (but see, e.g., Kadler et al. 2016).

In light of current developments of X-ray and γ-ray observatories with polarimetry capabilities, high-energy polarization may also be considered as a distinguishing diagnostic of leptonic and hadronic models. In particular, Zhang and Böttcher (2013) have shown that the X-ray and γ-ray emission from blazars in the case of a hadronic origin is expected to be polarized with polarization degrees similar to those observed in the optical band (i.e., $\Pi \lesssim$ a few tens of percent). On the other hand, in leptonic models, the X-ray emission of most (non-HBL) blazars is expected to be dominated by SSC emission, which is expected to exhibit a degree of polarization of about half that of the target photon field, i.e., $\Pi_X \sim \Pi_{\rm opt}/2$. In the case of EC dominated γ-ray emission, those γ-rays are expected to be essentially unpolarized, in stark contrast to the prediction of hadronic models. Thus, X-ray and γ-ray polarimetry of blazars may serve as a powerful diagnostic to distinguish leptonic and hadronic emission models.

3.2 One-Zone *vs* Multizone/Structured Jets

In the previous section we discussed the emission scenarios without specifying in detail the nature and the geometry of the emission region. Unfortunately, our knowledge of the geometry and even the composition of jets is still rather poor. Therefore in the modeling of blazars even the specific characteristics of the region responsible for the emission should be considered as a free parameters.

Blazar models can generally be divided in inhomogeneous (or jet) and homogeneous models. In the first class of models (Blandford and Königl 1979; Ghisellini et al. 1985) one assumes that the observed emission is produced in an extended volume along the jet and a particular geometry (e.g. conical, parabolical) for it (see also Sect. 2.1). Moreover one needs to specify the evolution of the main physical quantities (magnetic field, particle density, particle energy) along the jet: this necessarily introduces a great number of free parameters.

Most of the recent discussions in the literature consider a single, homogeneous emission region. The most direct support to these *one-zone* models comes from the correlated variability at different frequencies (Ulrich et al. 1997), although deviations from this behavior (e.g. the so-called orphan flares, Krawczynski et al. 2004) are sometimes observed. It is clear that this is an over-simplification of the actual geometry and physics involved in jets, but with the advantage to drastically reduce the unknown parameters of the model. This approach is valid to approximate the very compact region where high-energy radiation is produced. Consequently, such models do typically not reproduce the low-frequency radio emission from blazars, which is known to emerge in larger-scale ($\gg$ pc) jet structures, as evidenced by the drastically different radio variability time scales compared to optical through γ-ray variability, and consistent with the fact that such one-zone model configurations are optically thick to synchrotron self-absorption at radio frequencies.

Emission models typically assume a-priori that the jet contains high energy particles. An evident problem to consider is the identification of the physical processes able to energize particles to such high energies. Clearly, the problem of the nature of the emission region and that concerning the acceleration process(es) are coupled. A "paradigm" popular until recent years is the so-called *shock-in-jet* model (Marscher and Gear 1985; Spada et al. 2001). It is assumed that, due to instability in the flow, an internal shock can form in the jet. Particles are then accelerated at the shock front through diffusive shock acceleration (Fermi I) processes. This scenario is supported by the observations of traveling "knots" at VLBI ($\approx$ pc) scale. In this scheme the emitting region can be identified with the region downstream the shock, where particles injected from the front radiatively cool and emit (Kirk et al. 1998; Chiaberge and Ghisellini 1999; Sokolov et al. 2004; Böttcher and Dermer 2010; Joshi and Böttcher 2011; Chen et al. 2011).

However, several criticisms to this view have been advanced. In fact several observational and theoretical arguments seem to challenge the shock model, at least in its simplest form.

Among them, the observation of episodes of ultra-rapid variability in γ rays both in BL Lacs and FSRQ, with timescales down to few minutes (Aharonian et al. 2007; Albert et al. 2007; Arlen et al. 2013; Aleksić et al. 2011a), implies, through the causality argument, very compact emission regions ($R < ct_{\rm var} D \approx 10^{14}$ $[t_{\rm var}/1\ {\rm min}]$ $D/10$ cm) incompatible with the expected size of a shock. Also, detailed fits often reveal that a model with a unique region is not able to provide a good representation of the SED and more than one component is required (Aleksić et al. 2015; Barres de Almeida et al. 2014; Tavecchio et al. 2011). From a more fundamental perspective, it is expected that, at the typical distance of the "blazar zone", the flow is still dominated by the Poynting (i.e. magnetic) flux (e.g., Komissarov et al. 2007). In these circumstances diffusive particle acceleration by shocks is expected to be quite inefficient (Kennel and Coroniti 1984; Sironi et al. 2015).

All these issues bring stimulated new proposals for the structure of the emission region. Rapid variability suggests that (at least occasionally) the radiation is produced in very compact regions embedded in the jet. This have been identified with turbulent cells (Narayan and Piran 2012; Marscher 2014) or with large plasmoids resulting from efficient magnetic reconnection in a relativistic regime (Giannios et al. 2009; Giannios 2013; Sironi et al. 2015). In both cases, the compact regions are thought to move relativistically in the jet reference frame. The resulting beaming of the emitted radiation in the observer frame (which is the combination of the beaming factor of the compact region in the jet frame and that of the jet flow with respect to the observer) can be very large, with Doppler factors easily reaching $D \sim 50$.

In the framework of the Turbulent Extreme Multi-Zone (TEMZ) model (Marscher 2014), the turbulence also reproduces variations of the linear polarization angle (PA), possibly including large-angle PA swings occasionally observed during FSRQ monitoring (e.g., Marscher et al. 2010). An alternative explanation for such PA swings, correlated with multi-wavelength flares, was suggested by Zhang et al. (2014, 2015), in the context of an inhomogeneous shock-in-jet model. They have shown that such a model, in which the emission region is pervaded by a helical magnetic field, naturally produces PA swings of 180° (or multiples thereof in the case of multiple disturbances propagating through the jet) associated with multiwavelength flares, when carefully accounting for all light travel time effects, without the need for any asymmetric jet structures (such as bends or helical motions of plasmoids along the jet). In Zhang et al. (2015), this model was successfully applied to model, for the first time, snap-shot SEDs, multiwavelength light curve *and* time-dependent optical polarization signatures (Π and PA) of 3C279 during the correlated PA swing + γ-ray flare (Abdo et al. 2010) of 2009, in one coherent model (see Fig. 6). In a follow-up study, this model was more self-consistently coupled to the results of MHD simulations of relativistic shocks in blazar jets in order to constrain, from first principles, the particle energetics and magnetic-field topology (Zhang et al. 2016).

The magnetic reconnection scenario also appears rather promising. Dedicated Particle in Cell plasma simulations (Sironi and Spitkovsky 2014) indicate that the combined action of acceleration at X-points and the subsequent Fermi-II like acceleration from the magnetic islands could provide an efficient way to rapidly accelerate electrons (and ions). A possible problem for this scenario is that BL Lac jets (in which most of the ultra-fast events have been recorded) appear to be weakly magnetized when the SED is reproduced by standard emission models (Tavecchio and Ghisellini 2016). Even if in the process of magnetic reconnection, the original magnetic energy density is reduced in the reconnection zones, much of the radiative energy dissipation of accelerated particles is expected to occur in the remaining magnetic islands, in which the magnetization still has to be high for magnetic reconnection to occur.

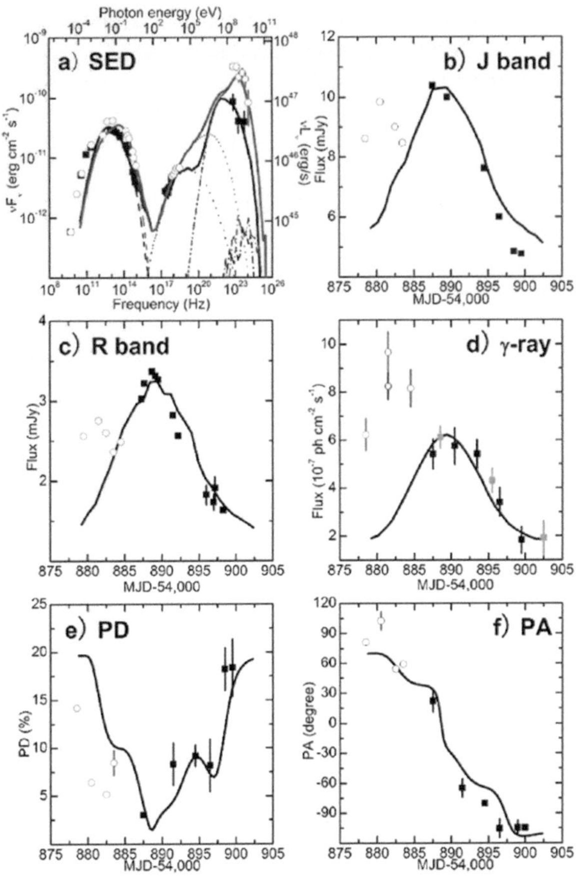

Fig. 6 Simultaneous fits to flux and polarization signatures of the FSRQ 3C279 during the correlated PA swing + multiwavelength flare of 2009 (Abdo et al. 2010), using the polarization-dependent shock-in-jet radiation transfer model of Zhang et al. (2014). (**a**) Snap-shot SEDs in quiescence and at the peak of the flare; (**b–d**) flux light curves in the IR (J-band), optical (R-band) and γ-rays (Fermi-LAT); (**e**) time-dependent optical polarization degree; (**f**) time-dependent polarization angle, reproducing the observed PA swing. From Zhang et al. (2015)

Another possible multi-zone model especially relevant for BL Lac jets is the so-called *spine-layer* model (Ghisellini et al. 2005; Tavecchio and Ghisellini 2008). In this picture, the jet is thought to be composed by two regions, an inner spine, moving with bulk Lorentz factor $\Gamma_s = 10$–20, surrounded by a slower ($\Gamma_l \simeq 5$) layer. This kind of structure could result naturally from the jet acceleration mechanism itself (McKinney 2006) or due to the interaction with the external environment (Rossi et al. 2008). If both structures produce radiation, the energy density of the radiation produced by one component is relativistically amplified in the rest frame of the other component by the square of the relative Lorentz factor, $\Gamma_{\rm rel} = \Gamma_s \Gamma_l (1 - \beta_s \beta_l)$. In this way the IC emission of the spine (which, having a much larger Lorentz factor, will preferentially dominate the emission at small viewing angles) receives a contribution from this kind of EC process. This structure could also be important in the possible production of high-energy neutrinos from BL Lac jets (Tavecchio et al. 2014) and to unify BL Lacs with the misaligned parent population of FRI radio galaxies (Chiaberge et al. 2000).

3.3 Location of the Emitting Region

Therefore the location of the emitting region in the jet is still quite hotly debated, especially for the case of γ rays. The problem is more acute for FSRQs for which the radiative environment of the jet changes significantly along the jet. As we have pointed out above, most of the models in the past assumed that the emission occurs within the BLR radius. Indeed, this is consistent with the variability timescale (down to a few hours, Tavecchio et al. 2010) and, importantly, the large radiative energy density ensures a large efficiency of the EC power, allowing one to minimize the total jet power (Ghisellini et al. 2014). However there is independent evidence that, at least occasionally, the emission occurs beyond the BLR radius, possibly even at several parsecs from the center (e.g., Agudo et al. 2011).

Important (almost model independent) evidence comes from the detection of VHE γ-ray photons from a handful of FSRQs (3C 279, PKS 1510-089, PKS 1222+216, PKS 1441+25, B0218+357) during active flaring states. The discovery of FSRQ as VHE sources was quite surprising since, in the standard scenario, a huge opacity for gamma rays with energies above few tens of GeV is expected (e.g., Donea and Protheroe 2003; Liu and Bai 2006). The UV radiation from the BLR (lines and continuum) is a target for $\gamma\gamma$ pair production, whose energy threshold is $E_{\gamma,\mathrm{th}} = 25/(E_{\mathrm{le}}/10\ \mathrm{eV})$ GeV. The optical depth starts to be important at an energy corresponding to the intense H Lyα line, $E_{\mathrm{le}} \sim 10$ eV, that is for $E_{\gamma} \sim 25$ GeV. The detection of photons with $E > 30$ GeV therefore, especially in powerful sources—for which the optical depth should be particularly large, naively increasing as $\tau \approx L_{\mathrm{disc}}^{0.5}$—immediately leads to the inference that the emission should occur beyond (or very close to) R_{BLR} (Pacciani et al. 2014; Boettcher and Els 2016; Coogan et al. 2016). Models for the VHE emission from FSRQs adopt this view, and generally assume that the observed high-energy radiation is produced through the IC scattering of the IR radiation field of the molecular torus (Tavecchio et al. 2011; Aleksić et al. 2011b, 2014; Pacciani et al. 2014; Ackermann et al. 2014; Ahnen et al. 2015).

For quiescent states there is no evidence of VHE emission from FSRQs (although this conclusion is clearly influenced by sensitivity limits). Therefore, it is not clear whether the quiescent emission should also originate outside the BLR. Through the analysis of long-term *Fermi*/LAT data of bright FSRQs, Poutanen and Stern (2010) claimed that the spectra show a well-defined and sharp break close to 6 GeV, which can be readily interpreted as induced by absorption through the interaction with the He II Lyα line/recombination continuum, expected from highly ionized clouds. On the other hand, such a break could also be intrinsic to the emitted spectrum, arising from the combination of two components, namely, the Compton-upscattered disk and BLR radiation fields (Finke and Dermer 2010), or from Klein-Nishina effects and a curving electron distribution (Cerruti et al. 2013). Further investigations showed that the amount $\Delta\alpha_{\gamma}$ of the break is variable, with also the indication of a possible anti-correlation between the column density of the He II Lyα continuum and the gamma-ray flux (Stern and Poutanen 2011). A possible way to interpret this complex phenomenology is to assume a stratified BLR, with lines of decreasing ionization produced at increasing distances from the central engine and with the emission region changing position with time (Stern and Poutanen 2014).

3.4 A Blazar Sequence?

Soon after the discovery of the intense gamma ray emission from blazars, Fossati et al. (1998) reported the existence of a trend between the radio luminosity (at 5 GHz) of a blazar and the position of the two SED peaks (see Fig. 7). This trend was dubbed the *blazar sequence*. According to the basic scheme, the SEDs of powerful sources (FSRQs) peak in the mm and MeV bands. Considering sources of lower luminosity, the two peaks progressively shift to higher frequency, reaching the X-ray band and the TeV band, respectively, in the case of the low-luminosity BL Lac objects. Together with this trend,

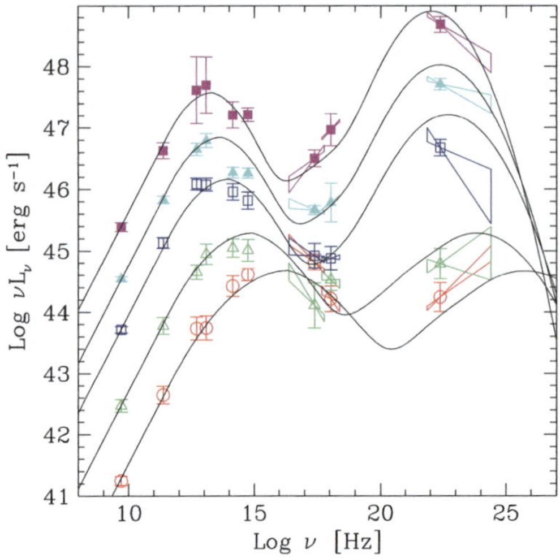

Fig. 7 The data-points show the average SED of blazars in bins of different radio (5 GHz) luminosity. The polynomial fitting curves (*black lines*) mark an inverse trend connecting the radio luminosity and the peak frequency of the two spectral component (the *blazar sequence*). From Donato et al. (2001)

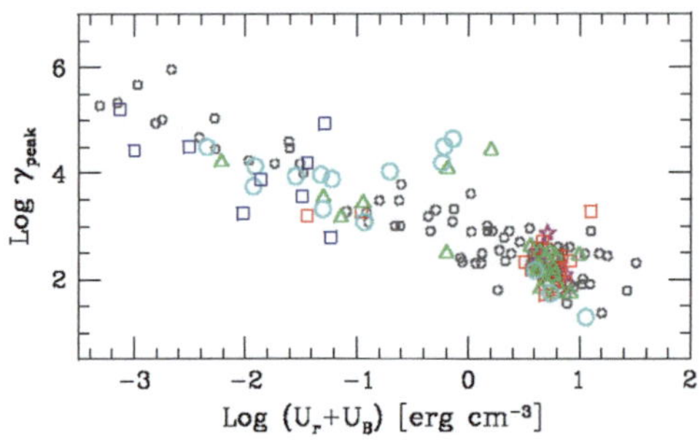

Fig. 8 Lorentz factor of relativistic electrons emitting at the SED peak versus the total energy density (magnetic plus radiative) for a sample of blazars. From Ghisellini et al. (2010)

one can also observe a decreasing *Compton dominance*, defined as the ratio of the luminosity of the high-energy bump with respect to the low energy one (see also Finke 2013).

The SED modeling (Ghisellini et al. 1998) shows that the *spectral* blazar sequence is accompanied by a correlation between two physical quantities, namely the energy (or the Lorentz factor) of the electrons emitting at the SED peaks $\gamma_{\rm p}$ and the sum of the energy density in magnetic field and radiation, $U = U_{\rm mag} + U_{\rm rad}$ of the form $\gamma_{\rm p} \propto U^{-1/2}$ (Fig. 8). This led Ghisellini et al. (1998) to heuristically propose that the main agent ruling the blazar sequence is the radiative cooling of the relativistic electrons. In fact, assuming a universal acceleration mechanism acting in all jets ensuring a characteristic acceleration rate per particle $\dot{\gamma}_{\rm acc}$, the equilibrium between acceleration gain and radiative losses $\dot{\gamma}_{\rm acc} = |\dot{\gamma}_{\rm cool}| \propto \gamma^2 U$ would naturally lead to the derived correlation. While the first work included only a few objects, with quite sparse data, subsequent studies using larger samples, more complete SED and better γ-ray data, confirmed the existence of the trend (Ghisellini et al. 2010; Ghisellini and Tavecchio 2015). The above *cooling scenario* has been further extended in Ghisellini and Tavecchio (2008), to include scaling laws between black hole mass, AGN luminosity, jet power and BLR radius. In this scheme the ultimate parameter determining the blazar appearance could be the accretion rate onto the central supermassive black hole in Eddington units (Ghisellini et al. 2009).

Still not clear is the connection between the general trend shown by the sequence (peaks shifting towards lower frequency for increasing emitted power) and the behavior of single sources which, instead, generally follow the opposite pattern between quiescent and flaring states. A possibility is that the evolution is still dictated by the equilibrium between cooling and acceleration rate, with the latter increasing during active states. This seems to be the case in the well studied BL Lac Mkn 421 (Mankuzhiyil et al. 2011), for which the SED fits of different states reveal that the magnetic field in the emitting region and the energy of the electrons emitting at the SED peaks are related by the expected relation $B \propto \gamma^{-2}$. On the other hand, for the other prototypical VHE BL Lac, Mkn 501, the evidence suggests that only the electron energy is changing (Tavecchio et al. 2001; Mankuzhiyil et al. 2012). In this case a possibility is that the parameter regulating the SED is the acceleration efficiency.

The actual existence of the blazar sequence is the subject of a lively debate. Most of the criticisms focus on the fact that observational biases could severely affect the selection of these sources, possibly introducing spurious trends. For instance, about 40–50 % of the known BL Lac objects lack a reliable redshift measurement (made difficult by the weakness of their emission lines) and this could hamper the identification of far, powerful BL Lacs, which would patently violate the SED sequence. As a way to circumvent this limitation, Finke (2013) suggested to use the Compton dominance instead of luminosity as the quantity to correlate with the SED peak frequencies, which does not require the knowledge of redshifts (with only a negligibly small error introduced by a lack of knowledge in the redshift of the peak frequencies). Using this new correlation discriminant, he found renewed evidence for the existence of a sequence, in agreement with the original Fossati et al. (1998) sequence for BL Lac objects, especially at high peak frequencies, but essentially no correlation for blazars with synchrotron peak frequencies below $\sim 10^{14}$ Hz, with Compton dominance parameters spreading over ~ 3 orders of magnitude without any correlation with the synchrotron peak frequency. The efforts devoted to investigate the sequence at low luminosities also start to show possible deviations from the expectations (e.g., Caccianiga and Marchã 2004; Padovani et al. 2007; Raiteri and Capetti 2016), although the situation is particularly delicate since low luminosity non-blazar AGNs, misaligned (hence less beamed) blazars or blazars with (relatively) low mass black holes could "pollute" the samples. These criticisms have been condensed by Giommi et al. (2012) (see also Giommi et al. 2013; Giommi and Padovani 2015) into a simplified view aiming at demonstrating that, even without any real trend in the intrinsic SED, selection effects alone can be responsible for the observed spectral sequence.

It is important to remark that, in discussing these issues, one should always clearly distinguish between the (observational) blazar sequence, in the sense of Fossati et al. (1998), and the (theoretical) cooling scenario (and its extensions). In fact, even if the spectral sequence could be affected by selection effects, the latter is a more general physical scheme which, in fact, can accommodate or even envisages the existence of possible outliers from the canonical blazar sequence (e.g., Ghisellini and Tavecchio 2008; Ghisellini et al. 2012, 2013). Note also that, according to the general scenario developed in Ghisellini and Tavecchio (2008), the observed sequence could just be the "tip of the icerberg" of the blazar population, representative only of the most massive and powerful blazars. Sources with smaller black hole masses or slightly misaligned jets are expected to produce low-luminosity jets with, however, a SED shape similar to that of the powerful FSRQ, thus appearing as outliers from the blazar sequence. In this sense the observed sequence could just be the upper edge of an envelope of progressively misaligned blazars (Meyer et al. 2011).

4 X-Ray Binaries—Microquasars

Over the last decades, it has become clear that many aspects of accretion physics are mirrored across the black hole mass scale, irregardless of the size difference and accretion environment. The extent of this 'mass-scaling' suggests the relative importance of mechanisms within the inner accretion flow rather than those dictated by fueling and/or outer disk geometry, and gives clues about the nature of jet launching. Although our theoretical understanding still lags the quality of the precision, broadband data in many cases, we are starting to find empirical trends that allow an urgently needed reduction in the number of free parameters.

In isolation, the modeling of jets and particle acceleration in XRBs is subject to the same uncertainties and avalanche of free parameters as discussed in the above Sections for AGN. However because XRBs evolve on much faster timescales, they offer a unique perspective, where an individual can traverse a wide variety of behaviors that only large samples of AGN can equally populate. This has opened a new window on one of the more persistent problems to solve in jet physics: the coupling between bulk dynamical properties (or macrophysics), and the microphysics of processes like particle acceleration and radiation. In this Section we will summarize some of the recent progress in this area and its extension to AGN.

XRBs come in several flavors, high and low mass (with respect to the companion star), and with black holes or neutron stars as the compact object. In order to compare with AGN we will be limiting ourselves to discussing black hole XRBs. The complicated phenomenology of XRBs has been covered in several papers and reviews (see e.g., McClintock and Remillard 2006; Fender 2006; Belloni 2010) to which we refer the interested reader. Here we will only summarize the most relevant properties, in order to focus the discussion on comparisons with jetted AGN with a focus on breaking model degeneracy and better understanding particle acceleration.

Low mass XRBs (LMXBs) are black holes with a main-sequence companion star whose mass is smaller than the compact object, typically on the order of a solar mass. The star fills its Roche lobe and thus feeds material to the black hole via its L1 Lagrangian point, which leads to a well defined entry point for material to populate an accretion disk in the orbital plane. LMXBs go through transient outbursts, likely driven by thermal instabilities in the accretion disk (e.g. Lasota 2001), which initiate a sequence of distinct states, originally classified in terms of X-ray and spectral timing properties. Figure 9 shows a schematic view of this pattern of cyclic behavior, using actual data from the Galactic LMXB GX339−4, which has been a great boon for our understanding because it experiences a full outburst cycle approximately every 1–2 years. The figure shows the outburst pattern in a commonly used phase space of X-ray luminosity (related to, not necessarily linearly) the accretion rate $\dot{M}$ onto the black hole, and the X-ray spectral hardness. The latter indicates the relative dominance of thermal (soft spectrum) versus nonthermal (hard power law) processes. This type of diagram is referred to a hardness-intensity (HID) diagram.

During a typical outburst, a LMXB rises out of quiescence (off the plot in the lower right corner in Fig. 9), a state which is quite difficult to study by definition because of the very low luminosity. The weakest sources seen in quiescence so far are XTE J1118+480, A0620−00, and Swift J1357.2−0933, all with $L_X \leq 10^{-8} L_{\rm Edd}$ (e.g. Gallo et al. 2007, 2014; Plotkin et al. 2016). At some point the rising X-rays trigger a monitoring instrument such as *SWIFT* or *MAXI* (in the past it was typically *RXTE*) and these facilities then send out alerts, allowing further multiwavelength coverage. At the point that a source is bright enough to trigger facilities, it is already in the so-called hard state. This state is associated with compact, steady jets which are synchrotron self-absorbed, resulting in the classical flat/inverted spectrum associated also with the compact cores of AGN (Blandford and Königl 1979). The hard state

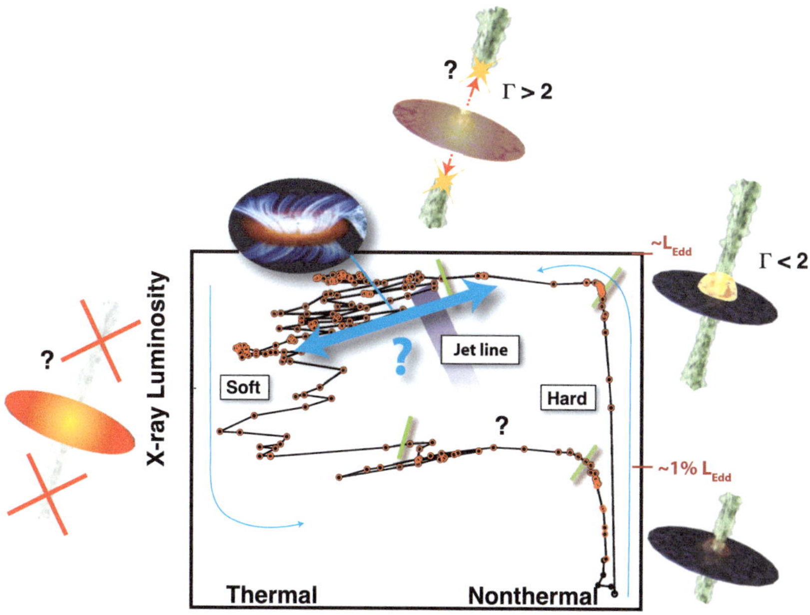

Fig. 9 The hardness-intensity diagram (HID) for an outburst of XRB GX339−4, with schematics of the corresponding inflow/outflow configuration per state. Only the two canonical states (hard and soft) are indicated

is the longest lasting of the LMXB states and, while quiescence has been argued to be a separate state, the latest observational evidence supports the persistence of jets smoothly down into quiescence, as well as other properties (see, e.g., Plotkin et al. 2015).

As the source luminosity increases (assumedly driven by a rising accretion rate), somewhere around $\sim 0.5 L_{\rm Edd}$ the spectrum begins to soften and the source moves counterclockwise around the upper right corner in the diagram, beginning a transition into a state often referred to as the Hard Intermediate State (HIMS). These transitions are harder to categorize based on radio or X-ray spectra, but are clearly demarcated via changes/features in the power density spectra (e.g., Belloni 2010), that we will not discuss in detail here, but that include the presence of quasi-periodic oscillations (QPOs) and various types of broadband variability. The hard-to-HIMS transition occurs very quickly and within a day or days, the jets are seen to radically transform from steady flow to a transient state with brighter, discrete ejecta at a distinct point often referred to as the "jet line" in the HID. The jets appear now as discrete optically thin blobs moving with superluminal motion; it was on this state that the original argument for the nickname "microquasar" was based (Mirabel and Rodríguez 1994). After the flaring ends, the source is in the Soft Intermediate State (SIMS) and the jets essentially deconstruct: the radio and IR synchrotron radiation vanishes and the source proceeds into the left side of the diagram, into a soft thermal state that is almost entirely disk dominated, with a multi-temperature black body. As the luminosity decreases (assumedly along with $\dot{M}$), the spectrum begins to harden again and the jets reform, at a luminosity around an order of magnitude or more below the hard-to-soft transition. This hysteresis effect is not well understood, but it seems clear that something besides the accretion rate, likely magnetic field topology, is an important driver. With renewed steady jets, the hard state recedes down into quiescence and the outburst is over. A typical outburst can last from months to $\sim$ a year.

In contrast to LMXBs, HMXBs tend to be persistent, with Cyg X-1 seen as the "prototypical" source of its class. Cyg X-1 does show state changes, but with a much lower range in luminosity, and because it does not go into quiescence it never undergoes the full hysteresis loop (e.g. Gleißner et al. 2004; Wilms et al. 2006). Mostly it varies between soft and hard states. In contrast to LMXBs, only a few HMXBs are known in our Galaxy (Tetarenko et al. 2016).

The two states that can be associated with particle acceleration via the presence of jets are thus the hard and HIMS. Interestingly, the discovery of jets in XRBs came over 30 years after their discovery as a class via X-ray observations, at which point a paradigm based entirely on accretion disk phenomena was already well established. The X-ray hard state SED is typically comprised of a nonthermal power law with $\Gamma \sim 1.5$–1.8, and an exponential cutoff around 100 keV. Sometimes there are also weak reflection features such as an iron fluorescent line and a "Compton bump" leading to hardening above $\sim$ 15 keV (Lightman and White 1988; Fabian et al. 1989). The hard state SED can be explained in terms of a moderately optically thick ($\tau \lesssim 1$), static region of hot electrons existing near the black hole and disk. The existence of this region, named the corona, was inferred in order to explain the hard X-ray power-law via unsaturated inverse Compton scattering. In this scenario, the corona upscatters the weaker disk thermal photons (not usually visible in the spectrum, but assumed present because of the clear thermal disk signature in the soft state). This type of scenario seems natural, because the temperature that gives both the spectral index (for some optical depth) and cutoff is consistent with the Virial temperature for a compact region within 10s of r_g from the black hole (see, e.g., Titarchuk 1994; Magdziarz and Zdziarski 1995).

As is common in astrophysics, however, increasing precision in the observations tends to provide challenges. More extensive monitoring after the launch of *RXTE* revealed significant variability in the cutoff over relatively short timescales (e.g., Rodriguez et al. 2003; Wilms et al. 2006), suggesting a more dynamic region than originally envisioned. Another complication was the relatively weak reflection signatures in many observations. In order to reduce the amount of coronal emission intercepted by the cooler disk, many variations of geometry were developed, including a recessed disk, a patchy corona, magnetic flares, etc. (see, e.g., Haardt 1997; Merloni and Fabian 2002). One paper in particular suggested that a magnetic corona could be accelerated to moderately relativistic speeds away from the disk, thus providing an explanation for the relatively weak reflection signatures in Cyg X-1 (Beloborodov 1999). This latter scenario is especially relevant in the context of integrating jets into the existing coronal picture.

The problem is that relativistic jets need dynamically important, ordered magnetic fields threading the disk near the black hole, in order to launch (e.g. McKinney 2006; Beckwith et al. 2008). The need for these fields creates two main complications to the original picture: first, the coronal electrons will want to follow the magnetic fields lines and the static picture breaks down. Secondly, the introduction of a strong magnetic field means that the electrons will also radiate (and thus cool) via synchrotron emission, and those photons can also be inverse Compton upscattered (SSC) along with those from the disk.

The presence of jets thus requires some rethinking of the original picture, and highlights the need to find a self-consistent link between the three types of flow: inflowing disk, corona, and outflowing jets. The problem is that multiple regions introduces degeneracy for the interpretation of the hard X-ray spectrum, which can now be from a thermal corona, a hybrid thermal/nonthermal corona, an isolated nonthermal corona or a magnetized corona comprising the base of the jets, as well as direct synchrotron emission (Markoff et al. 2005). In the

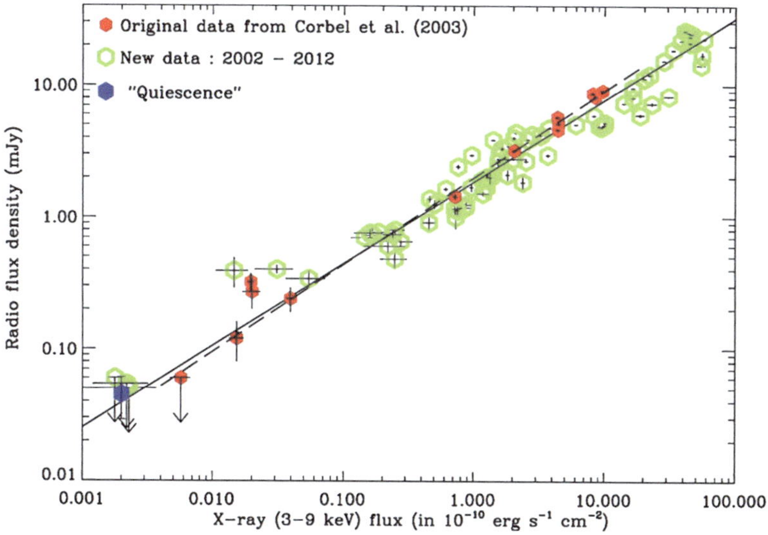

Fig. 10 Radio vs. X-ray correlation in the hard state of GX 339−4, with the original correlation shown in *red* from Corbel et al. (2003), and the newer outbursts. The *dashed line* illustrates the original radio/X-ray correlation from the 1997–1999 period, while the *solid line* corresponds to a fit to the new whole sample. The *blue point* shows the source close to quiescence. From Corbel et al. (2013)

meantime, more recent studies of reflection suggest that the data are more consistent with a source of disk illumination that lies on the axis above the disk, and changes its height ("lamppost" reflection, see, e.g., Miniutti and Fabian 2004; Dauser et al. 2013). This idea further supports a hybrid corona/jet base, where the effect of even mildly relativistic acceleration of the jets away from the disk means that IC/SSC from the jet base will dominate over synchrotron emission in the reflected component, unless particle acceleration starts within 10s of r_g of the black hole (see, e.g., Markoff and Nowak 2004).

In an attempt to rule out one or more of the various hard state emission scenarios, Nowak et al. (2011) carried out an extensive, simultaneous campaign on Cyg X-1 using *Suzaku*, *RXTE* and *Chandra-HETG*. Instead, they found that all main (Compton-dominated) scenarios provided equally good statistical descriptions of the new data sets. On the other hand, the fact that XRB jets clearly emit synchrotron radiation up to the mm/IR bands, depending on the observation (e.g. Fender 2006), presents a way to probe the connection between the components. For instance, coordinated multiwavelength observations can help to isolate the jet-related components, as well as build a picture of the jet properties in XRBs compared to AGN (content, particle acceleration, etc.).

Thus in parallel with some of the developments in ideas about the corona, the era of coordinated, quasi-simultaneous observations of XRBs really began to take off at the start of this century, particularly between the radio and X-ray bands (but see also Motch et al. 1982). Soon after, a tight, nonlinear correlation was discovered between the radio and X-ray luminosities of GX339−4, holding over many orders of magnitude, and repeated during different outbursts (Fig. 10 and, e.g., Corbel et al. 2000, 2003, 2013). This correlation was soon found to extend to other sources (e.g. Gallo et al. 2003), as well as between the IR and X-ray bands (Russell et al. 2006).

Because the different configurations of disk vs jet dominance in the HID are also seen among the various AGN classifications, the idea soon arose that AGN may be undergoing

XRB-like state changes, just over much longer timescales. This idea of trying to "pair" XRB states to AGN classes, often referred to as 'mass scaling' in accretion, has been the topic of much exploration in the last $\sim$ 13 years. The most successful pairings to date involve the longer-lasting hard and soft states. For instance, there is a compelling mass-dependent scaling between the power spectra of the soft/intermediate state and that of Seyferts (e.g., McHardy et al. 2006). However, in the context of jets and particle acceleration, the most interesting scaling is the so-called Fundamental Plane (FP) of black hole accretion.

The FP is an empirical relation, essentially a 2D plane linking all sub-Eddington, jetted black hole types in the 3D space of X-ray luminosity, radio luminosity and mass. After the discovery of the radio/X-ray correlation in individual XRBs, it was quickly realized that this correlation represents the ratio of the efficiencies for the two radiative processes. For instance Markoff et al. (2003) considered the scaling in the context of synchrotron emission from the jets. Using a Blandford-Königl-like jet model (Blandford and Königl 1979), one can show that the self-absorbed, flat/inverted radio-through-IR part of the spectrum will scale as $\sim \dot{M}^{17/12-2/3\alpha_{\rm RIR}}$, where $\alpha_{\rm RIR}$ is the spectral index of the self-absorbed part of the jet and is close to 0 (see also Falcke and Biermann 1995). The measured slope of the correlation is $L_{\rm X} \propto L_{\rm R}^{0.6-0.7}$ (Corbel et al. 2013), which constrains the dependence of $L_{\rm X} \propto \dot{M}^{2-2.3}$, ruling out radiatively efficient processes for the hard X-ray spectrum. This result was generalized by Heinz and Sunyaev (2003), expressing all physical scales in terms of $M_{\rm BH}$ (via r_g) and all powers in terms of $\dot{m} \sim L_{\rm Edd}/(c^2) \propto M_{\rm BH}$. This mass scaling effectively introduces a normalization term into the radio/IR/X-ray correlations. In other words, if compact jets and accretion disks are self-similar structures across the immense mass and power scales seen by "hard state-like" black holes, then all black holes should show a similar coupling of X-ray and radio luminosity as a function of relative accretion rate, once "corrected" by the mass.

This idea was explored and tested empirically by two different groups, using large samples of AGN Merloni et al. (2003), Falcke et al. (2004). If the AGN follow a similar correlation as hard state XRBs over much longer timescales, this trend would be revealed statistically by looking at snapshots of a large number of objects. Both groups, using different approaches, found that such a correlation indeed seems to extend to AGN, and in the subsequent years more discussion and exploration of the statistics has ensued. A good summary can be found in Plotkin et al. (2012b), as well as the most robust derivation of the FP, for the first time also including BL Lacs (see Fig. 11). While there is still some debate over exactly which AGN classes fit on the FP, it is pretty clear that sub-Eddington, steady jetted sources such as Low Luminosity AGN (LLAGN) or Low-Ionization Emission-Line Regions (LINERs), FRIs and BL Lacs do, and perhaps also Narrow-Line Seyfert-1 (NL Sy1) with radio detections (e.g. Gültekin et al. 2009).

The extension of the broadband correlations found in XRBs to AGN has therefore opened up a new pathway to study the disk/jet coupling in black holes. The X-rays probe the conditions near the black hole, in a corona or near the launchpoints of the jet, and its intimate connection to the transfer of energy into accelerated particles further out in the jets, responsible for the radio/IR parts of the FP correlation.

Having established that mass-scaling seems to hold in the disk/jet coupling, we can now consider how to use it in order to better understand the power and acceleration properties of the jets in the hard state (for all black hole masses).

4.1 Origin of the High-Energy Emission

Returning to the question of the origin of the hard state emission, the degeneracy between synchrotron and inverse Compton processes (and for the latter, also between different con-

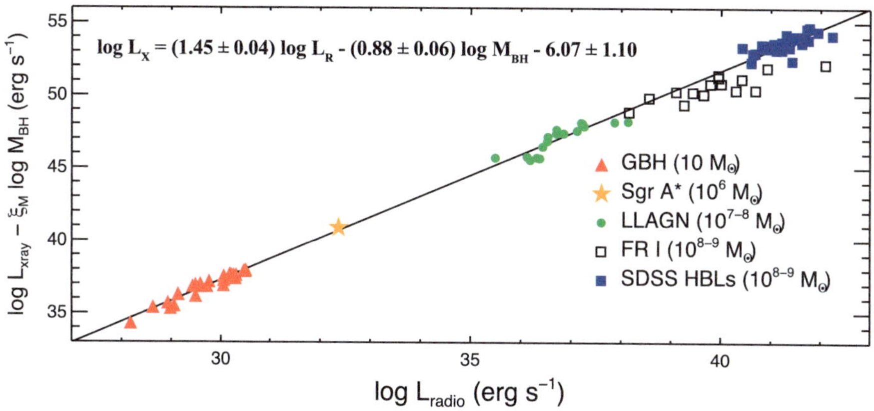

Fig. 11 A projection in mass of the Fundamental Plane for black hole accretion onto the radio and X-ray luminosity plane, using a Bayesian regression technique on data from hard state XRBs and a sample of sub-Eddington accreting AGN (Plotkin et al. 2012b)

tributing photon fields) exists at both ends of the mass range. In XRBs, the lower Lorentz factor of the jets limits EC to two contributions: the accretion disk or the companion star. In LMXBs the companion star is not bright enough to compete with the disk, reducing the possible source of the X-rays to direct synchrotron radiation, SSC emission, EC of disk photons, or some combination. In most scenarios all of these could give the correct scalings for the FP, thus the next step is to move beyond correlations and look at modeling the broadband SEDs.

The original suggestion that synchrotron could contribute significantly to the hard X-ray spectrum in XRBs came from Markoff et al. (2001a), for one of the first quasi-simultaneous broadband (radio through X-ray) SEDs compiled (Hynes et al. 2000). The hard X-ray spectrum shows little to no reflection, nor a high energy cutoff, in contrast to what would be expected for a thermal corona model. Furthermore, jets can account for all the nonthermal components in the spectrum, as long as the particle acceleration initiates somewhat offset from the black hole, at around $50r_g$. This requirement of having a transition from primarily thermal plasma in the jets (similar to Sgr A*, see, e.g., Markoff et al. 2001b) to a region where a powerlaw of particles is accelerated implies a break in the spectrum where this enhanced emission transitions from optically thick to thin. This feature is not always possible to detect because it often falls in the IR band where the companion star and/or disk can contribute, but it can be seen explicitly in GX339−4 (see Fig. 12, and, e.g. Corbel and Fender 2002), and detected or constrained in many other observations (Russell et al. 2010, 2013; Koljonen et al. 2015).

We will come back to the break in Sect. 4.2, but its location in the SED provides a hard limit on how much synchrotron emission can contribute to the hard X-ray spectrum. Unlike the SED for XTE J1118+480, most hard state XRBs, particularly at higher luminosity, do show evidence for some reflection and thus a contribution in SSC/EC from the corona, or jet base. However synchrotron emission can still contribute at the $\sim 10~\%$ level, which has implications for particle acceleration in XRBs. Even when the break is not visible, there are independent methods to gauge the jet synchrotron contribution. For instance Russell et al. (2010) studies the evolution of an outburst in the XRB XTE J1550−564 in three IR bands. During the decline from outburst in the disk-dominated soft state, a clear exponential decay is visible. Upon transition to the hard state, an excess emission above this monotonic

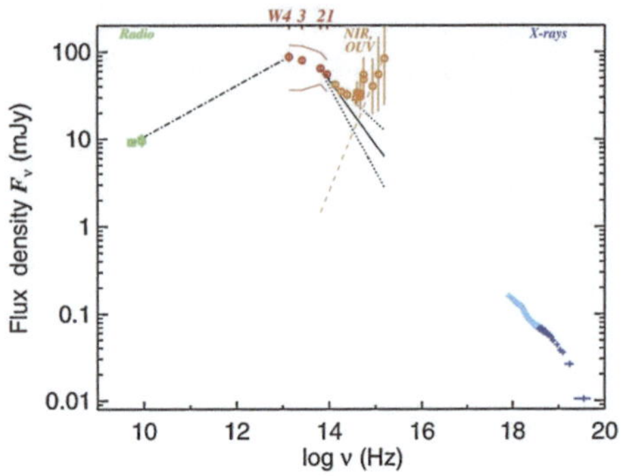

Fig. 12 Average dereddened SED of GX 339−4, including WISE data points for the first time (in *red*). *Red curves* represent the envelope of extreme variations over 13 WISE epochs. The near-infrared and optical/ultraviolet (OUV) points are plotted in *orange* and radio in *green*, and the break from optically thick to optically thin is apparent, see (Gandhi et al. 2011)

decrease appears, corresponding to the jets re-activating. Extracting this excess allows an independent determination of the jet synchrotron contribution to the IR band, and its spectral index, which is found to match that of the X-rays in both slope and normalization.

Another independent constraint on the contribution and geometry of the jets comes from fast IR/optical and X-ray variability studies. One of the first examples of such a spectro-timing study is presented for XTE J1118+480 in Hynes et al. (2003), where they extract the broadband SED of the rms variability in IR, UV and X-ray, and find it follows a $\nu^{-0.6}$ powerlaw, consistent with optically thin synchrotron emission. More recently, Kalamkar et al. (2016) detect the first IR QPO during a hard state of GX339−4 (at half the frequency of the X-ray QPO, and the same frequency as that found in optical/UV/Xray from Hynes et al. 2003 for XTE J1118), as well as strongly correlated sub-second variability, with the IR lagging the X-rays by 111 ms. The direction of the lag, as well as brightness temperature and size arguments, effectively rules out reprocessing or hot-inflow models, similar to argumentation in Hynes et al. (2003). The timescale for the lag corresponds to a light-travel distance of $3700 r_g$ for $\sim 6 M_\odot$, consistent with the distance of the IR-emitting region based on the location of the break in the spectrum of Gandhi et al. (2011) (Fig. 12). The first detection of a QPO in the IR further suggests a picture where oscillations rooted in the accretion flow are conveyed with the plasma into the jets, perhaps via ordered magnetic field lines.

Of all microquasars, Cygnus X-1 is the most extensively observed. This source has been the target of monitoring campaigns that allowed to estimate the parameters of the binary and to obtain detailed spectra at all wavelengths. Cygnus X-1 is located at 1.86 kpc from Earth. A high-mass stellar companion of spectral type O9.7 Iab and mass $\sim 20\ M_\odot$ and a black hole of 14.8 $M_\odot$ (Orosz et al. 2011) form the binary system.

In the X-ray band Cygnus X-1 switches between the typical hard and soft states of X-ray binaries. The soft state is characterized by a blackbody component of $kT \lesssim 0.5$ keV from an accretion disk, and a soft power-law with spectral photon index $\Gamma \sim 2$–3. The source, however, spends most of the time in its hard state, in which the spectral energy distribution is well described by a power-law of spectral index $\Gamma \sim 1.7$ that extends up to a high-energy cutoff at ~ 150 keV (e.g. Dove et al. 1997; Poutanen et al. 1997). The origin of this power-law is the Comptonization of disk photons by thermal electrons in the hot corona that partially covers the inner region of the disk. The detection of a Compton reflection bump and the Fe Kα line at ~ 6.4 keV support the presence of the corona during the low/hard state. Additionally, intermediate spectral states have also been reported (Belloni et al. 1996).

Persistent and transient jets have been resolved at radio wavelengths in Cygnus X-1 during the hard state (e.g. Stirling et al. 2001; Rushton et al. 2011). The outflow is extremely collimated (aperture angle $\sim 2°$, Stirling et al. 2001) and propagates at an angle of $\sim 29°$

with the line of sight (Orosz et al. 2011). The radio emission is modulated by the orbital period of the binary because of absorption in the wind of the companion star (e.g. Brocksopp et al. 2002).

Cygnus X-1 is one of the two confirmed MQs that is a gamma-ray source.[4] The first detection of soft gamma rays up to a few MeV was achieved with the instrument *COMPTEL* aboard the *Compton Gamma-Ray Observatory* (McConnell et al. 2002). Emission in the same energy range was later observed with *INTEGRAL* (e.g. Laurent et al. 2011). The *INTEGRAL* detections represented a breakthrough since it was found that the ~ 400 keV–2 MeV photons were highly polarized.

At higher energies Cygnus X-1 is fundamentally a transient source on timescales of 1–2 d. Episodes of gamma-ray emission have been detected with the satellite *AGILE* between 100 MeV and a few GeV in the hard state (Sabatini et al. 2010), and marginally during the hard-to-soft transition (Sabatini et al. 2013). The analysis of *Fermi*-Large Area Telescope (LAT) data at 0.1–10 GeV revealed weak flares (three of them quasi-simultaneous with *AGILE* detections; Bodaghee et al. 2013), and weak steady emission (Malyshev et al. 2013).

Finally, Cygnus X-1 has been observed in the very high energy band (≥ 100 GeV) with the Major Atmospheric Gamma Imaging Cherenkov (MAGIC) telescope during the hard state (Albert et al. 2007). During inferior conjunction a flare (duration of less than a day, rising time ~ 1 h) was likely detected with a significance of 4.0σ (3.2σ after trial correction). Only upper limits could be obtained for the steady emission.

The broadband spectral energy distribution (SED) of Cygnus X-1 in the hard state displays several components. The radio emission is synchrotron radiation of relativistic electrons accelerated in the jets. This component is observed up to its turnover at infrared frequencies, where the stellar continuum takes over. The emission of the disk/corona dominates up to the hard X-rays, but the origin of the MeV tail observed with *COMPTEL* and *INTEGRAL* is still disputed. Its high degree of polarization (> 65 %) suggests this component is emitted in an ordered magnetic field such as is expected to exist in the jets, a result supported by the fact that the polarized X-ray emission is only clearly detected during the hard state (Rodriguez et al. 2015). In this scenario the MeV tail would be the cutoff of the jet synchrotron spectrum, see for example the fits to the data obtained by Malyshev et al. (2013), Zdziarski et al. (2012, 2014), and Zhang et al. (2014). An alternative model was introduced by Romero et al. (2014), where the MeV tail is synchrotron radiation of secondary non-thermal electrons in the corona. This model predicts significant polarization also during intermediate spectral states, something that cannot be presently ruled out from the data (Rodriguez et al. 2015).

All known gamma-ray binaries host a high-mass donor star, a fact that points to a fundamental role played by the stellar wind and/or radiation field in the mechanisms that produce the high-energy photons. In leptonic models for MQs gamma rays are produced by inverse Compton (IC) scattering of the stellar radiation off relativistic electrons, whereas in hadronic models gamma-rays are generated by the decay of neutral pions injected in the interactions of non-thermal protons in the jets with cold protons of the stellar wind (Romero et al. 2003, 2005).

The inclusion of relativistic protons in the jets brings about a feature absent in purely leptonic models, namely the production of secondary particles (neutrinos, electron-positron pairs, muons, pions) in high-energy hadronic interactions. The cooling of charged secondaries may contribute to the radiative spectrum of the jets. In Cygnus X-1 the effects of the

[4] The other one is Cygnus X-3.

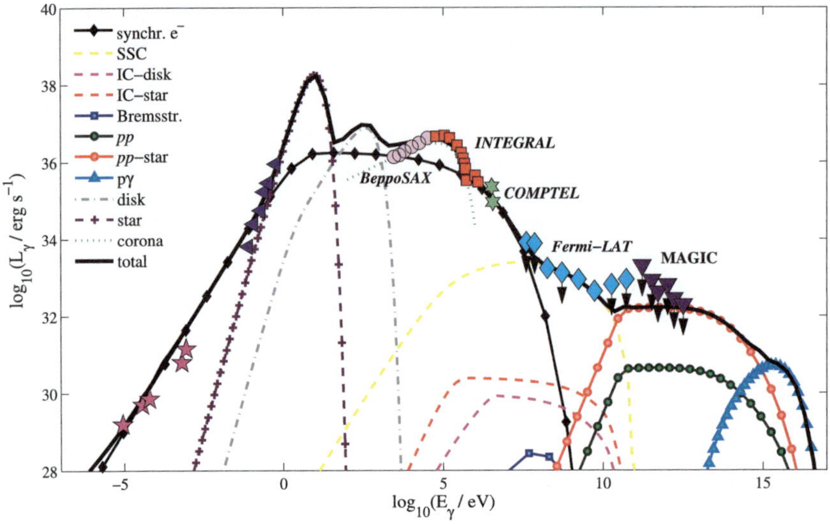

Fig. 13 Spectral energy distribution of the jet of the galactic microquasar Cygnus X-1. From Pepe et al. (2015)

impact of the jets in the interstellar medium suggest that they carry a significant amount of kinetic energy in cold protons (Gallo et al. 2005; Heinz 2006).

In Fig. 13 we show the results of the modeling of the broadband non-thermal emission reported by Pepe et al. (2015). In this model, the emission from 1 to ~ 150 keV is produced by Comptonization of disk photons in the coronae around the black hole (see, nevertheless, Georganopoulos et al. 2002). The synchrotron radiation extends up to soft gamma-rays. The high-energy and very high-energy gamma-rays are the result of a combination of SSC, *pp*, and $p\gamma$ interactions. The dominant source of thermal protons for *pp* is the wind of the star. Since these winds tend to be clumpy, flaring behavior is expected around 1 TeV (Romero et al. 2010).

4.2 Particle Acceleration

Thus due to a wide range of simultaneous broadband monitoring campaigns and spectro-timing campaigns over the last decades, a picture is emerging of a linked inflow/outflow system where both "sides" of the flow can contribute to the broadband spectrum. Contrary to the original picture of XRBs which focused exclusively on the accretion disk, these small accreting black holes are fully capable of launching relativistic jets, and those jets accelerate particles just as in blazars. Because of their relatively higher compactness, and thus stronger magnetic fields, jet synchrotron from moderately high-energy particles easily reaches the X-ray band and can compete with IC processes for the hard state spectrum. Finally, the detection of strong linear polarization in the soft gamma-rays from Cyg X-1 (Laurent et al. 2011), and more recently the *Fermi* detection of a hard power law (Zanin et al. 2016) seems to clinch the question about particle acceleration in XRB jets. Orbital modulations hint that for this source the GeV γ-ray emission occurs further out in the jets upon interaction with the stellar winds. A few detections of X-ray lines suggest hadronic content as well (e.g., Migliari et al. 2002; Díaz Trigo et al. 2013), though it is not clear if this is entrained or directly accelerated.

The question now becomes, can we learn something new from XRBs that will help us understand particle acceleration in jets in general, and help reduce the large number of free parameters in the various models? Many of the free parameters come from a lack of constraints on the jet bulk (M/HD driven) properties, and how they couple to particle acceleration properties (the macro-microphysics connection). XRBs offer a distinct advantage to AGN in this sense, given the realtime evolution displayed in this coupling. We will end this Section by discussing some new explorations of this connection.

The spectral break from the flat/inverted synchrotron spectrum to a power-law (e.g., Fig. 12) is the canonical signature of the transition from self-absorbed to optically thin regimes. The power-law at higher frequencies reveals an accelerated particle distribution, and the compactest part of the jet where particle acceleration is present will dominate the emission at the break itself. In AGN (see Sects. 3.2 and 3.3) this region is offset from the black hole, and often modeled de-facto by a single zone where dissipation of magnetic bulk energy into particle kinetic energy occurs via acceleration. In XRBs, the break location in the spectrum also corresponds to an offset from the black hole and may correspond to a similar dissipation zone.

Unlike AGN, XRB jets can barely be imaged with VLBI, however their smaller sizes allow for much more causality between the launch point and the outermost regions. Assuming a conservatively low plasma velocity of $0.3c$, the time for flow to travel from the black hole to the outermost regions of impact on the ISM at scales on the order of a pc (Corbel et al. 2002) is ~ 10 years, and most of the intrinsic broadband SED seems to be emitted within scales > 100 times smaller (e.g. Markoff et al. 2005). Thus an observing campaign covering the broadband emission from radio to X-ray can easily track the direct response of the steady jets to changes in the inner accretion flow in real time. However XRB jets are also seen to dismantle and reconfigure on timescales of weeks during state changes. During these times, multiwavelength monitoring reveals the buildup of the accelerated particle distribution, the transition from optically thin to self-absorbed (see Fig. 14), and the evolution in the spectrum including the spectral break, all as a function of changes in the accretion flow (Fig. 15).

Focusing now on the spectral break, if this feature originates in the *same* region for all sources (i.e., at some fixed number of r_g), then the dependence on $M_{\rm BH}$ and $\dot{m}$ can be derived for various standard accretion scenarios (see, e.g., Falcke and Biermann 1995; Markoff et al. 2003; Heinz and Sunyaev 2003). For the case of low-luminosity sources that fit on the FP, the slope of the correlation makes clear that these are in a radiatively inefficient state, as explained above. Briefly summarizing arguments presented in Heinz and Sunyaev (2003), this state is gas pressure dominated and the pressure is directly related to the particle density, which in turn is typically defined as $\rho = \dot{M}/(4\pi R H v)$. Since all physical scales for black holes can be expressed in terms of r_g, which is linear in mass, and $\dot{m} \equiv \dot{M}/M_{\rm BH}$, the pressure thus depends on the ratio $\dot{m}/M_{\rm BH}$. MRI-driven turbulence (Balbus and Hawley 1998) implies that the magnetic pressure is closely bound to the gas pressure, which means $B^2 \propto \dot{m}/M_{\rm BH}$ as well. If the jet is launched from the disk and shares the same scalings, one can solve for the photosphere where $\tau_{\rm SSA} = 1$, and show that for a fixed plasma β and a particle index of $p = 2$, the break frequency in the SED will scale as $\sim \dot{m}^{2/3}$, for objects of same mass and spin. *Thus in an individual XRB, this dependence on accretion rate can be isolated and tested.*

If the particle index p is significantly steeper, or the particle distribution is quasi-thermal, the scaling of $\dot{m}$ will be moderated slightly to $\sim 5/8$, while if the jet were launched from a non-advective disk (this seems ruled out by the FP, however for completeness we include it)

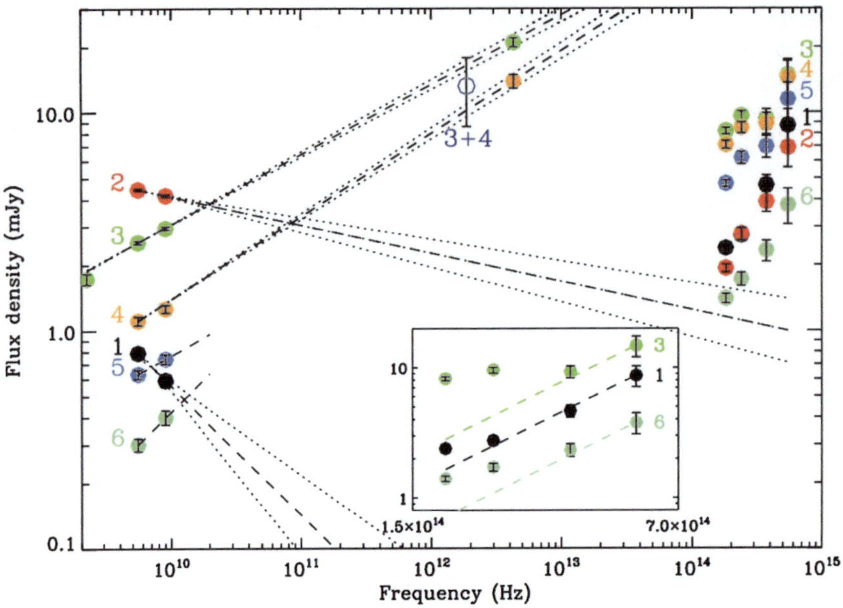

Fig. 14 Evolution of the radio to OIR SED during the soft to hard state transition as the jets reform, and become optically thick. The numbers (and associated *colors*) refer to sequential epochs, and the *dashed lines* correspond to the extrapolation of the radio spectra. *Inset*: zoom of the OIR spectra # 1, 3 and 6. From Corbel et al. (2013)

the scaling for all particle distributions is $\sim 1/2$. The point is that for any reasonable set of physical conditions for jet launching, the internal pressure provided by the gas and magnetic field would be expected to increase with $\dot{m}$ (for fixed spin and $M_{\rm BH}$), giving a positive scaling of the break frequency with $\dot{m}$ as the jet becomes optically thick at increasingly compacter scales.

However, already early monitoring of single sources like GX 339−4 indicated that the spectral break seems to move in the opposite direction expected from purely optical depth scalings as explained above (Markoff et al. 2003). By now there are a few well-observed outbursts from other sources that reveal the evolution of the SED (see, e.g., Fig. 15). These clearly show the spectral break increasing in frequency as the power in the thermal disk spectrum, and thus accretion rate, decreases. Broader compilations of multiple measurements from several XRBs Russell et al. (2013) show a similar trend in the location of the dissipation region.

The current best interpretation of these campaigns is that the location of the dissipation region that defines the break is itself moving, likely driven by changes in the jet itself. This evolution can also be seen in a correlation between break frequency and spectral index, indicating a blazar-sequence like relationship for XRBs (Koljonen et al. 2015). One important clue is that the break seems to be relatively stable within individual sources, and from source to source, for a given (relative) luminosity in Eddington units. Fits to multiple SEDs from many XRBs also show that the range of values for the break within a given source corresponds to a range of distance within the total length of the jets of ~ 10–$10^4\, r_g$.

Similar to other FP properties, the location of the break seems to be related more to the relative power (in, e.g., $\dot{m}$) than the size of the black hole. For instance in the few hard state-

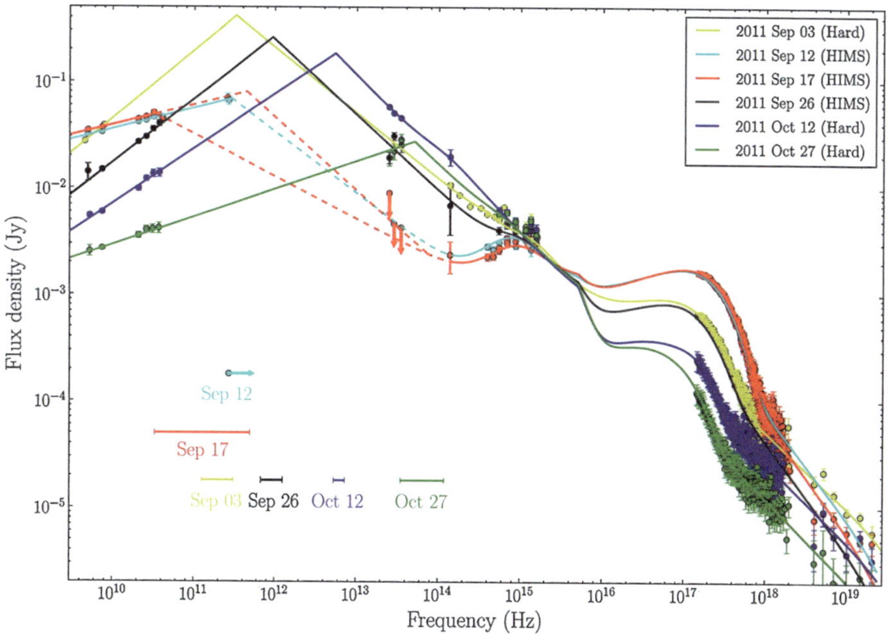

Fig. 15 Broadband radio to X-ray SED of MAXI J1836-194 taken during its 2011 outburst. *Dashed lines* indicate possible ranges of parameters, with the *horizontal bars* showing the uncertainty range for the optically thick-to-thin spectral break. Only four epochs of X-ray observational data are depicted to avoid crowding. The jet spectral break moves to higher frequency following the transition back to the hard state from the HIMS during a failed state transition, and is anticorrelated with accretion rate as indicated by the disk component. From Russell et al. (2014)

like LLAGN where quasi-simultaneous and/or high-spatial resolution data are available, the implied break occurs within a similar range as seen in XRBs (Markoff et al. 2008, and see Fig. 16). In fact, the same self-similar model (where all distances and powers are expressed in mass-scaling units of r_g and $L_{\rm Edd}$, or $\dot{m}$), can be used to fit the broadband SEDs of two black holes at opposite ends of the mass scale at similar accretion power, with all physical scales tied (see Fig. 17). Together these results suggest that the location of the particle acceleration zone is a direct consequence of conditions in the base of the jets, or in turn, the accretion flow. This opens the door to an exploration of whether the zone (and associated SED break) could be a predictive property of, e.g., an MHD outflow, providing a means to drastically reduce the number of free parameters currently employed.

In a series of papers, Polko et al. (2010, 2013, 2014) have derived a self-similar (following on prescriptions in Vlahakis et al. 2000 and Vlahakis and Königl 2003), relativistic MHD flow solution, including a pseudo-Newtonian treatment of gravity. These models can be used as intermediaries between special relativistic MHD and GRMHD simulations and observations, and also to explore potential drivers of the correlations observed in XRBs. For instance, if the break location is associated with particle acceleration at a disruption such as a recollimation shock (e.g., Marscher et al. 2008) or turbulent instabilities that form from shear once the bulk plasma velocity exceeds the fast magnetosonic speed, the scaling with other properties can be explored and used to rule out various launching scenarios. Thus the spectral breaks potentially provide an important new pivot for connecting inflow properties to the outflows, a longstanding issue in the field.

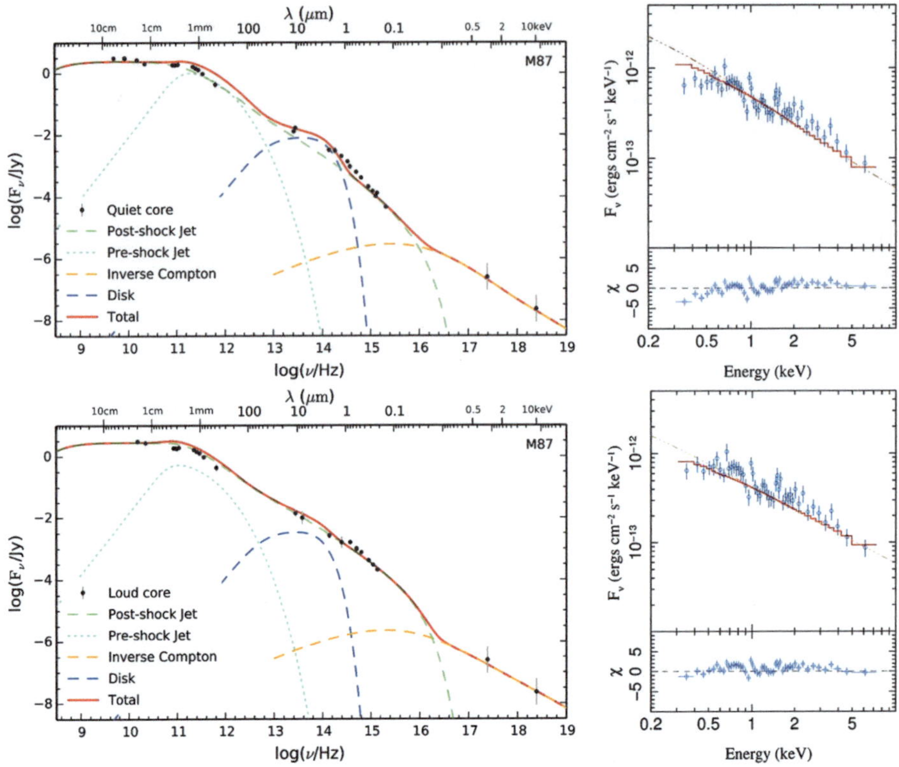

Fig. 16 An outflow-dominated model fit to the SED of M87, where all data are from within a 0.4 arcsec aperture radius (32 pc). The *top panel* shows the result for the quiescent state, the *bottom panel* for the active one. The X-ray spectrum used in the model for both states is the 2002 Chandra spectrum, the fit for which is shown as a separate plot. The data are fit using the model in Markoff et al. (2005) that is also applied to XRBs, and a location of the break implies an acceleration region of 10s of r_g, similar to XRBs as well. From Prieto et al. (2016)

4.3 Clues About Particle Acceleration Efficiency

XRBs have thus taught us that the location of particle acceleration may depend on the jet bulk properties, and that this property seems to extend to AGN, similar to the radio/X-ray correlations. Another indication from recent MWL campaigns is that there is a dependence of particle acceleration efficiency on accretion power. This phenomena was actually first noticed in Sgr A*, where the linearly polarized "submm bump" synchrotron feature in the spectrum (e.g., Bower et al. 2015) corresponds to emission from very close to the black hole and assumedly connects to a weak jet. No evidence for a spectral break or an optically thin power-law exists in the quiescent spectrum of Sgr A*, whose synchrotron and SSC spectrum can be explained by a mildly relativistic thermal distribution of particles in both the corona as well as jets. However Sgr A* also experiences a nonthermal flaring in the X-rays approximately daily, often accompanied by IR flaring, and models of this process suggest the appearance of sporadic and weak particle acceleration at the level of less than 1 % of the thermal peak (e.g., Yuan et al. 2003; Dibi et al. 2014, 2016).

Now that there are several observations of XRBs in a similar state of quiescence as Sgr A*, with $L_X < 10^{-8} L_{Edd}$, it seems that the lack of a strong power law in the emitting

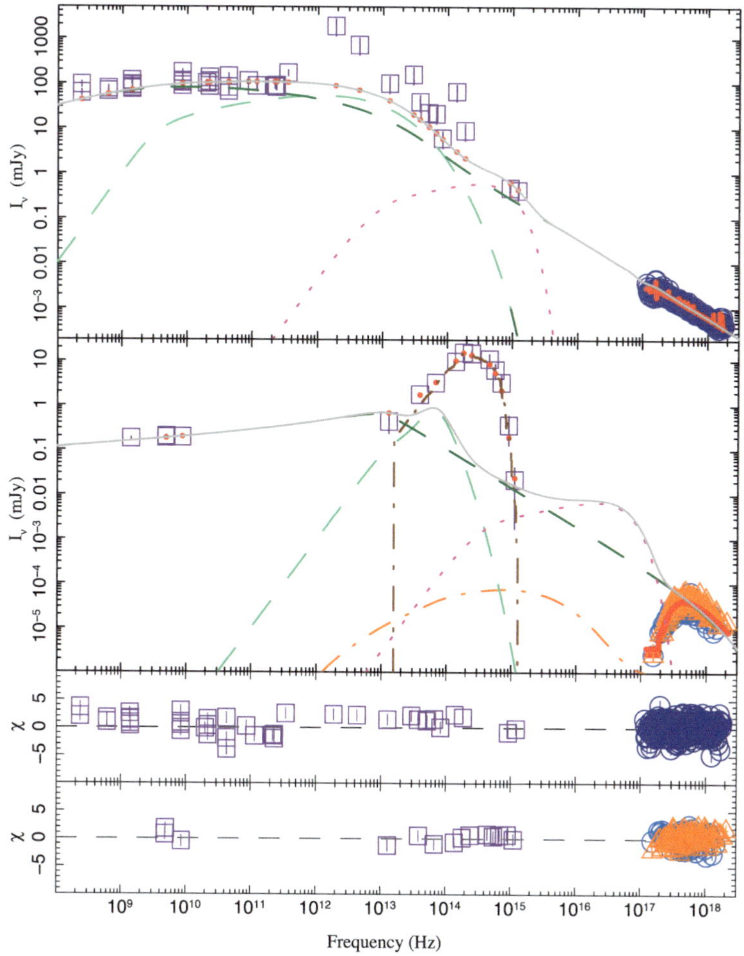

Fig. 17 Best joint-fit model with all physical free parameters tied to SEDs of similar accretion power (based on L_X/L_{Edd} from the LLAGN M81) (*top*) and hard state XRB V404 Cyg (*bottom*). *Lines* show the individual model components (*light green/dashed*: thermal synchrotron, *dark green/dashed*: non-thermal synchrotron, *orange/dashed–dotted*: synchrotron self-Compton, *magenta/dotted*: multicolor blackbody disk, and *gray/dashed–dotted*: stellar component in the case of V404), all absent absorption. The *gray/solid line* shows the total model, while the *red/solid line* and *dots* show the model after forward folding through detector space, including absorption. The IR/optical part of the M81 SED also includes a galactic stellar and dust contribution (not fitted). The spectral break in both sources implies a particle acceleration region in the jets starting at $300r_g$. From Markoff et al. (2015)

particles may be a universal hallmark of this state. For instance, the SEDs of A0620−00 (Gallo et al. 2007; Froning et al. 2011), XTE J1118+480 (Plotkin et al. 2015), and Swift J1357.2−0933 (Shahbaz et al. 2013) can all be explained with a dominant thermal synchrotron bump, and if any power-law is present it is consistent with being very weak and/or produced relatively far out in the jets. The physics of this dependence of particle acceleration on accretion rate is not yet understood, but may point to a jet power threshold below which the structures needed to accelerate particles cannot form or (as in the case of Sgr A*) be maintained.

One final note is that two distinct types of jets are seen in XRBs, at different accretion powers, and also radio-quiet states are seen. These can all occur in a single outburst of a source, thus ruling out spin as a driver for these states. The relationship between spin and jet power is also being explored for XRBs, but to date the jury is still out on this issue, due to disagreements about both how to measure the total jet power as well as the spin (e.g., Narayan and McClintock 2012; Russell et al. 2013).

In summary, XRBs are ideal testbeds for our understanding about black hole accretion and jet launching in general. Specifically they can reveal details about how particle acceleration is linked to the bulk plasma properties of the jets, in ways that cannot be as directly probed in AGN due to the longer timescales. XRB jets clearly accelerate particles, but the same questions about leptonic vs hadronic content still exist for these sources as for AGN. Similar to AGN, particularly blazars, there seems to be a dissipation/acceleration region offset from the black hole some distance down the jets, and a sequence in power and break frequency that may reveal the nature of the structures providing acceleration. The strong evidence for mass-scaling in the accretion physics around black holes makes XRBs valuable tools for probing these questions. The possibility that XRBs can accelerate hadrons to PeV energy is an interesting prospect, and something that near-future instruments like CTA will be able to test. Their energy budgets are smaller than supernovae, but there are many more of them, meaning that XRBs could be an important contributing source for the Galactic CR population. Some models, for instance the one by Pepe et al. (2015) some of whose predictions are illustrated in Fig. 13, can accommodate proton acceleration up to 10^{15} eV at the base of the jet and reach 10^{16} eV at distances $\sim 10^{12}$ cm from the compact object. These might be among the most energetic protons accelerated in the Galaxy.

5 Discussion

Galactic and extragalactic jets present many similarities, and also many differences. Both seem to be powered by accretion of magnetized matter with angular momentum onto a compact object; they both can display apparent superluminal motions; they seem to present a correlation of causal significance between the accretion and ejection processes; they display variability along the whole electromagnetic spectrum. These and other similarities are often stressed. The differences, however, are also important, and frequently overlooked. We emphasize the following ones:

- Galactic jets seem to be slower. In some cases, where the bulk motions are directly measured through emission lines, the plasma velocity is $\sim 0.3c$. There are some exceptional objects, however, such Sco X-1, with Lorentz factors up to 10.
- Galactic jets become dark not far from the central source. Typical lengths are ~ 1000 AU.
- They are heavy (have hadronic content) and produce thermal emission at their termination regions.
- They interact with stellar winds (in the case of microquasars with high-mass donor stars).
- The size of the accretion disk is constrained by the presence of the donor star.
- Gravitational and radiative effects of the donor star can be important to jet formation and propagation.
- The disk and corona are hotter than in active galactic nuclei; magnetic fields are also higher close to the jet base.

These differences can yield, however, opportunities for observational and theoretical studies of physical situations that can be applied, with suitable modifications, to extragalactic jets. In this review we have seen how a relatively simple model can easily be adapted to a quantitative description of both types of outflows.

Because of the proximity of Galactic jets, many physical processes that take place in mechanisms operating also in AGNs can be observationally probed in great detail and on vastly shorter timescales. The use of VLBI radio observations to resolve jets during simultaneous X-ray and gamma-ray monitoring of microquasars remains to be fully explored. The implementation of this kind of multiwavelength studies would facilitate the construction of refined models that can be later extrapolated to AGNs and larger scales. Testing the new predictions of such models in a very different scenario would strengthen our understanding of the underlying physics.

Acknowledgements We would like to thank the ISSI staff for providing an inspiring atmosphere favorable for intense discussions. GER thanks support from CONICET (PIP 2014-0338) and grant AYA2016-76012-C3-1-P (Ministro de Educación, Cultura y Deporte, Spain). The work of M.B. is supported by the South African Research Chairs Initiative (SARChI) of the Department of Science and Technology and the National Research Foundation[5] of South Africa, under SARChI Chair grant No. 64789.

References

M.G. Aartsen, M. Ackermann, J. Adams, J.A. Aguilar, M. Ahlers, M. Ahrens, D. Altmann, T. Anderson, C. Arguelles, T.C. Arlen, et al., Observation of high-energy astrophysical neutrinos in three years of IceCube data. Phys. Rev. Lett. **113**(10), 101101 (2014). doi:10.1103/PhysRevLett.113.101101

A.A. Abdo, M. Ackermann, I. Agudo, M. Ajello, H.D. Aller, M.F. Aller, E. Angelakis, A.A. Arkharov, M. Axelsson, U. Bach, et al., The spectral energy distribution of Fermi bright blazars. Astrophys. J. **716**, 30–70 (2010). doi:10.1088/0004-637X/716/1/30

G.O. Abell, B. Margon, A kinematic model for SS433. Nature **279**, 701–703 (1979). doi:10.1038/279701a0

M. Ackermann, M. Ajello, A. Allafort, E. Antolini, G. Barbiellini, D. Bastieri, R. Bellazzini, E. Bissaldi, E. Bonamente, J. Bregeon, et al., Multifrequency studies of the peculiar quasar 4C +21.35 during the 2010 flaring activity. Astrophys. J. **786**, 157 (2014). doi:10.1088/0004-637X/786/2/157

I. Agudo, S.G. Jorstad, A.P. Marscher, V.M. Larionov, J.L. Gómez, A. Lähteenmäki, M. Gurwell, P.S. Smith, H. Wiesemeyer, C. Thum, J. Heidt, D.A. Blinov, F.D. D'Arcangelo, V.A. Hagen-Thorn, D.A. Morozova, E. Nieppola, M. Roca-Sogorb, G.D. Schmidt, B. Taylor, M. Tornikoski, I.S. Troitsky, Location of γ-ray flare emission in the jet of the BL Lacertae object OJ287 more than 14 pc from the central engine. Astrophys. J. Lett. **726**, 13 (2011). doi:10.1088/2041-8205/726/1/L13

F.A. Aharonian, TeV gamma rays from BL Lac objects due to synchrotron radiation of extremely high energy protons. New Astron. **5**, 377–395 (2000). doi:10.1016/S1384-1076(00)00039-7

F.A. Aharonian, Proton-synchrotron radiation of large-scale jets in active galactic nuclei. Mon. Not. R. Astron. Soc. **332**, 215–230 (2002). doi:10.1046/j.1365-8711.2002.05292.x

F.A. Aharonian, *Very High Energy Cosmic Gamma Radiation: A Crucial Window on the Extreme Universe* 2004

F. Aharonian, A.G. Akhperjanian, A.R. Bazer-Bachi, B. Behera, M. Beilicke, W. Benbow, D. Berge, K. Bernlöhr, C. Boisson, O. Bolz, V. Borrel, T. Boutelier, I. Braun, E. Brion, A.M. Brown, R. Bühler, I. Büsching, T. Bulik, S. Carrigan, P.M. Chadwick, A.C. Clapson, L.-M. Chounet, G. Coignet, R. Cornils, L. Costamante, B. Degrange, H.J. Dickinson, A. Djannati-Ataï, W. Domainko, L.O. Drury, G. Dubus, J. Dyks, K. Egberts, D. Emmanoulopoulos, P. Espigat, C. Farnier, F. Feinstein, A. Fiasson, A. Förster, G. Fontaine, S. Funk, S. Funk, M. Füßling, Y.A. Gallant, B. Giebels, J.F. Glicenstein, B. Glück, P. Goret, C. Hadjichristidis, D. Hauser, M. Hauser, G. Heinzelmann, G. Henri, G. Hermann, J.A. Hinton, A. Hoffmann, W. Hofmann, M. Holleran, S. Hoppe, D. Horns, A. Jacholkowska, O.C. de Jager, E. Kendziorra, M. Kerschhaggl, B. Khélifi, N. Komin, K. Kosack, G. Lamanna, I.J. Latham, R. Le Gallou, A. Lemière, M. Lemoine-Goumard, J.-P. Lenain, T. Lohse, J.M. Martin, O. Martineau-Huynh, A. Marcowith, C. Masterson, G. Maurin, T.J.L. McComb, R. Moderski, E. Moulin, M. de Naurois, D. Nedbal, S.J. Nolan, J.-P. Olive, K.J. Orford, J.L. Osborne, M. Ostrowski, M. Panter, G. Pedaletti, G. Pelletier, P.-O. Petrucci,

[5] Any opinion, finding and conclusion or recommendation expressed in this material is hat of the authors, and the NRF does not accept any liability in this regard.

S. Pita, G. Pühlhofer, M. Punch, S. Ranchon, B.C. Raubenheimer, M. Raue, S.M. Rayner, M. Renaud, J. Ripken, L. Rob, L. Rolland, S. Rosier-Lees, G. Rowell, B. Rudak, J. Ruppel, V. Sahakian, A. Santangelo, L. Saugé, S. Schlenker, R. Schlickeiser, R. Schröder, U. Schwanke, S. Schwarzburg, S. Schwemmer, A. Shalchi, H. Sol, D. Spangler, Ł. Stawarz, R. Steenkamp, C. Stegmann, G. Superina, P.H. Tam, J.-P. Tavernet, R. Terrier, C. van Eldik, G. Vasileiadis, C. Venter, J.P. Vialle, P. Vincent, M. Vivier, H.J. Völk, F. Volpe, S.J. Wagner, M. Ward, A.A. Zdziarski, An exceptional very high energy gamma-ray flare of PKS 2155-304. Astrophys. J. Lett. **664**, 71–74 (2007). doi:10.1086/520635

M.L. Ahnen, S. Ansoldi, L.A. Antonelli, P. Antoranz, A. Babic, B. Banerjee, P. Bangale, U. Barres de Almeida, J.A. Barrio, W. Bednarek, et al., Very high energy γ-rays from the universe's middle age: Detection of the $z = 0.940$ blazar PKS 1441+25 with MAGIC. Astrophys. J. Lett. **815**, 23 (2015). doi:10.1088/2041-8205/815/2/L23

M. Ajello, R.W. Romani, D. Gasparrini, M.S. Shaw, J. Bolmer, G. Cotter, J. Finke, J. Greiner, S.E. Healey, O. King, W. Max-Moerbeck, P.F. Michelson, W.J. Potter, A. Rau, A.C.S. Readhead, J.L. Richards, P. Schady, The cosmic evolution of Fermi BL Lacertae objects. Astrophys. J. **780**, 73 (2014). doi:10.1088/0004-637X/780/1/73

J. Albert, E. Aliu, H. Anderhub, P. Antoranz, A. Armada, C. Baixeras, J.A. Barrio, H. Bartko, D. Bastieri, J.K. Becker, W. Bednarek, K. Berger, C. Bigongiari, A. Biland, R.K. Bock, P. Bordas, V. Bosch-Ramon, T. Bretz, I. Britvitch, M. Camara, E. Carmona, A. Chilingarian, J.A. Coarasa, S. Commichau, J.L. Contreras, J. Cortina, M.T. Costado, V. Curtef, V. Danielyan, F. Dazzi, A. De Angelis, C. Delgado, R. de los Reyes, B. De Lotto, E. Domingo-Santamaría, D. Dorner, M. Doro, M. Errando, M. Fagiolini, D. Ferenc, E. Fernández, R. Firpo, J. Flix, M.V. Fonseca, L. Font, M. Fuchs, N. Galante, R.J. García-López, M. Garczarczyk, M. Gaug, M. Giller, F. Goebel, D. Hakobyan, M. Hayashida, T. Hengstebeck, A. Herrero, D. Höhne, J. Hose, C.C. Hsu, P. Jacon, T. Jogler, R. Kosyra, D. Kranich, R. Kritzer, A. Laille, E. Lindfors, S. Lombardi, F. Longo, J. López, M. López, E. Lorenz, P. Majumdar, G. Maneva, K. Mannheim, O. Mansutti, M. Mariotti, M. Martínez, D. Mazin, C. Merck, M. Meucci, M. Meyer, J.M. Miranda, R. Mirzoyan, S. Mizobuchi, A. Moralejo, D. Nieto, K. Nilsson, J. Ninkovic, E. Oña-Wilhelmi, N. Otte, I. Oya, M. Panniello, R. Paoletti, J.M. Paredes, M. Pasanen, D. Pascoli, F. Pauss, R. Pegna, M. Persic, L. Peruzzo, A. Piccioli, E. Prandini, N. Puchades, A. Raymers, W. Rhode, M. Ribó, J. Rico, M. Rissi, A. Robert, S. Rügamer, A. Saggion, T. Saito, A. Sánchez, P. Sartori, V. Scalzotto, V. Scapin, R. Schmitt, T. Schweizer, M. Shayduk, K. Shinozaki, S.N. Shore, N. Sidro, A. Sillanpää, D. Sobczynska, A. Stamerra, L.S. Stark, L. Takalo, P. Temnikov, D. Tescaro, M. Teshima, D.F. Torres, N. Turini, H. Vankov, V. Vitale, R.M. Wagner, T. Wibig, W. Wittek, F. Zandanel, R. Zanin, J. Zapatero, Very high energy gamma-ray radiation from the stellar mass black hole binary Cygnus X-1. Astrophys. J. Lett. **665**, 51–54 (2007). doi:10.1086/521145

J. Aleksić, L.A. Antonelli, P. Antoranz, M. Backes, J.A. Barrio, D. Bastieri, J. Becerra González, W. Bednarek, A. Berdyugin, K. Berger, E. Bernardini, A. Biland, O. Blanch, R.K. Bock, A. Boller, G. Bonnoli, D. Borla Tridon, I. Braun, T. Bretz, A. Cañellas, E. Carmona, A. Carosi, P. Colin, E. Colombo, J.L. Contreras, J. Cortina, L. Cossio, S. Covino, F. Dazzi, A. De Angelis, E. De Cea del Pozo, B. De Lotto, C. Delgado Mendez, A. Diago Ortega, M. Doert, A. Domínguez, D. Dominis Prester, D. Dorner, M. Doro, D. Elsaesser, D. Ferenc, M.V. Fonseca, L. Font, C. Fruck, R.J. García López, M. Garczarczyk, D. Garrido, G. Giavitto, N. Godinović, D. Hadasch, D. Häfner, A. Herrero, D. Hildebrand, D. Höhne-Mönch, J. Hose, D. Hrupec, B. Huber, T. Jogler, S. Klepser, T. Krähenbühl, J. Krause, A. La Barbera, D. Lelas, E. Leonardo, E. Lindfors, S. Lombardi, M. López, E. Lorenz, M. Makariev, G. Maneva, N. Mankuzhiyil, K. Mannheim, L. Maraschi, M. Mariotti, M. Martínez, D. Mazin, M. Meucci, J.M. Miranda, R. Mirzoyan, H. Miyamoto, J. Moldón, A. Moralejo, D. Nieto, K. Nilsson, R. Orito, I. Oya, D. Paneque, R. Paoletti, S. Pardo, J.M. Paredes, S. Partini, M. Pasanen, F. Pauss, M.A. Perez-Torres, M. Persic, L. Peruzzo, M. Pilia, J. Pochon, F. Prada, P.G. Prada Moroni, E. Prandini, I. Puljak, I. Reichardt, R. Reinthal, W. Rhode, M. Ribó, J. Rico, S. Rügamer, A. Saggion, K. Saito, T.Y. Saito, M. Salvati, K. Satalecka, V. Scalzotto, V. Scapin, C. Schultz, T. Schweizer, M. Shayduk, S.N. Shore, A. Sillanpää, J. Sitarek, D. Sobczynska, F. Spanier, S. Spiro, A. Stamerra, B. Steinke, J. Storz, N. Strah, T. Surić, L. Takalo, F. Tavecchio, P. Temnikov, T. Terzić, D. Tescaro, M. Teshima, M. Thom, O. Tibolla, D.F. Torres, A. Treves, H. Vankov, P. Vogler, R.M. Wagner, Q. Weitzel, V. Zabalza, F. Zandanel, R. Zanin, MAGIC Collaboration, Y.T. Tanaka, D.L. Wood, S. Buson, MAGIC discovery of very high energy emission from the FSRQ PKS 1222+21. Astrophys. J. Lett. **730**, 8 (2011a). doi:10.1088/2041-8205/730/1/L8

J. Aleksić, L.A. Antonelli, P. Antoranz, M. Backes, J.A. Barrio, D. Bastieri, J. Becerra González, W. Bednarek, A. Berdyugin, K. Berger, E. Bernardini, A. Biland, O. Blanch, R.K. Bock, A. Boller, G. Bonnoli, D. Borla Tridon, I. Braun, T. Bretz, A. Cañellas, E. Carmona, A. Carosi, P. Colin, E. Colombo, J.L. Contreras, J. Cortina, L. Cossio, S. Covino, F. Dazzi, A. de Angelis, E. de Cea Del Pozo, B. de Lotto, C. Delgado Mendez, A. Diago Ortega, M. Doert, A. Domínguez, D. Dominis Prester, D. Dorner, M. Doro, D. Elsaesser, D. Ferenc, M.V. Fonseca, L. Font, C. Fruck, R.J. García López, M. Garczarczyk, D. Garrido, G. Giavitto, N. Godinović, D. Hadasch, D. Häfner, A. Herrero, D. Hildebrand, J. Hose,

D. Hrupec, B. Huber, T. Jogler, S. Klepser, T. Krähenbühl, J. Krause, A. La Barbera, D. Lelas, E. Leonardo, E. Lindfors, S. Lombardi, M. López, E. Lorenz, P. Majumdar, M. Makariev, G. Maneva, N. Mankuzhiyil, K. Mannheim, L. Maraschi, M. Mariotti, M. Martínez, D. Mazin, M. Meucci, J.M. Miranda, R. Mirzoyan, H. Miyamoto, J. Moldón, A. Moralejo, D. Nieto, K. Nilsson, R. Orito, I. Oya, R. Paoletti, S. Pardo, J.M. Paredes, S. Partini, M. Pasanen, F. Pauss, M.A. Perez-Torres, M. Persic, L. Peruzzo, M. Pilia, J. Pochon, F. Prada, P.G. Prada Moroni, E. Prandini, I. Puljak, I. Reichardt, R. Reinthal, W. Rhode, M. Ribó, J. Rico, S. Rügamer, M. Rüger, A. Saggion, K. Saito, T.Y. Saito, M. Salvati, K. Satalecka, V. Scalzotto, V. Scapin, C. Schultz, T. Schweizer, M. Shayduk, S.N. Shore, A. Sillanpää, J. Sitarek, D. Sobczynska, F. Spanier, S. Spiro, A. Stamerra, B. Steinke, J. Storz, N. Strah, T. Surić, L. Takalo, F. Tavecchio, P. Temnikov, T. Terzić, D. Tescaro, M. Teshima, M. Thom, O. Tibolla, D.F. Torres, A. Treves, H. Vankov, P. Vogler, R.M. Wagner, Q. Weitzel, V. Zabalza, F. Zandanel, R. Zanin, MAGIC observations and multiwavelength properties of the quasar 3C 279 in 2007 and 2009. Astron. Astrophys. **530**, 4 (2011b). doi:10.1051/0004-6361/201116497

J. Aleksić, S. Ansoldi, L.A. Antonelli, P. Antoranz, A. Babic, P. Bangale, U. Barres de Almeida, J.A. Barrio, J. Becerra González, W. Bednarek, et al., MAGIC gamma-ray and multi-frequency observations of flat spectrum radio quasar PKS 1510-089 in early 2012. Astron. Astrophys. **569**, 46 (2014). doi:10.1051/0004-6361/201423484

J. Aleksić, S. Ansoldi, L.A. Antonelli, P. Antoranz, A. Babic, P. Bangale, U. Barres de Almeida, J.A. Barrio, J. Becerra González, W. Bednarek, et al., Unprecedented study of the broadband emission of Mrk 421 during flaring activity in March 2010. Astron. Astrophys. **578**, 22 (2015). doi:10.1051/0004-6361/201424811

T. Arlen, T. Aune, M. Beilicke, W. Benbow, A. Bouvier, J.H. Buckley, V. Bugaev, A. Cesarini, L. Ciupik, M.P. Connolly, W. Cui, R. Dickherber, J. Dumm, M. Errando, A. Falcone, S. Federici, Q. Feng, J.P. Finley, G. Finnegan, L. Fortson, A. Furniss, N. Galante, D. Gall, S. Griffin, J. Grube, G. Gyuk, D. Hanna, J. Holder, T.B. Humensky, P. Kaaret, N. Karlsson, M. Kertzman, Y. Khassen, D. Kieda, H. Krawczynski, F. Krennrich, G. Maier, P. Moriarty, R. Mukherjee, T. Nelson, A. O'Faoláin de Bhróithe, R.A. Ong, M. Orr, N. Park, J.S. Perkins, A. Pichel, M. Pohl, H. Prokoph, J. Quinn, K. Ragan, L.C. Reyes, P.T. Reynolds, E. Roache, D.B. Saxon, M. Schroedter, G.H. Sembroski, D. Staszak, I. Telezhinsky, G. Tešić, M. Theiling, K. Tsurusaki, A. Varlotta, S. Vincent, S.P. Wakely, T.C. Weekes, A. Weinstein, R. Welsing, D.A. Williams, B. Zitzer, VERITAS Collaboration, S.G. Jorstad, N.R. MacDonald, A.P. Marscher, P.S. Smith, R.C. Walker, T. Hovatta, J. Richards, W. Max-Moerbeck, A. Readhead, M.L. Lister, Y.Y. Kovalev, A.B. Pushkarev, M.A. Gurwell, A. Lähteenmäki, E. Nieppola, M. Tornikoski, E. Järvelä, Rapid TeV gamma-ray flaring of BL Lacertae. Astrophys. J. **762**, 92 (2013). doi:10.1088/0004-637X/762/2/92

K. Asada, M. Nakamura, The structure of the M87 jet: A transition from parabolic to conical streamlines. Astrophys. J. Lett. **745**, 28 (2012). doi:10.1088/2041-8205/745/2/L28

A.M. Atoyan, C.D. Dermer, Neutral beams from blazar jets. Astrophys. J. **586**, 79–96 (2003). doi:10.1086/346261

S.A. Balbus, J.F. Hawley, Instability, turbulence, and enhanced transport in accretion disks. Rev. Mod. Phys. **70**(1), 1–53 (1998). doi:10.1103/RevModPhys.70.1

D.L. Band, J.E. Grindlay, Synchrotron and inverse Compton emission from expanding sources in jets—application to SS 433. Astrophys. J. **311**, 595–609 (1986). doi:10.1086/164799

M.G. Baring, Gamma-gamma attenuation and relativistic beaming in gamma-ray bursts. Astrophys. J. **418**, 391 (1993). doi:10.1086/173398

U. Barres de Almeida, F. Tavecchio, N. Mankuzhiyil, Polarimetric tomography of blazar jets. Mon. Not. R. Astron. Soc. **441**, 2885–2890 (2014). doi:10.1093/mnras/stu744

K. Beckwith, J.F. Hawley, J.H. Krolik, The influence of magnetic field geometry on the evolution of black hole accretion flows: Similar disks, drastically different jets. Astrophys. J. **678**, 1180–1199 (2008). doi:10.1086/533492

W. Bednarek, On the application of the mirror model for gamma-ray flare in 3C 279. Astron. Astrophys. **336**, 123–129 (1998)

M.C. Begelman, B. Rudak, M. Sikora, Consequences of relativistic proton injection in active galactic nuclei. Astrophys. J. **362**, 38–51 (1990). doi:10.1086/169241

M.C. Begelman, A.C. Fabian, M.J. Rees, Implications of very rapid TeV variability in blazars. Mon. Not. R. Astron. Soc. **384**, 19–23 (2008). doi:10.1111/j.1745-3933.2007.00413.x

T.M. Belloni, States and transitions in black hole binaries, in *The Jet Paradigm. From Microquasars to Quasars*, ed. by T. Belloni. Lecture Notes in Physics, vol. 794 (Springer, Berlin, 2010), p. 53

T. Belloni, M. Mendez, M. van der Klis, G. Hasinger, W.H.G. Lewin, J. van Paradijs, An intermediate state of Cygnus X-1. Astrophys. J. Lett. **472**, 107 (1996). doi:10.1086/310369

A.M. Beloborodov, Plasma ejection from magnetic flares and the X-ray spectrum of Cygnus X-1. Astrophys. J. **510**, 123–126 (1999)

V.S. Beskin, *MHD Flows in Compact Astrophysical Objects: Accretion, Winds and Jets*. Astronomy and Astrophysics Library (Springer, Berlin, 2009)
R.D. Blandford, A. Königl, Relativistic jets as compact radio sources. Astrophys. J. **232**, 34–48 (1979). doi:10.1086/157262
R.D. Blandford, A. Levinson, Pair cascades in extragalactic jets. 1: Gamma rays. Astrophys. J. **441**, 79–95 (1995). doi:10.1086/175338
R.D. Blandford, D.G. Payne, Hydromagnetic flows from accretion discs and the production of radio jets. Mon. Not. R. Astron. Soc. **199**, 883–903 (1982). doi:10.1093/mnras/199.4.883
R.D. Blandford, M.J. Rees, Some comments on radiation mechanisms in Lacertids, in *BL Lac Objects*, ed. by A.M. Wolfe, 1978, pp. 328–341
R.D. Blandford, R.L. Znajek, Electromagnetic extraction of energy from Kerr black holes. Mon. Not. R. Astron. Soc. **179**, 433–456 (1977). doi:10.1093/mnras/179.3.433
M. Błażejowski, M. Sikora, R. Moderski, G.M. Madejski, Comptonization of infrared radiation from hot dust by relativistic jets in quasars. Astrophys. J. **545**, 107–116 (2000). doi:10.1086/317791
D. Blinov, V. Pavlidou, I. Papadakis, S. Kiehlmann, G. Panopoulou, I. Liodakis, O.G. King, E. Angelakis, M. Baloković, H. Das, R. Feiler, L. Fuhrmann, T. Hovatta, P. Khodade, A. Kus, N. Kylafis, A. Mahabal, I. Myserlis, D. Modi, B. Pazderska, E. Pazderski, I. Papamastorakis, T.J. Pearson, C. Rajarshi, A. Ramaprakash, P. Reig, A.C.S. Readhead, K. Tassis, J.A. Zensus, RoboPol: First season rotations of optical polarization plane in blazars. Mon. Not. R. Astron. Soc. **453**, 1669–1683 (2015). doi:10.1093/mnras/stv1723
D. Blinov, V. Pavlidou, I.E. Papadakis, T. Hovatta, T.J. Pearson, I. Liodakis, G.V. Panopoulou, E. Angelakis, M. Baloković, H. Das, P. Khodade, S. Kiehlmann, O.G. King, A. Kus, N. Kylafis, A. Mahabal, A. Marecki, D. Modi, I. Myserlis, E. Paleologou, I. Papamastorakis, B. Pazderska, E. Pazderski, C. Rajarshi, A. Ramaprakash, A.C.S. Readhead, P. Reig, K. Tassis, J.A. Zensus, RoboPol: Optical polarization-plane rotations and flaring activity in blazars. Mon. Not. R. Astron. Soc. **457**, 2252–2262 (2016). doi:10.1093/mnras/stw158
G.R. Blumenthal, R.J. Gould, Bremsstrahlung, synchrotron radiation, and Compton scattering of high-energy electrons traversing dilute gases. Rev. Mod. Phys. **42**, 237–271 (1970). doi:10.1103/RevModPhys.42.237
A. Bodaghee, J.A. Tomsick, K. Pottschmidt, J. Rodriguez, J. Wilms, G.G. Pooley, Gamma-ray observations of the microquasars Cygnus X-1, Cygnus X-3, GRS 1915+105, and GX 339-4 with the Fermi Large Area Telescope. Astrophys. J. **775**, 98 (2013). doi:10.1088/0004-637X/775/2/98
M. Boettcher, P. Els, Gamma-gamma absorption in the broad line region radiation fields of gamma-ray blazars. arXiv:1603.02785 (2016)
G. Bonnoli, G. Ghisellini, L. Foschini, F. Tavecchio, G. Ghirlanda, The γ-ray brightest days of the blazar 3C 454.3. Mon. Not. R. Astron. Soc. **410**, 368–380 (2011). doi:10.1111/j.1365-2966.2010.17450.x
S. Bonometto, A. Saggion, Polarization in inverse Compton scattering of synchrotron radiation. Astron. Astrophys. **23**, 9 (1973)
V. Bosch-Ramon, G.E. Romero, J.M. Paredes, A broadband leptonic model for gamma-ray emitting microquasars. Astron. Astrophys. **447**, 263–276 (2006). doi:10.1051/0004-6361:20053633
M. Böttcher, Modeling intermediate BL Lac objects detected by Veritas. Int. J. Mod. Phys. D **19**, 873–878 (2010). doi:10.1142/S0218271810017135
M. Böttcher, J. Chiang, X-ray spectral variability signatures of flares in BL Lacertae objects. Astrophys. J. **581**, 127–142 (2002). doi:10.1086/344155
M. Böttcher, C.D. Dermer, On Compton scattering scenarios for blazer flares. Astrophys. J. Lett. **501**, 51–54 (1998). doi:10.1086/311458
M. Böttcher, C.D. Dermer, An evolutionary scenario for blazar unification. Astrophys. J. **564**, 86–91 (2002). doi:10.1086/324134
M. Böttcher, C.D. Dermer, Timing signatures of the internal-shock model for blazars. Astrophys. J. **711**, 445–460 (2010). doi:10.1088/0004-637X/711/1/445
M. Böttcher, R. Schlickeiser, The pair production spectrum from photon-photon annihilation. Astron. Astrophys. **325**, 866–870 (1997)
M. Böttcher, H. Mause, R. Schlickeiser, γ-Ray emission and spectral evolution of pair plasmas in AGN jets. I. General theory and a prediction for the GeV–TeV emission from ultrarelativistic jets. Astron. Astrophys. **324**, 395–409 (1997)
M. Böttcher, D.E. Harris, H. Krawczynski, *Relativistic Jets from Active Galactic Nuclei* (Wiley-VCH, Weinheim, 2012)
M. Böttcher, A. Reimer, K. Sweeney, A. Prakash, Leptonic and hadronic modeling of Fermi-detected blazars. Astrophys. J. **768**, 54 (2013). doi:10.1088/0004-637X/768/1/54
G.C. Bower, S. Markoff, J. Dexter, M.A. Gurwell, J.M. Moran, A. Brunthaler, H. Falcke, P.C. Fragile, D. Maitra, D. Marrone, A. Peck, A. Rushton, M.C.H. Wright, Radio and millimeter monitoring of Sgr A*:

Spectrum, variability, and constraints on the G2 encounter. Astrophys. J. **802**, 69 (2015). doi:10.1088/0004-637X/802/1/69
C. Brocksopp, R.P. Fender, G.G. Pooley, The orbital modulation in the radio emission of Cygnus X-1. Mon. Not. R. Astron. Soc. **336**, 699–704 (2002). doi:10.1046/j.1365-8711.2002.05813.x
G.R. Burbidge, On synchrotron radiation from Messier 87. Astrophys. J. **124**, 416 (1956). doi:10.1086/146237
B.J. Burn, On the depolarization of discrete radio sources by Faraday dispersion. Mon. Not. R. Astron. Soc. **133**, 67 (1966). doi:10.1093/mnras/133.1.67
A. Caccianiga, M.J.M. Marchã, The CLASS blazar survey: Testing the blazar sequence. Mon. Not. R. Astron. Soc. **348**, 937–954 (2004). doi:10.1111/j.1365-2966.2004.07415.x
A. Cavaliere, V. D'Elia, The blazar main sequence. Astrophys. J. **571**, 226–233 (2002). doi:10.1086/339778
M. Cerruti, C.D. Dermer, B. Lott, C. Boisson, A. Zech, Gamma-ray blazars near equipartition and the origin of the GeV spectral break in 3C 454.3. Astrophys. J. Lett. **771**, 4 (2013). doi:10.1088/2041-8205/771/1/L4
M. Cerruti, A. Zech, C. Boisson, S. Inoue, A hadronic origin for ultra-high-frequency-peaked BL Lac objects. Mon. Not. R. Astron. Soc. **448**, 910–927 (2015). doi:10.1093/mnras/stu2691
X. Chen, G. Fossati, E.P. Liang, M. Böttcher, Time-dependent simulations of multiwavelength variability of the blazar Mrk 421 with a Monte Carlo multizone code. Mon. Not. R. Astron. Soc. **416**, 2368–2387 (2011). doi:10.1111/j.1365-2966.2011.19215.x
X. Chen, M. Pohl, M. Böttcher, Particle diffusion and localized acceleration in inhomogeneous AGN jets—I. Steady-state spectra. Mon. Not. R. Astron. Soc. **447**, 530–544 (2015). doi:10.1093/mnras/stu2438
X. Chen, M. Pohl, M. Böttcher, S. Gao, Particle diffusion and localized acceleration in inhomogeneous AGN jets—Part II: Stochastic variation. Mon. Not. R. Astron. Soc. (2016). doi:10.1093/mnras/stw528
M. Chiaberge, G. Ghisellini, Rapid variability in the synchrotron self-Compton model for blazars. Mon. Not. R. Astron. Soc. **306**, 551–560 (1999). doi:10.1046/j.1365-8711.1999.02538.x
M. Chiaberge, A. Celotti, A. Capetti, G. Ghisellini, Does the unification of BL Lac and FR I radio galaxies require jet velocity structures? Astron. Astrophys. **358**, 104–112 (2000)
M.J. Chodorowski, A.A. Zdziarski, M. Sikora, Reaction rate and energy-loss rate for photopair production by relativistic nuclei. Astrophys. J. **400**, 181–185 (1992). doi:10.1086/171984
R.T. Coogan, A.M. Brown, P.M. Chadwick, Localizing the γ-ray emission region during the 2014 June outburst of 3C 454.3. Mon. Not. R. Astron. Soc. **458**, 354–365 (2016). doi:10.1093/mnras/stw199
S. Corbel, R.P. Fender, Near-infrared synchrotron emission from the compact jet of GX 339−4. Astrophys. J. **573**, 35–39 (2002)
S. Corbel, R.P. Fender, A.K. Tzioumis, M. Nowak, V. McIntyre, P. Durouchoux, R. Sood, Coupling of the X-ray and radio emission in the black hole candidate and compact jet source GX 339-4. Astron. Astrophys. **359**, 251–268 (2000)
S. Corbel, R.P. Fender, A.K. Tzioumis, J.A. Tomsick, J.A. Orosz, J.M. Miller, R. Wijnands, P. Kaaret, Large-scale, decelerating, relativistic X-ray jets from the microquasar XTE J1550−564. Science **298**, 196–199 (2002). doi:10.1126/science.1075857
S. Corbel, M. Nowak, R.P. Fender, A.K. Tzioumis, S. Markoff, Radio/X-ray correlations in the low/hard state of GX 339−4. Astron. Astrophys. **400**, 1007 (2003)
S. Corbel, M. Coriat, C. Brocksopp, A.K. Tzioumis, R.P. Fender, J.A. Tomsick, M.M. Buxton, C.D. Bailyn, The 'universal' radio/X-ray flux correlation: The case study of the black hole GX 339-4. Mon. Not. R. Astron. Soc. **428**, 2500–2515 (2013). doi:10.1093/mnras/sts215
H.D. Curtis, Descriptions of 762 nebulae and clusters photographed with the Crossley reflector. Publ. Lick. Obs. **13**, 9–42 (1918)
T. Dauser, J. Garcia, J. Wilms, M. Böck, L.W. Brenneman, M. Falanga, K. Fukumura, C.S. Reynolds, Irradiation of an accretion disc by a jet: General properties and implications for spin measurements of black holes. Mon. Not. R. Astron. Soc. **430**, 1694–1708 (2013). doi:10.1093/mnras/sts710
C.D. Dermer, G. Menon, *High Energy Radiation from Black Holes: Gamma Rays, Cosmic Rays, and Neutrinos* (2009)
C.D. Dermer, R. Schlickeiser, Model for the high-energy emission from blazars. Astrophys. J. **416**, 458 (1993). doi:10.1086/173251
M. Díaz Trigo, J.C.A. Miller-Jones, S. Migliari, J.W. Broderick, T. Tzioumis, Baryons in the relativistic jets of the stellar-mass black-hole candidate 4U1630-47. Nature **504**, 260–262 (2013). doi:10.1038/nature12672
S. Dibi, S. Markoff, R. Belmont, J. Malzac, N.M. Barrière, J.A. Tomsick, Exploring plasma evolution during Sagittarius A* flares. Mon. Not. R. Astron. Soc. **441**, 1005–1016 (2014). doi:10.1093/mnras/stu599
S. Dibi, S. Markoff, R. Belmont, J. Malzac, J. Neilsen, G. Witzel, Using infrared/X-ray flare statistics to probe the emission regions near the event horizon of Sgr A_*. Mon. Not. R. Astron. Soc. **461**(1), 552–559 (2016)

C. Diltz, M. Böttcher, Time dependent leptonic modeling of Fermi II processes in the jets of flat spectrum radio quasars. J. High Energy Astrophys. **1**, 63–70 (2014). doi:10.1016/j.jheap.2014.04.001

C. Diltz, M. Böttcher, G. Fossati, Time dependent hadronic modeling of flat spectrum radio quasars. Astrophys. J. **802**, 133 (2015). doi:10.1088/0004-637X/802/2/133

D. Donato, G. Ghisellini, G. Tagliaferri, G. Fossati, Hard X-ray properties of blazars. Astron. Astrophys. **375**, 739–751 (2001). doi:10.1051/0004-6361:20010675

L. Dondi, G. Ghisellini, Gamma-ray-loud blazars and beaming. Mon. Not. R. Astron. Soc. **273**, 583–595 (1995). doi:10.1093/mnras/273.3.583

A.-C. Donea, R.J. Protheroe, Radiation fields of disk, BLR and torus in quasars and blazars: Implications for γ-ray absorption. Astropart. Phys. **18**, 377–393 (2003). doi:10.1016/S0927-6505(02)00155-X

J.B. Dove, J. Wilms, M. Maisack, M.C. Begelman, Self-consistent thermal accretion disk corona models for compact objects. Astrophys. J. **487**, 759 (1997). doi:10.1086/304647

J.S. Dunlop, J.A. Peacock, The redshift cut-off in the luminosity function of radio galaxies and quasars. Mon. Not. R. Astron. Soc. **247**, 19 (1990)

A.C. Fabian, M.J. Rees, L. Stella, N.E. White, X-ray fluorescence from the inner disc in Cygnus X-1. Mon. Not. R. Astron. Soc. **238**, 729–736 (1989). doi:10.1093/mnras/238.3.729

H. Falcke, P.L. Biermann, The jet-disk symbiosis. I. Radio to X-ray emission models for quasars. Astron. Astrophys. **293**, 665–682 (1995)

H. Falcke, E. Körding, S. Markoff, A scheme to unify low-power accreting black holes. Astron. Astrophys. **414**, 895 (2004)

R. Fender, Jets from X-ray binaries, in *Compact Stellar X-ray Sources* (Cambridge University Press, Cambridge, 2006), pp. 381–419

C.E. Fichtel, D.L. Bertsch, J. Chiang, B.L. Dingus, J.A. Esposito, J.M. Fierro, R.C. Hartman, S.D. Hunter, G. Kanbach, D.A. Kniffen, P.W. Kwok, Y.C. Lin, J.R. Mattox, H.A. Mayer-Hasselwander, L. McDonald, P.F. Michelson, C. von Montigny, P.L. Nolan, K. Pinkau, H.-D. Radecke, H. Rothermel, P. Sreekumar, M. Sommer, E.J. Schneid, D.J. Thompson, T. Willis, The first energetic gamma-ray experiment telescope (EGRET) source catalog. Astrophys. J. Suppl. Ser. **94**, 551–581 (1994). doi:10.1086/192082

J.D. Finke, Compton dominance and the blazar sequence. Astrophys. J. **763**, 134 (2013). doi:10.1088/0004-637X/763/2/134

J.D. Finke, C.D. Dermer, On the break in the Fermi-Large Area Telescope spectrum of 3C 454.3. Astrophys. J. Lett. **714**, 303–307 (2010). doi:10.1088/2041-8205/714/2/L303

G. Fossati, L. Maraschi, A. Celotti, A. Comastri, G. Ghisellini, A unifying view of the spectral energy distributions of blazars. Mon. Not. R. Astron. Soc. **299**, 433–448 (1998). doi:10.1046/j.1365-8711.1998.01828.x

C.S. Froning, A.G. Cantrell, T.J. Maccarone, K. France, J. Khargharia, L.M. Winter, E.L. Robinson, R.I. Hynes, J.W. Broderick, S. Markoff, M.A.P. Torres, M. Garcia, C.D. Bailyn, J.X. Prochaska, J. Werk, C. Thom, S. Béland, C.W. Danforth, B. Keeney, J.C. Green, Multiwavelength observations of A0620−00 in quiescence. Astrophys. J. **743**, 26 (2011). doi:10.1088/0004-637X/743/1/26

E. Gallo, R.P. Fender, G.G. Pooley, A universal radio-X-ray correlation in low/hard state black hole binaries. Mon. Not. R. Astron. Soc. **344**, 60–72 (2003)

E. Gallo, R. Fender, C. Kaiser, D. Russell, R. Morganti, T. Oosterloo, S. Heinz, A dark jet dominates the power output of the stellar black hole Cygnus X-1. Nature **436**, 819–821 (2005). doi:10.1038/nature03879

E. Gallo, S. Migliari, S. Markoff, J.A. Tomsick, C.D. Bailyn, S. Berta, R. Fender, J.C.A. Miller-Jones, The spectral energy distribution of quiescent black hole X-ray binaries: New constraints from Spitzer. Astrophys. J. **670**, 600–609 (2007). doi:10.1086/521524

E. Gallo, J.C.A. Miller-Jones, D.M. Russell, P.G. Jonker, J. Homan, R.M. Plotkin, S. Markoff, B.P. Miller, S. Corbel, R.P. Fender, The radio/X-ray domain of black hole X-ray binaries at the lowest radio luminosities. Mon. Not. R. Astron. Soc. **445**, 290–300 (2014). doi:10.1093/mnras/stu1599

P. Gandhi, A.W. Blain, D.M. Russell, P. Casella, J. Malzac, S. Corbel, P. D'Avanzo, F.W. Lewis, S. Markoff, M. Cadolle Bel, P. Goldoni, S. Wachter, D. Khangulyan, A. Mainzer, A variable mid-infrared synchrotron break associated with the compact jet in GX 339-4. Astrophys. J. Lett. **740**, 13 (2011). doi:10.1088/2041-8205/740/1/L13

M. Georganopoulos, F.A. Aharonian, J.G. Kirk, External Compton emission from relativistic jets in Galactic black hole candidates and ultraluminous X-ray sources. Astron. Astrophys. **388**, 25–28 (2002). doi:10.1051/0004-6361:20020567

G. Ghisellini, P. Madau, On the origin of the gamma-ray emission in blazars. Mon. Not. R. Astron. Soc. **280**, 67–76 (1996). doi:10.1093/mnras/280.1.67

G. Ghisellini, F. Tavecchio, The blazar sequence: A new perspective. Mon. Not. R. Astron. Soc. **387**, 1669–1680 (2008). doi:10.1111/j.1365-2966.2008.13360.x

G. Ghisellini, F. Tavecchio, Canonical high-power blazars. Mon. Not. R. Astron. Soc. **397**, 985–1002 (2009). doi:10.1111/j.1365-2966.2009.15007.x
G. Ghisellini, F. Tavecchio, Fermi/LAT broad emission line blazars. Mon. Not. R. Astron. Soc. **448**, 1060–1077 (2015). doi:10.1093/mnras/stv055
G. Ghisellini, L. Maraschi, A. Treves, Inhomogeneous synchrotron-self-Compton models and the problem of relativistic beaming of BL Lac objects. Astron. Astrophys. **146**, 204–212 (1985)
G. Ghisellini, A. Celotti, G. Fossati, L. Maraschi, A. Comastri, A theoretical unifying scheme for gamma-ray bright blazars. Mon. Not. R. Astron. Soc. **301**, 451–468 (1998). doi:10.1046/j.1365-8711.1998.02032.x
G. Ghisellini, F. Tavecchio, M. Chiaberge, Structured jets in TeV BL Lac objects and radiogalaxies. Implications for the observed properties. Astron. Astrophys. **432**, 401–410 (2005). doi:10.1051/0004-6361:20041404
G. Ghisellini, L. Maraschi, F. Tavecchio, The Fermi blazars' divide. Mon. Not. R. Astron. Soc. **396**, 105–109 (2009). doi:10.1111/j.1745-3933.2009.00673.x
G. Ghisellini, F. Tavecchio, L. Foschini, G. Ghirlanda, L. Maraschi, A. Celotti, General physical properties of bright Fermi blazars. Mon. Not. R. Astron. Soc. **402**, 497–518 (2010). doi:10.1111/j.1365-2966.2009.15898.x
G. Ghisellini, F. Tavecchio, L. Foschini, T. Sbarrato, G. Ghirlanda, L. Maraschi, Blue Fermi flat spectrum radio quasars. Mon. Not. R. Astron. Soc. **425**, 1371–1379 (2012). doi:10.1111/j.1365-2966.2012.21554.x
G. Ghisellini, F. Tavecchio, L. Foschini, G. Bonnoli, G. Tagliaferri, The red blazar PMN J2345−1555 becomes blue. Mon. Not. R. Astron. Soc. **432**, 66–70 (2013). doi:10.1093/mnrasl/slt041
G. Ghisellini, F. Tavecchio, L. Maraschi, A. Celotti, T. Sbarrato, The power of relativistic jets is larger than the luminosity of their accretion disks. Nature **515**, 376–378 (2014). doi:10.1038/nature13856
D. Giannios, Reconnection-driven plasmoids in blazars: Fast flares on a slow envelope. Mon. Not. R. Astron. Soc. **431**, 355–363 (2013). doi:10.1093/mnras/stt167
D. Giannios, D.A. Uzdensky, M.C. Begelman, Fast TeV variability in blazars: Jets in a jet. Mon. Not. R. Astron. Soc. **395**, 29–33 (2009). doi:10.1111/j.1745-3933.2009.00635.x
V.L. Ginzburg, On the nature of the radio galaxies. Astron. Zh. **38**, 380 (1961)
P. Giommi, P. Padovani, A simplified view of blazars: Contribution to the X-ray and γ-ray extragalactic backgrounds. Mon. Not. R. Astron. Soc. **450**, 2404–2409 (2015). doi:10.1093/mnras/stv793
P. Giommi, P. Padovani, G. Polenta, S. Turriziani, V. D'Elia, S. Piranomonte, A simplified view of blazars: Clearing the fog around long-standing selection effects. Mon. Not. R. Astron. Soc. **420**, 2899–2911 (2012). doi:10.1111/j.1365-2966.2011.20044.x
P. Giommi, P. Padovani, G. Polenta, A simplified view of blazars: The γ-ray case. Mon. Not. R. Astron. Soc. **431**, 1914–1922 (2013). doi:10.1093/mnras/stt305
T. Gleißner, J. Wilms, G.G. Pooley, M.A. Nowak, K. Pottschmidt, S. Markoff, S. Heinz, M. Klein-Wolt, R.P. Fender, R. Staubert, Long term variability of Cygnus X-1. III. Radio-X-ray correlations. Astron. Astrophys. **425**, 1061–1068 (2004)
R.J. Gould, G.P. Schréder, Pair production in photon-photon collisions. Phys. Rev. **155**, 1404–1407 (1967). doi:10.1103/PhysRev.155.1404
K. Gültekin, E.M. Cackett, J.M. Miller, T. Di Matteo, S. Markoff, D.O. Richstone, The fundamental plane of accretion onto black holes with dynamical masses. Astrophys. J. **706**, 404–416 (2009). doi:10.1088/0004-637X/706/1/404
F. Guo, X. Li, H. Li, W. Daughton, B. Zhang, N. Lloyd-Ronning, Y.-H. Liu, H. Zhang, W. Deng, Efficient production of high-energy nonthermal particles during magnetic reconnection in a magnetically dominated ion-electron plasma. Astrophys. J. Lett. **818**, 9 (2016). doi:10.3847/2041-8205/818/1/L9
F. Haardt, Models for the X-ray emission from radio-quiet AGNs. Mem. Soc. Astron. Ital. **68**, 73 (1997)
S. Heinz, Composition, collimation, contamination: The jet of Cygnus X-1. Astrophys. J. **636**, 316–322 (2006). doi:10.1086/497954
S. Heinz, R.A. Sunyaev, The non-linear dependence of flux on black hole mass and accretion rate in core-dominated jets. Mon. Not. R. Astron. Soc. **343**, 59–64 (2003)
R.M. Hjellming, K.J. Johnston, Radio emission from conical jets associated with X-ray binaries. Astrophys. J. **328**, 600–609 (1988)
S. Hümmer, M. Rüger, F. Spanier, W. Winter, Simplified models for photohadronic interactions in cosmic accelerators. Astrophys. J. **721**, 630–652 (2010). doi:10.1088/0004-637X/721/1/630
R..I. Hynes, C..W. Mauche, C..A. Haswell, C..R. Shrader, W. Cui, S. Chaty, The X-ray transient xte j1118+480: Multiwavelength observations of a low-state minioutburst. Astrophys. J. **539**, 37 (2000)
R.I. Hynes, C.A. Haswell, W. Cui, C.R. Shrader, K. O'Brien, S. Chaty, D.R. Skillman, J. Patterson, K. Horne, The remarkable rapid X-ray, ultraviolet, optical and infrared variability in the black hole XTE J1118+480. Mon. Not. R. Astron. Soc. **345**, 292–310 (2003)
M. Joshi, M. Böttcher, Time-dependent radiation transfer in the internal shock model scenario for blazar jets. Astrophys. J. **727**, 21 (2011). doi:10.1088/0004-637X/727/1/21

M. Kadler, F. Krauß, K. Mannheim, R. Ojha, C. Müller, R. Schulz, G. Anton, W. Baumgartner, T. Beuchert, S. Buson, B. Carpenter, T. Eberl, P.G. Edwards, D. Eisenacher Glawion, D. Elsässer, N. Gehrels, C. Gräfe, H. Hase, S. Horiuchi, C.W. James, A. Kappes, A. Kappes, U. Katz, A. Kreikenbohm, M. Kreter, I. Kreykenbohm, M. Langejahn, K. Leiter, E. Litzinger, F. Longo, J.E.J. Lovell, J. McEnery, C. Phillips, C. Plötz, J. Quick, E. Ros, F.W. Stecker, T. Steinbring, J. Stevens, D.J. Thompson, J. Trüstedt, A.K. Tzioumis, J. Wilms, J.A. Zensus, Coincidence of a high-fluence blazar outburst with a PeV-energy neutrino event. ArXiv e-prints (2016)

M. Kalamkar, P. Casella, P. Uttley, K. O'Brien, D. Russell, T. Maccarone, M. van der Klis, F. Vincentelli, Detection of the first infra-red quasi periodic oscillation in a black hole X-ray binary. Mon. Not. R. Astron. Soc. **463**, 1822–1830 (2016)

M.M. Kaufman Bernadó, G.E. Romero, I.F. Mirabel, Precessing microblazars and unidentified gamma-ray sources. Astron. Astrophys. **385**, 10–13 (2002). doi:10.1051/0004-6361:20020251

S.R. Kelner, F.A. Aharonian, Energy spectra of gamma rays, electrons, and neutrinos produced at interactions of relativistic protons with low energy radiation. Phys. Rev. D **78**(3), 034013 (2008). doi:10.1103/PhysRevD.78.034013

S.R. Kelner, F.A. Aharonian, V.V. Bugayov, Energy spectra of gamma rays, electrons, and neutrinos produced at proton-proton interactions in the very high energy regime. Phys. Rev. D **74**(3), 034018 (2006). doi:10.1103/PhysRevD.74.034018

C.F. Kennel, F.V. Coroniti, Magnetohydrodynamic model of Crab nebula radiation. Astrophys. J. **283**, 710–730 (1984). doi:10.1086/162357

J.G. Kirk, P. Duffy, TOPICAL REVIEW: Particle acceleration and relativistic shocks. J. Phys. G, Nucl. Part. Phys. **25**, 163–194 (1999). doi:10.1088/0954-3899/25/8/201

J.G. Kirk, F.M. Rieger, A. Mastichiadis, Particle acceleration and synchrotron emission in blazar jets. Astron. Astrophys. **333**, 452–458 (1998)

K.I.I. Koljonen, D.M. Russell, J.A. Fernández-Ontiveros, S. Markoff, T.D. Russell, J.C.A. Miller-Jones, A.J. van der Horst, F. Bernardini, P. Casella, P.A. Curran, P. Gandhi, R. Soria, A connection between plasma conditions near black hole event horizons and outflow properties. Astrophys. J. **814**, 139 (2015). doi:10.1088/0004-637X/814/2/139

S.S. Komissarov, M.V. Barkov, N. Vlahakis, A. Königl, Magnetic acceleration of relativistic active galactic nucleus jets. Mon. Not. R. Astron. Soc. **380**, 51–70 (2007). doi:10.1111/j.1365-2966.2007.12050.x

H. Krawczynski, The polarization properties of inverse Compton emission and implications for blazar observations with the GEMS X-ray polarimeter. Astrophys. J. **744**, 30 (2012). doi:10.1088/0004-637X/744/1/30

H. Krawczynski, S.B. Hughes, D. Horan, F. Aharonian, M.F. Aller, H. Aller, P. Boltwood, J. Buckley, P. Coppi, G. Fossati, N. Götting, J. Holder, D. Horns, O.M. Kurtanidze, A.P. Marscher, M. Nikolashvili, R.A. Remillard, A. Sadun, M. Schröder, Multiwavelength observations of strong flares from the TeV blazar 1ES 1959+650. Astrophys. J. **601**, 151–164 (2004). doi:10.1086/380393

H.S. Krawczynski, D. Stern, F.A. Harrison, F.F. Kislat, A. Zajczyk, M. Beilicke, J. Hoormann, Q. Guo, R. Endsley, A.R. Ingram, H. Miyasaka, K.K. Madsen, K.M. Aaron, R. Amini, M.G. Baring, B. Beheshtipour, A. Bodaghee, J. Booth, C. Borden, M. Böttcher, F.E. Christensen, P.S. Coppi, R. Cowsik, S. Davis, J. Dexter, C. Done, L.A. Dominguez, D. Ellison, R.J. English, A.C. Fabian, A. Falcone, J.A. Favretto, R. Fernández, P. Giommi, B.W. Grefenstette, E. Kara, C.H. Lee, M. Lyutikov, T. Maccarone, H. Matsumoto, J. McKinney, T. Mihara, J.M. Miller, R. Narayan, L. Natalucci, F. Özel, M.J. Pivovaroff, S. Pravdo, D. Psaltis, T. Okajima, K. Toma, W.W. Zhang, X-ray polarimetry with the Polarization Spectroscopic Telescope Array (PolSTAR). Astropart. Phys. **75**, 8–28 (2016). doi:10.1016/j.astropartphys.2015.10.009

J.H. Krolik, *Active Galactic Nuclei: From the Central Black Hole to the Galactic Environment* (Princeton University Press, Princeton, 1999)

J.-P. Lasota, The disc instability model of dwarf novae and low-mass X-ray binary transients. New Astron. Rev. **45**, 449–508 (2001). doi:10.1016/S1387-6473(01)00112-9

P. Laurent, J. Rodriguez, J. Wilms, M. Cadolle Bel, K. Pottschmidt, V. Grinberg, Polarized gamma-ray emission from the galactic black hole Cygnus X-1. Science **332**, 438 (2011). doi:10.1126/science.1200848

A. Lazarian, G.L. Eyink, E.T. Vishniac, G. Kowal, Magnetic reconnection in astrophysical environments, in *Magnetic Fields in Diffuse Media*, ed. by A. Lazarian, E.M. de Gouveia Dal Pino, C. Melioli Astrophysics and Space Science Library, vol. 407, 2015, p. 311. doi:10.1007/978-3-662-44625-612

A. Levinson, R. Blandford, Pair cascades in extragalactic jets. II. The beamed X-ray spectrum. Astrophys. J. **449**, 86 (1995). doi:10.1086/176034

A. Levinson, R. Blandford, On the jets associated with galactic superluminal sources. Astrophys. J. Lett. **456**, 29 (1996). doi:10.1086/309851

A.P. Lightman, T.R. White, Effects of cold matter in active galactic nuclei—a broad hump in the X-ray spectra. Astrophys. J. **335**, 57–66 (1988). doi:10.1086/166905

K.R. Lind, R.D. Blandford, Semidynamical models of radio jets—relativistic beaming and source counts. Astrophys. J. **295**, 358–367 (1985). doi:10.1086/163380

P. Lipari, M. Lusignoli, D. Meloni, Flavor composition and energy spectrum of astrophysical neutrinos. Phys. Rev. D **75**(12), 123005 (2007). doi:10.1103/PhysRevD.75.123005

M.L. Lister, M.F. Aller, H.D. Aller, D.C. Homan, K.I. Kellermann, Y.Y. Kovalev, A.B. Pushkarev, J.L. Richards, E. Ros, T. Savolainen, MOJAVE. X. Parsec-scale jet orientation variations and superluminal motion in active galactic nuclei. Astron. J. **146**, 120 (2013). doi:10.1088/0004-6256/146/5/120

H.T. Liu, J.M. Bai, Absorption of 10–200 GeV gamma rays by radiation from broad-line regions in blazars. Astrophys. J. **653**, 1089–1097 (2006). doi:10.1086/509097

Y. Lyubarsky, Asymptotic structure of Poynting-dominated jets. Astrophys. J. **698**, 1570–1589 (2009). doi:10.1088/0004-637X/698/2/1570

P. Magdziarz, A.A. Zdziarski, Angle-dependent Compton reflection of X-rays and gamma-rays. Mon. Not. R. Astron. Soc. **273**, 837–848 (1995)

D. Malyshev, A.A. Zdziarski, M. Chernyakova, High-energy gamma-ray emission from Cyg X-1 measured by Fermi and its theoretical implications. Mon. Not. R. Astron. Soc. **434**, 2380–2389 (2013). doi:10.1093/mnras/stt1184

N. Mankuzhiyil, S. Ansoldi, M. Persic, F. Tavecchio, The environment and distribution of emitting electrons as a function of source activity in Markarian 421. Astrophys. J. **733**, 14 (2011). doi:10.1088/0004-637X/733/1/14

N. Mankuzhiyil, S. Ansoldi, M. Persic, E. Rivers, R. Rothschild, F. Tavecchio, Emitting electrons and source activity in Markarian 501. Astrophys. J. **753**, 154 (2012). doi:10.1088/0004-637X/753/2/154

K. Mannheim, The proton blazar. Astron. Astrophys. **269**, 67–76 (1993)

K. Mannheim, P.L. Biermann, Gamma-ray flaring of 3C 279—a proton-initiated cascade in the jet? Astron. Astrophys. **253**, 21–24 (1992)

L. Maraschi, F. Tavecchio, Spectral energy distributions of blazars: Facts and speculations, in *Blazar Demographics and Physics*, ed. by P. Padovani, C.M. Urry Astronomical Society of the Pacific Conference Series, vol. 227, 2001, p. 40

L. Maraschi, G. Ghisellini, A. Celotti, A jet model for the gamma-ray emitting blazar 3C 279. Astrophys. J. Lett. **397**, 5–9 (1992). doi:10.1086/186531

A. Marcowith, G. Henri, G. Pelletier, Gamma-ray emission of blazars by a relativistic electron-positron beam. Mon. Not. R. Astron. Soc. **277**, 681–699 (1995). doi:10.1093/mnras/277.2.681

S. Markoff, From multiwavelength to mass scaling: Accretion and ejection in microquasars and AGN, in *The Jet Paradigm. From Microquasars to Quasars*, ed. by T. Belloni Lecture Notes in Physics, vol. 794, 2010, p. 143. doi:10.1007/978-3-540-76937-8-6

S. Markoff, M.A. Nowak, Constraining X-ray binary jet models via reflection. Astrophys. J. **609**, 972–976 (2004)

S. Markoff, H. Falcke, R. Fender, A jet model for the broadband spectrum of XTE J1118+480. Synchrotron emission from radio to X-rays in the low/hard spectral state. Astron. Astrophys. **372**, 25–28 (2001a). doi:10.1051/0004-6361:20010420

S. Markoff, H. Falcke, F. Yuan, P.L. Biermann, The nature of the 10 kilosecond flare in Sgr A*. Astron. Astrophys. **379**, 13–16 (2001b)

S. Markoff, M. Nowak, S. Corbel, R. Fender, H. Falcke, Exploring the role of jets in the radio/X-ray correlations of GX 339−4. Astron. Astrophys. **397**, 645 (2003)

S. Markoff, M.A. Nowak, J. Wilms, Going with the flow: Can the base of jets subsume the role of compact accretion disk coronae? Astrophys. J. **635**, 1203–1216 (2005). doi:10.1086/497628

S. Markoff, M. Nowak, A. Young, H.L. Marshall, C.R. Canizares, A. Peck, M. Krips, G. Petitpas, R. Schödel, G.C. Bower, P. Chandra, A. Ray, M. Muno, S. Gallagher, S. Hornstein, C.C. Cheung, Results from an extensive simultaneous broadband campaign on the underluminous active nucleus M81*: Further evidence for mass-scaling accretion in black holes. Astrophys. J. **681**, 905–924 (2008). doi:10.1086/588718

S. Markoff, M.A. Nowak, E. Gallo, R. Hynes, J. Wilms, R.M. Plotkin, D. Maitra, C.V. Silva, S. Drappeau, As above, so below: Exploiting mass scaling in black hole accretion to break degeneracies in spectral interpretation. Astrophys. J. Lett. **812**, 25 (2015). doi:10.1088/2041-8205/812/2/L25

A.P. Marscher, Relativistic jets and the continuum emission in QSOs. Astrophys. J. **235**, 386–391 (1980a). doi:10.1086/157642

A.P. Marscher, Spontaneous formation of knots in relativistic flows—a model for variability in compact synchrotron sources. Astrophys. J. **239**, 296–304 (1980b). doi:10.1086/158110

A.P. Marscher, Turbulent, extreme multi-zone model for simulating flux and polarization variability in blazars. Astrophys. J. **780**, 87 (2014). doi:10.1088/0004-637X/780/1/87

A.P. Marscher, W.K. Gear, Models for high-frequency radio outbursts in extragalactic sources, with application to the early 1983 millimeter-to-infrared flare of 3C 273. Astrophys. J. **298**, 114–127 (1985). doi:10.1086/163592

A.P. Marscher, S.G. Jorstad, F.D. D'Arcangelo, P.S. Smith, G.G. Williams, V.M. Larionov, H. Oh, A.R. Olmstead, M.F. Aller, H.D. Aller, I.M. McHardy, A. Lähteenmäki, M. Tornikoski, E. Valtaoja, V.A. Hagen-Thorn, E.N. Kopatskaya, W.K. Gear, G. Tosti, O. Kurtanidze, M. Nikolashvili, L. Sigua, H.R. Miller, W.T. Ryle, The inner jet of an active galactic nucleus as revealed by a radio-to-γ-ray outburst. Nature **452**, 966–969 (2008). doi:10.1038/nature06895

A.P. Marscher, S.G. Jorstad, V.M. Larionov, M.F. Aller, H.D. Aller, A. Lähteenmäki, I. Agudo, P.S. Smith, M. Gurwell, V.A. Hagen-Thorn, T.S. Konstantinova, E.G. Larionova, L.V. Larionova, D.A. Melnichuk, D.A. Blinov, E.N. Kopatskaya, I.S. Troitsky, M. Tornikoski, T. Hovatta, G.D. Schmidt, F.D. D'Arcangelo, D. Bhattarai, B. Taylor, A.R. Olmstead, E. Manne-Nicholas, M. Roca-Sogorb, J.L. Gómez, I.M. McHardy, O. Kurtanidze, M.G. Nikolashvili, G.N. Kimeridze, L.A. Sigua, Probing the inner jet of the quasar PKS 1510-089 with multi-waveband monitoring during strong gamma-ray activity. Astrophys. J. Lett. **710**, 126–131 (2010). doi:10.1088/2041-8205/710/2/L126

E. Massaro, P. Giommi, C. Leto, P. Marchegiani, A. Maselli, M. Perri, S. Piranomonte, S. Sclavi, Roma-BZCAT: A multifrequency catalogue of blazars. Astron. Astrophys. **495**, 691–696 (2009). doi:10.1051/0004-6361:200810161

A. Mastichiadis, M. Petropoulou, S. Dimitrakoudis, Mrk 421 as a case study for TeV and X-ray variability in leptohadronic models. Mon. Not. R. Astron. Soc. **434**, 2684–2695 (2013). doi:10.1093/mnras/stt1210

J.E. McClintock, R.A. Remillard, Black hole binaries, in *Compact Stellar X-Ray Sources*, ed. by W. Lewin, M. van der Klis. Cambridge Astrophysics Series, vol. 39 (Cambridge University Press, Cambridge, 2006), pp. 157–213

J.E. McClintock, R. Narayan, J.F. Steiner, Black hole spin via continuum fitting and the role of spin in powering transient jets. Space Sci. Rev. **183**, 295–322 (2014). doi:10.1007/s11214-013-0003-9

M.L. McConnell, A.A. Zdziarski, K. Bennett, H. Bloemen, W. Collmar, W. Hermsen, L. Kuiper, W. Paciesas, B.F. Phlips, J. Poutanen, J.M. Ryan, V. Schönfelder, H. Steinle, A.W. Strong, The soft gamma-ray spectral variability of Cygnus X-1. Astrophys. J. **572**, 984–995 (2002). doi:10.1086/340436

I.M. McHardy, E. Koerding, C. Knigge, P. Uttley, R.P. Fender, Active galactic nuclei as scaled-up Galactic black holes. Nature **444**, 730–732 (2006). doi:10.1038/nature05389

J.C. McKinney, General relativistic magnetohydrodynamic simulations of the jet formation and large-scale propagation from black hole accretion systems. Mon. Not. R. Astron. Soc. **368**, 1561–1582 (2006). doi:10.1111/j.1365-2966.2006.10256.x

A. Merloni, A.C. Fabian, Coronal outflow dominated accretion discs: A new possibility for low-luminosity black holes? Mon. Not. R. Astron. Soc. **332**, 165–175 (2002)

A. Merloni, S. Heinz, T. di Matteo, A Fundamental Plane of black hole activity. Mon. Not. R. Astron. Soc. **345**, 1057–1076 (2003)

E.T. Meyer, G. Fossati, M. Georganopoulos, M.L. Lister, From the blazar sequence to the blazar envelope: Revisiting the relativistic jet dichotomy in radio-loud active galactic nuclei. Astrophys. J. **740**, 98 (2011). doi:10.1088/0004-637X/740/2/98

S. Migliari, R. Fender, M. Méndez, Iron emission lines from extended X-ray jets in SS 433: Reheating of atomic nuclei. Science **297**, 1673–1676 (2002). doi:10.1126/science.1073660

G. Miniutti, A.C. Fabian, A light bending model for the X-ray temporal and spectral properties of accreting black holes. Mon. Not. R. Astron. Soc. **349**, 1435–1448 (2004)

I.F. Mirabel, L.F. Rodríguez, A superluminal source in the Galaxy. Nature **371**, 46–48 (1994). doi:10.1038/371046a0

I.F. Mirabel, L.F. Rodriguez, B. Cordier, J. Paul, F. Lebrun, A double-sided radio jet from the compact Galactic Centre annihilator 1E1740.7-2942. Nature **358**, 215–217 (1992). doi:10.1038/358215a0

C. Motch, S.A. Ilovaisky, C. Chevalier, Discovery of fast optical activity in the X-ray source GX 339-4. Astron. Astrophys. **109**, 1–4 (1982)

A. Mücke, R.J. Protheroe, A proton synchrotron blazar model for flaring in Markarian 501. Astropart. Phys. **15**, 121–136 (2001). doi:10.1016/S0927-6505(00)00141-9

A. Mücke, R. Engel, J.P. Rachen, R.J. Protheroe, T. Stanev, Monte Carlo simulations of photohadronic processes in astrophysics. Comput. Phys. Commun. **124**, 290–314 (2000). doi:10.1016/S0010-4655(99)00446-4

A. Mücke, R.J. Protheroe, R. Engel, J.P. Rachen, T. Stanev, BL Lac objects in the synchrotron proton blazar model. Astropart. Phys. **18**, 593–613 (2003). doi:10.1016/S0927-6505(02)00185-8

R. Narayan, J.E. McClintock, Observational evidence for a correlation between jet power and black hole spin. Mon. Not. R. Astron. Soc. **419**, 69–73 (2012). doi:10.1111/j.1745-3933.2011.01181.x

R. Narayan, T. Piran, Variability in blazars: clues from PKS 2155-304. Mon. Not. R. Astron. Soc. **420**, 604–612 (2012). doi:10.1111/j.1365-2966.2011.20069.x

H. Netzer, Ionized gas in active galactic nuclei. New Astron. Rev. **52**, 257–273 (2008). doi:10.1016/j.newar.2008.06.009

M.A. Nowak, M. Hanke, S.N. Trowbridge, S.B. Markoff, J. Wilms, K. Pottschmidt, P. Coppi, D. Maitra, J.E. Davis, F. Tramper, Corona, jet, and relativistic line models for Suzaku/RXTE/Chandra-HETG observations of the Cygnus X-1 hard state. Astrophys. J. **728**, 13 (2011). doi:10.1088/0004-637X/728/1/13

J.A. Orosz, J.E. McClintock, J.P. Aufdenberg, R.A. Remillard, M.J. Reid, R. Narayan, L. Gou, The mass of the black hole in Cygnus X-1. Astrophys. J. **742**, 84 (2011). doi:10.1088/0004-637X/742/2/84

L. Pacciani, F. Tavecchio, I. Donnarumma, A. Stamerra, L. Carrasco, E. Recillas, A. Porras, M. Uemura, Exploring the blazar zone in high-energy flares of FSRQs. Astrophys. J. **790**, 45 (2014). doi:10.1088/0004-637X/790/1/45

A.G. Pacholczyk, Faraday rotation effect in extragalactic radio sources. Nature **200**, 765 (1963). doi:10.1038/200765a0

P. Padovani, P. Giommi, H. Landt, E.S. Perlman, The deep X-ray radio blazar survey. III. Radio number counts, evolutionary properties, and luminosity function of blazars. Astrophys. J. **662**, 182–198 (2007). doi:10.1086/516815

C. Pepe, G.S. Vila, G.E. Romero, Lepto-hadronic model for the broadband emission of Cygnus X-1. Astron. Astrophys. **584**, 95 (2015). doi:10.1051/0004-6361/201527156

M. Petropoulou, A. Mastichiadis, On proton synchrotron blazar models: The case of quasar 3C 279. Mon. Not. R. Astron. Soc. **426**, 462–472 (2012). doi:10.1111/j.1365-2966.2012.21720.x

M. Petropoulou, S. Dimitrakoudis, P. Padovani, A. Mastichiadis, E. Resconi, Photohadronic origin of γ-ray BL Lac emission: Implications for IceCube neutrinos. Mon. Not. R. Astron. Soc. **448**, 2412–2429 (2015). doi:10.1093/mnras/stv179

R.M. Plotkin, S.F. Anderson, W.N. Brandt, S. Markoff, O. Shemmer, J. Wu, The lack of torus emission from BL Lacertae objects: An infrared view of unification with WISE. Astrophys. J. Lett. **745**, 27 (2012a). doi:10.1088/2041-8205/745/2/L27

R.M. Plotkin, S. Markoff, B.C. Kelly, E. Körding, S.F. Anderson, Using the Fundamental Plane of black hole activity to distinguish X-ray processes from weakly accreting black holes. Mon. Not. R. Astron. Soc. **419**, 267–286 (2012b). doi:10.1111/j.1365-2966.2011.19689.x

R.M. Plotkin, E. Gallo, S. Markoff, J. Homan, P.G. Jonker, J.C.A. Miller-Jones, D.M. Russell, S. Drappeau, Constraints on relativistic jets in quiescent black hole X-ray binaries from broad-band spectral modelling. Mon. Not. R. Astron. Soc. **446**, 4098–4111 (2015). doi:10.1093/mnras/stu2385

R.M. Plotkin, E. Gallo, P.G. Jonker, J.C.A. Miller-Jones, J. Homan, T. Muñoz-Darias, S. Markoff, M. Armas Padilla, R. Fender, A.P. Rushton, D.M. Russell, M.A.P. Torres, A clean sightline to quiescence: multiwavelength observations of the high Galactic latitude black hole X-ray binary Swift J1357.2-0933. Mon. Not. R. Astron. Soc. **456**, 2707–2716 (2016). doi:10.1093/mnras/stv2861

M. Pohl, R. Schlickeiser, On the conversion of blast wave energy into radiation in AGN and GRBs, in *American Institute of Physics Conference Series*, ed. by B.L. Dingus, M.H. Salamon, D.B. Kieda American Institute of Physics Conference Series, vol. 515, 2000, pp. 249–252. doi:10.1063/1.1291375

P. Polko, D.L. Meier, S. Markoff, Determining the optimal locations for shock acceleration in magnetohydrodynamical jets. Astrophys. J. **723**, 1343–1350 (2010). doi:10.1088/0004-637X/723/2/1343

P. Polko, D.L. Meier, S. Markoff, Linking accretion flow and particle acceleration in jets—I. New relativistic magnetohydrodynamical jet solutions including gravity. Mon. Not. R. Astron. Soc. **428**, 587–598 (2013). doi:10.1093/mnras/sts052

P. Polko, D.L. Meier, S. Markoff, Linking accretion flow and particle acceleration in jets—II. Self-similar jet models with full relativistic MHD gravitational mass. Mon. Not. R. Astron. Soc. **438**, 959–970 (2014). doi:10.1093/mnras/stt2155

W.J. Potter, G. Cotter, Synchrotron and inverse-Compton emission from blazar jets—II. An accelerating jet model with a geometry set by observations of M87. Mon. Not. R. Astron. Soc. **429**, 1189–1205 (2013). doi:10.1093/mnras/sts407

J. Poutanen, Relativistic jets in blazars: Polarization of radiation. Astrophys. J. Suppl. Ser. **92**, 607–609 (1994). doi:10.1086/192024

J. Poutanen, B. Stern, GeV breaks in blazars as a result of gamma-ray absorption within the broad-line region. Astrophys. J. Lett. **717**, 118–121 (2010). doi:10.1088/2041-8205/717/2/L118

J. Poutanen, J.H. Krolik, F. Ryde, The nature of spectral transitions in accreting black holes—the case of CYG X-1. Mon. Not. R. Astron. Soc. **292**, 21–25 (1997). doi:10.1093/mnras/292.1.L21

M.A. Prieto, J.A. Fernández-Ontiveros, S. Markoff, D. Espada, O. González-Martín, The central parsecs of M87: jet emission and an elusive accretion disc. Mon. Not. R. Astron. Soc. **457**, 3801–3816 (2016). doi:10.1093/mnras/stw166

J.P. Rachen, P.L. Biermann, Extragalactic ultra-high energy cosmic-rays—Part One—Contribution from hot spots in Fr-II radio galaxies. Astron. Astrophys. **272**, 161 (1993)

C.M. Raiteri, A. Capetti, Testing the blazar sequence with the least luminous BL Lacertae objects. Astron. Astrophys. **587**, 8 (2016). doi:10.1051/0004-6361/201527679

T.A. Rector, J.T. Stocke, E.S. Perlman, S.L. Morris, I.M. Gioia, The properties of the X-ray-selected EMSS sample of BL Lacertae objects. Astron. J. **120**, 1626–1647 (2000). doi:10.1086/301587
M.J. Rees, Appearance of relativistically expanding radio sources. Nature **211**, 468–470 (1966). doi:10.1038/211468a0
M.M. Reynoso, G.E. Romero, Magnetic field effects on neutrino production in microquasars. Astron. Astrophys. **493**, 1–11 (2009). doi:10.1051/0004-6361:200811004
M.M. Reynoso, M.C. Medina, G.E. Romero, A lepto-hadronic model for high-energy emission from FR I radiogalaxies. Astron. Astrophys. **531**, 30 (2011). doi:10.1051/0004-6361/201014998
M.M. Reynoso, G.E. Romero, M.C. Medina, A two-component model for the high-energy variability of blazars. Application to PKS 2155-304. Astron. Astrophys. **545**, 125 (2012). doi:10.1051/0004-6361/201219873
J. Rodriguez, S. Corbel, J.A. Tomsick, Spectral evolution of the microquasar XTE J1550−564 over its entire 2000 outburst. Astrophys. J. **595**, 1032–1038 (2003). doi:10.1086/377478
J. Rodriguez, V. Grinberg, P. Laurent, M. Cadolle Bel, K. Pottschmidt, G. Pooley, A. Bodaghee, J. Wilms, C. Gouiffès, Spectral state dependence of the 0.4–2 MeV polarized emission in Cygnus X-1 seen with INTEGRAL/IBIS, and links with the AMI radio data. Astrophys. J. **807**, 17 (2015). doi:10.1088/0004-637X/807/1/17
G.E. Romero, G.S. Vila, The proton low-mass microquasar: High-energy emission. Astron. Astrophys. **485**, 623–631 (2008). doi:10.1051/0004-6361:200809563
G.E. Romero, G.S. Vila, *Introduction to Black Hole Astrophysics*. Lecture Notes in Physics, vol. 876 (Springer, Berlin, 2014). doi:10.1007/978-3-642-39596-3
G.E. Romero, J.A. Combi, H. Vucetich, Rapid variability in the southern blazar PKS 0521-365. Astrophys. Space Sci. **225**, 183–204 (1995). doi:10.1007/BF00613234
G.E. Romero, D.F. Torres, M.M. Kaufman Bernadó, I.F. Mirabel, Hadronic gamma-ray emission from windy microquasars. Astron. Astrophys. **410**, 1–4 (2003). doi:10.1051/0004-6361:20031314-1
G.E. Romero, H.R. Christiansen, M. Orellana, Hadronic high-energy gamma-ray emission from the microquasar LS I +61 303. Astrophys. J. **632**, 1093–1098 (2005). doi:10.1086/444446
G.E. Romero, F.L. Vieyro, G.S. Vila, Non-thermal processes around accreting galactic black holes. Astron. Astrophys. **519**, 109 (2010). doi:10.1051/0004-6361/200913663
G.E. Romero, F.L. Vieyro, S. Chaty, Coronal origin of the polarization of the high-energy emission of Cygnus X-1. Astron. Astrophys. **562**, 7 (2014). doi:10.1051/0004-6361/201323316
P. Rossi, A. Mignone, G. Bodo, S. Massaglia, A. Ferrari, Formation of dynamical structures in relativistic jets: The FRI case. Astron. Astrophys. **488**, 795–806 (2008). doi:10.1051/0004-6361:200809687
A. Rushton, J. Miller-Jones, Z. Paragi, T. Maccarone, G. Pooley, V. Tudose, R. Fender, R. Spencer, V. Dhawan, M. Garrett, VLBI constraints on the "jet-line" of Cygnus X-1. ArXiv e-prints (2011)
D.M. Russell, R.P. Fender, R.I. Hynes, C. Brocksopp, J. Homan, P.G. Jonker, M.M. Buxton, Global optical/infrared-X-ray correlations in X-ray binaries: Quantifying disc and jet contributions. Mon. Not. R. Astron. Soc. **371**, 1334–1350 (2006). doi:10.1111/j.1365-2966.2006.10756.x
D.M. Russell, D. Maitra, R.J.H. Dunn, S. Markoff, Evidence for a compact jet dominating the broad-band spectrum of the black hole accretor XTE J1550−564. Mon. Not. R. Astron. Soc. **405**, 1759–1769 (2010). doi:10.1111/j.1365-2966.2010.16547.x
D.M. Russell, E. Gallo, R.P. Fender, Observational constraints on the powering mechanism of transient relativistic jets. Mon. Not. R. Astron. Soc. **431**, 405–414 (2013). doi:10.1093/mnras/stt176
D.M. Russell, S. Markoff, P. Casella, A.G. Cantrell, R. Chatterjee, R.P. Fender, E. Gallo, P. Gandhi, J. Homan, D. Maitra, J.C.A. Miller-Jones, K. O'Brien, T. Shahbaz, Jet spectral breaks in black hole X-ray binaries. Mon. Not. R. Astron. Soc. **429**, 815–832 (2013). doi:10.1093/mnras/sts377
T.D. Russell, R. Soria, J.C.A. Miller-Jones, P.A. Curran, S. Markoff, D.M. Russell, G.R. Sivakoff, The accretion-ejection coupling in the black hole candidate X-ray binary MAXI J1836-194. Mon. Not. R. Astron. Soc. **439**, 1390–1402 (2014). doi:10.1093/mnras/stt2498
G.B. Rybicki, A.P. Lightman, *Radiative Processes in Astrophysics* (1979)
S. Sabatini, M. Tavani, E. Striani, A. Bulgarelli, V. Vittorini, G. Piano, E. Del Monte, M. Feroci, F. de Pasquale, M. Trifoglio, F. Gianotti, A. Argan, G. Barbiellini, P. Caraveo, P.W. Cattaneo, A.W. Chen, F. D'Ammando, E. Costa, G. De Paris, G. Di Cocco, I. Donnarumma, Y. Evangelista, A. Ferrari, M. Fiorini, F. Fuschino, M. Galli, A. Giuliani, M. Giusti, C. Labanti, F. Lazzarotto, P. Lipari, F. Longo, M. Marisaldi, S. Mereghetti, E. Morelli, E. Moretti, A. Morselli, L. Pacciani, A. Pellizzoni, F. Perotti, P. Picozza, M. Pilia, G. Pucella, M. Prest, M. Rapisarda, A. Rappoldi, A. Rubini, E. Scalise, P. Soffitta, A. Trois, E. Vallazza, S. Vercellone, A. Zambra, D. Zanello, C. Pittori, F. Verrecchia, P. Santolamazza, P. Giommi, S. Colafrancesco, L.A. Antonelli, L. Salotti, Episodic transient gamma-ray emission from the microquasar Cygnus X-1. Astrophys. J. Lett. **712**, 10–15 (2010). doi:10.1088/2041-8205/712/1/L10

S. Sabatini, M. Tavani, P. Coppi, G. Pooley, M. Del Santo, R. Campana, A. Chen, Y. Evangelista, G. Piano, A. Bulgarelli, P.W. Cattaneo, S. Colafrancesco, E. Del Monte, A. Giuliani, M. Giusti, F. Longo, A. Morselli, A. Pellizzoni, M. Pilia, E. Striani, M. Trifoglio, S. Vercellone, Gamma-ray observations of Cygnus X-1 above 100 MeV in the hard and soft states. Astrophys. J. **766**, 83 (2013). doi:10.1088/0004-637X/766/2/83
R.M. Sambruna, L. Maraschi, C.M. Urry, On the spectral energy distributions of blazars. Astrophys. J. **463**, 444 (1996). doi:10.1086/177260
R. Schlickeiser, Models of high-energy emission from active galactic nuclei. Astron. Astrophys. Suppl. Ser. **120**, 481–489 (1996)
R. Schlickeiser, *Cosmic Ray Astrophysics* (2002)
M. Schmidt, 3C 273: A star-like object with large red-shift. Nature **197**, 1040 (1963). doi:10.1038/1971040a0
T. Shahbaz, D.M. Russell, C. Zurita, J. Casares, J.M. Corral-Santana, V.S. Dhillon, T.R. Marsh, Evidence for quiescent synchrotron emission in the black hole X-ray transient Swift J1357.2-0933. Mon. Not. R. Astron. Soc. **434**, 2696–2706 (2013). doi:10.1093/mnras/stt1212
I.S. Shklovskii, Radio galaxies. Sov. Astron. **4**, 885 (1961)
M. Sikora, G. Madejski, On pair content and variability of subparsec jets in quasars. Astrophys. J. **534**, 109–113 (2000). doi:10.1086/308756
M. Sikora, M.C. Begelman, M.J. Rees, Comptonization of diffuse ambient radiation by a relativistic jet: The source of gamma rays from blazars? Astrophys. J. **421**, 153–162 (1994). doi:10.1086/173633
L. Sironi, A. Spitkovsky, Relativistic reconnection: An efficient source of non-thermal particles. Astrophys. J. Lett. **783**, 21 (2014). doi:10.1088/2041-8205/783/1/L21
L. Sironi, M. Petropoulou, D. Giannios, Relativistic jets shine through shocks or magnetic reconnection? Mon. Not. R. Astron. Soc. **450**, 183–191 (2015). doi:10.1093/mnras/stv641
P. Soffitta, X. Barcons, R. Bellazzini, J. Braga, E. Costa, G.W. Fraser, S. Gburek, J. Huovelin, G. Matt, M. Pearce, J. Poutanen, V. Reglero, A. Santangelo, R.A. Sunyaev, G. Tagliaferri, M. Weisskopf, R. Aloisio, E. Amato, P. Attiná, M. Axelsson, L. Baldini, S. Basso, S. Bianchi, P. Blasi, J. Bregeon, A. Brez, N. Bucciantini, L. Burderi, V. Burwitz, P. Casella, E. Churazov, M. Civitani, S. Covino, R.M. Curado da Silva, G. Cusumano, M. Dadina, F. D'Amico, A. De Rosa, S. Di Cosimo, G. Di Persio, T. Di Salvo, M. Dovciak, R. Elsner, C.J. Eyles, A.C. Fabian, S. Fabiani, H. Feng, S. Giarrusso, R.W. Goosmann, P. Grandi, N. Grosso, G. Israel, M. Jackson, P. Kaaret, V. Karas, M. Kuss, D. Lai, G.L. Rosa, J. Larsson, S. Larsson, L. Latronico, A. Maggio, J. Maia, F. Marin, M.M. Massai, T. Mineo, M. Minuti, E. Moretti, F. Muleri, S.L. O'Dell, G. Pareschi, G. Peres, M. Pesce, P.-O. Petrucci, M. Pinchera, D. Porquet, B. Ramsey, N. Rea, F. Reale, J.M. Rodrigo, A. Różańska, A. Rubini, P. Rudawy, F. Ryde, M. Salvati, V.A. de Santiago, S. Sazonov, C. Sgró, E. Silver, G. Spandre, D. Spiga, L. Stella, T. Tamagawa, F. Tamborra, F. Tavecchio, T. Teixeira Dias, M. van Adelsberg, K. Wu, S. Zane, XIPE: The X-ray imaging polarimetry explorer. Exp. Astron. **36**, 523–567 (2013). doi:10.1007/s10686-013-9344-3
A. Sokolov, A.P. Marscher, I.M. McHardy, Synchrotron self-Compton model for rapid nonthermal flares in blazars with frequency-dependent time lags. Astrophys. J. **613**, 725–746 (2004). doi:10.1086/423165
M. Spada, G. Ghisellini, D. Lazzati, A. Celotti, Internal shocks in the jets of radio-loud quasars. Mon. Not. R. Astron. Soc. **325**, 1559–1570 (2001). doi:10.1046/j.1365-8711.2001.04557.x
H.C. Spruit, Theory of magnetically powered jets, in *The Jet Paradigm. From Microquasars to Quasars*, ed. by T. Belloni Lecture Notes in Physics, vol. 794 (Springer, Berlin, 2010), p. 233. doi:10.1007/978-3-540-76937-8
B.E. Stern, J. Poutanen, Variation of the $\gamma\gamma$ opacity by the He II Lyman continuum constrains the location of the γ-ray emission region in the blazar 3C 454.3. Mon. Not. R. Astron. Soc. **417**, 11–15 (2011). doi:10.1111/j.1745-3933.2011.01107.x
B.E. Stern, J. Poutanen, The mystery of spectral breaks: Lyman continuum absorption by photon-photon pair production in the Fermi GeV spectra of bright blazars. Astrophys. J. **794**, 8 (2014). doi:10.1088/0004-637X/794/1/8
A.M. Stirling, R.E. Spencer, C.J. de la Force, M.A. Garrett, R.P. Fender, R.N. Ogley, A relativistic jet from Cygnus X-1 in the low/hard X-ray state. Mon. Not. R. Astron. Soc. **327**, 1273–1278 (2001). doi:10.1046/j.1365-8711.2001.04821.x
E.J. Summerlin, M.G. Baring, Diffusive acceleration of particles at oblique, relativistic, magnetohydrodynamic shocks. Astrophys. J. **745**, 63 (2012). doi:10.1088/0004-637X/745/1/63
M. Tavani, V. Vittorini, A. Cavaliere, An emerging class of Gamma-ray flares from blazars: Beyond one-zone models. Astrophys. J. **814**, 51 (2015). doi:10.1088/0004-637X/814/1/51
F. Tavecchio, G. Ghisellini, Spine-sheath layer radiative interplay in subparsec-scale jets and the TeV emission from M87. Mon. Not. R. Astron. Soc. **385**, 98–102 (2008). doi:10.1111/j.1745-3933.2008.00441.x
F. Tavecchio, G. Ghisellini, On the magnetization of BL Lac jets. Mon. Not. R. Astron. Soc. **456**, 2374–2382 (2016). doi:10.1093/mnras/stv2790

F. Tavecchio, L. Maraschi, G. Ghisellini, Constraints on the physical parameters of TeV blazars. Astrophys. J. **509**, 608–619 (1998). doi:10.1086/306526

F. Tavecchio, L. Maraschi, E. Pian, L. Chiappetti, A. Celotti, G. Fossati, G. Ghisellini, E. Palazzi, C.M. Raiteri, R.M. Sambruna, A. Treves, C.M. Urry, M. Villata, A. Djannati-Ataï, Theoretical implications from the spectral evolution of Markarian 501 observed with BeppoSAX. Astrophys. J. **554**, 725–733 (2001). doi:10.1086/321394

F. Tavecchio, G. Ghisellini, G. Bonnoli, G. Ghirlanda, Constraining the location of the emitting region in Fermi blazars through rapid γ-ray variability. Mon. Not. R. Astron. Soc. **405**, 94–98 (2010). doi:10.1111/j.1745-3933.2010.00867.x

F. Tavecchio, J. Becerra-Gonzalez, G. Ghisellini, A. Stamerra, G. Bonnoli, L. Foschini, L. Maraschi, On the origin of the γ-ray emission from the flaring blazar PKS 1222+216. Astron. Astrophys. **534**, 86 (2011). doi:10.1051/0004-6361/201117204

F. Tavecchio, G. Ghisellini, D. Guetta, Structured jets in BL Lac objects: Efficient PeV neutrino factories? Astrophys. J. Lett. **793**, 18 (2014). doi:10.1088/2041-8205/793/1/L18

A. Tchekhovskoy, J.C. McKinney, R. Narayan, Simulations of ultrarelativistic magnetodynamic jets from gamma-ray burst engines. Mon. Not. R. Astron. Soc. **388**, 551–572 (2008). doi:10.1111/j.1365-2966.2008.13425.x

A. Tchekhovskoy, R. Narayan, J.C. McKinney, Magnetohydrodynamic simulations of gamma-ray burst jets: Beyond the progenitor star. New Astron. **15**, 749–754 (2010). doi:10.1016/j.newast.2010.03.001

B.E. Tetarenko, G.R. Sivakoff, C.O. Heinke, J.C. Gladstone, WATCHDOG: A comprehensive all-sky database of galactic black hole X-ray binaries. Astrophys. J. Suppl. Ser. **222**, 15 (2016). doi:10.3847/0067-0049/222/2/15

L. Titarchuk, Generalized Comptonization models and application to the recent high-energy observations. Astrophys. J. **434**, 570–586 (1994). doi:10.1086/174760

M.-H. Ulrich, L. Maraschi, C.M. Urry, Variability of active galactic nuclei. Annu. Rev. Astron. Astrophys. **35**, 445–502 (1997). doi:10.1146/annurev.astro.35.1.445

C.M. Urry, P. Padovani, Unified schemes for radio-loud active galactic nuclei. Publ. Astron. Soc. Pac. **107**, 803 (1995). doi:10.1086/133630

G.S. Vila, G.E. Romero, Leptonic/hadronic models for electromagnetic emission in microquasars: The case of GX 339-4. Mon. Not. R. Astron. Soc. **403**, 1457–1468 (2010). doi:10.1111/j.1365-2966.2010.16208.x

G.S. Vila, G.E. Romero, N.A. Casco, An inhomogeneous lepto-hadronic model for the radiation of relativistic jets. Application to XTE J1118+480. Astron. Astrophys. **538**, 97 (2012). doi:10.1051/0004-6361/201118106

N. Vlahakis, A. Königl, Relativistic magnetohydrodynamics with application to gamma-ray burst outflows. I. Theory and semianalytic trans-Alfvénic solutions. Astrophys. J. **596**, 1080–1103 (2003). doi:10.1086/378226

N. Vlahakis, K. Tsinganos, C. Sauty, E. Trussoni, A disc-wind model with correct crossing of all magnetohydrodynamic critical surfaces. Mon. Not. R. Astron. Soc. **318**, 417–428 (2000). doi:10.1046/j.1365-8711.2000.03703.x

M. Weidinger, F. Spanier, A self-consistent and time-dependent hybrid blazar emission model. Properties and application. Astron. Astrophys. **573**, 7 (2015). doi:10.1051/0004-6361/201424159

J. Wilms, M.A. Nowak, K. Pottschmidt, G.G. Pooley, S. Fritz, Long term variability of Cygnus X-1. IV. Spectral evolution 1999–2004. Astron. Astrophys. **447**, 245–261 (2006). doi:10.1051/0004-6361:20053938

D. Yan, L. Zhang, Understanding the TeV emission from a distant blazar PKS 1424+240 in a lepto-hadronic jet model. Mon. Not. R. Astron. Soc. **447**, 2810–2816 (2015). doi:10.1093/mnras/stu2551

F. Yuan, E. Quataert, R. Narayan, Nonthermal electrons in radiatively inefficient accretion flow models of Sagittarius A*. Astrophys. J. **598**, 301–312 (2003)

R. Zanin, A. Fernández-Barral, E. de Oña-Wilhelmi, F. Aharonian, O. Blanch, V. Bosch-Ramon, D. Galindo, Detection of gamma rays of likely jet origin in Cygnus X-1. ArXiv e-prints (2016)

A.A. Zdziarski, M. Böttcher, Hadronic models of blazars require a change of the accretion paradigm. Mon. Not. R. Astron. Soc. **450**, 21–25 (2015). doi:10.1093/mnrasl/slv039

A.A. Zdziarski, P. Lubiński, M. Sikora, The MeV spectral tail in Cyg X-1 and optically thin emission of jets. Mon. Not. R. Astron. Soc. **423**, 663–675 (2012). doi:10.1111/j.1365-2966.2012.20903.x

A.A. Zdziarski, Ł. Stawarz, P. Pjanka, M. Sikora, Jet models for black hole binaries in the hard spectral state. Mon. Not. R. Astron. Soc. **440**, 2238–2254 (2014). doi:10.1093/mnras/stu420

J.A. Zensus, T.J. Pearson (eds.), Superluminal radio sources, in *Superluminal Radio Sources. Proceedings of the Workshop*, Pasadena, CA, Oct. 28–30, 1986 (1987)

H. Zhang, M. Böttcher, X-ray and gamma-ray polarization in leptonic and hadronic jet models of blazars. Astrophys. J. **774**, 18 (2013). doi:10.1088/0004-637X/774/1/18

H. Zhang, X. Chen, M. Böttcher, Synchrotron polarization in blazars. Astrophys. J. **789**, 66 (2014). doi:10.1088/0004-637X/789/1/66

H. Zhang, X. Chen, M. Böttcher, F. Guo, H. Li, Polarization swings reveal magnetic energy dissipation in blazars. Astrophys. J. **804**, 58 (2015). doi:10.1088/0004-637X/804/1/58

H. Zhang, W. Deng, H. Li, M. Böttcher, Polarization signatures of relativistic magnetohydrodynamic shocks in the blazar emission region. I. Force-free helical magnetic fields. Astrophys. J. **817**, 63 (2016). doi:10.3847/0004-637X/817/1/63

Space Sci Rev (2017) 207:63–86
DOI 10.1007/s11214-017-0366-4

Gamma-Ray Bursts and Fast Transients

Multi-wavelength Observations and Multi-messenger Signals

R. Willingale[1] · P. Mészáros[2]

Received: 13 September 2016 / Accepted: 11 April 2017 / Published online: 25 April 2017

Abstract The current status of observations and theoretical models of gamma-ray bursts and some other related transients, including ultra-long bursts and tidal disruption events, is reviewed. We consider the impact of multi-wavelength data on the formulation and development of theoretical models for the prompt and afterglow emission including the standard fireball model utilizing internal shocks and external shocks, photospheric emission, the role of the magnetic field and hadronic processes. In addition, we discuss some of the prospects for non-photonic multi-messenger detection and for future instrumentation, and comment on some of the outstanding issues in the field.

Keywords Gamma-rays: bursts · Black holes · Neutron stars · Shocks · Neutrinos · Gravitational waves

1 Introduction

Gamma-ray bursts (GRBs) and other one-shot, fast, high energy transients are likely to produce UHE Cosmic rays, VHE neutrinos and Gravitational waves in addition to radiation across the entire electromagnetic spectrum from radio through to VHE gamma-rays. Multi-wavelength observations have been and will, without doubt, continue to be crucial in the study of GRBs and the detection of multi-messenger signals, along side the high energy photon prompt emission, will broaden our understanding of the extreme physical conditions and processes which give rise to such high energy phenomena.

Many of the major breakthroughs and results in GRB studies to date have arisen from multi-wavelength observations. The realisation that all GRBs are extra galactic was initiated by the first X-ray afterglow detection made by the *BeppoSAX* soft X-ray narrow field

✉ R. Willingale
zrw@le.ac.uk

P. Mészáros
nnp@psu.edu

1 Department of Physics and Astronomy, University of Leicester, Leicester, UK

2 Pennsylvania State University, 525 Davey Lab, University Park, PA 16802, USA

instruments for the GRB 970228 (Costa et al. 1997). This provided a position with a 90% confidence error radius of 50 arc seconds which lead to the identification of the optical emission (Van Paradijs et al. 1997) and subsequent deep observations with the Keck telescopes gave a redshift of $z = 0.695$ (Bloom et al. 2001). The exceptionally bright GRB 030329 in 2003 was detected by all three instruments on the High Energy Transient Explorer satellite (HETE) across a wide photon energy range: Fregate 7–400 keV; the WXM 2–25 keV; SXC 2–10 keV. Rapid communication of the $2'$ error box 73 minutes after the burst allowed the detection of a counterpart in optical and radio emission (Vanderspek et al. 2004). This multi-wavelength coverage lead to the measurement of a redshift, $z = 0.1685$, the discovery of a supernova associated with the burst and established that long GRBs can be produced by the explosion of some extreme type Ic SNe.

The redshift distribution of GRBs is a moving target with new results being published all the time. A good up to date summary can be found at the web page maintained by Greiner.[1] The redshift measurements span the range 0.0085-9.4 with a mean of $\sim$2.5. This distribution is the product of intense multi-wavelength observation activity and is subject to many selection biases and constraints, e.g. Coward et al. (2013). HST imaging of the error boxes of long GRBs reveals that they occur predominantly in UV bright star forming regions of their host galaxies, e.g. Fruchter et al. (2006).

The detection and accurate positioning of GRBs combined with a rapid alert system and fast slewing response to initiate near immediate follow-up observations was pioneered by the *Swift* satellite (Gehrels et al. 2004). This has provided a wealth of new and intriguing results including the first accurate localizations and redshifts for short GRBs (Gehrels et al. 2005), the identification of the highest redshift GRBs (Cucchiara et al. 2011), a population of low luminosity GRBs often associated with supernovae (Stanek et al. 2003; Zhang et al. 2012), the detection of the brightest optical GRB, visible with the naked eye (Bloom et al. 2009) and detailed multi-wavelength light curves and spectra from the prompt through to the afterglow emission. With the rapid development of new facilities like *LIGO*, *Virgo*, *IceCube* etc. there is now the possibility of the simultaneous detection of multi-messenger signals, including gravitational wave and neutrinos, from GRBs. The likelihood of such detections is, of course, dependant on a multitude of factors such as the local rate of mergers of compact objects and the beaming angle of GRB jets but it is clear that a rapid response and rapid alert enabling multi-wavelength and multi-messenger observations from the prompt through to late afterglow will be mandatory if we hope to gain a detailed understanding of GRBs and other fast, high energy transients.

2 The Fireball Model Then and Now

The theoretical paradigm underpinning most investigations of GRBs is the standard fireball shock model (see Mészáros 2002, and references therein). At early times, and close to the central engine, internal shocks within the relativistically expanding fireball produce the highly variable prompt hard X-ray and Gamma-ray emission we see as the GRB (Rees and Mészáros 1994) and as the fireball is decelerated by interaction with the low-density circumburst medium an external shock forward shock develops to produce the longer lasting afterglow emission seen initially from the soft X-ray band through to the IR (Rees and Mészáros 1992). The predominant emission mechanism is expected to be synchrotron radiation from relativistic electrons accelerated in the shocks, both internal and external, gyrating in the magnetic field wound up by compression and turbulence in the same shocks. In

[1] http://www.mpe.mpg.de/~jcg/grbgen.html.

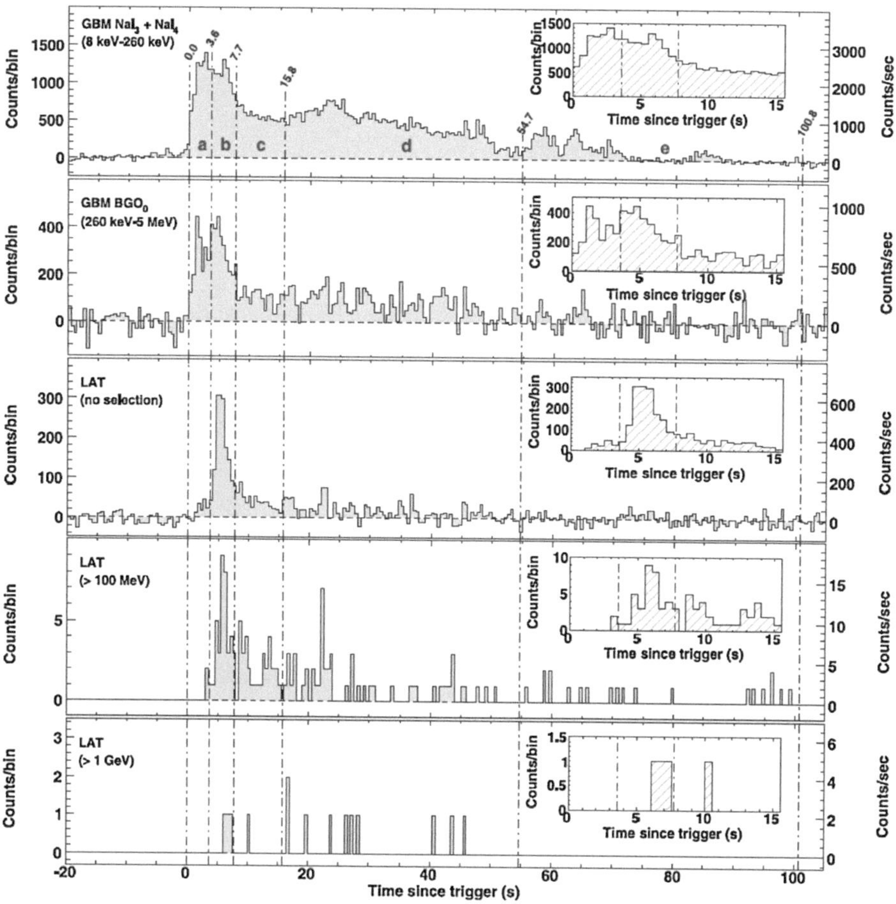

Fig. 1 The Fermi GRM and LAT lightcurves for GRB 080916C (Abdo et al. 2009). The first pulse, 0.0–3.6 seconds is bright in the GBM but absent from the LAT

addition, inverse Compton up-scattering of low energy synchrotron photons by the relativistic electron population can produce a high energy component or tail to the expected synchrotron spectrum such that the broadband spectrum observed is described by a so-called Synchrotron-Self-Compton (SSC) model. When energy injection from the central engine ceases the forward shock emission enters a final power-law decay and produces a characteristic synchrotron radiation afterglow spectrum stretching from radio through to X-ray (Sari et al. 1998; Wijers and Galama 1999).

With the launch of *Fermi* late 2008, the detection of GRBs with both the Large Area Telescope (LAT, 20 MeV to $>$300 GeV) and the Gamma-ray Burst Monitor (GBM, 8 keV to 40 MeV) has given us exceptional broadband spectra of the prompt emission over more than seven decades of energy. An interesting and unexpected finding is that for many GRBs the GeV emission starts with a noticeable delay after the MeV emission. For example, the first pulse of GRB 080916C is visible only in the MeV while the second pulse, that appears $\sim$4 seconds later, is seen strongly in the GeV band by the LAT, Fig. 1 (Abdo et al. 2009). For GRB 080916C there is no evidence for a second spectral component; the spectrum is fitted by a single component that evolves from soft to hard and back to soft again. Furthermore,

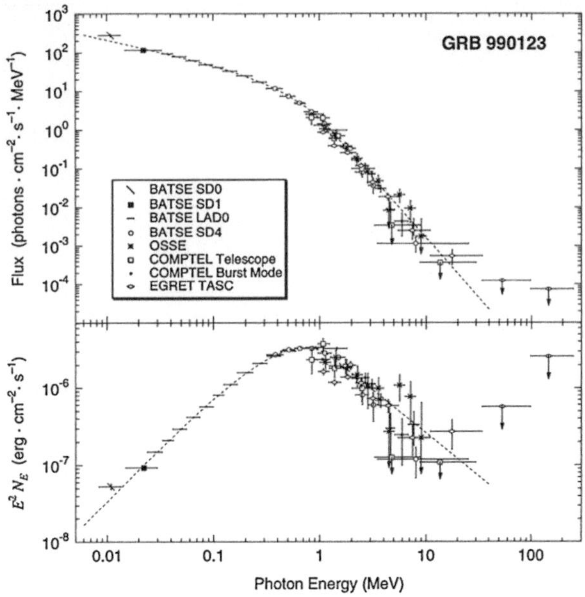

Fig. 2 The Band function illustrated by spectrum of GRB 990123 with $E_{peak} = 720$ keV (Briggs et al. 1999)

for this and other LAT bursts the GeV emission persists into the afterglow phase >1000 s after the initial trigger. However, for some bursts, for example GRB 090902B (Abdo et al. 2009) and GRB 090926A (Ackermann et al. 2011), a distinct second hard spectral component is observed. Explaining the GeV–MeV time delays and the second distinct hard spectral component is difficult in the context of a one-zone (internal shocks) SSC mechanism (Ackermann and the Fermi collab. 2011). The initial GeV emission is obviously associated with the prompt pulse structure but the GeV emission extending into the afterglow, seen after the main prompt pulses, is unlikely to come from internal shocks and maybe SSC arising from the forward shock as the jet interacts with the surrounding circumburst medium.

Recent *Fermi* observations of the early prompt spectra of a few bursts have revealed a second mid-energy range spectral component in the emission again calling into question the validity of the one-zone internal shock/SSC radiation model for the prompt phase (Veres and Mészáros 2012). The new paradigm which may solve this problem evokes photospheric emission to explain at least part if not all of the prompt emission in the energy band 20 keV to 100 GeV.

3 The Prompt Emission

GRBs are primarily detected by virtue of the initial short burst of gamma-rays. A succession of instruments have provided GRB triggers, e.g. BATSE on *CGRO*, the Konus-Wind Experiment, FREGATE on *HETE-2*, IBIS on *INTEGRAL*, BAT on *Swift*, GRID on *AGILE* and the LAT and GRM on *Fermi*. The spectrum of the prompt emission is usually observed as a broken powerlaw characterised by the so-called Band function (Band et al. 1993), as shown in Fig. 2. The peak of the $E.F(E)$ spectrum, E_{peak}, represents the characteristic photon energy for the GRB. In order to get a reliable estimate of E_{peak} broad energy band coverage is required.

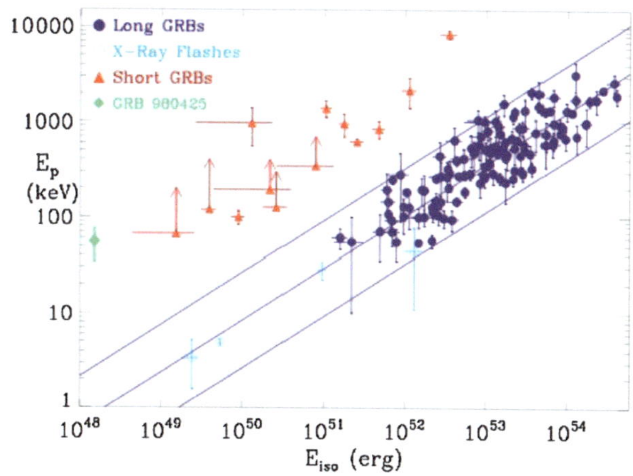

Fig. 3 The Amati Relation—a correlation between the characteristic photon peak energy, E_p, in the rest frame of the burst and the equivalent isotropic energy, E_{iso}. Long GRBs and X-ray Flashes lie in the correlation band which is ~1 dex wide in energy. All short GRBs are outliers and form a band of similar width but offset from the main correlation by ~2 dex in energy

There is a well known correlation, called the Amati Relation, between the E_{peak} energy expressed in the rest frame of the GRB and the isotropic energy, E_{iso}, estimated from the total fluence from the burst and the redshift (Amati et al. 2002; Amati 2006). This is illustrated in Fig. 3. The correlation is different for short ($T_{90} < 2$ seconds) and long ($T_{90} > 2$ seconds) with higher E_{peak} values for a given E_{iso} value for the short bursts. For both long and short bursts the correlation is not tight with a spread of $\sim$ factor 10. There has been much discussion as to whether the correlation is an intrinsic property of the GRB emission or an observational or selection artifact (Butler et al. 2007). It may be that the relation is a signature of something fundamental about the GRB emission but better broad energy band spectral observations are required, coupled with more accurate determination of other burst properties, to reveal the true nature of the relationship.

The prompt emission of a few bright GRBs observed by *Fermi* have hard X-ray and Gamma-ray spectra which, as expected, are dominated by a Band function, usually attributed to non-thermal emission from internal shocks, but also include a highly significant component fitted by a thermal spectrum, e.g. GRB 100724B (Guiriec et al. 2011). Simultaneous observation across a broad energy band 10–10^7 keV is required to unambiguously confirm the presence of two components in the prompt spectra, see Fig. 4. Including the BB spectral component over the brightest part of the burst (T0 $-$ 1.024 s to T0 $+$ 83.969 seconds) shows a significant improvement in the fit compared to a Band function by itself however, the second component is weak compared to the dominant Band function and there is no guarantee that the additional component is thermal in origin and uniquely fitted by a Planck function. In GRB 100724B the two components seem to vary independently which favours the interpretation that the thermal component is of photospheric origin close to the central engine while the non-thermal component occurs at larger radii.

The challenge to any theoretical model is to produce a Band-function-like broadband spectrum for the prompt emission with a characteristic peak energy in the 0.2–1 MeV range (Gehrels et al. 2009), with spectral evolution soft-to-hard-to-soft or more simply hard-to-soft and to include a (delayed) GeV tail component and possibly a lower energy Planck-function-like second component. Within the SSC model the peak is produced by a combination of the underlying relativistic electron spectrum, synchrotron cooling, synchrotron self absorption and up-scattering of the synchrotron photons by inverse Compton from the relativistic electrons. The observed energy peak is dependent on the bulk Lorentz factor of the outflow. Whether it is possible to produce the range of observed Band function parameters by

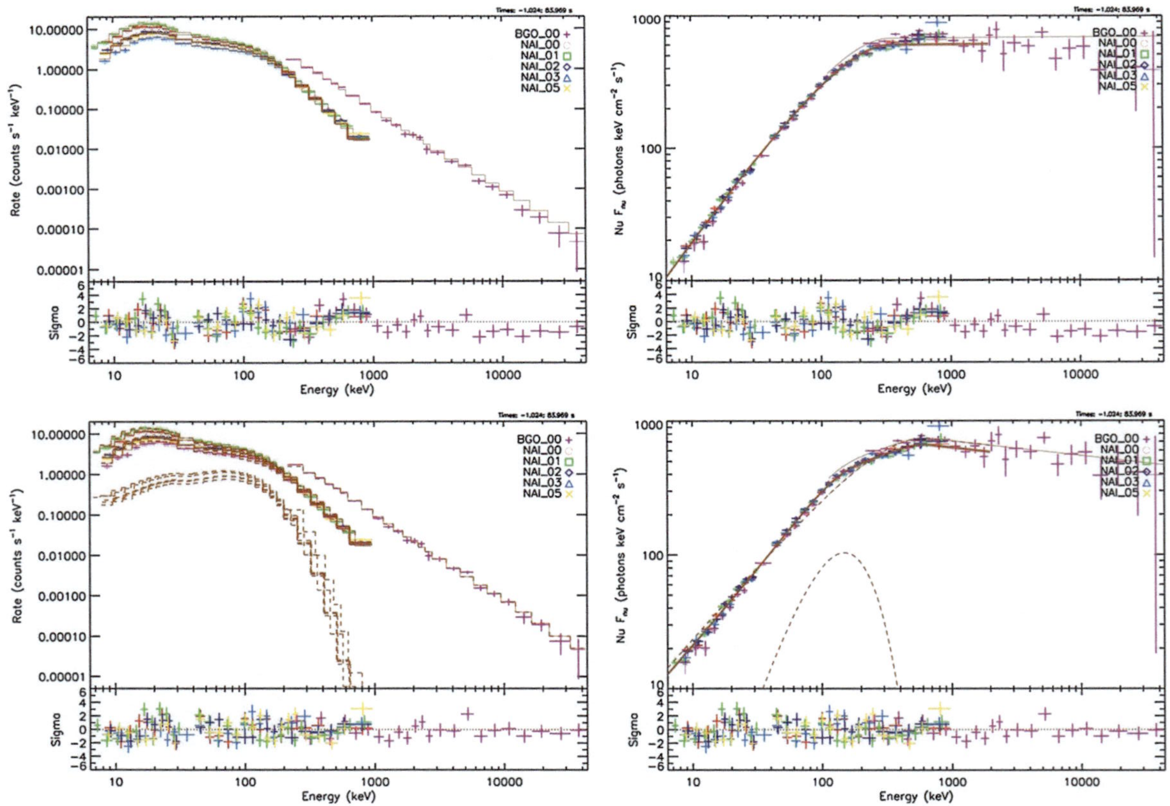

Fig. 4 Time-integrated spectrum of GRB 100724B (Guiriec et al. 2011). *Top panels*: Band function only. *Bottom panels*: Band function plus Black Body. *Left panels*: count spectra. *Right panels*: deconvolved $\nu F\nu$ spectra. The data points appear as *colour crosses*. *Dashed lines* indicate the individual spectral functions, and *solid lines* show the summed model fit

SSC has been questioned by many authors (e.g. Ghisellini et al. 2000; Preece et al. 2000) and the radiation efficiency of internal shocks is low, 5%–10%, unless the colliding shells have widely differing Lorentz factors (Beloborodov 2000). An alternative is to associate the energy peak to a blackbody at the comoving pair combination temperature in the fireball's photosphere (Eichler and Levinson 2000). The spectrum below the peak energy can then be attributed to the Rayleigh–Jeans part of the photospheric spectrum but the high energy power law tail and the GeV emission require an additional explanation (Meszaros and Rees 2000). If dissipation due to shocks or other processes occurs at or below the photosphere then a high efficiency is achievable and the thermal peak energies occur in the observed Band function range and Compton equilibrium of internal shock electrons or pairs with photospheric photons lead to spectra with a break at the right preferred energy and steep low energy slopes (Rees and Mészáros 2005; Pe'er et al. 2005, 2006). The photospheric model also provides a possible physical explanation for the Amati relation (or Ghirlanda relation, see below) (Thompson et al. 2007).

The magnetic field in the outflow plays a vital role. It is necessary for the production of synchrotron radiation and is certainly crucial in the acceleration of the relativistic charged particle distribution, by a Fermi mechanism, that ultimately produces the high energy tail of the spectrum via inverse Compton up-scattering. However, if the magnetic field is strong enough the magnetic stresses will be dominant and the whole dynamics of the jet will be driven by the magnetic field and must be modelled using magnetohydrodynamics (MHD). Magnetically dominated (or so-called Poynting dominated) GRB jets fall into two categories, baryon-free in which the baryons are absent or dynamically negligible (e.g. Drenkhahn 2002) and those with a non-negligible baryon load (e.g. Thompson 2006). In both cases the Alfvén (sound) speed in the ejecta is very high and therefore internal shocks maybe absent or weaker than in the standard hydrodynamics case. Similarly the reverse shock may be weaker although an external shock and associated forward blast wave will still be produced (Mészáros and Rees 1997). Any internal or reverse shocks that do form will have different radiation characteristics dependent on the magnetization parameter σ. Along side or in place of internal shocks "internal dissipation regions" formed by magnetic reconnection will lead to particle acceleration via electric fields produced by magnetic reconnection, rather than a Fermi mechanism, and a high radiative efficiency is likely. In order to get the isotropically equivalent luminosities of $L_\gamma \geq 10^{52}\ \mathrm{erg\,s^{-1}}$ seen in GRBs the magnetic fields at the base of the jet must be $B \sim 10^{15}$ G. The baryon-free Poynting jet models are like pulsar wind models. In a similar way to the purely hydro baryon-loaded models the initially high magnetic stresses, involving pairs and photons, lead to acceleration and growth of the Lorentz factor, $\Gamma \propto r$ up to a pair annihilation photosphere which provides the first prompt emission peaking in the hard X-ray to MeV with up-scattering adding a high energy power law tail. In the absence of internal shocks the external shock must provide photons in the GeV–TeV range via inverse Compton scattering. On the other hand, baryon-loaded magnetically dominated jets behave differently with acceleration leading to $\Gamma \propto r^{1/3}$ or similar intermediate power law dependencies (McKinney and Uzdensky 2012; Metzger et al. 2011). The spectral signatures expected from Poynting dominated jets have not been extensively studied. Calculations made using a one-zone steady state approximation for a simple magnetised outflow photosphere (Giannios and Spruit 2007) indicate that a Band-type spectrum can be produced. The time delays and spectral components of a simple baryon-loaded magnetic model have be calculated (Bošnjak and Kumar 2012) giving results in the range observed by *Fermi*.

In GRB jets with a significant baryonic content the charged baryons will be co-accelerated with the electrons in any shocks or magnetic reconnection zones, and hadronic

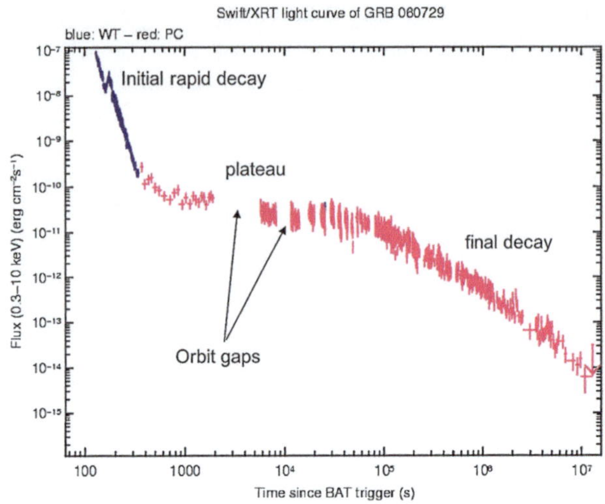

Fig. 5 The canonical GRB X-ray afterglow. The prompt emission typically ceases after ∼100 seconds and ends in a rapid fall with temporal decay index >2. In the vast majority of bursts this rapid decay is halted by a late emission component which forms a plateau that decays slowly, temporal index ∼0.5. Finally, the plateau turns down to give a final afterglow decay with index ∼1.5. The duration and decay rate of each of the three phases vary considerably from burst to burst

processes may lead to secondary high energy photon emission and neutrinos. For example, the *Fermi* LAT observations of the short GRB 090510 were modelled by Asano et al. (2009) as electron synchrotron for the MeV component and photohadronic cascade radiation for the distinct GeV power law component. Furthermore, since the acceleration and cascade development take time, a one-zone model including hadronic effects might explain the GeV–MeV photon delays seen by *Fermi* (Razzaque 2010). The hadronic interactions might also modify the low energy branch of the Band spectrum resulting in a GRB optical flash (Asano et al. 2010). The spectral signatures and GeV–MeV time delay can be produced by combining a magnetically dominated, baryon loaded model with a hadronic component involving nuclear collisions as calculated by Mészáros and Rees (2011). Murase et al. (2012) attempt to self consistently produce the GeV radiation, the MeV Band component and the low energy optical power law using a hadronic model. Protons accelerated by shocks or magnetic reconnection regions give hadronic cascades which in produce both the GeV and optical power laws. Cooled electron secondaries are then re-accelerated via a Fermi second order mechanism in the turbulent MHD waves produced by the same shocks, or reconnection regions, to yield a MeV Band spectrum.

More simultaneous, early, broad energy band/multi-wavelength observations with high signal to noise are required to reveal the detailed temporal behaviour of the components contributing to the prompt phase and to discriminate between the many theoretical possibilities.

4 The Canonical GRB X-Ray Afterglow

Before the arrival of the first *Swift XRT* observations of the X-ray afterglow of GRBs, starting only ∼100 seconds after the prompt trigger from the Burst Alert Telescope (BAT), it was expected that the X-ray (and optical) afterglows would fall as a single simple power law. On the contrary, the vast majority of X-ray afterglows have the canonical form shown in Fig. 5 (Nousek et al. 2006; Evans et al. 2007). The afterglow seems to arise from two components. The fading emission from the tails of the prompt GRB pulses seen as the initial rapid decay and the plateau which turns over into a final powerlaw decay, typically 10^4–10^5 seconds after the prompt emission has died, which most likely arises from the external shock formed

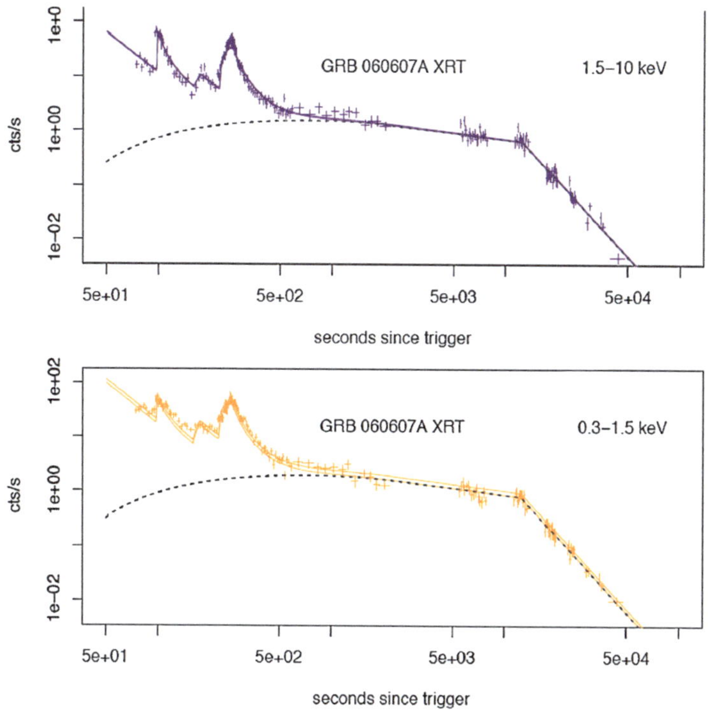

Fig. 6 Late X-ray flares not seen in the *Swift* BAT. The lightcurves are shown in two XRT bands 0.3–1.5 keV and 1.5–10 keV. The ratio of the fluxes in these bands gives us a measure of the soft X-ray spectra of the features seen. The afterglow component is indicated by the *dotted line*

as jetted material sweeps up the circumburst medium (O'Brien et al. 2006; Willingale et al. 2007).

The slow decay over the plateau phase was unexpected and is difficult to explain. It requires some form of energy injection, probably from the central engine, although the initial fast decay would suggest that the central engine has already switched off. In some cases the energy to sustain the plateau may come from the spin-down of a magnetar formed during the initial collapse (Zhang and Mészáros 2001; Rowlinson et al. 2010), but in most instances it is probably continued low level activity of the central engine. For a few GRBs the afterglow plateau persists for $>10^4$ seconds in which case the required total late energy injection would have to be large and comparable to the prompt energy. Another unexpected feature in the X-ray afterglows is the appearance of late flares, seen in the *Swift* XRT but not in the BAT (Chincarini et al. 2007), see Fig. 6. These flares are probably indicative that the central engine can and does remain active after the prompt Gamma-ray and hard X-ray emission has ceased. Figure 6 shows the afterglow decay in two XRT energy bands, 0.3–1.5 keV and 1.5–10 keV. In all bursts the ratio of the afterglow fluxes in these bands, over the plateau and in the final decay, remains constant indicating that there is no spectral evolution and that the temporal break from the end of the plateau into the final decay is achromatic.

If the relativistic outflow is collimated and slowing down, as the circumburst medium is swept up, we expect an achromatic break in the afterglow decay when the edge of the jet becomes visible as the relativistic beaming angle opens up. The standard formulation for so-called jet-breaks is given by Sari et al. (1999) and reference therein. Temporal breaks

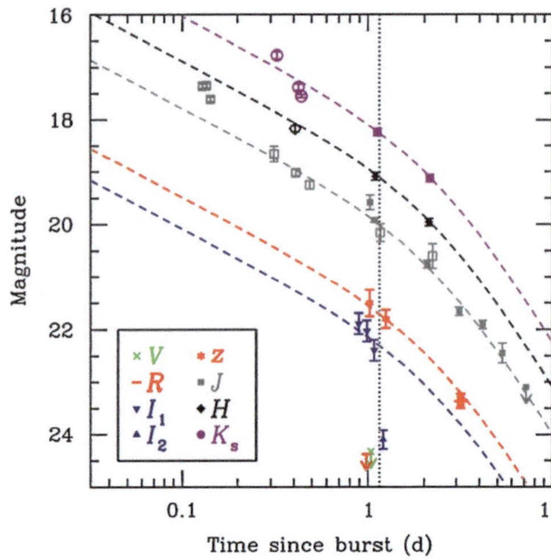

Fig. 7 The optical/IR break seen in the afterglow of GRB 050904 (Tagliaferri et al. 2005). The profile of the break is the same in the z, J, H and K bands indicating that the break is achromatic. When combined with the redshift the break time can be used to estimate the collimation angle of the jet

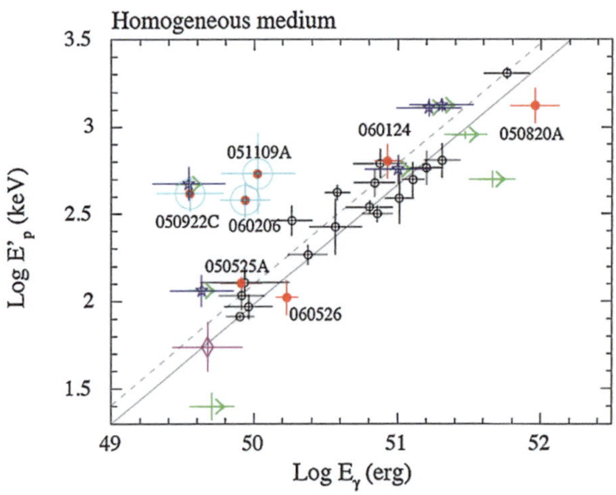

Fig. 8 The Ghirlanda relation (Ghirlanda et al. 2004) between the photon energy E_{peak} and the collimated energy estimated from E_{iso} and the jet collimation angle (Campana et al. 2007; McBreen et al. 2010). The unlabelled *black points* are the original GRBs used by Ghirlanda. The other values are for *Swift* and *Fermi* bursts which don't all conform to the original tight correlation

identified as jet-breaks in the afterglow decay are often seen in the optical and IR, e.g. Tagliaferri et al. (2005), Fig. 7, and when combined with a redshift measurement and an estimate of the total equivalent isotropic energy released the break times can be used to estimate the collimation angle of the jet. However, although late breaks are sometimes seen in the X-ray afterglows they rarely coincide with a break at other wavelengths so the simple achromatic break scenario is probably complicated by other factors.

Using an estimate of the jet collimation angle, derived from the assumed jet-break time observed in the optical/IR, we can correct the isotropic energy E_{iso} to a true collimated energy value E_γ. An original analysis performed by Ghirlanda et al. (2004) using 18 bursts showed that the correlation of E_{peak} with E_γ (known as the Ghirlanda Relation) was significantly tighter than the Amati Relation for the same bursts. However, more recent analysis using *Swift* and *Fermi* bursts, illustrated in Fig. 8, are not so conclusive (Campana et al. 2007; McBreen et al. 2010). This again brings into question whether the optical late breaks are really jet-breaks and whether or not the E_γ estimates are valid. More high quality multi-wavelength observations are required to tell us if the breaks are chromatic or achromatic to resolve this issue.

5 Prompt to Afterglow

The prompt Gamma-ray and hard X-ray emission from GRBs is very variable, both spectrally and temporally, but it can be characterised as a stochastic sequence of pulses each with a distinctive shape and spectral signature including a fast rise, slower decay and spectral softening. Several authors have attempted to model these pulses, e.g. Norris et al. (2005) using an empirical mathematical model and Willingale et al. (2010) using a physically motivated formulation. It is supposed that each pulse is associated with internal shocks within a thin ultra-relativistically expanding shell. The time profile we observe is produced by the relativistic aberrations and the Equal Arrival Time Surfaces (EATS) arising from the large Lorentz factor of the relativistic jet pointing towards us and the spherical geometry of the shell. Figure 9 shows the typical model pulse profile and an example of a GRB multi-energyband light curve fitted as a sequence of pulses and an afterglow component. The fitting is only possible because of the multi-energy band coverage and nearly continuous temporal coverage from the trigger to very late in the afterglow phase. When combined with a redshift the model provides rest frame timing and pulse/afterglow duration estimates and luminosity estimates for all the components.

On rare occasions the prompt emission is heralded by a precursor pulse which is bright enough to trigger the *Swift* BAT. In these cases the XRT and UVOT telescopes and alerted ground based instruments may be able to detect the latter stages of the prompt in which case we can get better multi-band coverage across the entire burst including the prompt phase as shown in Fig. 10. For GRB 080810 the burst also triggered the *Fermi* GBM so we also have simultaneous hard Gamma-ray coverage (Page et al. 2009) and for both GRB 061124 and GRB 080810 the afterglow flux was also detected in the radio as indicated on Fig. 10. Using data like these we can set up a temporally evolving Spectral Energy Distribution (SED) for both the prompt and afterglow emission. For GRB 061124 both the prompt and afterglow emission appear in the optical and the SED has a break between the optical and X-ray bands. The afterglow exhibits spectral evolution and decays too slowly compared with the standard synchrotron models. In addition there is evidence for a late break which might be a jet-break. For GRB 080810 the peak photon frequency in the SED, E_{peak}, evolves as a function of time and the afterglow optical-X-ray SED initially hardens but subsequently softens. The early rise and peak of the optical afterglow is also seen.

The rise and peak of the soft X-ray afterglow component is very rarely seen because it is hidden by the X-ray tail of the prompt hard X-ray and Gamma-ray emission. However, the early optical/IR afterglow peak can be observed (as in the case of GRB 080810 above). The optical emission most likely arises from the reverse shock which is expected to produce photons of lower energy than the forward shock which is responsible for the X-ray afterglow. The optical/IR afterglow peak can be assumed to correspond to the deceleration radius of the expanding fireball. The Lorentz factor of the outflow at the time of the peak, t_d, is half the initial value, Γ_0 (Sari and Piran 1999). From observations of t_d and using estimates of the isotropic energy released, E_{iso}, we can estimate the value of $\Gamma_0/2$. For both GRB 060418 and GRB 060607A the inferred deceleration radius is $R_{dec} \approx 10^{17}$ cm and $\Gamma_0 \sim 400$ giving us direct confirmation of the highly relativistic nature of the GRB fireballs (Molinari et al. 2007). Using values of the initial Lorentz factor derived in this way Liang et al. (2010) showed a correlation between Γ_0 and E_{iso}, Fig. 11. More recently the same authors published

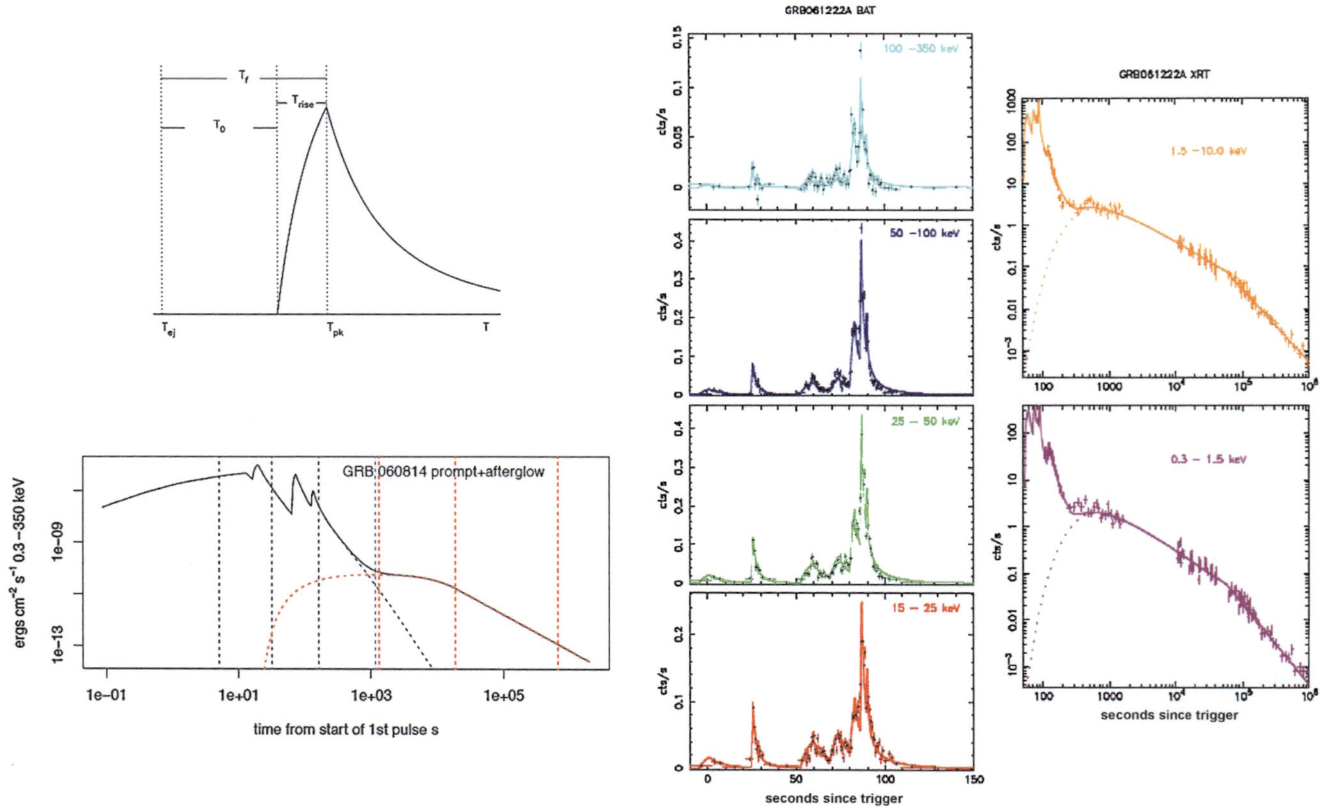

Fig. 9 *Top left*: the model profile for prompt pulses. *Bottom left*: the combination of prompt pulses and an afterglow representing the entire light curve emission (in this case from GRB 060814). *Right*: the prompt and afterglow emission from GRB 061222A seen in a combination of the energy bands from the *Swift* BAT and XRT

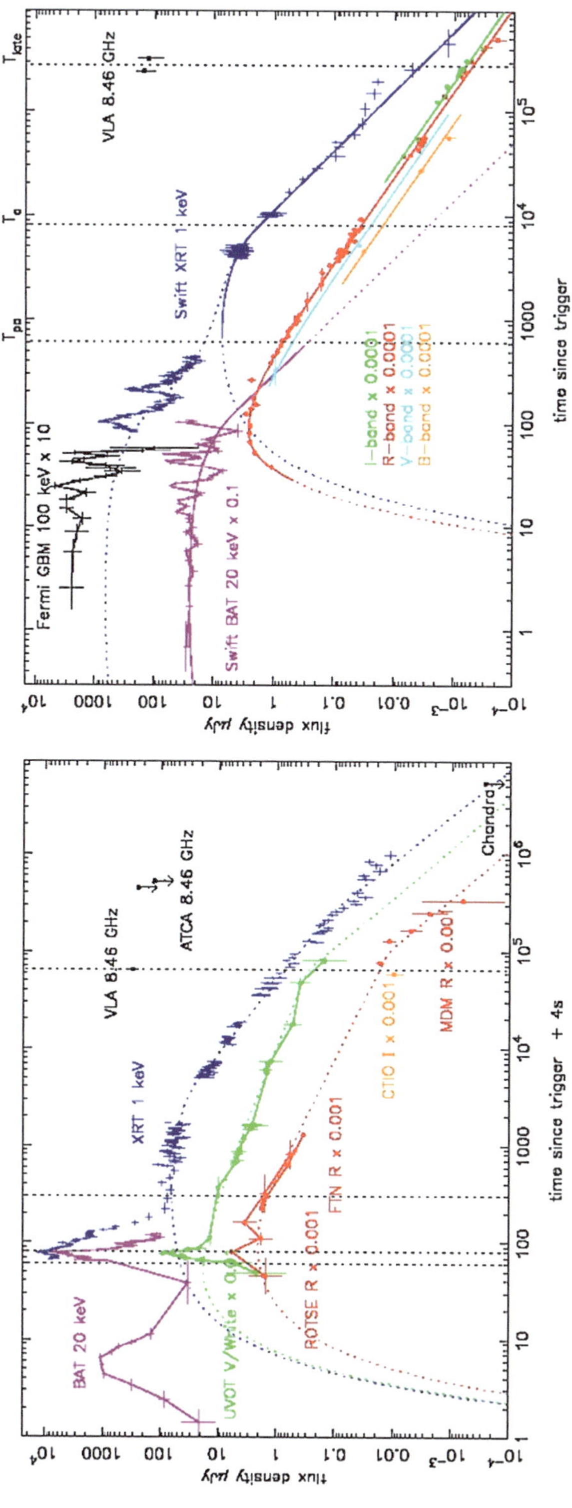

Fig. 10 Multi-band observations of the prompt and afterglow for *left*: GRB 061124 (Page et al. 2007) and *right*: GRB 080810 (Page et al. 2009)

a 3-parameter correlation between the isotropic luminosity, L_{iso}, E_{peak} and Γ_0 (Liang et al. 2015). Clearly, multi-wavelength observations which cover the prompt emission and the early afterglow have a great potential to reveal the key parameters which are driving the burst fireball.

6 Ultra-long GRBs and Tidal Disruption Events

Recently it has been noted that a few GRBs have a prompt emission phase which lasts an extraordinarily long time, measured in kiloseconds (Levan et al. 2014). Figure 12 shows an example of such a burst monitored both by the *Swift* BAT and the *Konus-Wind* Experiment, GRB 130925A (Evans et al. 2014). There are currently just four bursts which have a prompt duration (T_{90}) that is an order of magnitude longer than the main population of long GRBs (GRB 101225A, GRB 111209A, GRB 121027A, GRB 130925A). It may be that the progenitor and emission mechanism that gives rise to such persistent bursts is distinct from the majority but at the moment we don't have a detailed explanation for these ultra-long bursts. In addition there are one or two events detected by *Swift* which have repeated outbursts over a period of many days. For example, GRB 110328A was originally designated as a GRB but produced several BAT triggers and it soon became clear that the light curve didn't look like a conventional GRB, showing repeated flaring that eventually lasted for greater than 5×10^6 seconds. It was established that it occurred in a host galaxy at a redshift of $z = 0.351$ and was most probably a Tidal Disruption Event (TDE) in which a star was ripped apart by tidal forces when it wandered too close to a massive galactic black hole (Cannizzo et al. 2011). This event was subsequently re-labelled as Sw 1644+57. We require continued operation of *Swift* and future satellites like *SVOM* (Cordier et al. 2015) coupled with extended multi-wavelength observations to find and characterise more of these oddball events which clearly don't match the paradigm of conventional GRBs.

7 Follow-up of Gravitational Wave and Neutrino Triggers

A flux of neutrinos is expected from gamma-ray bursts (Asano and Mészáros 2014), but so far searching retrospectively through neutrino data from *IceCube* (IceCube Collaboration 2012; Aartsen et al. 2015), and *ANTARES* (Adrián-Martínez et al. 2013), looking for matches corresponding to the time and location of GRB triggers has not yielded any positive results. Alternatively, we can choose to follow-up neutrino triggers with an instrument like the *Swift* XRT (Evans et al. 2015). The XRT field of view is 24 arc minutes in diameter and the error radius of *IceCube* triggers is typically 1 degree at 50% confidence so this can be covered by a tiling of 7 XRT pointings as shown in Fig. 13. Each pointing has an exposure of 1–2 ksec depending on the satellite orbit conditions at the time of the trigger. The delay from neutrino trigger to the tiling of X-ray observations is typically 1–3 hours. The *Swift* XRT has a sensitivity of 5×10^{-13} $\mathrm{erg\,cm^{-2}\,s^{-1}}$ in 1 ksec in the 0.3–10 keV band and can get source positions with a 90% confidence radius of 3.5 arcsec or better for brighter sources (Evans et al. 2009).

Gravitational waves are expected from a range of transients, in particular the merger of two neutron stars which is also expected to produce a short GRB. Targeted searches for gravitational waves from short GRBs have so far failed to produce detections (Abadie et al. 2010). Similarly, *Swift* XRT follow-up of gravitational wave triggers has so far also drawn a

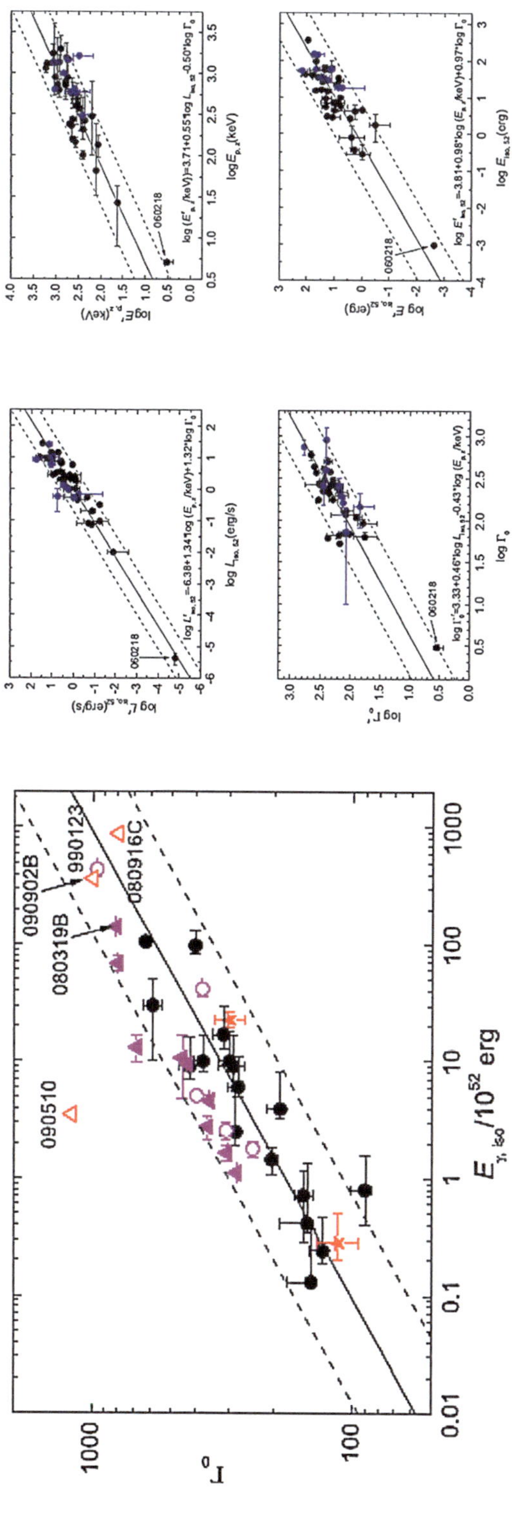

Fig. 11 *Left*: the correlation between the initial Lorentz factor and the isotropic energy. *Right*: a 3-parameter correlation between the isotropic luminosity, E_{peak} and Γ_0

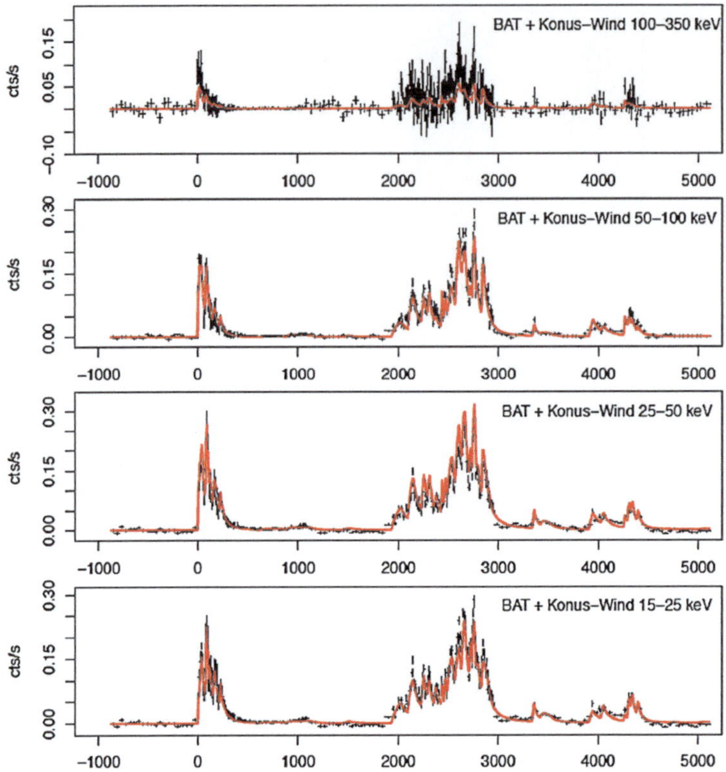

Fig. 12 The combined *Swift* BAT and *Konus-Wind* prompt light curve of GRB 130925A with a prompt duration of ∼5000 seconds

Fig. 13 Exposure map of a 7-tile *Swift* XRT observation of an IceCube trigger. The *blank lines* and *dots* are bad columns and pixels on the CCD

blank (Evans et al. 2012), although this is to be expected because the error boxes associated with LIGO-Virgo triggers have to date been very large, many tens of square degrees.

The *Swift* XRT is probably the best instrument currently available for neutrino and gravitational wave follow-up, looking for the high energy transient counterparts, because of the reasonable sensitivity of the XRT in the 0.3–10 keV band and the rapid repointing response capability. However, it has a rather small field of view so tiling is required and the response time of a few hours means that any short transient may have faded significantly by the time the observations begin. A high sensitivity soft X-ray transient monitor with a wide field of many square degrees is required to increase the probability of finding any fast transients associated with these triggers.

8 The Lobster Eye X-Ray Transient Monitor

The wide field detectors used to find and position GRBs and similar transients, listed in Sect. 3, all use collimation and/or coded shadow mask techniques to detect and estimate the position of hard X-ray/Gamma-ray transients. Such devices have the required large fields of view but the sensitivity is limited by the multiplex nature of the positioning capability. If the X-ray flux could be focused in some way, so that multiplexing of the detector imaging elements could be reduced or eliminated, the sensitivity would be greatly improved. The lobster eye X-ray focusing geometry originally described by Angel (1979) provides just that. In a lobster eye the image is produced by grazing incidence reflections off the inside walls of microscopic square pores which are arranged over a spherical surface all pointing towards a common centre of curvature. The image is formed on a spherical focal surface with half the radius of curvature of the square pore optic. Unlike a fly's eye optic, which uses an array of refracting lenses, the lobster eye produces a single continuous image of the sky. The field of view is limited only by the size of the optic and size of the detector and if an optic (and detector) could be constructed across a complete sphere the whole sky would be captured in a single continuous image.

The idea of using the lobster eye principle to produce a soft X-ray all-sky monitor was first proposed by Priedhorsky et al. (1996) and a method of manufacturing the square pore array has been described by several groups (Chapman et al. 1991; Wilkins et al. 1989; Fraser et al. 1993). Recent proposals for soft X-ray wide field transient monitors using lobster eye optics include *ISS Lobster* submitted to NASA December 2014 and the *Theseus* SXT submitted the ESA M4 call at about the same time. Figure 14 shows a schematic view of a typical module and the flux sensitivity it can achieve. Each module has a field of view of $\sim 1/6$ steradian so an array of 6 modules can view 1 steradian. An angular resolution of ~ 5 arc minutes is currently achievable using the available square pore micro-channel plate technology. The sensitivity is $\sim 2 \times 10^{-10}\ \mathrm{erg\,cm^{-2}\,s^{-1}}$ in 100 sec in the 0.3–5 keV energy band. This is a factor of ~ 100 higher than the *Swift* XRT but the lobster eye transient monitor will see transients when they happen rather than 1–2 hours later, in response to a trigger. It is predicted that such a monitor would see greater than 90% of the GRBs seen by the *Swift* BAT and provide positions <1 minute for all transients. A lobster eye X-ray transient monitor is the ideal instrument to hunt for and position fast soft X-ray transients or look for the X-ray counterparts to neutrino and gravitational wave triggers.

9 Discussion and Conclusion

The study of GRBs is entering an exciting new phase. After great advances obtained through multi-wavelength studies made possible by *CGRO*, *Beppo-SAX*, *HETE-2* and *Swift*, GRB studies have been extended well into the GeV range by the *Fermi* mission (Bissaldi et al. 2015a), and higher energies reaching the TeV range are being probed by the High Altitude Water Cherenkov Observatory (HAWC) (Lennarz et al. 2015) and the High Energy Stereoscopic System (HESS https://www.mpi-hd.mpg.de/hfm/HESS/). The next instruments after that are expected to be, at X-rays to MeV energies, the SVOM satellite (Cordier et al. 2015),

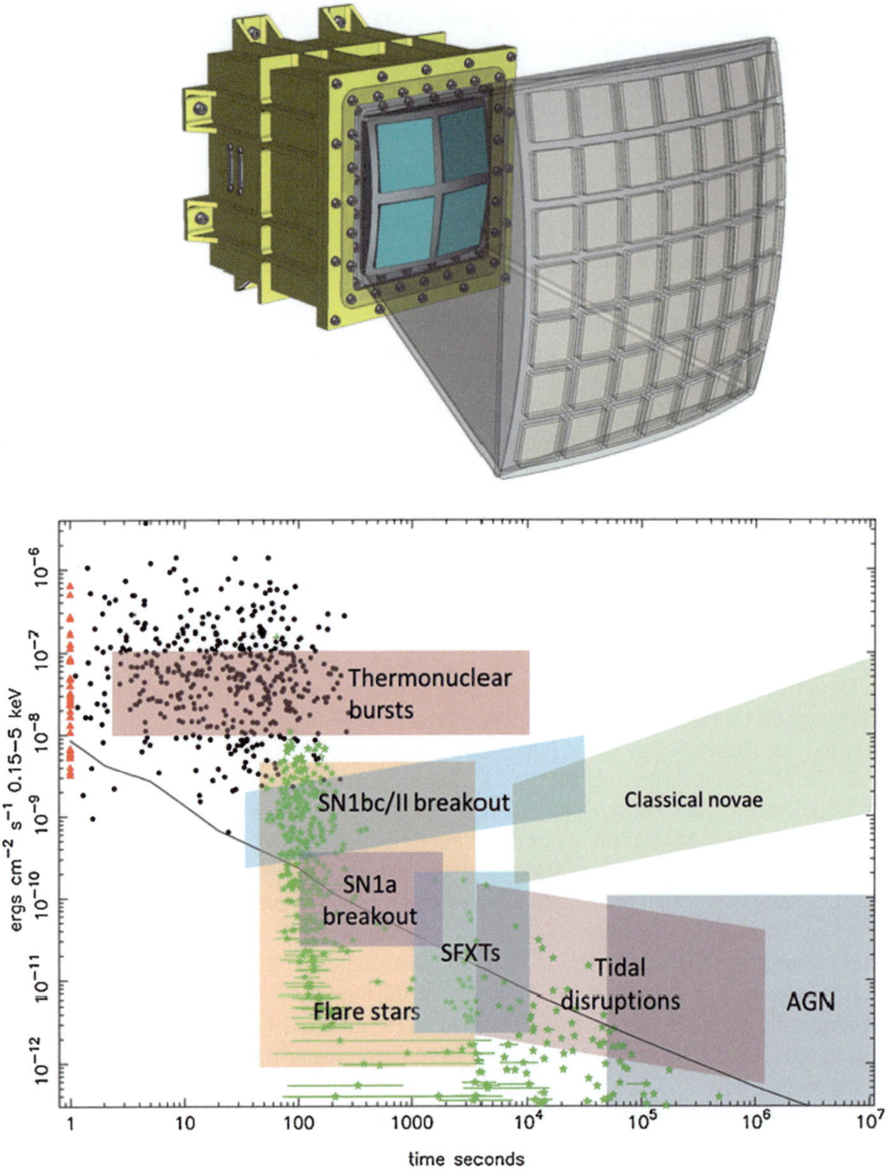

Fig. 14 *Top*: a typical lobster eye module with a focal length of 300 mm and total mass of ~20 kg. *Bottom*: the flux sensitivity of the module compared to typical flux ranges and timescales associated with types of transients. The *black dots* indicate the prompt flux from *Swift* GRBs detected by the BAT and the *green dots* are the afterglow flux levels seen in the afterglows by the XRT. The *black curve* is the module flux sensitivity in the energy band 0.15–5 keV as a function of integration time

and in the TeV to multi-TeV range the ground-based (air) Cherenkov Telescope Array (CTA) (Bissaldi et al. 2015b).

Non-photonic messengers are also being probed, most notably high energy neutrinos, with *IceCube* setting strong constraints against classical (Swift-detected) GRBs providing more than 1% of the observed neutrino background (Aartsen et al. 2015). One possibility is that the gamma-rays arise at lower radii, e.g. in the photosphere, while cosmic rays are only accelerated in shocks further out, where the efficiency for photo-hadronic interaction producing neutrinos is low (Asano and Mészáros 2016). For a discussion of the physics and gamma-ray emission from GRB photospheres, see Beloborodov and Mészáros, 2016, elsewhere in this volume.

Following the recent announcement of the first detection of gravitational waves from the merger of stellar mass binary black holes by LIGO (Abbott et al. 2016) results from non-photonic GW searches are now eagerly awaited. Neutron star or neutron star-black hole mergers are also expected to produce gravitational waves detectable by LIGO and VIRGO, and their detection in coincidence with a short GRB gamma-ray signal is the smoking gun signal being looked for in order to confirm the progenitor's identity.

Gamma-ray bursts continue to surprise us, and it is clear that the array of major new facilities that are being brought to bear on these sources will play a major role in solving some of the many questions that they pose.

References

M.G. Aartsen, M. Ackermann, J. Adams, J.A. Aguilar, M. Ahlers, M. Ahrens, D. Altmann, T. Anderson, C. Arguelles, T.C. Arlen et al., Search for prompt neutrino emission from gamma-ray bursts with IceCube. Astrophys. J. Lett. **805**, 5 (2015). doi:10.1088/2041-8205/805/1/L5

J. Abadie, B. Abbott, R. Abbott, T. Accadia, F. Acernese, R. Adhikari, P. Ajith, B. Allen, G. Allen, E.A. Ceron et al., Search for gravitational-wave inspiral signals associated with short gamma-ray bursts during ligo's fifth and virgo's first science run. Astrophys. J. **715**(2), 1453 (2010)

B.P. Abbott, R. Abbott, T.D. Abbott, M.R. Abernathy, F. Acernese, K. Ackley, C. Adams, T. Adams, P. Addesso, R.X. Adhikari et al., Observation of gravitational waves from a binary black hole merger. Phys. Rev. Lett. **116**(6), 061102 (2016). doi:10.1103/PhysRevLett.116.061102

A.A. Abdo, M. Ackermann, M. Arimoto, K. Asano, W.B. Atwood, M. Axelsson, L. Baldini, J. Ballet, D.L. Band, G. Barbiellini et al., Fermi observations of high-energy gamma-ray emission from GRB 080916C. Science **323**, 1688 (2009). doi:10.1126/science.1169101

A.A. Abdo, M. Ackermann, M. Ajello, K. Asano, W.B. Atwood et al., Astrophys. J. Lett. **706**, 138 (2009)

M. Ackermann, the Fermi collab., Detection of a spectral break in the extra hard component of GRB 090926A. Astrophys. J. **729**, 114 (2011). doi:10.1088/0004-637X/729/2/114

M. Ackermann, M. Ajello, K. Asano, M. Axelsson, L. Baldini, J. Ballet, G. Barbiellini, M. Baring, D. Bastieri, K. Bechtol et al., Detection of a spectral break in the extra hard component of GRB 090926a. Astrophys. J. **729**(2), 114 (2011)

S. Adrián-Martínez, A. Albert, I. Al Samarai, M. André, M. Anghinolfi, G. Anton, S. Anvar, M. Ardid, T. Astraatmadja, J.-J. Aubert et al., Search for muon neutrinos from gamma-ray bursts with the ANTARES neutrino telescope using 2008 to 2011 data. Astron. Astrophys. **559**, 9 (2013)

L. Amati, The Ep,i-Eiso correlation in gamma-ray bursts: updated observational status, re-analysis and main implications. Mon. Not. R. Astron. Soc. **372**, 233–245 (2006). doi:10.1111/j.1365-2966.2006.10840.x

L. Amati, F. Frontera, M. Tavani, J.J.M. in't Zand, A. Antonelli, E. Costa, M. Feroci, C. Guidorzi, J. Heise, N. Masetti, E. Montanari, L. Nicastro, E. Palazzi, E. Pian, L. Piro, P. Soffitta, Intrinsic spectra and energetics of BeppoSAX gamma-ray bursts with known redshifts. Astron. Astrophys. **390**, 81–89 (2002). doi:10.1051/0004-6361:20020722

J.R.P. Angel, Lobster eyes as X-ray telescopes. Astrophys. J. **233**, 364–373 (1979). doi:10.1086/157397

K. Asano, P. Mészáros, Ultrahigh-energy cosmic ray production by turbulence in gamma-ray burst jets and cosmogenic neutrinos. Phys. Rev. D **94**(2), 023005 (2016). doi:10.1103/PhysRevD.94.023005

K. Asano, P. Mészáros, Neutrino and cosmic-ray release from gamma-ray bursts: time-dependent simulations. Astrophys. J. **785**(1), 54 (2014)

K. Asano, S. Guiriec, P. Mészáros, Hadronic models for the extra spectral component in the short GRB 090510. Astrophys. J. Lett. **705**(2), 191 (2009)

K. Asano, S. Inoue, P. Mészáros, Prompt x-ray and optical excess emission due to hadronic cascades in gamma-ray bursts. Astrophys. J. Lett. **725**(2), 121 (2010)

J.-L. Atteia, N. Kawai, R. Vanderspek, G. Pizzichini, G.R. Ricker, C. Barraud, M. Boer, J. Braga, N. Butler, T. Cline, G.B. Crew, J.-P. Dezalay, T.Q. Donaghy, J. Doty, E.E. Fenimore, M. Galassi, C. Graziani, K. Hurley, J.G. Jernigan, D.Q. Lamb, A. Levine, R. Manchanda, F. Martel, M. Matsuoka, E. Morgan, Y. Nakagawa, J.-F. Olive, G. Prigozhin, T. Sakamoto, R. Sato, Y. Shirasaki, M. Suzuki, K. Takagishi, T. Tamagawa, K. Torii, J. Villasenor, S.E. Woosley, M. Yamauchi, A. Yoshida, HETE-2 observation of two gamma-ray bursts at $z > 3$. Astrophys. J. **626**, 292–297 (2005). doi:10.1086/430057

D. Band, J. Matteson, L. Ford, B. Schaefer, D. Palmer, B. Teegarden, T. Cline, M. Briggs, W. Paciesas, G. Pendleton, G. Fishman, C. Kouveliotou, C. Meegan, R. Wilson, P. Lestrade, BATSE observations of gamma-ray burst spectra. I—Spectral diversity. Astrophys. J. **413**, 281–292 (1993). doi:10.1086/172995

A.M. Beloborodov, On the efficiency of internal shocks in gamma-ray bursts. Astrophys. J. Lett. **539**(1), 25 (2000)

E. Bissaldi, F. Longo, N. Omodei, G. Vianello, A. von Kienlin, Gamma-ray burst observations with Fermi (2015a). ArXiv e-prints arXiv:1507.03478

E. Bissaldi, T. Di Girolamo, F. Longo, P. Vallania, C. Vigorito, Prospects for gamma-ray bursts detection by the Cherenkov Telescope Array (2015b). ArXiv e-prints arXiv:1509.01438

J.S. Bloom, S.G. Djorgovski, S.R. Kulkarni, The redshift and the ordinary host galaxy of GRB 970228. Astrophys. J. **554**(2), 678 (2001). http://stacks.iop.org/0004-637X/554/i=2/a=678

J.S. Bloom, D.A. Perley, W. Li, N.R. Butler, A.A. Miller, D. Kocevski, D.A. Kann, R.J. Foley, H.-W. Chen, A.V. Filippenko, D.L. Starr, B. Macomber, J.X. Prochaska, R. Chornock, D. Poznanski, S. Klose, M.F. Skrutskie, S. Lopez, P. Hall, K. Glazebrook, C.H. Blake, Observations of the naked-eye GRB 080319B: implications of nature's brightest explosion. Astrophys. J. **691**, 723–737 (2009). doi:10.1088/0004-637X/691/1/723

Ž. Bošnjak, P. Kumar, Magnetic jet model for GRBs and the delayed arrival of >100 MeV photons. Mon. Not. R. Astron. Soc. Lett. **421**(1), 39–43 (2012)

M. Briggs, D. Band, R. Kippen, R. Preece, C. Kouveliotou, J. Van Paradijs, G. Share, R. Murphy, S. Matz, A. Connors et al., Observations of GRB 990123 by the Compton gamma ray observatory. Astrophys. J. **524**(1), 82 (1999)

N.R. Butler, D. Kocevski, J.S. Bloom, J.L. Curtis, A complete catalog of Swift gamma-ray burst spectra and durations: demise of a physical origin for pre-Swift high-energy correlations. Astrophys. J. **671**, 656–677 (2007). doi:10.1086/522492

S. Campana, C. Guidorzi, G. Tagliaferri, G. Chincarini, A. Moretti, D. Rizzuto, P. Romano, Are Swift gamma-ray bursts consistent with the Ghirlanda relation? Astron. Astrophys. **472**, 395–401 (2007). doi:10.1051/0004-6361:20066984

J.K. Cannizzo, E. Troja, G. Lodato, GRB 110328A/Swift J164449.3+573451: the tidal obliteration of a deeply plunging star? Astrophys. J. **742**, 32 (2011). doi:10.1088/0004-637X/742/1/32

H. Chapman, K. Nugent, S. Wilkins, X-ray focusing using square channel-capillary arrays. Rev. Sci. Instrum. **62**(6), 1542–1561 (1991)

G. Chincarini, A. Moretti, P. Romano, A.D. Falcone, D. Morris, J. Racusin, S. Campana, S. Covino, C. Guidorzi, G. Tagliaferri, D.N. Burrows, C. Pagani, M. Stroh, D. Grupe, M. Capalbi, G. Cusumano, N. Gehrels, P. Giommi, V. La Parola, V. Mangano, T. Mineo, J.A. Nousek, P.T. O'Brien, K.L. Page, M. Perri, E. Troja, R. Willingale, B. Zhang, The first survey of X-ray flares from gamma-ray bursts observed by Swift: temporal properties and morphology. Astrophys. J. **671**, 1903–1920 (2007). doi:10.1086/521591

B. Cordier, J. Wei, J.-L. Atteia, S. Basa, A. Claret, F. Daigne, J. Deng, Y. Dong, O. Godet, A. Goldwurm, D. Götz, X. Han, A. Klotz, C. Lachaud, J. Osborne, Y. Qiu, S. Schanne, B. Wu, J. Wang, C. Wu, L. Xin, B. Zhang, S.-N. Zhang, The SVOM gamma-ray burst mission (2015). ArXiv e-prints arXiv:1512.03323

E. Costa, F. Frontera, J. Heise, M. Feroci, J.I. Zand, F. Fiore, M.N. Cinti, D.D. Fiume, L. Nicastro, M. Orlandini et al., Discovery of the x-ray afterglow of the gamma-ray burst of February 28 1997 (1997). ArXiv preprint arXiv:astro-ph/9706065

D.M. Coward, E.J. Howell, M. Branchesi, G. Stratta, D. Guetta, B. Gendre, D. Macpherson, The Swift gamma-ray burst redshift distribution: selection biases and optical brightness evolution at high z? Mon. Not. R. Astron. Soc. **432**(3), 2141–2149 (2013). doi:10.1093/mnras/stt537. http://mnras.oxfordjournals.org/content/432/3/2141.abstract

A. Cucchiara, A.J. Levan, D.B. Fox, N.R. Tanvir, T.N. Ukwatta, E. Berger, T. Krühler, A. Küpcü Yoldaş, X.F. Wu, K. Toma, J. Greiner, F.E. Olivares, A. Rowlinson, L. Amati, T. Sakamoto, K. Roth, A. Stephens, A. Fritz, J.P.U. Fynbo, J. Hjorth, D. Malesani, P. Jakobsson, K. Wiersema, P.T. O'Brien, A.M. Soderberg, R.J. Foley, A.S. Fruchter, J. Rhoads, R.E. Rutledge, B.P. Schmidt, M.A. Dopita, P. Pod-

siadlowski, R. Willingale, C. Wolf, S.R. Kulkarni, P. D'Avanzo, A photometric redshift of $z \sim 9.4$ for GRB 090429B. Astrophys. J. **736**, 7 (2011). doi:10.1088/0004-637X/736/1/7
G. Drenkhahn, Acceleration of GRB outflows by Poynting flux dissipation. Astron. Astrophys. **387**(2), 714–724 (2002)
D. Eichler, A. Levinson, A compact fireball model of gamma-ray bursts. Astrophys. J. **529**(1), 146 (2000)
P.A. Evans, A.P. Beardmore, K.L. Page, L.G. Tyler, J.P. Osborne, M.R. Goad, P.T. O'Brien, L. Vetere, J. Racusin, D. Morris, D.N. Burrows, M. Capalbi, M. Perri, N. Gehrels, P. Romano, An online repository of Swift/XRT light curves of γ-ray bursts. Astron. Astrophys. **469**, 379–385 (2007). doi:10.1051/0004-6361:20077530
P.A. Evans, R. Willingale, J.P. Osborne, P.T. O'Brien, N.R. Tanvir, D.D. Frederiks, V.D. Pal'shin, D.S. Svinkin, A. Lien, J. Cummings, S. Xiong, B.-B. Zhang, D. Götz, V. Savchenko, H. Negoro, S. Nakahira, K. Suzuki, K. Wiersema, R.L.C. Starling, A.J. Castro-Tirado, A.P. Beardmore, R. Sánchez-Ramírez, J. Gorosabel, S. Jeong, J.A. Kennea, D.N. Burrows, N. Gehrels, GRB 130925A: an ultralong gamma ray burst with a dust-echo afterglow, and implications for the origin of the ultralong GRBs. Mon. Not. R. Astron. Soc. **444**, 250–267 (2014). doi:10.1093/MNRAS/stu1459
P.A. Evans, J.P. Osborne, J.A. Kennea, M. Smith, D.M. Palmer, N. Gehrels, J.M. Gelbord, A. Homeier, M. Voge, N.L. Strotjohann, D.F. Cowen, S. Böser, M. Kowalski, A. Stasik, Swift follow-up of IceCube triggers, and implications for the advanced-LIGO era. Mon. Not. R. Astron. Soc. **448**, 2210–2223 (2015). doi:10.1093/mnras/stv136
P. Evans, A. Beardmore, K. Page, J. Osborne, P. O'Brien, R. Willingale, R. Starling, D. Burrows, O. Godet, L. Vetere et al., Methods and results of an automatic analysis of a complete sample of Swift-XRT observations of GRBs. Mon. Not. R. Astron. Soc. **397**(3), 1177–1201 (2009)
P. Evans, J. Fridriksson, N. Gehrels, J. Homan, J. Osborne, M. Siegel, A. Beardmore, P. Handbauer, J. Gelbord, J. Kennea et al., Swift follow-up observations of candidate gravitational-wave transient events. Astrophys. J. Suppl. Ser. **203**(2), 28 (2012)
G. Fraser, A. Brunton, J. Lees, J. Pearson, W. Feller, X-ray focusing using square-pore microchannel plates first observation of cruxiform image structure. Nucl. Instrum. Methods Phys. Res., Sect. A, Accel. Spectrom. Detect. Assoc. Equip. **324**(1), 404–407 (1993)
A.S. Fruchter, A.J. Levan, L. Strolger, P.M. Vreeswijk, S.E. Thorsett, D. Bersier, I. Burud, J.M. Castro Cerón, A.J. Castro-Tirado, C. Conselice, T. Dahlen, H.C. Ferguson, J.P.U. Fynbo, P.M. Garnavich, R.A. Gibbons, J. Gorosabel, T.R. Gull, J. Hjorth, S.T. Holland, C. Kouveliotou, Z. Levay, M. Livio, M.R. Metzger, P.E. Nugent, L. Petro, E. Pian, J.E. Rhoads, A.G. Riess, K.C. Sahu, A. Smette, N.R. Tanvir, R.A.M.J. Wijers, S.E. Woosley, Long γ-ray bursts and core-collapse supernovae have different environments. Nature **441**, 463–468 (2006). doi:10.1038/nature04787
N. Gehrels, E. Ramirez-Ruiz, D.B. Fox, Gamma-ray bursts in the Swift era. Annu. Rev. Astron. Astrophys. **47**, 567–617 (2009). doi:10.1146/annurev.astro.46.060407.145147
N. Gehrels, G. Chincarini, P. Giommi, K.O. Mason, J.A. Nousek, A.A. Wells, N.E. White, S.D. Barthelmy, D.N. Burrows, L.R. Cominsky, K.C. Hurley, F.E. Marshall, P. Mészáros, P.W.A. Roming, L. Angelini, L.M. Barbier, T. Belloni, S. Campana, P.A. Caraveo, M.M. Chester, O. Citterio, T.L. Cline, M.S. Cropper, J.R. Cummings, A.J. Dean, E.D. Feigelson, E.E. Fenimore, D.A. Frail, A.S. Fruchter, G.P. Garmire, K. Gendreau, G. Ghisellini, J. Greiner, J.E. Hill, S.D. Hunsberger, H.A. Krimm, S.R. Kulkarni, P. Kumar, F. Lebrun, N.M. Lloyd-Ronning, C.B. Markwardt, B.J. Mattson, R.F. Mushotzky, J.P. Norris, J. Osborne, B. Paczynski, D.M. Palmer, H.-S. Park, A.M. Parsons, J. Paul, M.J. Rees, C.S. Reynolds, J.E. Rhoads, T.P. Sasseen, B.E. Schaefer, A.T. Short, A.P. Smale, I.A. Smith, L. Stella, G. Tagliaferri, T. Takahashi, M. Tashiro, L.K. Townsley, J. Tueller, M.J.L. Turner, M. Vietri, W. Voges, M.J. Ward, R. Willingale, F.M. Zerbi, W.W. Zhang, The Swift gamma-ray burst mission. Astrophys. J. **611**, 1005–1020 (2004). doi:10.1086/422091
N. Gehrels, C.L. Sarazin, P.T. O'Brien, B. Zhang, L. Barbier, S.D. Barthelmy, A. Blustin, D.N. Burrows, J. Cannizzo, J.R. Cummings, M. Goad, S.T. Holland, C.P. Hurkett, J.A. Kennea, A. Levan, C.B. Markwardt, K.O. Mason, P. Meszaros, M. Page, D.M. Palmer, E. Rol, T. Sakamoto, R. Willingale, L. Angelini, A. Beardmore, P.T. Boyd, A. Breeveld, S. Campana, M.M. Chester, G. Chincarini, L.R. Cominsky, G. Cusumano, M. de Pasquale, E.E. Fenimore, P. Giommi, C. Gronwall, D. Grupe, J.E. Hill, D. Hinshaw, J. Hjorth, D. Hullinger, K.C. Hurley, S. Klose, S. Kobayashi, C. Kouveliotou, H.A. Krimm, V. Mangano, F.E. Marshall, K. McGowan, A. Moretti, R.F. Mushotzky, K. Nakazawa, J.P. Norris, J.A. Nousek, J.P. Osborne, K. Page, A.M. Parsons, S. Patel, M. Perri, T. Poole, P. Romano, P.W.A. Roming, S. Rosen, G. Sato, P. Schady, A.P. Smale, J. Sollerman, R. Starling, M. Still, M. Suzuki, G. Tagliaferri, T. Takahashi, M. Tashiro, J. Tueller, A.A. Wells, N.E. White, R.A.M.J. Wijers, A short γ-ray burst apparently associated with an elliptical galaxy at redshift $z = 0.225$. Nature **437**, 851–854 (2005). doi:10.1038/nature04142
G. Ghirlanda, G. Ghisellini, D. Lazzati, The collimation-corrected gamma-ray burst energies correlate with the peak energy of their νF_ν spectrum. Astrophys. J. **616**, 331–338 (2004). doi:10.1086/424913

G. Ghisellini, A. Celotti, D. Lazzati, Constraints on the emission mechanisms of gamma-ray bursts. Mon. Not. R. Astron. Soc. **313**, 1–5 (2000). doi:10.1046/j.1365-8711.2000.03354.x

D. Giannios, H.C. Spruit, Spectral and timing properties of a dissipative gamma-ray burst photosphere. Astron. Astrophys. **469**(1), 1–9 (2007)

S. Guiriec, V. Connaughton, M.S. Briggs, M. Burgess, F. Ryde, F. Daigne, P. Mészáros, A. Goldstein, J. McEnery, N. Omodei, P.N. Bhat, E. Bissaldi, A. Camero-Arranz, V. Chaplin, R. Diehl, G. Fishman, S. Foley, M. Gibby, M.M. Giles, J. Greiner, D. Gruber, A. von Kienlin, M. Kippen, C. Kouveliotou, S. McBreen, C.A. Meegan, W. Paciesas, R. Preece, A. Rau, D. Tierney, A.J. van der Horst, C. Wilson-Hodge, Detection of a thermal spectral component in the prompt emission of GRB 100724b. Astrophys. J. **727**(2), 33 (2011). http://stacks.iop.org/2041-8205/727/i=2/a=L33

IceCube Collaboration, An absence of neutrinos associated with cosmic-ray acceleration in [ggr]-ray bursts. Nature **484**(7394), 351–354 (2012)

D. Lennarz, I. Taboada, for the HAWC Collaboration, First results from HAWC on GRBs (2015). ArXiv e-prints arXiv:1508.07325

A.J. Levan, N.R. Tanvir, R.L.C. Starling, K. Wiersema, K.L. Page, D.A. Perley, S. Schulze, G.A. Wynn, R. Chornock, J. Hjorth, S.B. Cenko, A.S. Fruchter, P.T. O'Brien, G.C. Brown, R.L. Tunnicliffe, D. Malesani, P. Jakobsson, D. Watson, E. Berger, D. Bersier, B.E. Cobb, S. Covino, A. Cucchiara, A. de Ugarte Postigo, D.B. Fox, A. Gal-Yam, P. Goldoni, J. Gorosabel, L. Kaper, T. Krühler, R. Karjalainen, J.P. Osborne, E. Pian, R. Sánchez-Ramírez, B. Schmidt, I. Skillen, G. Tagliaferri, C. Thöne, O. Vaduvescu, R.A.M.J. Wijers, B.A. Zauderer, A new population of ultra-long duration gamma-ray bursts. Astrophys. J. **781**, 13 (2014). doi:10.1088/0004-637X/781/1/13

E.-W. Liang, S.-X. Yi, J. Zhang, H.-J. Lü, B.-B. Zhang, B. Zhang, Constraining gamma-ray burst initial Lorentz factor with the afterglow onset feature and discovery of a tight $\Gamma_0 - E_{\gamma,iso}$ correlation. Astrophys. J. **725**, 2209–2224 (2010). doi:10.1088/0004-637X/725/2/2209

E.-W. Liang, T.-T. Lin, J. Lü, R.-J. Lu, J. Zhang, B. Zhang, A tight $L_{iso} - E_{p,z} - \Gamma_0$ correlation of gamma-ray bursts. Astrophys. J. **813**, 116 (2015). doi:10.1088/0004-637X/813/2/116

S. McBreen, T. Krühler, A. Rau, J. Greiner, D.A. Kann, S. Savaglio, P. Afonso, C. Clemens, R. Filgas, S. Klose, A. Küpcü Yoldaş, F. Olivares E., A. Rossi, G.P. Szokoly, A. Updike, A. Yoldaş, Optical and near-infrared follow-up observations of four Fermi/LAT GRBs: redshifts, afterglows, energetics, and host galaxies. Astron. Astrophys. **516**, 71 (2010). doi:10.1051/0004-6361/200913734

J.C. McKinney, D.A. Uzdensky, A reconnection switch to trigger gamma-ray burst jet dissipation. Mon. Not. R. Astron. Soc. **419**(1), 573–607 (2012)

P. Mészáros, Theories of gamma-ray bursts. Annu. Rev. Astron. Astrophys. **40**, 137–169 (2002). doi:10.1146/annurev.astro.40.060401.093821

P. Mészáros, M. Rees, Poynting jets from black holes and cosmological gamma-ray bursts. Astrophys. J. Lett. **482**(1), 29 (1997). http://stacks.iop.org/1538-4357/482/i=1/a=L29

P. Mészáros, M. Rees, GeV emission from collisional magnetized gamma-ray bursts. Astrophys. J. Lett. **733**(2), 40 (2011)

P. Meszaros, M. Rees, Steep slopes and preferred breaks in gamma-ray burst spectra: the role of photospheres and comptonization. Astrophys. J. **530**(1), 292 (2000)

B. Metzger, D. Giannios, T. Thompson, N. Bucciantini, E. Quataert, The protomagnetar model for gamma-ray bursts. Mon. Not. R. Astron. Soc. **413**(3), 2031–2056 (2011)

E. Molinari, S.D. Vergani, D. Malesani, S. Covino, P. D'Avanzo, G. Chincarini, F.M. Zerbi, L.A. Antonelli, P. Conconi, V. Testa, G. Tosti, F. Vitali, F. D'Alessio, G. Malaspina, L. Nicastro, E. Palazzi, D. Guetta, S. Campana, P. Goldoni, N. Masetti, E.J.A. Meurs, A. Monfardini, L. Norci, E. Pian, S. Piranomonte, D. Rizzuto, M. Stefanon, L. Stella, G. Tagliaferri, P.A. Ward, G. Ihle, L. Gonzalez, A. Pizarro, P. Sinclaire, J. Valenzuela, REM observations of GRB 060418 and GRB 060607A: the onset of the afterglow and the initial fireball Lorentz factor determination. Astron. Astrophys. **469**, 13–16 (2007). doi:10.1051/0004-6361:20077388

K. Murase, K. Asano, T. Terasawa, P. Mészáros, The role of stochastic acceleration in the prompt emission of gamma-ray bursts: application to hadronic injection. Astrophys. J. **746**(2), 164 (2012)

J.P. Norris, J.T. Bonnell, D. Kazanas, J.D. Scargle, J. Hakkila, T.W. Giblin, Long-lag, wide-pulse gamma-ray bursts. Astrophys. J. **627**, 324–345 (2005). doi:10.1086/430294

J.A. Nousek, C. Kouveliotou, D. Grupe, K.L. Page, J. Granot, E. Ramirez-Ruiz, S.K. Patel, D.N. Burrows, V. Mangano, S. Barthelmy, A.P. Beardmore, S. Campana, M. Capalbi, G. Chincarini, G. Cusumano, A.D. Falcone, N. Gehrels, P. Giommi, M.R. Goad, O. Godet, C.P. Hurkett, J.A. Kennea, A. Moretti, P.T. O'Brien, J.P. Osborne, P. Romano, G. Tagliaferri, A.A. Wells, Evidence for a canonical gamma-ray burst afterglow light curve in the Swift XRT data. Astrophys. J. **642**, 389–400 (2006). doi:10.1086/500724

P.T. O'Brien, R. Willingale, J. Osborne, M.R. Goad, K.L. Page, S. Vaughan, E. Rol, A. Beardmore, O. Godet, C.P. Hurkett, A. Wells, B. Zhang, S. Kobayashi, D.N. Burrows, J.A. Nousek, J.A. Kennea, A. Falcone,

D. Grupe, N. Gehrels, S. Barthelmy, J. Cannizzo, J. Cummings, J.E. Hill, H. Krimm, G. Chincarini, G. Tagliaferri, S. Campana, A. Moretti, P. Giommi, M. Perri, V. Mangano, V. LaParola, The early X-ray emission from GRBs. Astrophys. J. **647**, 1213–1237 (2006). doi:10.1086/505457

K.L. Page, R. Willingale, J.P. Osborne, J.P. Norris, P.T. O'Brien, V. Pal'shin, B. Zhang, O. Godet, F.E. Marshall, A. Melandri, E. Rol, P. Romano, R.L.C. Starling, P. Schady, S.A. Yost, S.D. Barthelmy, A.P. Beardmore, GRB 061121: broadband spectral evolution through the prompt and afterglow phases of a bright burst. Astrophys. J. **663**, 1125–1138 (2007). doi:10.1086/518821

K.L. Page, R. Willingale, E. Bissaldi, A.D.U. Postigo, S.T. Holland, S. McBreen, P.T. O'Brien, J.P. Osborne, J.X. Prochaska, E. Rol, E.S. Rykoff, R.L.C. Starling, N.R. Tanvir, A.J. van der Horst, K. Wiersema, B. Zhang, F.J. Aceituno, C. Akerlof, A.P. Beardmore, M.S. Briggs, D.N. Burrows, A.J. Castro-Tirado, V. Connaughton, P.A. Evans, J.P.U. Fynbo, N. Gehrels, C. Guidorzi, A.W. Howard, J.A. Kennea, C. Kouveliotou, C. Pagani, R. Preece, D. Perley, I.A. Steele, F. Yuan, Multiwavelength observations of the energetic GRB 080810: detailed mapping of the broad-band spectral evolution. Mon. Not. R. Astron. Soc. **400**, 134–146 (2009). doi:10.1111/j.1365-2966.2009.15462.x

A. Pe'er, P. Mészáros, M.J. Rees, Peak energy clustering and efficiency in compact objects. Astrophys. J. **635**(1), 476 (2005)

A. Pe'er, P. Mészáros, M.J. Rees, The observable effects of a photospheric component on GRB and XRF prompt emission spectrum. Astrophys. J. **642**(2), 995 (2006)

R.D. Preece, M.S. Briggs, R.S. Mallozzi, G.N. Pendleton, W. Paciesas, D.L. Band, The BATSE gamma-ray burst spectral catalog. I. High time resolution spectroscopy of bright bursts using high energy resolution data. Astrophys. J. Suppl. Ser. **126**(1), 19 (2000)

W.C. Priedhorsky, A.G. Peele, K.A. Nugent, An X-ray all-sky monitor with extraordinary sensitivity. Mon. Not. R. Astron. Soc. **279**, 733–750 (1996)

S. Razzaque, A leptonic–hadronic model for the afterglow of gamma-ray burst 090510. Astrophys. J. Lett. **724**(1), 109 (2010)

M.J. Rees, P. Mészáros, Relativistic fireballs—energy conversion and time-scales. Mon. Not. R. Astron. Soc. **258**, 41–43 (1992). doi:10.1093/mnras/258.1.41P

M.J. Rees, P. Mészáros, Unsteady outflow models for cosmological gamma-ray bursts. Astrophys. J. **430**, 93–96 (1994). doi:10.1086/187446

M.J. Rees, P. Mészáros, Dissipative photosphere models of gamma-ray bursts and x-ray flashes. Astrophys. J. **628**(2), 847 (2005)

A. Rowlinson, P.T. O'Brien, N.R. Tanvir, B. Zhang, P.A. Evans, N. Lyons, A.J. Levan, R. Willingale, K.L. Page, O. Onal, D.N. Burrows, A.P. Beardmore, T.N. Ukwatta, E. Berger, J. Hjorth, A.S. Fruchter, R.L. Tunnicliffe, D.B. Fox, A. Cucchiara, The unusual X-ray emission of the short Swift GRB 090515: evidence for the formation of a magnetar? Mon. Not. R. Astron. Soc. **409**, 531–540 (2010). doi:10.1111/j.1365-2966.2010.17354.x

R. Sari, T. Piran, Predictions for the very early afterglow and the optical flash. Astrophys. J. **520**, 641–649 (1999). doi:10.1086/307508

R. Sari, T. Piran, J.P. Halpern, Jets in gamma-ray bursts. Astrophys. J. **519**, 17–20 (1999). doi:10.1086/312109

R. Sari, T. Piran, R. Narayan, Spectra and light curves of gamma-ray burst afterglows. Astrophys. J. **497**, 17–20 (1998). doi:10.1086/311269

K.Z. Stanek, T. Matheson, P.M. Garnavich, P. Martini, P. Berlind, N. Caldwell, P. Challis, W.R. Brown, R. Schild, K. Krisciunas, M.L. Calkins, J.C. Lee, N. Hathi, R.A. Jansen, R. Windhorst, L. Echevarria, D.J. Eisenstein, B. Pindor, E.W. Olszewski, P. Harding, S.T. Holland, D. Bersier, Spectroscopic discovery of the supernova 2003dh associated with GRB 030329. Astrophys. J. **591**, 17–20 (2003). doi:10.1086/376976

G. Tagliaferri, L.A. Antonelli, G. Chincarini, A. Fernández-Soto, D. Malesani, M. Della Valle, P. D'Avanzo, A. Grazian, V. Testa, S. Campana, S. Covino, F. Fiore, L. Stella, A.J. Castro-Tirado, J. Gorosabel, D.N. Burrows, M. Capalbi, G. Cusumano, M.L. Conciatore, V. D'Elia, P. Filliatre, D. Fugazza, N. Gehrels, P. Goldoni, D. Guetta, S. Guziy, E.V. Held, K. Hurley, G.L. Israel, M. Jelínek, D. Lazzati, A. López-Echarri, A. Melandri, I.F. Mirabel, M. Moles, A. Moretti, K.O. Mason, J. Nousek, J. Osborne, L.J. Pellizza, R. Perna, S. Piranomonte, L. Piro, A. de Ugarte Postigo, P. Romano, GRB 050904 at redshift 6.3: observations of the oldest cosmic explosion after the Big Bang. Astron. Astrophys. **443**, 1–5 (2005). doi:10.1051/0004-6361:200500196

C. Thompson, P. Mészáros, M. Rees, Thermalization in relativistic outflows and the correlation between spectral hardness and apparent luminosity in gamma-ray bursts. Astrophys. J. **666**(2), 1012 (2007)

C. Thompson, Deceleration of a relativistic, photon-rich shell: end of preacceleration, damping of magnetohydrodynamic turbulence, and the emission mechanism of gamma-ray bursts. Astrophys. J. **651**(1), 333 (2006)

J. Van Paradijs, P. Groot, T. Galama, C. Kouveliotou, R. Strom, J. Telting, R. Rutten, G. Fishman, C. Meegan, M. Pettini et al., Transient optical emission from the error box of the γ-ray burst of 28 February 1997. Nature **386**(6626), 686–689 (1997)

R. Vanderspek, T. Sakamoto, C. Barraud, T. Tamagawa, C. Graziani, M. Suzuki, Y. Shirasaki, G. Prigozhin, J. Villasenor, J. Jernigan et al., Hete observations of the gamma-ray burst GRB 030329: evidence for an underlying soft x-ray component. Astrophys. J. **617**(2), 1251 (2004)

P. Veres, P. Mészáros, Single- and two-component gamma-ray burst spectra in the Fermi GBM-LAT energy range. Astrophys. J. **755**(1), 12 (2012). http://stacks.iop.org/0004-637X/755/i=1/a=12

R.A.M.J. Wijers, T.J. Galama, Physical parameters of GRB 970508 and GRB 971214 from their afterglow synchrotron emission. Astrophys. J. **523**, 177–186 (1999). doi:10.1086/307705

S.W. Wilkins, A.W. Stevenson, K.A. Nugent, H. Chapman, S. Steenstrup, On the concentration, focusing, and collimation of x-rays and neutrons using microchannel plates and configurations of holes. Rev. Sci. Instrum. **60**(6), 1026–1036 (1989)

R. Willingale, P.T. O'Brien, J.P. Osborne, O. Godet, K.L. Page, M.R. Goad, D.N. Burrows, B. Zhang, E. Rol, N. Gehrels, G. Chincarini, Testing the standard fireball model of gamma-ray bursts using late X-ray afterglows measured by Swift. Astrophys. J. **662**, 1093–1110 (2007). doi:10.1086/517989

R. Willingale, F. Genet, J. Granot, P.T. O'Brien, The spectral-temporal properties of the prompt pulses and rapid decay phase of gamma-ray bursts. Mon. Not. R. Astron. Soc. **403**, 1296–1316 (2010). doi:10.1111/j.1365-2966.2009.16187.x

B. Zhang, P. Mészáros, Gamma-ray burst afterglow with continuous energy injection: signature of a highly magnetized millisecond pulsar. Astrophys. J. **552**, 35–38 (2001). doi:10.1086/320255

B.-B. Zhang, Y.-Z. Fan, R.-F. Shen, D. Xu, F.-W. Zhang, D.-M. Wei, D.N. Burrows, B. Zhang, N. Gehrels, GRB 120422A: a low-luminosity gamma-ray burst driven by a central engine. Astrophys. J. **756**, 190 (2012). doi:10.1088/0004-637X/756/2/190

Space Sci Rev (2017) 207:87–110
DOI 10.1007/s11214-017-0348-6

Photospheric Emission of Gamma-Ray Bursts

A.M. Beloborodov[1] · **P. Mészáros**[2,3]

Received: 11 October 2016 / Accepted: 11 March 2017 / Published online: 28 March 2017

Abstract We review the physics of GRB production by relativistic jets that start highly opaque near the central source and then expand to transparency. We discuss dissipative and radiative processes in the jet and how radiative transfer shapes the observed nonthermal spectrum released at the photosphere. A comparison of recent detailed models with observations gives estimates for important parameters of GRB jets, such as the Lorentz factor and magnetization. We also discuss predictions for GRB polarization and neutrino emission.

Keywords Gamma-ray bursts · Radiative processes

1 Introduction

The emission mechanism of gamma-ray bursts (GRBs) has been a puzzle since their discovery half a century ago. Following the discovery of GRB afterglows in the end of 1990s, it has been established that GRB emission arises in ultra-relativistic jets from extremely powerful cosmic explosions at cosmological distances. The existing emission models are mainly guided by the observed spectrum of the jet radiation and its fast variability.

The main phase of a GRB, usually called "prompt" emission, typically lasts seconds or minutes, with huge luminosities, sometimes reaching $L_\gamma \sim 10^{54}$ erg s^{-1} in isotropic equivalent (before correcting for the beaming of radiation from a collimated relativistic jet). The

✉ A.M. Beloborodov
amb@phys.columbia.edu

P. Mészáros
nnp@psu.edu

1 Physics Department and Columbia Astrophysics Laboratory, Columbia University, 538 West 120th Street, New York, NY 10027, USA

2 Center for Particle and Gravitational Astrophysics, Dept. of Astronomy & Astrophysics, Pennsylvania State University, University Park, PA 16802, USA

3 Center for Particle and Gravitational Astrophysics, Dept. of Physics, Pennsylvania State University, University Park, PA 16802, USA

prompt phase should be distinguished from the GRB "afterglow"—the long-lasting and weaker emission with a broad spectrum extending from gamma-rays to radio waves.

The spectrum of prompt emission has a well defined, sharp peak at an energy $E_{\rm pk}$ that varies around 1 MeV, after correcting by $1+z$ for the cosmological redshift (Kaneko et al. 2006; Goldstein et al. 2012). For most bursts the spectrum around the peak can be approximately described by a simple Band function (Band et al. 2009)—two power laws that are smoothly connected at $E_{\rm pk}$. Bursts of higher luminosity are observed to have higher $E_{\rm pk}$. An approximate correlation $E_{\rm pk} \approx 0.3 L_{\gamma,52}^{1/2}$ MeV was reported (e.g. Wei and Gao 2003; Yonetoku et al. 2004; Ghirlanda et al. 2012), where $L_{\gamma,52}$ is the burst luminosity (isotropic equivalent) in units of 10^{52} erg s^{-1}.

The short durations of $\sim$0.1–10 s and fast variability of GRB emission imply that it is produced by outflows or jets with Lorentz factors $\Gamma \gtrsim 10^2$. The Doppler effect compresses the typical observed timescale of radiation released at a radius r, from $\sim r/c$ to $\sim r/c\Gamma^2$. In addition, a high Γ allows the radiation to decouple from the jet at smaller radii; typically, the jet becomes transparent to MeV radiation at radii somewhere between 10^{12} and 10^{14} cm. The characteristic scattering optical depth seen by photons propagating from a radius r to infinity is given by

$$\tau_{\rm T} = \frac{Z_\pm \dot{M}_p \sigma_{\rm T}}{4\pi r \Gamma^2 m_p c}, \tag{1}$$

where $\dot{M}_p$ is the isotropic equivalent of the proton mass outflow rate, and $Z_\pm$ is the number of $e^\pm$ per proton. The optically thin zone of the jet roughly corresponds to $\tau_{\rm T} < 1$, and the sub-photospheric region is where $\tau_{\rm T} > 1$. Strong energy dissipation must occur in the jet to generate the observed non-thermal spectra. This dissipation can occur before and after the jet becomes transparent to radiation.

1.1 Optically Thin Synchrotron Emission?

A simple phenomenological GRB model posits that we observe Doppler-shifted synchrotron radiation, similar to blazar jets. The model assumes that a nonthermal electron population is injected in the jet by some dissipative process in an optically thin outflow. It gives a radiation spectrum that peaks at the photon energy

$$E_{\rm pk} \approx E_{\rm syn} = 0.4\Gamma\, \gamma_{\rm pk}^2\, \hbar \frac{eB}{m_e c}. \tag{2}$$

Here B is the magnetic field measured in the rest frame of the jet ("fluid frame"), $\gamma_{\rm pk}$ is the Lorentz factor at which the injected electron distribution peaks (also measured in the fluid frame), and Γ is the bulk Lorentz factor of the jet itself. If the electron distribution at $\gamma_e > \gamma_{\rm pk}$ is a power-law $dN_e/d\gamma_e \propto \gamma_e{}^{-p}$ then the synchrotron spectrum has a high-energy power-law tail, $dN_{\rm ph}/dE \propto E^{-p/2-1}$ at $E > E_{\rm pk}$.

One possibility for the injection of high-energy electrons is associated with internal shocks (Rees and Mészáros 1994; Kobayashi et al. 1997; Daigne and Mochkovitch 1998). A mildly relativistic electron-ion shock produces an electron distribution with $\gamma_{\rm pk} \sim \epsilon_e(m_p/m_e)$, where ϵ_e can be a significant fraction of unity. This gives

$$E_{\rm pk} \approx 1\, r_{12}^{-1} \epsilon_B^{1/2} L_{52}^{1/2} \left(\frac{\epsilon_e}{0.3}\right)^2 \text{MeV}, \tag{3}$$

where r_{12} is radius in units in 10^{12} cm, L_{52} is the isotropic equivalent of the jet power in units of 10^{52} erg s^{-1}, and ϵ_B is the fraction of jet energy that is carried by the magnetic field. If the shock radius happens to be $r \sim 10^{13}\epsilon_B^{1/2}\epsilon_e^2$ cm then $E_{\rm pk}$ would be consistent with observations. A Band-type spectrum is emitted if the electron energy distribution extends as a power law at $\gamma_e > \gamma_{\rm pk}$ and cuts off at $\gamma_e < \gamma_{\rm pk}$, with a negligible Maxwellian component (e.g. Burgess et al. 2011). This is a problematic requirement. In observed shocks, e.g. in the solar system or in supernovae, as well as in computer-simulated shocks, most of the dissipated energy is given to Maxwellian particles, with only a small fraction of particles populating the high-energy nonthermal tail.

Strong magnetization $\varepsilon_{\rm B} \gtrsim 0.1$ offers an alternative mechanism for particle acceleration: the dissipation of magnetic energy, in particular through magnetic reconnection (Drenkhahn and Spruit 2002). Recent dedicated numerical studies of relativistic reconnection show that it gives a broad flat electron distribution, with no cutoff at low energies (Kagan et al. 2015). Synchrotron emission from this electron distribution does not resemble the pronounced spectral peak observed in GRBs.

The synchrotron model may be viewed more broadly as a phenomenological model which does not specify how the desired electron distribution is created. In this form the model still faces a few problems:

(1) Thousands of GRBs have been observed, and most of them have $E_{\rm pk}$ near 1 MeV (Goldstein et al. 2012). Few bursts have $E_{\rm pk}$ above 10 MeV and no bursts are known with $E_{\rm pk} > 20$ MeV. This is at odds with the naive model prediction $E_{\rm pk} \propto \Gamma\gamma_{\rm pk}^2 B$, which should give a broad distribution of $E_{\rm pk}$ due to variations in the local B, $\gamma_{\rm pk}$, and Γ.
(2) High-energy electrons quickly cool to $\gamma_e < \gamma_{\rm pk}$ (which makes the process radiatively efficient) and should emit radiation at $E < E_{\rm pk}$ with a photon index $\alpha = -3/2$. A typical low-energy index observed in GRBs is $\alpha \sim -1$, and many bursts have even harder slopes $\alpha > 0$ (Kaneko et al. 2006). Such hard slopes are in conflict with the synchrotron model.
(3) The synchrotron spectrum should have a rather broad peak, even when the source is assumed to have a uniform magnetic field B and the power-law electron distribution is assumed to cut off at $\gamma_e < \gamma_{\rm pk}$. Attempts to reproduce GRB spectra with such idealized models gave acceptable fits for a small number of bursts (Burgess et al. 2011, 2014; Preece and The Fermi collab. 2014) and were found incompatible with observed spectra in the dominant majority of GRBs (Axelsson and Borgonovo 2015; Yu et al. 2015).

The peak sharpness problem is illustrated in Fig. 1 where a synchrotron spectrum is compared with the Band fit of a bright burst with a typical spectrum, GRB 990123. In the realistic fast-cooling regime, the minimum width of the synchrotron peak at half maximum exceeds 1.5 orders in photon energy, even with a single value of B in the emitting source and the narrow injected electron distribution (single-temperature Maxwellian). It is significantly broader than the Band fit to the *time-average* of the GRB spectrum (Briggs et al. 1999). Uncertainties in the observed spectrum due to the detector response and limited photon statistics allow some room for stretching the measured peak width; however it can hardly be made consistent with the synchrotron model, especially when the inevitable broadening due to variable magnetic fields, electron injection and the Doppler factor are taken into account. Note also that synchrotron spectra with sharp peaks consistent with a Band function are not observed in any other astrophysical objects. One example is provided by blazars—their synchrotron spectra have the half-maximum width of several orders of magnitude, broader than in GRBs (e.g. Ghisellini 2006).

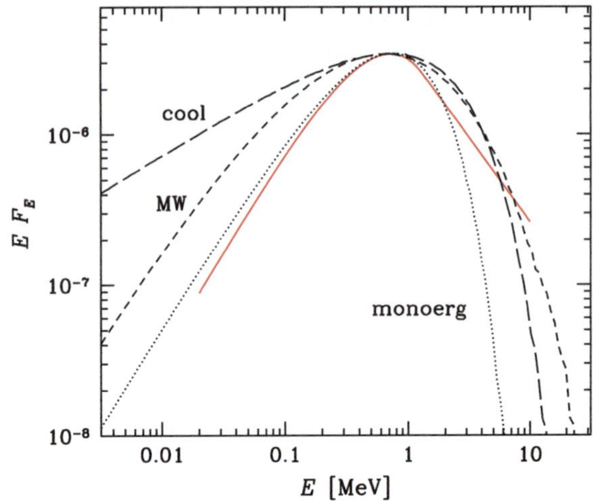

Fig. 1 Synchrotron spectra from an optically thin spherical shell. Three models are shown with different electron distributions: mono-energetic (*dotted line*), Maxwellian (*short-dashed line*), and fast-cooling Maxwellian (*long-dashed line*). For comparison the Band fit of GRB 990123 is shown by the red curve. (From Vurm and Beloborodov 2016b)

1.2 Photospheric Emission

The sharp MeV spectral peak provides evidence for thermalization of radiation at early, opaque stages of the GRB explosion (Paczýnski 1986; Goodman 1986; Thompson 1994; Mészáros and Rees 2000b). The inheritance of the spectral peak from an initial thermalization stage is also supported by the observed distribution of E_{pk}, which cuts off above $\sim$10 MeV in agreement with a theoretical maximum (Beloborodov 2013).

Accepting that the MeV peak forms inside an opaque jet leads to so-called "photospheric" emission models: GRB radiation is released where the jet becomes transparent to scattering. Several versions of the photospheric model have been discussed (Thompson 1994; Eichler and Levinson 2000; Rees and Mészáros 2005; Giannios and Spruit 2007; Beloborodov 2010; Levinson 2012; Thompson and Gill 2014). All share a key feature: the jet is *dissipative*, i.e. significantly heated as it propagates away from the central engine. This heating modifies the emitted photospheric radiation from simple blackbody emission. The resulting spectrum was investigated using dedicated numerical simulations and found to have a nonthermal shape that closely resembles the phenomenological Band function (Pe'er et al. 2006; Giannios 2008; Beloborodov 2010; Vurm et al. 2011; Gill and Thompson 2014; Vurm and Beloborodov 2016b). It was proposed that the dissipative photosphere provides the best description for the observed spectra (Ryde et al. 2011).

Previous observational search for photospheric emission mainly looked for a Planck component or its modifications (e.g. Ryde et al. 2011; Guiriec et al. 2011, 2015, see Pe'er 2016, for a recent review). In contrast, the detailed theoretical calculations suggest that photospheric emission has a broad Band-like spectrum that dominates the prompt GRB rather than adds a blackbody component to it.

2 Sub-Photospheric Dissipation

Internal bulk motions and magnetic fields provide energy reservoirs available for dissipation in GRB jets. Three mechanisms can make photospheric emission nonthermal:

1. Dissipation of magnetic energy.
2. Collisions of drifting neutrons.
3. Internal shocks.

The details of magnetic dissipation are complicated and still poorly understood. The other two dissipation mechanisms can be examined from first principles and give predictions that can be compared with observations; we will discuss them is some detail below.

A neutron component is expected in GRB explosions (Derishev et al. 1999; Bahcall and Mészáros 2000; Mészáros and Rees 2000a; Beloborodov 2003; Rossi et al. 2006). Free neutrons enter the jet from the neutronized central engine and are also produced by spallation reactions in the jet itself. They begin to drift with respect to the proton flow in the sub-photospheric region and collide with the protons. These nuclear collisions create pions, whose decay leads to the production of $e^{\pm}$ pairs with Lorentz factors $\gamma_e \sim m_\pi/m_e \sim 300$ in the jet rest frame and to the emission of neutrinos with energy $\sim\Gamma m_\pi c^2$ in the observer frame. These multi-GeV neutrinos are detectable with IceCube (Bartos et al. 2013; Murase et al. 2013) and can provide an important test for GRB models. The photospheric spectrum generated by neutron collisions has been calculated by Beloborodov (2010) and Vurm et al. (2011), and found to be consistent with GRB spectra.

Internal shocks are generally expected to develop in GRB jets, as the jets are highly variable. The central engine of the explosion is likely unsteady, and additional variability is induced as the jet breaks out of the progenitor star (Lazzati et al. 2009, 2013; Ito et al. 2015). This breakout is accompanied by multiple internal and recollimation shocks. Shock heating is expected to occur in an extended range of radii and in an extended range of timescales, which is consistent with the broad power spectrum of variability observed in GRB light curves (Beloborodov et al. 2000; Morsony et al. 2010).

As long as the flow is opaque and radiation dominates its energy density, photon diffusion plays a leading role in shaping the shock front. Its thickness is then comparable to the photon mean free path. The radiation-mediated shock (RMS) is not capable of electron acceleration by the standard Fermi mechanism, since the electron radiates its energy much faster than it can cross the shock. Photons can experience significant energy gains by crossing the shock back and forth multiple times (Blandford and Payne 1981; Riffert 1988; Levinson 2012). This "bulk Comptonization" upscatters photons up to the MeV band (in the fluid frame). Upscattering to higher energies is hindered by the energy loss due to electron recoil in scattering. Photon upscattering beyond $\sim$1 MeV is also stopped by the absorption reaction $\gamma + \gamma \to e^- + e^+$, and the produced mildly relativistic $e^{\pm}$ pairs quickly cool down due to Coulomb and inverse Compton (IC) losses.

All this would suggest that sub-photospheric shocks are inefficient in producing particles with energies $E \gg m_e c^2$ in the fluid frame. However, a more realistic picture of sub-photospheric shocks differs from the simple RMS, in particular in outflows that carry magnetic fields and free neutrons in addition to the plasma and radiation. The shock wave is capable of generating ultra-relativistic electrons in two ways:

- A strong collisionless subshock forms in the RMS. This is inevitable (even deep below the photosphere) if the flow is sufficiently magnetized (Beloborodov 2016). A mildly relativistic collisionless subshock heats the electrons to an ultra-relativistic temperature T_e, as they receive a significant fraction of the ion kinetic energy (e.g. Sironi and Spitkovsky 2011). One can show that the inverse Compton (IC) emission from the heated electrons breeds $e^{\pm}$ pairs in the upstream, and this process regulates the subshock temperature to $kT_e \sim 10 m_e c^2$.

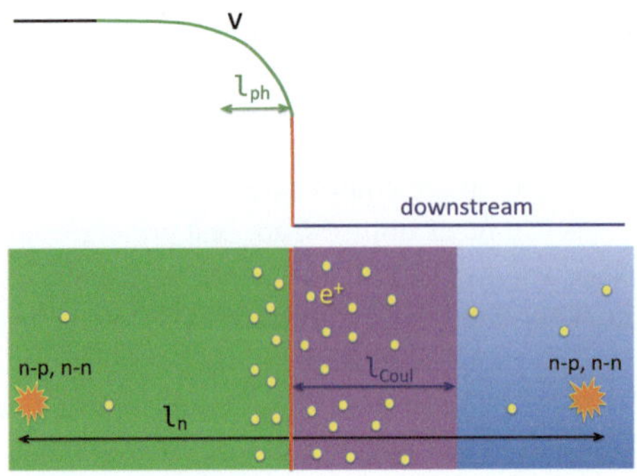

Fig. 2 Structure of a relativistic internal shock in a hot, opaque, magnetized outflow. The *green part* of the velocity profile is shaped by radiation pressure on a scale comparable to the photon mean free path to scattering, $l_{\rm ph}$. The *vertical red part* is the collisionless subshock, which heats ions to a mildly relativistic temperature and electrons to an ultra-relativistic temperature. The ultra-relativistic electrons are quickly cooled, producing synchrotron and inverse Compton (IC) photons, which create $e^{\pm}$ pairs ahead and behind the shock. The ion cooling occurs on a longer scale $l_{\rm Coul}$, which exceeds the RMS thickness if the subshock is relativistic. In the presence of a free neutron component in the flow, the shock wave is partially shaped by neutron migration between the upstream and downstream. The neutron mean free path ℓ_n is the longest scale in the shock structure. The migrating neutrons are stopped by nuclear collisions. Inelastic nuclear collisions result in injection of $e^{\pm}$ with Lorentz factors $\gamma_e \sim m_\pi/m_e \sim 300$, distributed over a broad region $\sim\ell_n$. The nuclear collisions also emit neutrinos; they escape the outflow with energies $\sim\Gamma m_\pi c^2$ where Γ is the outflow Lorentz factor. (From Beloborodov 2016)

- Inelastic nuclear collisions inject $e^{\pm}$ pairs with Lorentz factors $\sim m_\pi/m_e \sim 300$ in the fluid frame. This mechanism becomes particularly efficient if the outflow carries free neutrons, as they can migrate across the RMS, making the shock wave partially mediated by neutrons (Fig. 2).

The structure of sub-photospheric shocks can be studied using detailed numerical simulations. One method is to search for a steady-state propagating solution of the kinetic equations (e.g. Levinson and Bromberg 2008; Budnik et al. 2010), which may be found by iterations. Alternatively, one can use direct time-dependent simulations of shock formation. The simulation can be set up to follow the evolution of a compressive wave in the outflow, which leads to formation of a pair of shocks and their subsequent quasi-steady propagation. Then the shock structure is obtained from first principles by calculating the time-dependent radiative transfer through the moving plasma. A detailed first-principle simulation involves tracking of a large number of individual photons and their interaction with the moving plasma (using Monte-Carlo techniques).

The shock structure obtained with this method is shown in Fig. 3, for two cases—with and without magnetic fields. The simulations demonstrate the formation of a strong collisionless subshock in the magnetized RMS. In GRB jets, a moderate magnetization ~ 0.1 is sufficient to generate a strong subshock.

Electron-positron pair creation was expected in dissipative GRB jets since the early works in the 1990s (e.g. Rees and Mészáros 1994; Thompson 1994; Mészáros and Rees 2000b). The efficiency of this process is described by the pair loading factor $Z_\pm = n_\pm/n_p$ where n_p is the proton density and $n_\pm$ is the $e^{\pm}$ density. One can show that $Z_\pm$ reaches a peak of $\sim 10^2$ inside the shock front and decreases behind it due to pair annihilation (Beloborodov 2016). The pairs are produced in collisions between MeV photons, which are generated by two mechanisms: (1) An RMS without a strong collisionless subshock produces MeV photons due to bulk Comptonization by a relatively cold flow (the flow is everywhere in local

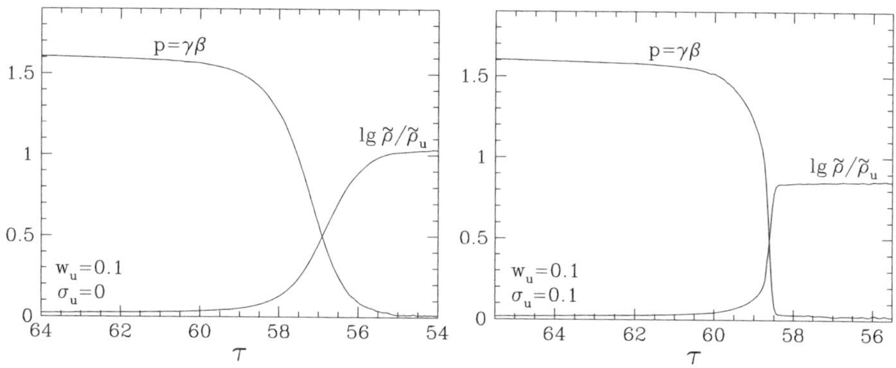

Fig. 3 Snapshot of a shock propagating in the flow with upstream radiation enthalpy $w_u = 0.1$ and upstream magnetization $\sigma_u = 0$ (*left*) and 0.1 (*right*). The parameters w and σ are defined in the fluid frame as $w = (4/3)U_{\rm rad}/\tilde{\rho}c^2$ and $\sigma = B^2/4\pi\tilde{\rho}c^2 = 2\varepsilon_{\rm B}$, where $U_{\rm rad}$ is the radiation energy density and $\tilde{\rho}$ is the proper mass density of the fluid. The shock is propagating to the left and the Thomson optical depth τ is measured from the site of shock formation (caustic of the initial supersonic wave). The *solid curves* show the profiles of momentum $p = \gamma\beta$ and proper density $\tilde{\rho}$ (normalized to the upstream proper density $\tilde{\rho}_u$). A strong subshock has formed in the magnetized case; it is highlighted by the *gray strip*. The electron temperature has a strong and narrow peak at the subshock. The subshock is resolved (not a discontinuous jump) due to a finite viscosity employed in the simulation of the plasma dynamics. The radiation is everywhere simulated directly as a large collection of individual photons whose propagation and scattering is followed using the Monte-Carlo technique. (From Beloborodov 2016)

Compton equilibrium with radiation at $kT \ll m_ec^2$). (2) In the presence of a collisionless jump, MeV photons are produced by IC cooling of $e^\pm$ heated to the subshock temperature $kT_e \sim 10m_ec^2$.

The shock "breakout" at the photosphere[1] occurs through the growth of the collisionless jump in the RMS until radiation completely decouples from the plasma. Eventually the entire velocity jump becomes mediated by collective plasma processes, regardless of magnetization. This somewhat resembles the shock breakout in non-relativistic supernova explosions (Waxman and Loeb 2001; Giacinti and Bell 2015). However, the photospheric shock breakout in GRBs has a special feature: $e^\pm$ pair creation. The shock sustains a local optical depth $\tau_{\rm T} \gtrsim 1$ even after the background electron-ion plasma becomes transparent. Effectively, the shock carries the photosphere with it until it expands by an additional factor $\sim$30–100, continually producing photospheric emission (Beloborodov 2016). This emission should be observed as a prominent pulse of nonthermal radiation in the GRB light curve. In the observer frame, the emission from the $e^\pm$-dressed shocks extends up to the GeV band, where γ–γ absorption shapes a break in the spectrum.

The high-energy photospheric pulses contribute to the prompt GRB emission. This emission is observed to overlap (in arrival time) with GeV afterglow, which is associated with the external blast wave from the explosion (Ackermann and The Fermi Collaboration 2013). The early peak of GeV afterglow is explained by the $e^\pm$-loading of the blast wave (Beloborodov et al. 2014; Hascoët et al. 2015), and its overlap with the high-energy photospheric pulses can explain the observed variable component of GeV emission at early times.

Internal shocks are the main heating mechanism for jets that are not dominated by magnetic fields. This is a likely situation in GRBs, as comparison of detailed models of photo-

[1] The photospheric shock breakout should not be confused with the jet breakout from the progenitor star. The jet breakout occurs at smaller radii and huge optical depths.

spheric radiation with observed GRB spectra suggests a moderate magnetization in the subphotospheric region, $\sigma \sim 10^{-2}$–10^{-1} (see Sect. 4 below). This does not exclude a stronger magnetization closer to the central engine, allowing for magnetic dissipation that reduces σ as the jet expands. In this scenario, the early heating at very large optical depths may be dominated by magnetic dissipation. Constraints on the jet magnetization at small radii were recently estimated using the jet breakout time by Bromberg et al. (2015); they find that the jet is not dominated by magnetic fields.

Neutron migration across the shock front leads to nuclear collisions which generate additional $e^{\pm}$ pairs and produce neutrino emission. The typical observed energy of neutrinos produced by this mechanism is $\sim\Gamma m_{\pi} c^2 \gtrsim 10$ GeV. Sub-photospheric internal shocks were also proposed to emit ultra-high-energy neutrinos (Mészáros and Waxman 2001; Razzaque et al. 2003). This proposal assumes efficient ion acceleration by the Fermi diffusive mechanism, which does not operate in an RMS and requires a collisionless (sub)shock. Diffusive acceleration is, however, suppressed in shocks with a transverse magnetic field, which advects the particles downstream before they have a chance to cross the shock many times (Sironi and Spitkovsky 2011). Obliqueness of the magnetic field could help the ion acceleration. Another possible way to accelerate ions is the converter mechanism proposed by Derishev et al. (2003). Numerical simulations of Kashiyama et al. (2013) suggest that the converter mechanism becomes efficient for ultra-relativistic shocks, with amplitude $\gamma_0 \gtrsim 4$. This mechanism is, however, quite slow as it relies on neutron-to-proton and proton-to-neutron conversion in (inelastic) nuclear collisions, which occur with a large free path $l_n \sim (n_p \sigma_n)^{-1}$ due to the modest nuclear cross section $\sigma_n \approx \sigma_T/20$.

3 Peak Position of the Photospheric Spectrum $E_{\rm pk}$

Photons deep below the photosphere keep interacting with the plasma, and their spectrum forms a sharp Wien peak. The simplest model of photospheric emission assumes a freely expanding radiation-dominated outflow with no baryon loading or magnetic fields (Paczýnski 1986; Goodman 1986). In this case, dissipation weakly influences the radiation spectrum and it is not far from a Planck shape. The peak energy of the Planck spectrum (defined as the peak of $dL/d\ln E$) is related to the average photon energy $\bar{E}$ by $E_{\rm pk} \approx 1.45\bar{E}$. In the ideal radiation-dominated outflow $\bar{E}$ remains constant and equal to its value near the central engine E_0.

This simple dissipationless model, however, fails to explain the observed spectra. Although the Planck spectrum may appear in the time-resolved emission of some bursts (e.g. Ryde 2004; Ryde et al. 2011), GRB spectra are typically nonthermal, with an extended high-energy tail.

3.1 Effects of Collimation and Early Dissipation

The average photon energy at the explosion center, E_0, is a useful parameter of more general jets, which may be dissipative, magnetized, and collimated. E_0 may be expressed in terms of the jet power L_0 and an initial radius r_0 (comparable to the size of the central compact object) using the relation $\epsilon_0 L_0 \approx 4\pi r_0^2 a T_0^4 c$,

$$E_0 \approx 10\,\epsilon_0^{1/4} L_{0,52}^{1/4} r_{0,6}^{-1/2}\ \text{MeV}. \tag{4}$$

Here ϵ_0 is the initial thermal fraction of the power L_0 (radiation-dominated jets with sub-dominant magnetic energy have $\epsilon_0 \approx 1$). Collimation of the flow within a small angle θ_c implies a smaller true jet power L_0 compared with its apparent isotropic equivalent L_γ,

$$L_0 \approx \frac{\theta_c^2}{2} L_\gamma, \tag{5}$$

and hence E_0 is reduced as $\theta_c^{1/2}$. Achromatic breaks in GRB afterglow light curves are often interpreted as evidence for jet collimation, with a typical opening angle of 5–10°. Collimation helps explain the extremely high apparent luminosities L_γ, which reach 10^{54} erg s^{-1} in some GRBs. This collimation can be the result of the pressure confinement of the jet by the progenitor star and the breakout cocoon, or by a dense wind from the outer regions of the central accretion disk. Collimation may also be assisted by strong magnetic fields (Komissarov et al. 2009).

If collimation is not accompanied by significant dissipation, the expanding jet can be described as an ideal relativistic flow confined by a wall that determines the cross section of the jet. Dissipationless collimation conserves entropy, and hence does not change the photon-to-baryon ratio n_γ/n. This implies conservation of the photon number carried by the jet and a constant energy per photon $\bar{E}$. Collimation boosts the isotropic equivalent of the luminosity L_γ and the isotropic equivalent of the photon flux $\dot{N}_\gamma$ by the same factor $\sim\theta_c^{-2}$, and their ratio $\bar{E} = L_\gamma/\dot{N}_\gamma$ remains unchanged from its value at r_0, $\bar{E} = E_0$.

In contrast, dissipative collimation (e.g. involving re-collimation shocks) generates entropy and hence increases n_γ/n. Then the total photon number carried by the jet is increased by a factor of $Q > 1$ and hence $\bar{E}$ (jet energy per photon) is reduced as Q^{-1}, giving a GRB with a reduced $E_{\rm pk}$.

If there is a relation between θ_c and Q, it leads to a correlated variation of L and $E_{\rm pk}$ (with θ_c being the varying parameter). Thompson et al. (2007) considered the possibility that θ_c always tends to its maximum value $\theta_c \sim \Gamma^{-1}$ allowed by the condition of causal contact across the jet. They pointed out that this gives $\bar{E} \propto \theta_c^{-1}$ and hence $E_{\rm pk} \propto L^{1/2}$, similar to the observed trend. This simple explanation of the 1/2 slope of the $E_{\rm pk}$-L_γ correlation is however problematic, as it invokes huge variations in θ_c to cover the observed correlation range, not only from burst to burst but also within a single burst (e.g. Ghirlanda et al. 2012). It also assumes that GRBs of various apparent luminosities $L \sim L_0\theta_c^{-2}$ have approximately the same true power $L_0 \sim 10^{50}$ erg s^{-1}, which implies a central temperature $kT_0 \approx 1\, r_{0,6}^{-1/2}$ MeV. Then the brightest bursts would have the highest $E_{\rm pk} \approx 3kT_0 \approx 3r_{0,6}^{-1/2}$ MeV, which falls short of the observed highest $E_{\rm pk} \sim 15$ MeV.

Although the dissipative collimation is not the only process regulating the observed $E_{\rm pk}$ (as will be further discussed below), it plays an important role illustrated in Fig. 4: beaming/collimation increases the apparent luminosity L_γ while photon production due to dissipation reduces the observed $E_{\rm pk}$. Reasonable beaming factors $L/L_0 \sim 10^2$ (suggested by the burst energetics and the afterglow data analysis) together with the expected photon production in a dissipative jet naturally explains the location of observed GRBs on the L_γ-$E_{\rm pk}$ diagram.

Evidence for dissipation at small radii and large optical depths is provided by the observed *photon number* emitted in GRBs. In many GRBs, the blackbody central engine is unable to provide the observed photon number, so additional photons must be produced in the expanding jet.

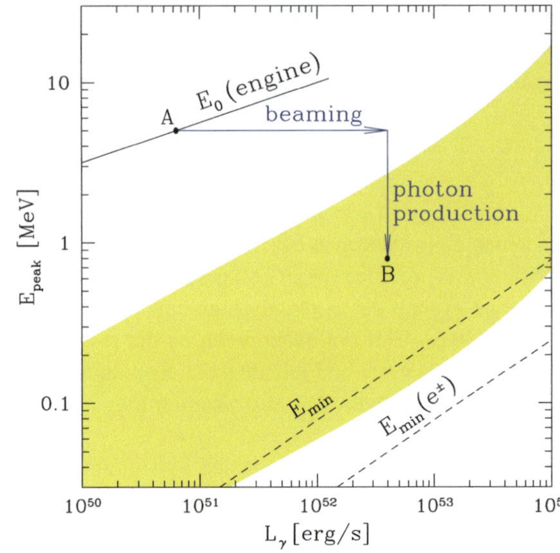

Fig. 4 L_γ-E_{pk} diagram. Point A is an example of the initial condition near the central engine; the jet starts with E_{pk} close to E_0 (4). As the jet expands, its apparent luminosity L_γ is increased by beaming and E_{pk} is reduced by photon production (dissipation offsets adiabatic cooling). Point B shows the resulting photospheric emission. The approximate region populated by observed GRBs is shown in *yellow*. The observed E_{pk} should not violate the lower bound E_{min} that is set by the effective blackbody temperature of the photospheric radiation. The *dashed lines* show E_{min} with and without pair enrichment of the photosphere. (From Beloborodov 2013)

3.2 Planck, Wien, and Nonthermal Zones Below Photosphere

Observations also require that dissipation in many GRBs continues at least to the photospheric radius, so that the released spectrum has a nonthermal tail. The effect of dissipation on E_{pk} differs in the following three sub-photospheric zones shown in Fig. 5 (Beloborodov 2013).

1. Planck zone ($r \lesssim 10^{10}$ cm, $\tau_T \gtrsim 10^5$): the density of the jet is sufficiently high to maintain blackbody radiation in detailed equilibrium with the thermalized plasma. The thermal plasma efficiently produces photons through double Compton scattering.
2. Wien zone ($\tau_T \gtrsim 10^2$): the dissipated heat is thermalized into a Bose-Einstein photon distribution with a *finite chemical potential*. Kinetic equilibrium between the photons and the plasma is maintained due to the large Kompaneets parameter of the thermal plasma, $y = 4(kT/m_e c^2)\tau_T \gg 1$. The number of photons accumulated in the Wien peak saturates near the Wien radius R_W where y drops to $\sim$1 and Comptonization becomes unable to bring new generated photons to the spectral peak.
3. Unsaturated Comptonization zone ($1 \lesssim \tau_T \lesssim 10^2$): heating maintains a Compton parameter $y \sim 1$. The final non-thermal shape of the spectrum is produced in this region. Depending on the details of the dissipation mechanism, the spectrum can develop a high-energy tail up to the GeV band. In the presence of a soft photon source (synchrotron emission) unsaturated Comptonization also produces a low-energy photon index $\alpha \sim 1$.

Dissipation in the Planck zone reduces E_{pk}. Dissipation in the Wien zone increases E_{pk} if no new photons are produced; strong synchrotron emission in the Wien zone can reduce E_{pk}. In the non-thermal zone E_{pk} remains practically unchanged.

A blackbody photosphere would have the temperature $T = T_{eff}$ defined by

$$\frac{4}{3} a T_{eff}^4 \Gamma^2 4\pi R_\star^2 c = L_\gamma. \tag{6}$$

The corresponding E_{pk} would be given by $E_{pk} \approx 4\Gamma k T_{eff}$. This estimate should be viewed as a robust lower limit on E_{pk} rather than its true value. One could also define the effective

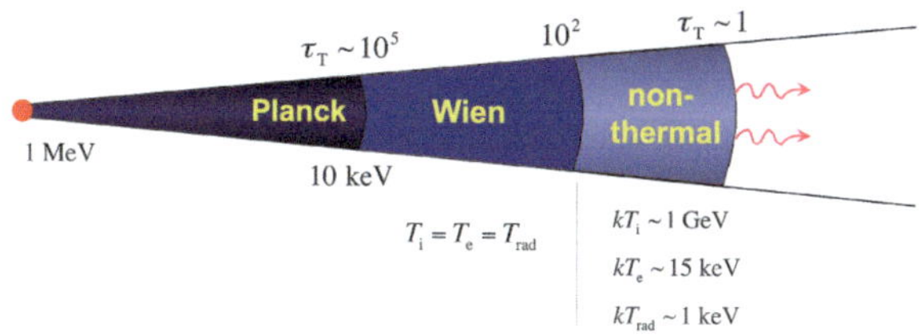

Fig. 5 Three zones in a GRB jet: (1) *Planck zone* where a blackbody spectrum is enforced; this zone ends where the Thomson optical depth decreases to $\tau_T \approx 10^5$. (2) *Wien zone* with Kompaneets parameter $y \gg 1$ where radiation has a Bose-Einstein spectrum, and (3) Comptonization ("*nonthermal*") zone where the radiation spectrum develops the high-energy tail. In the *Planck and Wien zones* the ion and electron temperatures are equal to the photon temperature, $T_i = T_e = T_{\rm rad}$. Outside the *Wien zone* $T_i \gg T_e \gg T_{\rm rad}$. The approximate values of the temperatures are indicated in the figure; these characteristic values weakly depend on how the jet is heated. The radiation spectrum is nonthermal in the *Comptonization zone* and here $T_{\rm rad}$ is defined as the Compton temperature, $T_{\rm rad} = T_{\rm C}$ (a measure of the mean photon energy)

blackbody temperature at the Wien radius, replacing $R_\star$ with $R_{\rm W}$ in (6) (Giannios 2012). This would still give $E_{\rm pk}$ lower than its true value. A more realistic model should not make the blackbody assumptions at any radii. Instead, it should explicitly follow the production of photons (and their Comptonization) in the expanding jet. Such simulations have been performed recently; their results are described below.

Radiative transfer simulations also give a detailed description of radiation decoupling from the expanding jet. There is no unique sphere of last scattering, and the notion of photosphere is quite fuzzy. $R_\star$ is defined as the radius where $\tau_T = 1$ (see (1)). This gives a characteristic decoupling radius, however the location of last scattering is random and surprisingly broadly distributed around $R_\star$ (Pe'er 2008; Beloborodov 2011). 2/3 of photons diffusing from large optical depths are last scattered between $0.3R_\star$ and $3R_\star$, and 1/3—outside this interval. Another surprising transfer effect is the significant anisotropy of radiation (measured in the jet rest frame) developing at small radii $r \sim 0.1R_\star$ (Beloborodov 2011).

4 GRB Spectrum from Radiative Transfer Simulations

Radiation emerging at the photosphere carries information about the entire expansion history of the jet, and the observed spectrum may be used to reconstruct dissipative processes hidden in the opaque region behind the photosphere. This can be achieved by solving a carefully formulated radiative transfer problem, which must be solved consistently with the flow dynamics and heating. A global simulation starting at a sufficiently small radius will show how the jet radiation deviates from blackbody and becomes nonthermal, shaping the observed GRB spectrum.

The effect of subphotospheric heating on GRB radiation has been studied with four different numerical codes (Pe'er et al. 2006; Giannios 2008; Beloborodov 2010; Vurm et al. 2011), consistently giving Band-type spectra. These calculations explored how the spectrum develops the high-energy tail at optical depths $\tau_T \lesssim 30$ as a result of thermal and nonthermal Comptonization. Recently, Vurm and Beloborodov (2016b) have performed global simulations of radiative transfer starting from very high optical depths $\tau_T > 10^3$ and including photon production and absorption processes.

4.1 Formulation of the Transfer Problem

A convenient starting point of transfer simulations is the radius R_c where the collimation process ends and the jet begins to expand conically. The jet is still accelerating at this stage. The acceleration can be self-consistently determined using the dynamical equation (Beloborodov 2011),

$$\frac{d\Gamma}{dr} = \frac{\sigma_T Z_\pm}{m_p c^3} F, \tag{7}$$

where F is the radiation flux measured in the plasma rest frame, and $Z_\pm$ is the number of electrons and positrons per proton. Equation (7) assumes that radiation pressure dominates the jet acceleration, neglecting the effect of magnetic forces; this approximation may be reasonable even when the jet is strongly magnetized, as long as the jet is hot (Vlahakis and Königl 2003; Russo and Thompson 2013).

Radiation carried by the dissipative jet is influenced by electron heating. Two types of heating should be distinguished: thermal and nonthermal:

- Thermal heating is the energy deposition into the Maxwellian $e^\pm$ plasma. In particular, in a two-temperature plasma $T_i \gg T_e$ below the photosphere the electrons are efficiently heated by Coulomb collisions with the hot ions (Beloborodov 2010). Thermal heating tends to increase the $e^\pm$ temperature above the local Compton temperature of radiation T_C, and the decoupling $T_e > T_C$ becomes significant at $\tau_T \lesssim 10^2$.
- Nonthermal heating is the injection of ultra-relativistic particles with a Lorentz factor γ_{inj}, e.g. by inelastic nuclear collisions ($\gamma_{inj} \sim m_\pi/m_e \sim 300$) or by other dissipation mechanisms discussed in Sect. 2. The energetic particles can generate an inverse Compton cascade and their energy is processed into secondary $e^\pm$ pairs of lower energies (Svensson 1987). The value of γ_{inj} is only important for synchrotron emission, which dominates the observed spectrum at $E \ll E_{pk}$ (see below) and has little relevance to the main MeV peak. Like thermal heating, the main parameter of nonthermal heating is its power.

To keep the number of parameters to a minimum, Vurm and Beloborodov (2016b) employed a simple model of continuous internal dissipation, approximating the thermal and nonthermal dissipation rates as power laws of radius,

$$\frac{dL_{th}}{d\ln r} = \varepsilon_{0,th} L \left(\frac{r}{R_c}\right)^{k_{th}}, \qquad \frac{dL_{nth}}{d\ln r} = \varepsilon_{0,nth} L \left(\frac{r}{R_c}\right)^{k_{nth}}, \tag{8}$$

where L is the total energy flow rate in the jet. The power law may be a crude approximation to jet heating by multiple internal shocks or reconnection, however it allows one to study the global picture of the evolution of radiative processes with radius in the expanding jet. The formation of photospheric radiation extends over several decades in radius, and the global simulation allows one to see all stages of this process.

Radiation is described by its intensity $I_\nu(r, \theta, \nu)$ in the fluid frame, where θ is the photon angle with respect to the radial direction. The radiative transfer equation in ultra-relativistic outflows ($\Gamma > 10$) is given by Beloborodov (2011),

$$\frac{1}{r^2 \Gamma} \frac{\partial}{\partial \ln r}\left[(1+\mu) r^2 \Gamma I_\nu\right] = \frac{r}{\Gamma}(j_\nu - \kappa_\nu I_\nu) + (1+\mu)(1-g\mu)\frac{\partial I_\nu}{\partial \ln \nu} - \frac{\partial}{\partial \mu}\left[(1-\mu^2)(1+\mu) g I_\nu\right], \tag{9}$$

where $\mu = \cos\theta$, κ_ν is the opacity, j_ν is the emissivity, and

$$g = 1 - \frac{d\ln\Gamma}{d\ln r}. \tag{10}$$

Equation (9) describes a steady jet, i.e. a simplified model where variability and internal shocks are averaged out and replaced by dissipation terms. At very large optical depths below the photosphere, $\tau_T > 100$, the radiation is approximately isotropic in the fluid frame, which simplifies the transfer problem—it is reduced to advection of isotropic radiation by the expanding outflow. However, a new feature in this region is the importance of induced scattering. Therefore, at $\tau_T > 100$ (9) should be replaced by the Kompaneets equation (see Vurm et al. 2013, Vurm and Beloborodov 2016b). At optical depths $\tau_T \ll 100$ the radiation becomes significantly anisotropic, and the full transfer equation (9) should be used; the induced scattering rate is small at these radii and can be neglected.

The radiative transfer is coupled to the evolution of the electron/positron distribution in the expanding jet, which is described by the kinetic equation (Vurm et al. 2011),

$$\begin{aligned}\frac{1}{r^2\Gamma}\frac{\partial}{\partial\ln r}\left[r^2\Gamma\, n_\pm(p)\right] &= \frac{r}{c\Gamma}\left(j_\pm - c\kappa_\pm n_\pm(p)\right)\\ &\quad - \frac{\partial}{\partial p}\left\{\frac{r}{c\Gamma}\left[\dot{p}\, n_\pm(p) - \frac{1}{2}\frac{\partial}{\partial\gamma}\left[D n_\pm(p)\right]\right] - \frac{3-g}{3}\, p\, n_\pm(p)\right\}.\end{aligned} \tag{11}$$

Here $p = \sqrt{\gamma^2 - 1}$ is the electron/positron momentum in units of $m_e c$, and $n_\pm(p)$ is the $e^\pm$ momentum distribution. The terms $\dot{p}$ and D account for heating/cooling and diffusion in the momentum space due to radiative processes and Coulomb collisions. The emission term $j_\pm$ includes the injection of new pairs due to non-thermal dissipation and photon-photon absorption $\gamma + \gamma \to e^+ + e^-$, and $c\kappa_\pm n_\pm(p)$ is the pair annihilation rate.

Thermal radiative processes include Comptonization of photons by the thermal plasma, induced Compton scattering at low frequencies, cyclotron emission/absorption, bremsstrahlung, and double Compton scattering. Nonthermal radiative processes result from the injection of relativistic leptons. This leads to the IC $e^\pm$ cascade and synchrotron emission, which is an important source of photons. All these processes have been followed in the expanding dissipative jet by Vurm and Beloborodov (2016b) using the exact cross sections and rates and solving numerically the transfer problem described by the coupled equations (7), (9), (11).

4.2 Main Features of Sub-Photospheric Radiative Transfer

Figure 6 shows the evolution of the radiation spectrum within a dissipative jet as the jet expands from the highly opaque region $\tau_T \sim 3 \times 10^3$ to transparency. There are two main stages of the spectral evolution:

(1) Generation of photons below the Wien radius ($\tau_T \gtrsim 10^2$) and their Comptonization to the Wien peak. A major source of photons is the synchrotron emission from nonthermal leptons. If the magnetization is relatively weak, e.g. $\varepsilon_B \sim 10^{-2}$, then the synchrotron power is modest, however the *number* of generated photons is substantial, because they are emitted with low energies. Below the Wien radius R_W many of them are Comptonized to the Wien peak and eventually dominate the peak, significantly increasing its photon number and shifting E_{pk} to lower energies. The decrease of E_{pk} due to the continuing photon supply to the peak ends when τ_T is reduced below $\sim 10^2$, which approximately corresponds to R_W.

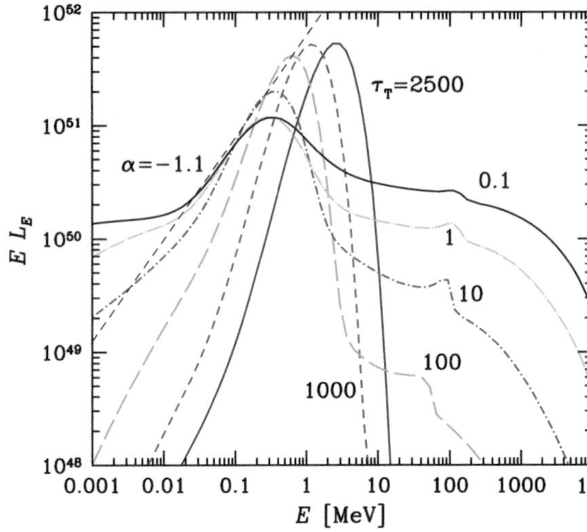

Fig. 6 Evolution of the radiation spectrum carried by the jet from high to low optical depths τ_T in a sample model with $L = 2 \times 10^{52}$ erg/s, $\eta = 190$, and $\varepsilon_B = \sigma/2 = 10^{-2}$. The simulation started at radius $R_c = 10^{11}$ cm with $\Gamma_c = 20$, and the jet was continually heated with $\epsilon_{th} = \epsilon_{nth} = 0.025$. The spectra are measured in the rest frame of the central engine and not corrected for a cosmological redshift. The *black curve* ($\tau_T = 0.1$) shows the observed (escaping) spectrum that result from this evolution. (From Vurm and Beloborodov 2016b)

(2) Broadening of the spectrum by unsaturated Comptonization outside the Wien radius, leading to a nonthermal shape of the spectrum. The thermal Comptonization gradually switches to the unsaturated regime at $r \sim R_W$ and then the Compton y-parameter stays near unity. The production of synchrotron photons continues at $r > R_W$ and results in the soft "excess" seen below a few $\times 10$ keV in the emitted spectrum (Fig. 6). More importantly, unsaturated Comptonization of these photons plays a key role in determining the spectral slope α below the peak. The slope begins to soften near the Wien radius and attains its final value $\alpha \sim -1$ near the photosphere. The value of α depends on the heating history at $r > R_W$, in particular on the nonthermal dissipation channel, and is expected to vary, however $\alpha \sim -1$ is a characteristic signature of unsaturated Comptonization.

The high-energy tail of the spectrum develops mostly near the photosphere, due to two effects, thermal and nonthermal. Outside the Wien radius the electron temperature rises significantly above E_{pk}/Γ and thermal Comptonization begins to produce photons with energies $E > E_{pk}$. As the optical depth drops, the nonthermal high-energy component becomes increasingly prominent, especially in weakly magnetized jets with the dominant IC cooling. Then the overlapping thermal and nonthermal Comptonization components together form an extended high-energy spectrum, which may superficially appear as a single emission component.

As seen in Fig. 6, the spectrum has a distinctly nonthermal shape already well below the photosphere (at $\tau_T \gtrsim 10$). In particular, the low-energy slope is significantly softer than the Planck (or Wien) spectrum. Even if dissipation stopped completely at $\tau_T \sim 10$, a thermal-looking prompt emission would not be expected; instead, the emerging spectrum would resemble an exponentially cutoff power law.

The dependence of the photospheric spectrum on the jet magnetization ε_B is shown in Fig. 7. Very weak magnetization $\varepsilon_B < 10^{-3}$ is disfavored for typical GRBs, as it leads to a hard slope α at low energies—then practically the entire spectrum is dominated by the Comptonized photons advected from very large optical depths, with little photon production at moderate τ_T. The preferred magnetization is in the range $10^{-3} < \varepsilon_B < 0.1$, which gives α varying around -1. At the low-ε_B end, the nonthermal $e^{\pm}$ cascades are particularly efficient, and the pairs outnumber protons on average by a factor of $\gtrsim 10$.

A large fraction of produced pairs "freeze out" in the expanding jet, so the jet remains forever dominated by $e^{\pm}$ pairs. This should affect the afterglow emission produced by the

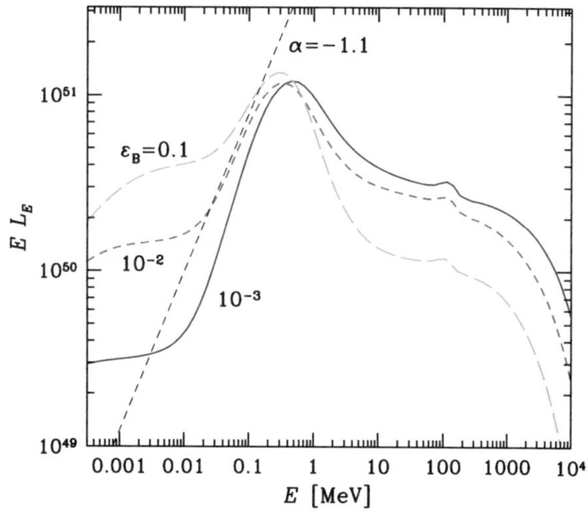

Fig. 7 Effect of varying magnetization on the emerging spectrum (Vurm and Beloborodov 2016b). If variations of ε_B are not resolved, a superposition of the spectra will be observed

reverse shock when the jet interacts with the external medium. Thus, the reverse-shock component in a GRB afterglow could serve as a probe of pair loading.

In strongly magnetized jets with $\varepsilon_B \gtrsim 1$, the pair cascade and nonthermal Comptonization are suppressed, leading to a different radiation spectrum at high energies (Vurm et al. 2011). Radiation from magnetically dominated jets was recently considered by Gill and Thompson (2014), Thompson and Gill (2014) and Bégué and Pe'er (2015). Gill and Thompson envision a two-stage evolution of a jet that starts out baryon-free and dominated by the Poynting flux. In their scenario, baryon loading occurs at a large radius, and the jet opacity is dominated by electron-positron pairs generated by dissipation, which takes place in two separate episodes. Despite the complicated details, the evolution of radiative processes in their picture resembles that shown in Fig. 6—basically, generation of synchrotron photons at large optical depths is followed by Comptonization into a Band-like spectrum.

4.3 Comparison with Observed Spectra

A continuously heated and moderately magnetized jet generates a Band-type photospheric spectrum, with spectral slopes α, β, and peak position E_{pk} consistent with observations. The radiative transfer simulations show a moderate dependence of E_{pk} on the parameters of the problem, and no fine-tuning is needed to bring the peak into the observed range around 1 MeV.

A typical burst with $L \approx 10^{52}$ erg/s, a canonical Band spectrum, and $E_{pk} \lesssim 1$ MeV is reproduced by the model if:

1. The jet magnetization is in the range $10^{-3} < \varepsilon_B < 0.1$. Very weak magnetization increases E_{pk} by suppressing synchrotron emission; strong magnetization softens the spectrum both below and above the peak, and generates a prominent soft "excess" below a few tens of keV.
2. The jet Lorentz factor $\Gamma(R_c) = 10$–100 at radii comparable to that of the stellar progenitor. Jets with a low $\Gamma(R_c)$ are more "photon-rich" and produce GRBs with low E_{pk}.
3. Heating has a nonthermal component that injects relativistic leptons into the jet. The subphotopsheric nonthermal particles play a key role by providing soft synchrotron photons as seeds for Comptonization. This impacts the formation of both E_{pk} (at $\tau_T \gtrsim 10^2$) and

Table 1 Jet parameters for three fitted GRBs (from Vurm and Beloborodov 2016b)

GRB	L_{54}	Γ_c	Γ_f	ε_B	$\epsilon_{0,\text{th}}$	$\epsilon_{0,\text{nth}}$
990123	0.44	35	500	0.018	0.099	0.021
090902B	2.3	70	1220	0.012	0.055	0.065
130427A	0.47	64	300	0.046	0.093	0.067

slope α (at $1 < \tau < 10^2$). The absence of nonthermal leptons would lead to hard spectra with high E_{pk}.

The detailed radiative transfer models have been applied to three well-studied bright bursts, GRB 990123, GRB 090902B, and GRB 130427A, which show different prompt spectra, and successful fits have been found in all three cases (Vurm and Beloborodov 2016b). The fits gave estimates for the main parameters of the GRB jets (Table 1). In particular, the best-fit magnetization varies between 0.01 and 0.1, and the final Lorentz factor Γ_f between 300 and 1200. The corresponding photospheric radius $R_\star$ varies around 10^{13} cm.

The obtained Γ_f can be compared with the results of another, independent method of estimating the jet Lorentz factor. This method is based on the reconstruction of the GeV+optical flash produced by the jet at much larger radii $r \gtrsim 10^{16}$ cm where it drives the external blast wave (Beloborodov et al. 2014; Vurm et al. 2014; Hascoët et al. 2015). Remarkably, the prompt emission fit for GRB 130427A gave $\Gamma_{\text{f}} \approx 300$, close to the ejecta Lorentz factor $\Gamma_{\text{ej}} \approx 350$ obtained from the GeV+optical flash reconstruction (Vurm et al. 2014). The fit to GRB 090902B gave $\Gamma_{\text{f}} \approx 1200$, which is higher than $\Gamma_{\text{ej}} = 600$–900 used to fit the GeV flash. However, in this case there is no significant inconsistency, because the reverse shock in the blast wave from GRB 090902B is relativistic; in this case the flash modeling only gives a lower limit $\Gamma_{\text{ej}} > 600$ (Hascoët et al. 2015). A similar $\Gamma \approx 1000$ was proposed by Pe'Er et al. (2012) based on a different phenomenological model for GRB 090902B (a multicolor blackbody for the photosphere and nonthermal radiation from dissipation at a large radius).

Besides the values of Γ and ε_{B}, the radiative transfer models of the prompt emission give insights into the heating history of the GRB jets. The fit to GRB 130427A requires that the nonthermal heating be significantly deep below the photosphere and become weak well before the jet expands to transparency; the observed Band-like spectrum is mainly shaped by thermal Comptonization. In GRB 990123, thermal heating appears to dominate at all radii, and the nonthermal synchrotron source is relatively weak. The model still predicts a moderate excess below a few tens of keV due to synchrotron emission. A hint of such excess is indeed seen in the data (Fig. 2 in Briggs et al. 1999; R. Preece, private communication).

In contrast, nonthermal dissipation in GRB 090902B is found to be strong up to the photosphere and beyond. It explains the observed high-energy component and the soft excess, which were previously modeled as a power law of unknown origin (Abdo and The Fermi Collaboration 2009). The two models are compared in Fig. 8. The radiative transfer model shows that the soft excess is the synchrotron emission and the high-energy component is IC emission from nonthermal electrons.[2] They are not parts of a single power law, however they are produced by the same nonthermal electron population. The high-energy component of the prompt emission is also expected to contaminate (at early times) the GeV flash observed in GRB 090902B by *Fermi* LAT.

[2] A different hadronic shock model explanation was also proposed by Asano et al. (2010).

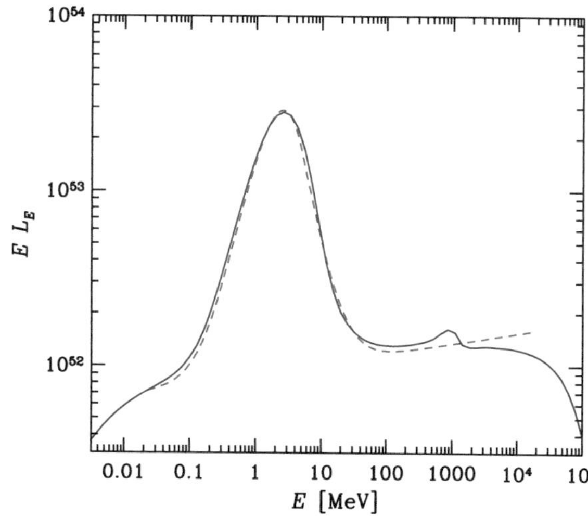

Fig. 8 Radiative transfer model for GRB 090902B (*solid curve*). The *dashed curve* shows the Band+power-law fit by Abdo and The Fermi Collaboration (2009), in time bin b. (From Vurm and Beloborodov 2016b)

5 Polarization

Linear polarization measurements in the hard X-ray band have been reported for GRB 041219A (Kalemci et al. 2007; McGlynn et al. 2007; Götz et al. 2009), GRB 061122 (McGlynn et al. 2009; Götz et al. 2013), GRB 100826A (Yonetoku et al. 2011), GRB 110301A, GRB 110721A (Yonetoku et al. 2012), and GRB 140206A (Götz et al. 2014). However, the detection significance for all these bursts is modest, because of a low sensitivity, and the measurements have poorly known systematic uncertainties. Future detectors can significantly improve the polarization information, which will be highly valuable for GRB modeling.

Radiative transfer in the jet occurs in two linear polarization states with different intensities, however this gives no observable polarization as long as the emission is symmetric about the line of sight (Beloborodov 2011). Due to the radial Doppler beaming of radiation, the observed GRB radiation is expected to come from a narrow region around the line of sight, of angular size $\delta\theta \sim \Gamma^{-1}$. Its symmetry can be broken in two ways.

(1) A preferred direction for polarization appears if there is a strong transverse gradient in the jet parameters at the photosphere (e.g. a gradient in luminosity, Lorentz factor, etc.). Then observable polarization emerges after the last scattering around $R_\star$, which was investigated by Lundman et al. (2014) and Ito et al. (2014). They predicted polarization degrees up to $\Pi \sim 40\%$ if the jet has significant structure on angular scales $\delta\theta \sim \Gamma^{-1}$. In particular, a strong transverse gradient is expected near the jet edge $\theta = \theta_c$. If the collimation angle θ_c is not much larger than Γ^{-1} then observers detecting a bright burst (but not exactly on the jet axis) will see the jet cross section $\sim\pi r^2\theta_c^2$ at some inclination. Then the line of sight and the jet axis form a preferred plane and a strong linear polarization should show up in the main peak of the GRB spectrum. Lundman et al. (2014) also showed that for a jet with $\theta_c \gg \Gamma^{-1}$ the probability to observe its edge (and hence to break the symmetry of observed emission) is $\sim 4(\Gamma\theta_c)^{-1}$. For instance, if $\theta_c \sim 0.1$ and $\Gamma \approx 400$ the edge should be visible in roughly 10% of detected GRBs. The actual distribution of θ_c is however unknown; attempts to infer this distribution from the analysis of so-called "jet breaks" in afterglow light curves have been inconclusive.

(2) If part of the GRB spectrum is dominated by synchrotron radiation, a preferred direction for polarization can be set by the magnetic field where this radiation is produced. The transfer models discussed in Sect. 4 show that the main MeV peak is dominated by the Comptonized radiation advected from large optical depths. The Comptonized peak is unpolarized as long as there is no significant structure on angular scales $\delta\theta \sim \Gamma^{-1}$. However, at sufficiently low energies $E \ll E_{\rm pk}$ the spectrum is dominated by synchrotron emission. This leads to a unique polarization signature: a rise in GRB polarization toward lower energies (Lundman et al. 2016).

Synchrotron emission from relativistic electrons in a uniform magnetic field **B** is linearly polarized in the plane perpendicular to **B**. The polarization degree for an isotropic electron distribution is $\Pi_{\rm syn} = (p+1)/(p+7/3)$, where $p = d\ln N/d\ln E_e$ is the slope of the electron spectrum (e.g. Rybicki and Lightman 1979). This standard result is somewhat modified when the observed region is a spherical patch in a relativistic outflow, carrying an ordered transverse magnetic field, which gives $\Pi_{\rm syn}$ varying around 50%, depending on p (Lyutikov et al. 2003). Synchrotron radiation produced well below the photosphere will lose its polarization after a few scatterings, however a significant fraction of synchrotron photons produced around and above the photosphere will escape without scattering and preserve their polarization. The observed GRB spectrum $L(E)$ can be viewed as the sum of two contributions: photons escaping the jet after their last Compton scattering, $L_{\rm sc}(E)$, and photons escaping directly after their emission by the synchrotron mechanism, with no scattering, $L_{\rm nsc}(E)$. The observed polarization is then given by

$$\Pi(E) = f_{\rm nsc}(E)\Pi_{\rm syn}, \qquad f_{\rm nsc}(E) = \frac{L_{\rm nsc}(E)}{L(E)}. \tag{12}$$

Using the transfer models of Vurm and Beloborodov (2016b), Lundman et al. (2016) calculated $f_{\rm nsc}(E)$ and the expected polarization degree $\Pi(E)$ for GRBs 990123, 090902B, and 110721A. Their results demonstrate that strong polarization is expected in the X-ray band under two conditions: (1) significant dissipation extends to $r \gtrsim R_\star$ and is accompanied by injection of nonthermal particles, and (2) the jet magnetization is significant, $\varepsilon_{\rm B} \gtrsim 10^{-2}$. These conditions are well satisfied in GRB 090902B (Sect. 4). Then the GRB spectrum has a prominent soft excess emerging above the Band spectrum at low energies, and a significant fraction of this excess is formed by unscattered synchrotron photons. This leads to the strong polarization rise toward low energies (Fig. 9).

6 Possible Future Developments

The detailed radiative transfer simulations can be systematically applied to all available spectra of prompt GRB emission, including those resolved in time. Such fits would reconstruct the jet heating and magnetization for a large sample of GRBs and may reveal important correlations. This could shed light on how the jets are produced and heated, and what role is played in this process by the magnetic field.

The analysis of prompt emission can be combined with that of GRB afterglow, especially its early phase. The GeV+optical flash predicted and observed at the onset of the afterglow (Beloborodov et al. 2014) provides particularly useful data that recently helped to disentangle the main parameters of the blast wave in seven GRBs (Hascoët et al. 2015). The initial Lorentz factor of the blast wave reconstructed by this method can be compared with the jet Lorentz factor inferred from the transfer simulations of the prompt emission, as has been

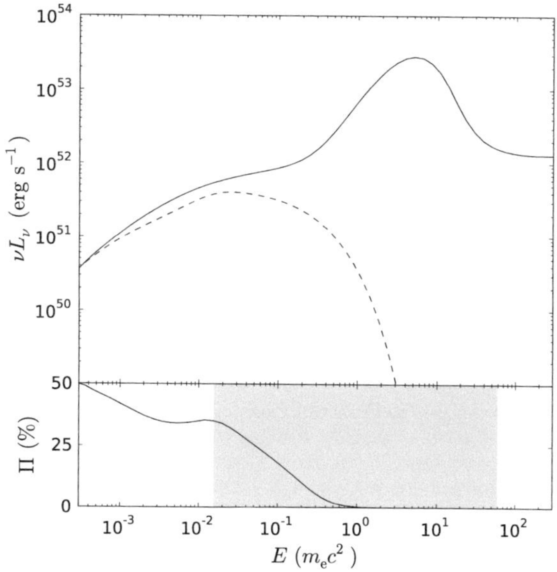

Fig. 9 *Top panel*: the simulated spectrum of GRB 090902B (*solid curve*) and the contribution of unscattered synchrotron emission (*dashed curve*). Their ratio gives the unscattered fraction $f_{\rm nsc}(E)$. *Bottom panel*: expected polarization as a function of photon energy E. The *shaded region* shows the Fermi GBM band (NaI + BGO detectors, 8 keV to 30 MeV). (From Lundman et al. 2016)

done for GRB 130427A. An additional test would be provided by future detections of the predicted TeV counterpart of the GeV+optical flash, which should rise with some delay (Hascoët et al. 2015). It is detectable by current ground-based Cherenkov telescopes and should certainly be detected by the future Cherenkov Telescope Array (CTA) (Vurm and Beloborodov 2016a).

Spectra of prompt GRB emission are poorly known at low energies, from soft X-rays to the optical band. While the main MeV peak is dominated by the Comptonized radiation advected from large optical depths, the prompt optical radiation must be dominated by synchrotron emission outside the scattering photosphere. The low-frequency emission is predicted to be affected by synchrotron self-absorption, which tends to give a flat spectral slope (Vurm et al. 2011). While prompt optical emission has been detected in some GRBs, there is currently no information on its spectral slope. A color measurement and better sampling of the optical light curve would provide useful information about dissipation outside the photosphere and inside of the external blast wave. This is a rather broad range of radii, even taking into account that the photosphere can be "carried" forward by the $e^{\pm}$-dressed internal shocks up to two decades in radius (Sect. 2). Radiation produced outside the photosphere is expected to have a broad nonthermal spectrum with no sharp MeV peak and a slower variability on timescales $\sim r/c\Gamma^2$.

Important developments may result from future observations of polarization of the prompt GRB emission and its variation across the burst spectrum. An unpolarized MeV peak and a rise in linear polarization toward low photon energies would confirm the synchrotron contribution at $E \ll E_{\rm pk}$ predicted by the radiative transfer simulations (Sects. 4 and 5). Bursts similar to GRB 090902B are the most promising targets for such observations. Observations of polarization of the main MeV peak of photospheric emission would provide evidence for a transverse structure of the jet on angular scales $\delta\theta \sim \Gamma^{-1}$.

Detection of neutrinos from GRB explosions is possible in the near future. The detection (or strong upper limits) would test the ideas for the dissipation mechanism in the relativistic jet, including that in the opaque region well below the photosphere. In particular, the expected presence of a free neutron component leads to strong nuclear collisional dissipation, which produces 10–100 GeV neutrinos detectable by IceCube DeepCore.

Acknowledgements AMB's research is supported by NSF grant AST-1412485, NASA grant NNX15AE26G, and a grant from the Simons Foundation (#446228, Andrei Beloborodov). PM's research is supported by NASA grant NNX13AH50G.

References

A.A. Abdo (The Fermi Collaboration), Fermi observations of GRB 090902B: a distinct spectral component in the prompt and delayed emission. Astrophys. J. Lett. **706**, 138–144 (2009). doi:10.1088/0004-637X/706/1/L138

M. Ackermann (The Fermi Collaboration), Multiwavelength observations of GRB 110731A: GeV emission from onset to afterglow. Astrophys. J. **763**, 71 (2013). doi:10.1088/0004-637X/763/2/71

K. Asano, S. Inoue, P. Mészáros, Prompt X-ray and optical excess emission due to hadronic cascades in gamma-ray bursts. Astrophys. J. Lett. **725**, 121–125 (2010). doi:10.1088/2041-8205/725/2/L121

M. Axelsson, L. Borgonovo, The width of gamma-ray burst spectra. Mon. Not. R. Astron. Soc. **447**, 3150–3154 (2015). doi:10.1093/mnras/stu2675

J.N. Bahcall, P. Mészáros, 5–10 GeV Neutrinos from gamma-ray burst fireballs. Phys. Rev. Lett. **85**, 1362–1365 (2000)

D.L. Band, M. Axelsson, L. Baldini, G. Barbiellini, M.G. Baring, D. Bastieri, M. Battelino, R. Bellazzini, E. Bissaldi, G. Bogaert, J. Bonnell, J. Chiang, J. Cohen-Tanugi, V. Connaughton, S. Cutini, F. de Palma, B.L. Dingus, E. do Couto e Silva, G. Fishman, A. Galli, N. Gehrels, N. Giglietto, J. Granot, S. Guiriec, R.E. Hughes, T. Kamae, N. Komin, F. Kuehn, M. Kuss, F. Longo, P. Lubrano, R.M. Kippen, M.N. Mazziotta, J.E. McEnery, S. McGlynn, E. Moretti, T. Nakamori, J.P. Norris, M. Ohno, M. Olivo, N. Omodei, V. Pelassa, F. Piron, R. Preece, M. Razzano, J.J. Russell, F. Ryde, P.M. Saz Parkinson, J.D. Scargle, C. Sgrò, T. Shimokawabe, P.D. Smith, G. Spandre, P. Spinelli, M. Stamatikos, B.L. Winer, R. Yamazaki, Prospects for GRB science with the Fermi Large Area Telescope. Astrophys. J. **701**, 1673–1694 (2009). doi:10.1088/0004-637X/701/2/1673

I. Bartos, A.M. Beloborodov, K. Hurley, S. Márka, Detection prospects for GeV neutrinos from collisionally heated gamma-ray bursts with IceCube/DeepCore. Phys. Rev. Lett. **110**(24), 241101 (2013). doi:10.1103/PhysRevLett.110.241101

D. Bégué, A. Pe'er, Poynting-flux-dominated jets challenged by their photospheric emission. Astrophys. J. **802**, 134 (2015). doi:10.1088/0004-637X/802/2/134

A.M. Beloborodov, Nuclear composition of gamma-ray burst fireballs. Astrophys. J. **588**, 931–944 (2003). doi:10.1086/374217

A.M. Beloborodov, Collisional mechanism for gamma-ray burst emission. Mon. Not. R. Astron. Soc. **407**, 1033–1047 (2010). doi:10.1111/j.1365-2966.2010.16770.x

A.M. Beloborodov, Radiative transfer in ultrarelativistic outflows. Astrophys. J. **737**, 68 (2011). doi:10.1088/0004-637X/737/2/68

A.M. Beloborodov, Regulation of the spectral peak in gamma-ray bursts. Astrophys. J. **764**, 157 (2013). doi:10.1088/0004-637X/764/2/157

A.M. Beloborodov, Sub-photospheric shocks in relativistic explosions. ArXiv e-prints (2016). 1604.02794

A.M. Beloborodov, R. Hascoët, I. Vurm, On the origin of GeV emission in gamma-ray bursts. Astrophys. J. **788**, 36 (2014). doi:10.1088/0004-637X/788/1/36

A.M. Beloborodov, B.E. Stern, R. Svensson, Power density spectra of gamma-ray bursts. Astrophys. J. **535**, 158–166 (2000). doi:10.1086/308836

R.D. Blandford, D.G. Payne, Compton scattering in a converging fluid flow— part two—radiation dominated shock. Mon. Not. R. Astron. Soc. **194**, 1041 (1981). doi:10.1093/mnras/194.4.1041

M.S. Briggs, D.L. Band, R.M. Kippen, R.D. Preece, C. Kouveliotou, J. van Paradijs, G.H. Share, R.J. Murphy, S.M. Matz, A. Connors, C. Winkler, M.L. McConnell, J.M. Ryan, O.R. Williams, C.A. Young, B. Dingus, J.R. Catelli, R.A.M.J. Wijers, Observations of GRB 990123 by the Compton Gamma Ray Observatory. Astrophys. J. **524**, 82–91 (1999). doi:10.1086/307808

O. Bromberg, A. Levinson, Recollimation and radiative focusing of relativistic jets: applications to blazars and M87. Astrophys. J. **699**, 1274–1280 (2009). doi:10.1088/0004-637X/699/2/1274

O. Bromberg, J. Granot, T. Piran, On the composition of GRBs' collapsar jets. Mon. Not. R. Astron. Soc. **450**, 1077–1084 (2015). doi:10.1093/mnras/stv226

R. Budnik, B. Katz, A. Sagiv, E. Waxman, Relativistic radiation mediated shocks. Astrophys. J. **725**, 63–90 (2010). doi:10.1088/0004-637X/725/1/63

J.M. Burgess, R.D. Preece, M.G. Baring, M.S. Briggs, V. Connaughton, S. Guiriec, W.S. Paciesas, C.A. Meegan, P.N. Bhat, E. Bissaldi, V. Chaplin, R. Diehl, G.J. Fishman, G. Fitzpatrick, S. Foley, M. Gibby, M. Giles, A. Goldstein, J. Greiner, D. Gruber, A.J. van der Horst, A. von Kienlin, M. Kippen, C. Kouveliotou, S. McBreen, A. Rau, D. Tierney, C. Wilson-Hodge, Constraints on the synchrotron shock model for the Fermi GRB 090820A observed by gamma-ray burst monitor. Astrophys. J. **741**, 24 (2011). doi:10.1088/0004-637X/741/1/24

J.M. Burgess, R.D. Preece, F. Ryde, P. Veres, P. Mészáros, V. Connaughton, M. Briggs, A. Pe'er, S. Iyyani, A. Goldstein, M. Axelsson, M.G. Baring, P.N. Bhat, D. Byrne, G. Fitzpatrick, S. Foley, D. Kocevski, N. Omodei, W.S. Paciesas, V. Pelassa, C. Kouveliotou, S. Xiong, H.-F. Yu, B. Zhang, S. Zhu, An observed correlation between thermal and non-thermal emission in gamma-ray bursts. Astrophys. J. Lett. **784**, 43 (2014). doi:10.1088/2041-8205/784/2/L43

F. Daigne, R. Mochkovitch, Gamma-ray bursts from internal shocks in a relativistic wind: temporal and spectral properties. Mon. Not. R. Astron. Soc. **296**, 275–286 (1998). doi:10.1046/j.1365-8711.1998.01305.x

E.V. Derishev, V.V. Kocharovsky, V.V. Kocharovsky, The neutron component in fireballs of gamma-ray bursts: dynamics and observable imprints. Astrophys. J. **521**, 640–649 (1999). doi:10.1086/307574

E.V. Derishev, F.A. Aharonian, V.V. Kocharovsky, V.V. Kocharovsky, Particle acceleration through multiple conversions from a charged into a neutral state and back. Phys. Rev. D **68**(4), 043003 (2003). doi:10.1103/PhysRevD.68.043003

G. Drenkhahn, H.C. Spruit, Efficient acceleration and radiation in Poynting flux powered GRB outflows. Astron. Astrophys. **391**, 1141–1153 (2002). doi:10.1051/0004-6361:20020839

D. Eichler, A. Levinson, A compact fireball model of gamma-ray bursts. Astrophys. J. **529**, 146–150 (2000). doi:10.1086/308245

G. Ghirlanda, L. Nava, G. Ghisellini, A. Celotti, D. Burlon, S. Covino, A. Melandri, Gamma-ray bursts in the comoving frame. Mon. Not. R. Astron. Soc. **420**, 483–494 (2012). doi:10.1111/j.1365-2966.2011.20053.x

G. Ghisellini, Blazars and gamma ray bursts, in *VI Microquasar Workshop: Microquasars and Beyond* (2006), pp. 1–27

G. Giacinti, A.R. Bell, Collisionless shocks and TeV neutrinos before supernova shock breakout from an optically thick wind. Mon. Not. R. Astron. Soc. **449**, 3693–3699 (2015). doi:10.1093/mnras/stv561

D. Giannios, Prompt GRB emission from gradual energy dissipation. Astron. Astrophys. **480**, 305–312 (2008). doi:10.1051/0004-6361:20079085

D. Giannios, The peak energy of dissipative gamma-ray burst photospheres. Mon. Not. R. Astron. Soc. **422**, 3092–3098 (2012). doi:10.1111/j.1365-2966.2012.20825.x

D. Giannios, H.C. Spruit, Spectral and timing properties of a dissipative gamma-ray burst photosphere. Astron. Astrophys. **469**, 1–9 (2007). doi:10.1051/0004-6361:20066739

R. Gill, C. Thompson, Non-thermal gamma-ray emission from delayed pair breakdown in a magnetized and photon-rich outflow. Astrophys. J. **796**, 81 (2014). doi:10.1088/0004-637X/796/2/81

A. Goldstein, J.M. Burgess, R.D. Preece, M.S. Briggs, S. Guiriec, A.J. van der Horst, V. Connaughton, C.A. Wilson-Hodge, W.S. Paciesas, C.A. Meegan, A. von Kienlin, P.N. Bhat, E. Bissaldi, V. Chaplin, R. Diehl, G.J. Fishman, G. Fitzpatrick, S. Foley, M. Gibby, M. Giles, J. Greiner, D. Gruber, R.M. Kippen, C. Kouveliotou, S. McBreen, S. McGlynn, A. Rau, D. Tierney, The Fermi GBM gamma-ray burst spectral catalog: the first two years. Astrophys. J. Suppl. Ser. **199**, 19 (2012). doi:10.1088/0067-0049/199/1/19

J. Goodman, Are gamma-ray bursts optically thick? Astrophys. J. Lett. **308**, 47–50 (1986). doi:10.1086/184741

D. Götz, P. Laurent, F. Lebrun, F. Daigne, Ž. Bošnjak, Variable polarization measured in the prompt emission of GRB 041219A using IBIS on board INTEGRAL. Astrophys. J. Lett. **695**, 208–212 (2009). doi:10.1088/0004-637X/695/2/L208

D. Götz, S. Covino, A. Fernández-Soto, P. Laurent, Ž. Bošnjak, The polarized gamma-ray burst GRB 061122. Mon. Not. R. Astron. Soc. **431**, 3550–3556 (2013). doi:10.1093/mnras/stt439

D. Götz, P. Laurent, S. Antier, S. Covino, P. D'Avanzo, V. D'Elia, A. Melandri, GRB 140206A: the most distant polarized gamma-ray burst. Mon. Not. R. Astron. Soc. **444**, 2776–2782 (2014). doi:10.1093/mnras/stu1634

S. Guiriec, V. Connaughton, M.S. Briggs, M. Burgess, F. Ryde, F. Daigne, P. Mészáros, A. Goldstein, J. McEnery, N. Omodei, P.N. Bhat, E. Bissaldi, A. Camero-Arranz, V. Chaplin, R. Diehl, G. Fishman, S. Foley, M. Gibby, M.M. Giles, J. Greiner, D. Gruber, A. von Kienlin, M. Kippen, C. Kouveliotou, S. McBreen, C.A. Meegan, W. Paciesas, R. Preece, A. Rau, D. Tierney, A.J. van der Horst, C. Wilson-Hodge, Detection of a thermal spectral component in the prompt emission of GRB 100724B. Astrophys. J. Lett. **727**, 33 (2011). doi:10.1088/2041-8205/727/2/L33

S. Guiriec, R. Mochkovitch, T. Piran, F. Daigne, C. Kouveliotou, J. Racusin, N. Gehrels, J. McEnery, GRB 131014A: a laboratory for studying the thermal-like and non-thermal emissions in gamma-ray bursts, and the new L_i^{nTh}-$E_{peak,i}^{nTh,rest}$ relation. Astrophys. J. **814**, 10 (2015). doi:10.1088/0004-637X/814/1/10

R. Hascoët, I. Vurm, A.M. Beloborodov, Measuring ambient densities and Lorentz factors of gamma-ray bursts from GeV and optical observations. Astrophys. J. **813**, 63 (2015). doi:10.1088/0004-637X/813/1/63

H. Ito, S. Nagataki, J. Matsumoto, S.-H. Lee, A. Tolstov, J. Mao, M. Dainotti, A. Mizuta, Spectral and polarization properties of photospheric emission from stratified jets. Astrophys. J. **789**, 159 (2014). doi:10.1088/0004-637X/789/2/159

H. Ito, J. Matsumoto, S. Nagataki, D.C. Warren, M.V. Barkov, Photospheric emission from collapsar jets in 3D relativistic hydrodynamics. Astrophys. J. Lett. **814**, 29 (2015). doi:10.1088/2041-8205/814/2/L29

D. Kagan, L. Sironi, B. Cerutti, D. Giannios, Relativistic magnetic reconnection in pair plasmas and its astrophysical applications. Space Sci. Rev. **191**, 545–573 (2015). doi:10.1007/s11214-014-0132-9

E. Kalemci, S.E. Boggs, C. Kouveliotou, M. Finger, M.G. Baring, Search for polarization from the prompt gamma-ray emission of GRB 041219a with SPI on INTEGRAL. Astrophys. J. Suppl. Ser. **169**, 75–82 (2007). doi:10.1086/510676

Y. Kaneko, R.D. Preece, M.S. Briggs, W.S. Paciesas, C.A. Meegan, D.L. Band, The complete spectral catalog of bright BATSE gamma-ray bursts. Astrophys. J. Suppl. Ser. **166**, 298–340 (2006). doi:10.1086/505911

K. Kashiyama, K. Murase, P. Mészáros, Neutron-proton-converter acceleration mechanism at subphotospheres of relativistic outflows. Phys. Rev. Lett. **111**(13), 131103 (2013). doi:10.1103/PhysRevLett.111.131103

S. Kobayashi, T. Piran, R. Sari, Can internal shocks produce the variability in gamma-ray bursts? Astrophys. J. **490**, 92 (1997)

S.S. Komissarov, N. Vlahakis, A. Königl, M.V. Barkov, Magnetic acceleration of ultrarelativistic jets in gamma-ray burst sources. Mon. Not. R. Astron. Soc. **394**, 1182–1212 (2009). doi:10.1111/j.1365-2966.2009.14410.x

D. Lazzati, B.J. Morsony, M.C. Begelman, Very high efficiency photospheric emission in long-duration γ-ray bursts. Astrophys. J. Lett. **700**, 47–50 (2009). doi:10.1088/0004-637X/700/1/L47

D. Lazzati, M. Villeneuve, D. Lopez-Camara, B. Morsony, R. Perna, On the observed duration distribution of gamma-ray bursts from collapsars. ArXiv e-prints (2013)

A. Levinson, Observational signatures of sub-photospheric radiation-mediated shocks in the prompt phase of gamma-ray bursts. Astrophys. J. **756**, 174 (2012). doi:10.1088/0004-637X/756/2/174

A. Levinson, O. Bromberg, Relativistic photon mediated shocks. Phys. Rev. Lett. **100**(13), 131101 (2008). doi:10.1103/PhysRevLett.100.131101

C. Lundman, A. Pe'er, F. Ryde, Polarization properties of photospheric emission from relativistic, collimated outflows. Mon. Not. R. Astron. Soc. **440**, 3292–3308 (2014). doi:10.1093/mnras/stu457

C. Lundman, I. Vurm, A.M. Beloborodov, Polarization of gamma-ray bursts in the dissipative photosphere model. ArXiv e-prints (2016)

M. Lyutikov, V.I. Pariev, R.D. Blandford, Polarization of prompt gamma-ray burst emission: evidence for electromagnetically dominated outflow. Astrophys. J. **597**, 998–1009 (2003). doi:10.1086/378497

S. McGlynn, D.J. Clark, A.J. Dean, L. Hanlon, S. McBreen, D.R. Willis, B. McBreen, A.J. Bird, S. Foley, Polarisation studies of the prompt gamma-ray emission from GRB 041219a using the spectrometer aboard INTEGRAL. Astron. Astrophys. **466**, 895–904 (2007). doi:10.1051/0004-6361:20066179

S. McGlynn, S. Foley, B. McBreen, L. Hanlon, S. McBreen, D.J. Clark, A.J. Dean, A. Martin-Carrillo, R. O'Connor, High energy emission and polarisation limits for the INTEGRAL burst GRB 061122. Astron. Astrophys. **499**, 465–472 (2009). doi:10.1051/0004-6361/200810920

P. Mészáros, M.J. Rees, Multi-GEV neutrinos from internal dissipation in gamma-ray burst fireballs. Astrophys. J. Lett. **541**, 5–8 (2000a). doi:10.1086/312894

P. Mészáros, M.J. Rees, Steep slopes and preferred breaks in gamma-ray burst spectra: the role of photospheres and comptonization. Astrophys. J. **530**, 292–298 (2000b). doi:10.1086/308371

P. Mészáros, E. Waxman, TeV neutrinos from successful and choked gamma-ray bursts. Phys. Rev. Lett. **87**(17), 171102 (2001)

B.J. Morsony, D. Lazzati, M.C. Begelman, The origin and propagation of variability in the outflows of long-duration gamma-ray bursts. Astrophys. J. **723**, 267–276 (2010). doi:10.1088/0004-637X/723/1/267

K. Murase, K. Kashiyama, P. Mészáros, Subphotospheric neutrinos from gamma-ray bursts: the role of neutrons. Phys. Rev. Lett. **111**(13), 131102 (2013). doi:10.1103/PhysRevLett.111.131102

B. Paczýnski, Gamma-ray bursters at cosmological distances. Astrophys. J. Lett. **308**, 43–46 (1986). doi:10.1086/184740

A. Pe'er, Temporal evolution of thermal emission from relativistically expanding plasma. Astrophys. J. **682**, 463–473 (2008). doi:10.1086/588136

A. Pe'er, Photospheric emission in gamma-ray bursts, in *41st COSPAR Scientific Assembly*. COSPAR Meeting, vol. 41 (2016)

A. Pe'er, P. Mészáros, M.J. Rees, The observable effects of a photospheric component on GRB and XRF prompt emission spectrum. Astrophys. J. **642**, 995–1003 (2006). doi:10.1086/501424

A. Pe'Er, B.-B. Zhang, F. Ryde, S. McGlynn, B. Zhang, R.D. Preece, C. Kouveliotou, The connection between thermal and non-thermal emission in gamma-ray bursts: general considerations and GRB 090902B as a case study. Mon. Not. R. Astron. Soc. **420**, 468–482 (2012). doi:10.1111/j.1365-2966.2011.20052.x

R. Preece (The Fermi Collaboration), The first pulse of the extremely bright GRB 130427A: a test lab for synchrotron shocks. Science **343**, 51–54 (2014). doi:10.1126/science.1242302

S. Razzaque, P. Mészáros, E. Waxman, Neutrino tomography of gamma ray bursts and massive stellar collapses. Phys. Rev. **D68**, 083001 (2003)

M.J. Rees, P. Mészáros, Unsteady outflow models for cosmological gamma-ray bursts. Astrophys. J. Lett. **430**, 93–96 (1994). doi:10.1086/187446

M.J. Rees, P. Mészáros, Dissipative photosphere models of gamma-ray bursts and X-ray flashes. Astrophys. J. **628**, 847–852 (2005). doi:10.1086/430818

H. Riffert, A self-consistent shock solution for radiation-dominated flows. Astrophys. J. **327**, 760–771 (1988). doi:10.1086/166234

E.M. Rossi, A.M. Beloborodov, M.J. Rees, Neutron-loaded outflows in gamma-ray bursts. Mon. Not. R. Astron. Soc. **369**, 1797–1807 (2006). doi:10.1111/j.1365-2966.2006.10417.x

M. Russo, C. Thompson, Hot electromagnetic outflows. I. acceleration and spectra. Astrophys. J. **767**, 142 (2013). doi:10.1088/0004-637X/767/2/142

G.B. Rybicki, A.P. Lightman, *Radiative Processes in Astrophysics* (1979)

F. Ryde, The cooling behavior of thermal pulses in gamma-ray bursts. Astrophys. J. **614**, 827–846 (2004). doi:10.1086/423782

F. Ryde, A. Pe'Er, T. Nymark, M. Axelsson, E. Moretti, C. Lundman, M. Battelino, E. Bissaldi, J. Chiang, M.S. Jackson, S. Larsson, F. Longo, S. McGlynn, N. Omodei, Observational evidence of dissipative photospheres in gamma-ray bursts. Mon. Not. R. Astron. Soc. **415**, 3693–3705 (2011). doi:10.1111/j.1365-2966.2011.18985.x

L. Sironi, A. Spitkovsky, Particle acceleration in relativistic magnetized collisionless electron-ion shocks. Astrophys. J. **726**, 75 (2011). doi:10.1088/0004-637X/726/2/75

R. Svensson, Nonthermal pair production in compact X-ray sources—first-order Compton cascades in soft radiation fields. Mon. Not. R. Astron. Soc. **227**, 403–451 (1987). doi:10.1093/mnras/227.2.403

C. Thompson, A model of gamma-ray bursts. Mon. Not. R. Astron. Soc. **270**, 480 (1994)

C. Thompson, R. Gill, Hot electromagnetic outflows. III. Displaced fireball in a strong magnetic field. Astrophys. J. **791**, 46 (2014). doi:10.1088/0004-637X/791/1/46

C. Thompson, P. Mészáros, M.J. Rees, Thermalization in relativistic outflows and the correlation between spectral hardness and apparent luminosity in gamma-ray bursts. Astrophys. J. **666**, 1012–1023 (2007). doi:10.1086/518551

N. Vlahakis, A. Königl, Relativistic magnetohydrodynamics with application to gamma-ray burst outflows. I. Theory and semianalytic trans-Alfvénic solutions. Astrophys. J. **596**, 1080–1103 (2003). doi:10.1086/378226

I. Vurm, A.M. Beloborodov, On the prospects of gamma-ray burst detection in the TeV band. ArXiv e-prints (2016a)

I. Vurm, A.M. Beloborodov, Radiative transfer models for gamma-ray bursts. Astrophys. J. **831**, 175 (2016b). doi:10.3847/0004-637X/831/2/175

I. Vurm, A.M. Beloborodov, J. Poutanen, Gamma-ray bursts from magnetized collisionally heated jets. Astrophys. J. **738**, 77 (2011). doi:10.1088/0004-637X/738/1/77

I. Vurm, Y. Lyubarsky, T. Piran, On thermalization in gamma-ray burst jets and the peak energies of photospheric spectra. Astrophys. J. **764**, 143 (2013). doi:10.1088/0004-637X/764/2/143

I. Vurm, R. Hascoët, A.M. Beloborodov, Pair-dominated GeV-optical flash in GRB 130427A. Astrophys. J. Lett. **789**, 37 (2014). doi:10.1088/2041-8205/789/2/L37

E. Waxman, A. Loeb, TeV neutrinos and GeV photons from shock breakout in supernovae. Phys. Rev. Lett. **87**(7), 071101 (2001). doi:10.1103/PhysRevLett.87.071101

D.M. Wei, W.H. Gao, Are there cosmological evolution trends on gamma-ray burst features? Mon. Not. R. Astron. Soc. **345**, 743–746 (2003). doi:10.1046/j.1365-8711.2003.06971.x

D. Yonetoku, T. Murakami, T. Nakamura, R. Yamazaki, A.K. Inoue, K. Ioka, Gamma-ray burst formation rate inferred from the spectral peak energy-peak luminosity relation. Astrophys. J. **609**, 935–951 (2004). doi:10.1086/421285

D. Yonetoku, T. Murakami, S. Gunji, T. Mihara, K. Toma, T. Sakashita, Y. Morihara, T. Takahashi, N. Toukairin, H. Fujimoto, Y. Kodama, S. Kubo (IKAROS Demonstration Team), Detection of gamma-ray polarization in prompt emission of GRB 100826A. Astrophys. J. Lett. **743**, 30 (2011). doi:10.1088/2041-8205/743/2/L30

D. Yonetoku, T. Murakami, S. Gunji, T. Mihara, K. Toma, Y. Morihara, T. Takahashi, Y. Wakashima, H. Yonemochi, T. Sakashita, N. Toukairin, H. Fujimoto, Y. Kodama, Magnetic structures in gamma-ray burst jets probed by gamma-ray polarization. Astrophys. J. Lett. **758**, 1 (2012). doi:10.1088/2041-8205/758/1/L1

H.-F. Yu, H.J. van Eerten, J. Greiner, R. Sari, P. Narayana Bhat, A. von Kienlin, W.S. Paciesas, R.D. Preece, The sharpness of gamma-ray burst prompt emission spectra. Astron. Astrophys. **583**, 129 (2015). doi:10.1051/0004-6361/201527015

Space Sci Rev (2017) 207:111–136
DOI 10.1007/s11214-016-0315-7

Electrodynamics of Pulsar Magnetospheres

Benoît Cerutti[1] · Andrei M. Beloborodov[2]

Received: 1 August 2016 / Accepted: 9 November 2016 / Published online: 6 December 2016

Abstract We review electrodynamics of rotating magnetized neutron stars, from the early vacuum model to recent numerical experiments with plasma-filled magnetospheres. Significant progress became possible due to the development of global particle-in-cell simulations which capture particle acceleration, emission of high-energy photons, and electron-positron pair creation. The numerical experiments show from first principles how and where electric gaps form, and promise to explain the observed pulsar activity from radio waves to gamma-rays.

Keywords Pulsars: general · Acceleration of particles · Radiation mechanisms: non-thermal · Magnetic reconnection · Relativistic processes · Stars: winds, outflows

1 Introduction

The canonical radio pulsar is an isolated, rotating, magnetized neutron star. Its radius is $r_\star = 10$–13 km and its mass is $M_\star \approx 1.4 M_\odot$. The rotation periods of pulsars vary from milliseconds to seconds and their surface magnetic fields—from 10^9 G to 10^{14} G. Pulsars have been a great puzzle since their discovery in 1967 (Hewish et al. 1968). It is established that the canonical pulsars are powered by the kinetic energy of their rotation. Their powerful electromagnetic winds extract the rotational energy of the star and create bright nebulae, e.g. the famous Crab nebula. More surprisingly, pulsars themselves are bright sources of broadband radiation, from radio waves to gamma-rays, whose mechanisms have been debated for almost five decades.

✉ B. Cerutti
benoit.cerutti@univ-grenoble-alpes.fr

A.M. Beloborodov
amb@phys.columbia.edu

1 Univ. Grenoble Alpes, CNRS, IPAG, 38000 Grenoble, France

2 Physics Department and Columbia Astrophysics Laboratory, Columbia University, 538 West 120th Street, New York, NY 10027, USA

The precise measurements of the pulsar rotation period P and its time derivative $\dot{P}$ provide accurate knowledge of the energy reservoir and the total released power. A significant fraction of the spindown power (typically 1–10 %) is converted to high-energy gamma-rays, as clearly demonstrated by the *Fermi*-LAT observations in the 0.1–10 GeV band (Abdo et al. 2010, 2013). Thus pulsars must be efficient particle accelerators. Understanding their mechanism is a difficult task, because it involves a complex interplay between electrodynamics, high-energy radiative processes, and electron-positron pair creation.

The basic picture of pulsar magnetosphere needs to be established before addressing the many puzzles posed by pulsar observations—the extremely high number of $e^{\pm}$ ejected by the Crab pulsar, the drifting radio sub-pulses, the giant radio pulses, intermittency, gamma-ray pulse profiles, the spectrum shape etc. In this introductory review, we summarize the attempts to understand the basic picture of the pulsar magnetosphere over the past five decades. The topic has been discussed in many reviews, e.g. by Arons (1979), Michel (2004) and recently by Arons (2009, 2012), Spitkovsky (2011), Bühler and Blandford (2014), Beskin et al. (2015), Grenier and Harding (2015), Pétri (2016). We focus on the pulsar itself, i.e. the region inside and around the light cylinder where the pulsar wind forms; an excellent theoretical review of pulsar-wind nebulae is found in Kirk et al. (2009).

2 Vacuum Dipole Magnetosphere

In the early model of pulsar magnetosphere, the star is surrounded by vacuum (Pacini 1967, 1968; Ostriker and Gunn 1969). The initial argument behind this assumption was that the strong gravity of the neutron star keeps any free particles at the stellar surface and does not allow them to fill the magnetosphere (Hoyle et al. 1964). This model neglects the lifting of particles by strong induced electric fields and pair creation in the magnetosphere, which will be discussed in the next sections.

The pulsar is described by its magnetic moment $\boldsymbol{\mu}$ and angular velocity $\boldsymbol{\Omega}$, and the light cylinder radius is defined by $R_{\rm LC} = c/\Omega$. Let us first consider the "aligned rotator" $\boldsymbol{\mu} \parallel \boldsymbol{\Omega}$ with the magnetic axis passing through the center of the star. Clearly, this "pulsar" would not pulsate, as the configuration is symmetric about the rotation axis. However, understanding this simplest configuration plays an important role in the development of pulsar theory.

The neutron star is an excellent conductor, and therefore the electric field measured in the corotating frame inside the star nearly vanishes, $\mathbf{E}'_{\rm int} = \mathbf{0}$. This condition determines the internal electric field in the fixed lab frame $E_{\rm int}$,

$$\mathbf{E}'_{\rm int} = \mathbf{E}_{\rm int} + \frac{\mathbf{V} \times \mathbf{B}_{\rm int}}{c} = \mathbf{0}, \quad \mathbf{V} = \boldsymbol{\Omega} \times \mathbf{r}. \tag{1}$$

It implies that the star is polarized by rotation, very much like a Faraday disk (the unipolar induction effect). Electrons re-distribute themselves inside the star to create the electric field $\mathbf{E}_{\rm int}$ that balances the Lorentz force $\mathbf{V} \times \mathbf{B}_{\rm int}/c$, where $\mathbf{V} = c\mathbf{E}_{\rm int} \times \mathbf{B}_{\rm int}/B^2_{\rm int} = \boldsymbol{\Omega} \times \mathbf{r}$ is the drift speed of the electrons. This polarization creates an excess of negative charges at the poles and positive charges at the equator.

The internal electric field, in particular its θ-component $E^{\rm int}_{\theta} = -V_{\phi} B^{\rm int}_r / c$ (we use spherical coordinates r, θ, ϕ) determines the electrostatic potential drop with latitude on the stellar surface $\Phi(r_\star, \theta) - \Phi(r_\star, 0) = -\int E^{\rm int}_{\theta} r_\star d\theta$. Using the surface dipole magnetic field $B_r = 2\mu \cos\theta / r_\star^3$ and $V_\phi = \Omega r_\star \sin\theta$ one finds

$$\Phi(r_\star, \theta) = \Phi_p + \frac{\Omega \mu}{c r_\star} \sin^2\theta, \quad \Phi_p \equiv \Phi(r_\star, 0). \tag{2}$$

It is straightforward to solve Laplace equation $\nabla^2\Phi = 0$ in the vacuum domain outside the star with the boundary condition given in Eq. (2) (using that $\sin^2\theta - 2/3 = -(2/3)P_2(\theta)$ is an eigen function of the θ-part of the Laplacian). The solution is given by (Davis 1947; Deutsch 1955; Hones and Bergeson 1965)

$$\Phi(r,\theta) = \left(\Phi_p + \frac{2\Omega\mu}{3cr_\star}\right)\frac{r_\star}{r} + \frac{\Omega\mu}{cr_\star}\left(\frac{r_\star}{r}\right)^3\left(\sin^2\theta - \frac{2}{3}\right), \quad r \geq r_\star. \tag{3}$$

Φ_p determines the monopole contribution to Φ. It is related to the net charge of the star $Q = -r_\star\Phi_p - 2\Omega\mu/3c$, as seen from the Gauss theorem applied to a sphere of radius $r \gg r_\star$. If the neutron star is formed with zero charge and no particles are lifted from its surface then $Q = 0$ and $\Phi_p = -(2\Omega\mu/3cr_\star)$.

The electric field outside the star is found from $\mathbf{E} = -\nabla\Phi$. Note that while Φ is continuous at the surface $r = r_\star$, its derivative is not. Thus, a continuous dipolar magnetic field implies a discontinuous E_r i.e. there is a surface charge $4\pi\Sigma = E_r - E_r^{\rm int}$. The star also possesses an interior charge, which is found from the Gauss theorem applied to a sphere just inside $r_\star$ using $E_r^{\rm int}(r_\star,\theta) = (V_\phi/c)B_\theta^{\rm int} = \Omega\mu\sin^2\theta/cr_\star^2$, which gives $Q_{\rm int} = 2\Omega\mu/3c$ (Michel and Li 1999). The net charge Q is the sum of the interior and surface charges.

The dipole magnetic field remains a valid vacuum solution for the aligned rotator, as if there were no rotation. In summary, the vacuum electromagnetic field of the aligned dipole rotator with $Q = 0$ is given by[1]

$$(B_r, B_\theta, B_\phi) = B_\star\left(\frac{r_\star}{r}\right)^3(2\cos\theta, \sin\theta, 0) \tag{4}$$

$$(E_r, E_\theta, E_\phi) = \frac{\Omega r_\star}{c}B_\star\left(\frac{r_\star}{r}\right)^4\left(1 - 3\cos^2\theta, -\sin 2\theta, 0\right) \tag{5}$$

where $B_\star = \mu/r_\star^3$ is the magnetic field strength at the stellar equator. The vacuum electric field has a monopole component if $Q \neq 0$; otherwise it is a pure quadrupole (see Fig. 1).

The net Poynting flux from the aligned rotator (integrated over a sphere of radius $r > r_\star$) vanishes, $L_{\rm vac} = 0$. Therefore, the rotating star does not lose energy and hence does not spin down. The vacuum solution can be generalized to the case of an inclined rotator, where the magnetic moment is directed at an angle χ relative to the rotation axis. In this case, the net Poynting flux becomes (Pacini 1968; Ostriker and Gunn 1969)

$$L_{\rm vac} = \frac{2}{3}\frac{\mu^2\Omega^4}{c^3}\sin^2\chi. \tag{6}$$

3 The Electrosphere

Particles are strongly magnetized near the neutron star, so that their velocities perpendicular to the magnetic field lines are limited to the drift velocity $\mathbf{V}_D = \mathbf{E}\times\mathbf{B}/B^2$. However, particles can freely slide along the magnetic field, in particular if there is a parallel electric field component $\mathbf{E}_\parallel = \mathbf{E}\cdot\mathbf{B}/|\mathbf{B}| \neq 0$. From the vacuum solution, one immediately realizes that there is a strong unscreened $E_\parallel$, in contrast to the interior of the star where $\mathbf{E}\cdot\mathbf{B} = 0$.

[1] Generalization of the Deutsch solution to an arbitrary multipolar order is discussed by Bonazzola et al. (2015), Pétri (2015).

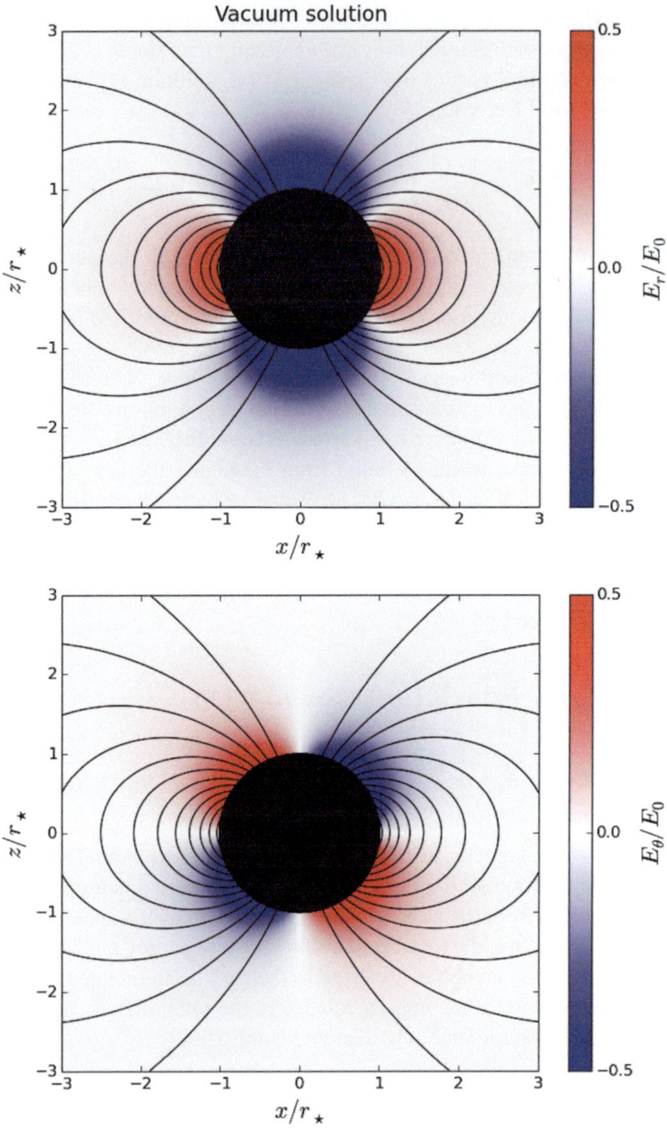

Fig. 1 Vacuum electromagnetic field of an aligned dipole rotator, $\boldsymbol{\Omega} \parallel \boldsymbol{\mu}$ (both along the z-axis in the figure). *Black curves* show the magnetic field lines. *Color* shows the electric field outside the star (*top*: E_r, *bottom*: E_θ) induced by the rotation of the perfectly conducting neutron star with no net charge ($Q = 0$). The electric field is shown in units of $E_0 = \Omega\mu/r_\star^2 c = \Omega r_\star B_\star/c$

Goldreich and Julian (1969) pointed out that the electric force greatly exceeds gravity of the star and will lift electrons in the polar region and ions in the equatorial region. The work function W at the condensed surface of a neutron star crust is negligible for electrons and may be significant for ions (Ruderman 1974). However, detailed calculations of W (Jones 1986; Medin and Lai 2007) show that the typical temperatures of pulsars provide sufficient surface emission of ions; these free ions are picked up by the electric field and lifted to the magnetosphere.

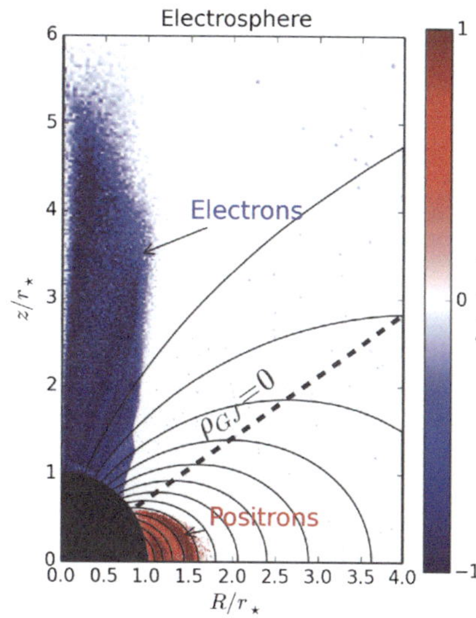

Fig. 2 Fully charge-separated solution (electrosphere) of the aligned rotator obtained with a 2D axisymmetric PIC simulation after one rotation period. Electrons (*blue*) form a dome above the magnetic pole while positive charges (here positrons, in *red*) form the equatorial torus. Both species are confined well within the light-cylinder radius, here set at $R_{LC}=6r_\star$. Charge densities are normalized by the surface Goldreich-Julian density at the pole. *Solid curves* show the magnetic field lines and the *oblique dashed line* shows the null line (where $\rho_{GJ}=0$, i.e., $\theta\approx 55^\circ$)

Thus the entire surface charge of the aligned rotator must be lifted and distributed in the magnetosphere. There exists an axisymmetric electrostatic equilibrium for the lifted charge (Jackson 1979; Krause-Polstorff and Michel 1985a,b): the electrons lifted in the polar region form a dome above the star, and the ions lifted in the equatorial region form a torus. In both dome and torus $\mathbf{E}\cdot\mathbf{B}=0$. There is a vacuum gap between (and around) the dome and torus, i.e. the lifted charge does not fill the entire magnetosphere. The equilibrium solution for the lifted charge implies no outflow, no poloidal electric current, and no toroidal magnetic field B_ϕ. This "electrosphere" produces no net Poynting flux and the pulsar is dead—there is no spindown power.

In the corotating regions (stellar interior, dome, and part of the torus) the condition $\mathbf{E}=-\mathbf{V}\times\mathbf{B}/c$ determines charge density ρ by $\nabla\cdot\mathbf{E}=4\pi\rho$, which is often called "corotation density" or Goldreich-Julian density (Hones and Bergeson 1965; Goldreich and Julian 1969),

$$\rho_{GJ}\approx-\frac{\boldsymbol{\Omega}\cdot\mathbf{B}}{2\pi c}=\frac{B_\star}{2\pi c}\left(\frac{r_\star}{r}\right)^3\left(3\cos^2\theta-1\right). \tag{7}$$

The electrosphere configuration was obtained by several groups using iterative or PIC numerical simulations (Krause-Polstorff and Michel 1985a,b; Shibata 1989; Neukirch 1993; Thielheim and Wolfsteller 1994; Smith et al. 2001; Pétri et al. 2002b; Spitkovsky and Arons 2002; McDonald and Shearer 2009; Wada and Shibata 2011; Philippov and Spitkovsky 2014; Cerutti et al. 2015). In almost all cases, the simulations are initiated with the vacuum solution and zero work function at the stellar surface. Then, as expected, the electric field extracts charges from the star; they fill the magnetosphere and form the charge-separated dome + torus structure. The charge distribution obtained by an axisymmetric PIC simulation (Cerutti et al. 2015) is shown in Fig. 2.

The equilibrium electrosphere model is however incomplete, because it turns out unstable to non-axisymmetric perturbations (Pétri et al. 2002a; Spitkovsky and Arons 2002). The magnetosphere rotates with the drift velocity $\mathbf{E}\times\mathbf{B}/B^2$, and the electrosphere solution implies a strong velocity shear. This causes the diocotron instability of the torus, an analog of the Kelvin-Helmholtz instability in neutral fluids. This instability is captured only by 3D simulations as it grows from non-axisymmetric modes. It induces an expansion of the torus, and can even produce an outflow of charge through the light cylinder. However, this outflow is too weak to give a significant spindown power. Furthermore, Pétri (2007) showed that the diocotron instability is suppressed by relativistic effects that become important near the light cylinder.

In summary, the electrosphere formed by particles lifted from the surface can hardly explain the spindown and the magnetospheric activity of pulsars. The model is, however, useful for old inactive neutron stars.

4 The Force-Free Magnetosphere: The Plasma-Filled Solution

Observations of pulsar wind nebulae indicate that the wind is heavily loaded with $e^{\pm}$ plasma, which must be created in the pulsar magnetosphere (e.g., Kirk et al. 2009). Creation of $e^{\pm}$ pairs is also expected theoretically, due to strong electric fields that must develop in plasma-starved regions—so-called "gaps." For instance, the electrosphere solution has a gap between the dome and the torus, and a seed electron placed in the gap will be accelerated to enormous Lorentz factors. The electron is accelerated by $\mathbf{E}_{\parallel}$ along the curved field lines and emits high-energy photons (curvature radiation) which convert to $e^{\pm}$ pairs in the strong magnetic field (Erber 1966; Harding and Lai 2006). This process ignites an electromagnetic cascade of gamma rays and pairs until the density of the plasma is high enough to screen the accelerating electric field, so that $\mathbf{E}\cdot\mathbf{B}=0$ becomes nearly satisfied.

Although the existence of gaps is required to fill the magnetosphere with pair plasma, as a first approximation it makes sense to study "force-free" magnetospheres with $\mathbf{E}\cdot\mathbf{B}\approx 0$. This model assumes that small deviations from the screening condition are sufficient to fill the magnetosphere with plasma. The model also assumes that the inertial mass density of the plasma is much smaller than $B^2/8\pi c^2$. This limit is called "force-free electrodynamics" (FFE). It satisfies the equation,

$$\rho\mathbf{E}+\frac{\mathbf{J}\times\mathbf{B}}{c}=\mathbf{0}, \tag{8}$$

where ρ and $\mathbf{J}$ are the charge and current densities. The FFE describes the behavior of the electromagnetic field, however does not provide any information about the plasma creation and dynamics, except that it sustains the electric current $\mathbf{J}$ and charge density $\rho=\nabla\cdot\mathbf{E}/4\pi$ demanded by the electromagnetic field.

4.1 Axisymmetric "Pulsar Equation"

Since the ϕ-derivative vanishes for axisymmetric configurations, the condition $\nabla\cdot\mathbf{B}=0$ gives $\nabla\cdot\mathbf{B}_{\mathrm{P}}=0$. This condition implies that there is only one degree of freedom in the poloidal magnetic field $\mathbf{B}_{\mathrm{P}}=(B_r, B_\theta)$. It is convenient to use the so-called "poloidal flux function" $\Psi(r,\theta)$ or $\Psi(R,z)$ in cylindrical coordinates R, z. It is defined so that $2\pi\Psi$ equals the magnetic flux through the circle of radius $R=r\sin\theta$ around the rotation axis at height $z=r\cos\theta$. The flux function $\Psi(R,z)$ is related to $\mathbf{B}_{\mathrm{P}}$ by

$$\mathbf{B}_{\mathrm{P}}=\frac{\nabla\Psi\times\mathbf{e}_\phi}{R}. \tag{9}$$

The poloidal electric current through the same circle, $2\pi I$, is related to the toroidal magnetic field B_ϕ according to the Stokes' theorem,

$$B_\phi=\frac{I}{R}. \tag{10}$$

The two scalar functions—the flux function $\Psi(R, z)$ and the current function $I(R, z)$—completely describe an axisymmetric magnetic field.

In a steady state, it is possible to translate the force-free condition into the so-called "pulsar equation" for Ψ and I (Scharlemann and Wagoner 1973; Michel 1973b). This equation reads,

$$\left(1-\frac{R^2}{R_{\rm LC}^2}\right)\left(\frac{\partial^2\Psi}{\partial R^2}+\frac{\partial^2\Psi}{\partial z^2}\right)-\left(1+\frac{R^2}{R_{\rm LC}^2}\right)\frac{1}{R}\frac{\partial\Psi}{\partial R}+I(\Psi)\frac{\partial I}{\partial\Psi}=0. \tag{11}$$

No analytical solution to the pulsar equation is known for a rotating dipole. However, an exact solution was found for a rotating monopole (Michel 1973b). Some features of this solution are shared by the aligned dipole rotator, in particular in the wind zone beyond the light cylinder. Below we first describe the monopole solution and then discuss the dipole rotator.

4.2 (Split) Monopole: Michel's Solution

Michel's solution is given by (in spherical coordinates)

$$B_{\rm r}=B_\star\left(\frac{r_\star}{r}\right)^2 \tag{12}$$

$$B_\theta=0 \tag{13}$$

$$B_\phi=-B_\star\left(\frac{r_\star}{R_{\rm LC}}\right)\left(\frac{r_\star}{r}\right)\sin\theta \tag{14}$$

$$E_{\rm r}=0 \tag{15}$$

$$E_\theta=B_\phi \tag{16}$$

$$E_\phi=0. \tag{17}$$

The solution is remarkably simple, as the poloidal magnetic field $B_P=B_r$ is unchanged from the normal monopole. Rotation creates the additional toroidal field $B_\phi=(V_\phi/c)B_r$ (where $V_\phi=\Omega r\sin\theta$ is the co-rotation velocity) and $E_\theta=B_\phi$. The corresponding current density is purely radial,

$$\mathbf{J}=\frac{c}{4\pi}(\nabla\times\mathbf{B})=-\frac{\boldsymbol{\Omega}\cdot\mathbf{B}}{2\pi}\,\mathbf{e}_r=c\rho\,\mathbf{e}_r, \tag{18}$$

where $\rho=\nabla\cdot\mathbf{E}/4\pi$ is the charge density. According to this force-free solution, the current can be carried by charge-separated plasma extracted from the star and moving with the speed of light. Since magnetic monopoles are forbidden by $\nabla\cdot\mathbf{B}=0$, it is better to consider a "split monopole," which is antisymmetric about the equatorial plane. Then the above solution describes one hemisphere, and the solution in the opposite hemisphere is obtained by the transformation $\mathbf{B}\to-\mathbf{B}$, $\mathbf{E}\to-\mathbf{E}$, and $\mathbf{J}\to-\mathbf{J}$. The reversal of the magnetic field across the equator implies the presence of an equatorial current sheet, which is not included in Eq. (18). It carries the return current and ensures that the net current through the stellar surface vanishes, so the stellar charge does not grow.

The relation $B_\phi/B_r=r\sin\theta/R_{\rm LC}$ shows that the magnetic field lines are nearly radial close to the star ($r\sin\theta\ll R_{\rm LC}$) and start winding up around the pulsar with increasing distance (see Fig. 3, top panel). At the light-cylinder, in the equatorial plane, the poloidal and

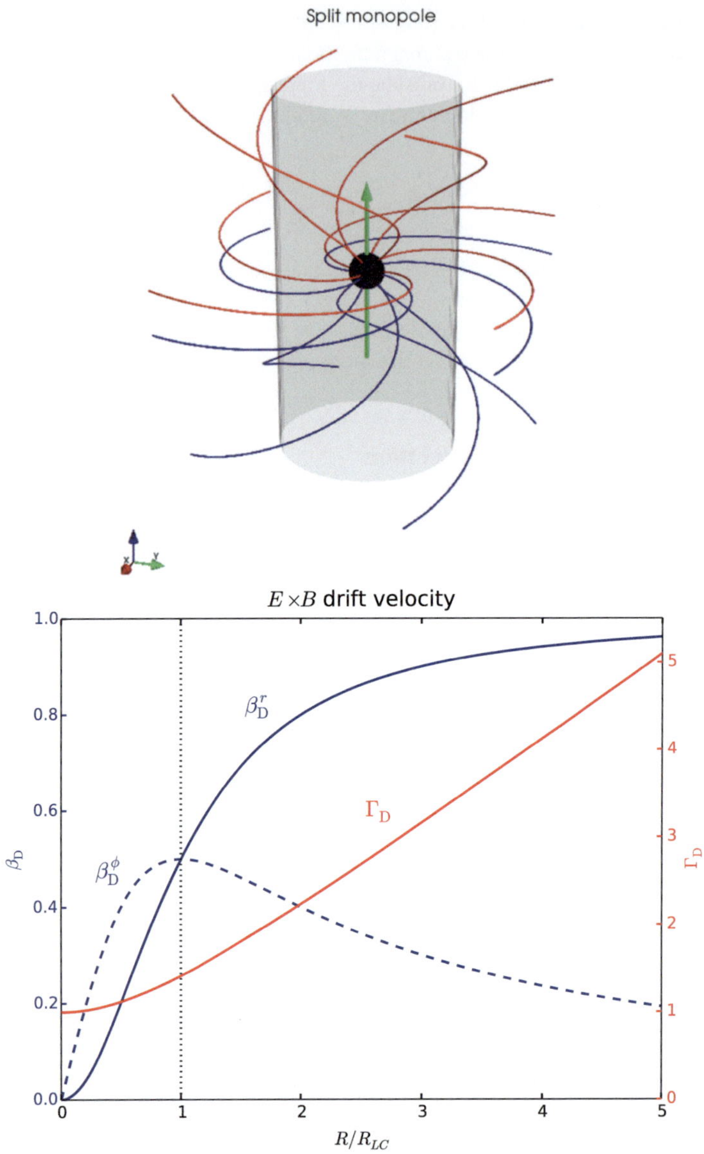

Fig. 3 Analytical solution of the aligned split monopole (Michel 1973b). *Top*: Magnetic field line structure winding up around the star (*black sphere*) due to the stellar rotation. The *green arrow* is the rotation axis of the star. *Red lines* represent outgoing field lines while *blue lines* show the incoming field lines. The field changes its polarity across the equator. The magnetosphere contained within the light-cylinder radius is shown in *grey*. *Bottom*: $\mathbf{E} \times \mathbf{B}$ drift velocity (*solid line*: $\beta_D^r = V_D^r/c$, *dashed line*: $\beta_D^\phi = V_D^\phi/c$) and Lorentz factor, $\Gamma_D = 1/\sqrt{1-\beta_D^2} = \sqrt{1+(R/R_{LC})^2}$, as a function of the cylindrical radius $R = r\sin\theta$

toroidal magnetic components are equal, i.e. the field line makes a 45° angle with respect to the radial direction. Far outside the light cylinder, the magnetic field is almost purely toroidal. The Poynting flux $\mathbf{\Pi} = c\mathbf{E} \times \mathbf{B}/4\pi$ integrated over any sphere of radius $r > r_\star$

determines the spindown power extracted from the star,

$$L_{\rm mono} = \int \boldsymbol{\Pi} \cdot d\mathbf{S} = \frac{2cB_\star^2 r_\star^4}{3R_{\rm LC}^2}. \tag{19}$$

This result turns out also valid for any oblique split monopole, i.e. it does not depend on the inclination angle χ between $\boldsymbol{\Omega}$ and the normal to the antisymmetry plane of $\mathbf{B}$ (Bogovalov 1999).

The drift velocity of the magnetic field lines is given by (Fig. 3, bottom panel)

$$\mathbf{V}_{\rm D} = c\frac{\mathbf{E}\times\mathbf{B}}{\mathbf{B}^2} = \frac{c\,\mathbf{e}_{\rm r}}{1+(\frac{R_{\rm LC}}{R})^2} + \frac{R\Omega\mathbf{e}_\phi}{1+(\frac{R}{R_{\rm LC}})^2}. \tag{20}$$

This velocity does not describe the true motion of the charge-separated plasma in the monopole magnetosphere. The force-free result $\mathbf{J} = c\rho\mathbf{e}_r$ implies that the charge-separated plasma everywhere flows radially with the speed of light, which is inconsistent with $\mathbf{V}_{\rm D}$. The true plasma velocity can only be found beyond the realm of force-free models. The plasma is co-rotating with the star at the stellar surface and starts radial motion into the magnetosphere with a negligible v_r and a very high charge density $\rho \gg J/c$. The Lorentz factor of the radial outflow is controlled by $E_\parallel \neq 0$ which requires a self-consistent calculation (Michel 1974). The outflow Lorentz factor quickly grows with altitude and saturates at $\gamma \sim \sigma^{1/2}\cos\theta$, where $\sigma = e\Omega B_\star r_\star^2/m_e c^3$. While $E_\parallel \neq 0$ is essential for the plasma acceleration, it is much smaller than E_θ, B_r, B_ϕ when $\sigma \gg 1$. In this limit, the electromagnetic field around the star approaches the force-free model with $E_\parallel = 0$.

The drift velocity profile given in Eq. (20) becomes relevant to plasma motion if the plasma has a high multiplicity of charges instead of being charge separated. In particular, the force-free pulsar wind of a high multiplicity (discussed below) approaches a split-monopole structure at $R \gg R_{\rm LC}$. In this case, the plasma flies away with $\mathbf{V} \approx \mathbf{V}_{\rm D}$ until it reaches the fast magnetosonic point. It follows from Eq. (20) that the bulk Lorentz factor of the flow is given by

$$\Gamma_{\rm D} = \sqrt{1+\left(\frac{R}{R_{\rm LC}}\right)^2}, \tag{21}$$

so the pulsar wind accelerates linearly with cylindrical radius beyond the light cylinder (Buckley 1977; Contopoulos and Kazanas 2002).

4.3 Dipole

The dipole magnetic flux emerging from the stellar surface is described by the flux function

$$\Psi(r_\star,\theta) = \frac{\mu\sin^2\theta}{r_\star}. \tag{22}$$

As in the monopole case, rotation of the star $\boldsymbol{\Omega}$ distorts the external dipole field. The force-free magnetosphere of the aligned rotator ($\boldsymbol{\Omega} \parallel \boldsymbol{\mu}$) is described by the poloidal flux function $\Psi(r,\theta)$ that obeys the pulsar equation (Eq. (11)) with the boundary condition Eq. (22). This equation was solved numerically by Contopoulos et al. (1999) (see also Goodwin et al. 2004, Gruzinov 2005, Contopoulos 2005, Timokhin 2006). The solution gives the poloidal magnetic field $\mathbf{B}_P$ (which determines the charge density $\rho_{\rm GJ} = -\boldsymbol{\Omega}\cdot\mathbf{B}_P/2\pi c$) and the self-consistent poloidal current $\mathbf{J}_P$ (which determines B_ϕ).

Fig. 4 Sketch of the ideal force-free magnetosphere of the aligned pulsar. The main elements are: (i) The closed field line region (*grey*, and *black field lines*) lying between the star surface and the light cylinder. This zone is dead and does not participate to the pulsar activity. (ii) The open field line region (*red* and *blue field lines*) extending beyond the light cylinder. The open field-line bundle carries the outflowing electric current, Poynting flux and the relativistic pulsar wind. (iii) The equatorial current sheet (*green*) between the opposite magnetic fluxes in the wind zone. It splits at the light cylinder into two separatrix current sheets that go around the closed zone, between the last open and the first closed field lines

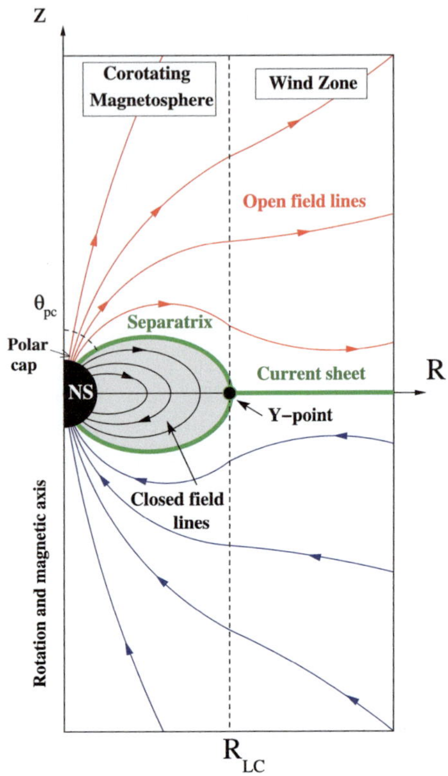

The force-free magnetosphere possesses the following three basic features depicted schematically in Fig. 4: (i) an open zone with $\mathbf{J}_P \neq 0$ and $B_\phi \neq 0$, resembling the monopole solution, especially far outside the light cylinder, (ii) a closed zone with $\mathbf{J}_P = 0$ and $B_\phi = 0$, and (iii) a Y-shaped current sheet. The current sheet supports the jump in B_ϕ between the open and closed zones inside the light cylinder (the separatrix) and the jump $\mathbf{B} \to -\mathbf{B}$ across the equatorial plane in the wind zone. The current sheet is positively charged outside the light cylinder and negatively charged inside, as predicted by Lyubarskii (1990). The presence of the closed zone and the Y-point in the magnetic field topology is an essential difference from the monopole magnetosphere. The closed magnetic field lines are not much different from a normal dipole configuration. However, the closed zone is not vacuum—it is filled with charge density $\rho_{\rm GJ}$, similar to the corotating part of the torus in the electrosphere solution. It co-rotates with the star, creating current $J_\phi = V_\phi \rho_{\rm GJ}$ which increases the effective dipole moment of the star, slightly inflating the poloidal magnetic field lines.

The field lines in the open zone would be closed if there were no rotation; rotation pushes them to infinity. This effect may be thought of as a result of the effective inertial mass of the magnetic field $B^2/8\pi c^2$ (Michel 1973a), as there is no plasma inertia in the force-free magnetosphere. Like the monopole case, the field outside the light cylinder becomes increasingly dominated by the toroidal component, and the field energy flows out with nearly speed of light. The amount of open flux $\Psi_{\rm pc}$ may be estimated approximating the closed zone by the dipole field. Then the last closed field line crosses the stellar surface at the polar angle

θ_{pc} given by

$$\sin^2\theta_{pc} = \frac{r_\star}{R_{LC}}, \qquad \Psi_{pc} = \frac{\mu}{R_{LC}}. \tag{23}$$

The angle θ_{pc} defines the "polar cap"—the footprint of the open field-line bundle on the star, carrying magnetic flux Ψ_{pc}. Its accurate value is slightly larger, because of the poloidal inflation of magnetic field lines caused by $J_\phi > 0$. The open field-line bundle carries the polar-cap current and the Poynting flux into the wind zone. This is the active part of the magnetosphere which spins down the star.

The spindown power may be estimated using the fact that $B_\phi \sim B_P$ and $E_\theta \sim B_P$ near the light cylinder (similar to the monopole magnetosphere). The net radial Poynting flux may be estimated as $L \approx 4\pi R_{LC}^2 E_\theta B_\phi / 4\pi c$, which gives

$$L \approx \frac{\mu^2 \Omega^4}{c^3} \quad \text{(aligned rotator)}. \tag{24}$$

This simple expression turns out quite accurate, within a numerical factor close to unity.

4.4 Time-Dependent Force-Free Simulations

In the standard force-free solution for the aligned rotator, the last closed field line touches the light-cylinder radius, i.e. the Y-point is at $r = R_{LC}$. Timokhin (2006) pointed out that this is not required by the pulsar equation, and in fact there is a continuous sequence of steady solutions with the Y-point located at $r \leq R_{LC}$. This degeneracy is resolved by direct *time-dependent* simulations of the force-free magnetosphere, which show how the steady state is established (Spitkovsky 2006; McKinney 2006; Parfrey et al. 2012).

The force-free electrodynamic equations can be formulated exclusively in terms of $\mathbf{E}$ and $\mathbf{B}$ with no reference to the plasma. Combining Eq. (8) with $\partial(\mathbf{E}\cdot\mathbf{B})/\partial t = 0$, one obtains (Blandford 2002)

$$\mathbf{J} = c\rho\left(\frac{\mathbf{E}\times\mathbf{B}}{\mathbf{B}^2}\right) + \frac{c}{4\pi}(\mathbf{B}\cdot\nabla\times\mathbf{B} - \mathbf{E}\cdot\nabla\times\mathbf{E})\frac{\mathbf{B}}{\mathbf{B}^2}. \tag{25}$$

Using $\rho = \nabla\cdot\mathbf{E}/4\pi$ and substituting Eq. (25) into the Maxwell equation $\partial\mathbf{E}/\partial t = \nabla\times\mathbf{B} - (4\pi/c)\mathbf{J}$ one can express $\partial\mathbf{E}/\partial t$ in terms of $\mathbf{E}$, $\mathbf{B}$, and their spatial derivatives. Together with $\partial\mathbf{B}/\partial t = -c\nabla\times\mathbf{E}$, this gives a closed dynamical system of field equations that can be evolved in time. For example, one can start with a non-rotating star with a vacuum dipole magnetosphere, then spin it up to a desired $\boldsymbol{\Omega}$, and let it relax to the quasi-steady state. For the aligned rotator, these simulations reproduced the steady-state solution with the Y-point located close to the light cylinder.

Three-dimensional time dependent simulations were then performed for the oblique rotator (right panel in Fig. 5, Spitkovsky 2006, Kalapotharakos and Contopoulos 2009, Kalapotharakos et al. 2012, Pétri 2012). These simulations showed how the spindown power depends on the angle χ between $\boldsymbol{\Omega}$ and $\boldsymbol{\mu}$ (Fig. 6). The result is well approximated by a simple formula proposed by Spitkovsky (2006),

$$L \approx \frac{\mu^2\Omega^4}{c^3}\left(1 + \sin^2\chi\right). \tag{26}$$

In contrast to the oblique split monopole, the dipole spindown power depends on χ. The term proportional to $\sin^2\chi$ is similar to the vacuum model, except for a different numerical

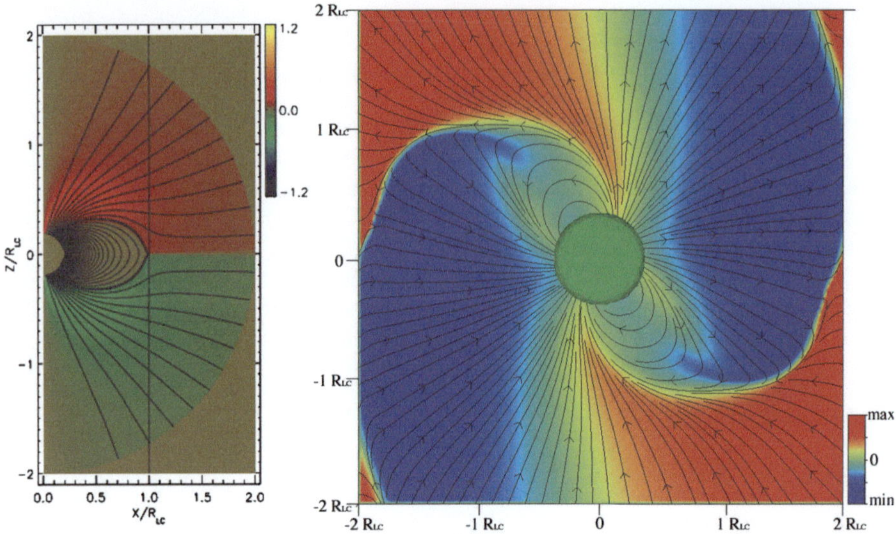

Fig. 5 Snapshots of time-dependent force-free simulations of the aligned (*left*) and oblique (*right*) rotators (from Spitkovsky 2006). The oblique rotator magnetosphere is shown in the $\mathbf{\Omega} - \boldsymbol{\mu}$ plane; the inclination angle is $\chi = 60°$. *Solid lines* represent magnetic field lines, and *color* shows the strength of the magnetic field component perpendicular to the plane of the figure (the toroidal field in the aligned rotator case)

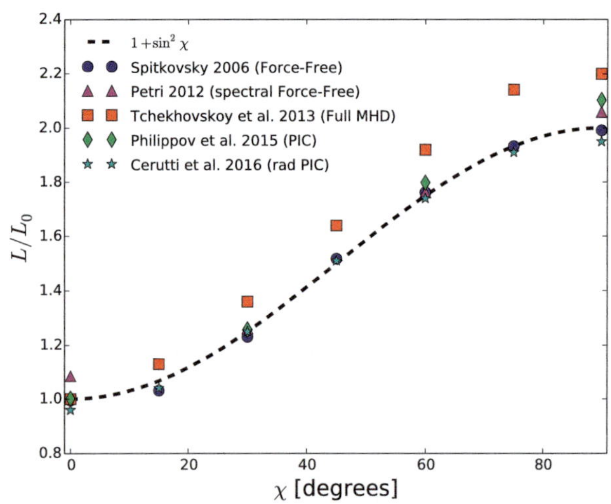

Fig. 6 Compilation of the spindown power for the oblique plasma-filled magnetosphere reported in the literature in different numerical models: force-free (Spitkovsky 2006), spectral force-free (Pétri 2012), full MHD (Tchekhovskoy et al. 2013), PIC (Philippov et al. 2015a), and radiative PIC (Cerutti et al. 2016b). The spindown power is normalized by $L_0 = \mu^2 \Omega^4 / c^3$. The *black dashed line* shows the approximate analytical fit $L \approx L_0(1 + \sin^2 \chi)$

prefactor (1 instead of 2/3). Note also that L does not vanish in the aligned case $\chi = 0$. The origin of the variation $L(\chi)$ was discussed by Tchekhovskoy et al. (2016). Part of it comes from the increasing open magnetic flux $\Psi_{\rm pc}(\chi)$ (explaining 40 % of the increase in L), and the remaining part is caused by the increasing concentration of the open magnetic flux toward the equator. Apart from the moderate dependence $L(\chi)$, the wind from the dipole rotator is similar to that from a split monopole with equal open magnetic flux.

Force-free simulations have also verified the existence of the current sheet in the dipole magnetosphere. Its presence is important as it carries most of the return current, which makes the net current through each polar cap equal to zero. In the wind zone, the current sheet separates the two opposite magnetic fluxes. For the aligned rotator, the current sheet is flat outside the light cylinder and lies in the equatorial plane. For an oblique rotator, the current sheet outside the light cylinder has the shape of a ballerina skirt ("striped wind", a relativistic analog of the heliospheric current sheet). It oscillates in θ around $\pi/2$ with the amplitude χ and a wavelength of $2\pi R_{\rm LC}$.

A key feature of the equatorial current sheet is that it is the place where the force-free approximation certainly breaks down. The magnetic field flips across the sheet, i.e. it must pass through $B=0$, making the plasma unmagnetized (infinite Larmor radius) and violating the force-free condition $B^2/8\pi \gg U_{\rm pl}$ where $U_{\rm pl}$ is the plasma energy density including its rest-mass energy. A complete model of the current sheet should include the plasma inertia and pressure, a finite resistivity and localized dissipation, which promotes magnetic reconnection. In numerical force-free models, the singular behavior of the current sheet outside the light cylinder is regularized by introducing artificial resistivity (Li et al. 2012; Parfrey et al. 2012; Kalapotharakos et al. 2012). Alternatively, the full MHD approach can be employed to model the plasma inertia and pressure in the sheet (Komissarov 2006; Tchekhovskoy et al. 2013).

5 Toward a Realistic Magnetosphere: Pair Creation, Particle Acceleration and Radiation

Observations show that the spindown power is partially dissipated in the magnetosphere, resulting in bright X-ray and gamma-ray emission. Radiation is produced by relativistic particles and a self-consistent model is needed that would demonstrate how and where the particles are created and accelerated. A traditional approach to pulsar modeling was based on guessing the location of electric gaps inside the light cylinder—places where a strong electric field $E_\parallel$ accelerates particles to relativistic energies. Various gaps were proposed: polar-cap gap (Sturrock 1971; Ruderman and Sutherland 1975), slot gap (Arons 1983; Muslimov and Harding 2004), and outer gap (Cheng et al. 1986). The approach to this problem significantly evolved in the last decade, as described below.

5.1 Hints from the Force-Free Model

The force-free model shows one obvious candidate for strong emission—the equatorial current sheet outside the light cylinder, as anticipated in some earlier work (Mestel and Shibata 1994; Lyubarskii 1996). In addition, the force-free model provides guidance to possible locations of the particle accelerator inside the light cylinder, since it shows the current density $\mathbf{J}$ and charge density ρ desired by the magnetosphere. Using the dimensionless parameter $\alpha = J/c\rho$, one can formulate a simple necessary condition for particle acceleration (Beloborodov 2008):

$$\alpha^{-1} \le 1, \qquad \alpha \equiv \frac{J}{c\rho}, \tag{27}$$

which corresponds to $\alpha \ge 1$ or $\alpha < 0$. If the opposite condition $0 \le \alpha < 1$ is satisfied along the magnetic field line, the charge-separated flow extracted from the star is pushed to velocity $\mathbf{v} = \mathbf{J}/\rho$ and provides the desired J and ρ with insignificant acceleration. The condition $0 \le \alpha < 1$ corresponds to $0 \le v < c$, making the velocity $v = J/\rho$ accessible to the flow, which is sufficient for avoiding gaps. In contrast, if $\alpha^{-1} \ge 1$, the charge-separated outflow either fails to conduct the current (and then $E_\parallel$ is generated by the induction effect) or fails to sustain ρ. In this case, the magnetosphere may approach the desired configuration only if the field lines are loaded with $e^\pm$ pairs created in the magnetosphere—otherwise a gap with a strong unscreened $E_\parallel$ must form.

This criterion can be applied to the polar cap of the aligned dipole rotator using α calculated for the force-model by Timokhin (2006), see also Parfrey et al. (2012). On the polar-cap surface, $\alpha(\theta)$ decreases from unity at $\theta = 0$ and becomes negative near the edge of the polar cap $\theta_{\rm pc}$. The return current sheet at the edge also has $\alpha < 0$ (recall that the force-free magnetosphere requires the current sheet inside the light cylinder both to be negatively charged and to conduct $J_r > 0$).

The criterion in Eq. (27) then says that the polar-cap accelerator is suppressed in most of the polar-cap interior—here the extracted charge-separated flow with $v = \alpha c$ screens $E_\parallel$. More exactly, a moderate $E_\parallel$ is sustained to move the plasma from the surface, and the flow velocity v oscillates around αc (Mestel et al. 1985; Beloborodov 2008). The resulting oscillating flow has the maximum dimensionless momentum $p = \gamma\beta$ given by

$$p_{\rm max} = \frac{2\alpha}{1-\alpha^2} \quad (0 < \alpha < 1). \tag{28}$$

This result was confirmed by one-dimensional particle-in-cell (PIC) simulations of the charge-separated flow above the polar cap (Chen and Beloborodov 2013; Timokhin and Arons 2013).

The fact that $\alpha < 0$ near the edge of the polar cap implies strong particle acceleration, in particular in the current sheet at the boundary with the closed magnetosphere. One can show that $e^\pm$ creation in the accelerator with $\alpha^{-1} \leq 1$ must be time-dependent and follow a limit-cycle pattern (Beloborodov 2008). The cyclic discharge is demonstrated by one-dimensional PIC simulations that track $e^\pm$ creation (Timokhin and Arons 2013). This is a rather general property of $e^\pm$ discharge; similar one-dimensional PIC simulations of pair creation in the twisted closed magnetospheres of magnetars (which have $\alpha \gg 1$) also exhibit quasi-periodic behavior (Beloborodov and Thompson 2007).

5.2 Global Numerical Experiments: Particle-in-Cell (PIC) Simulations

The full problem of gap formation and particle acceleration is a global problem, because particles and pair-creating photons can flow through the magnetosphere. For instance, above the polar cap, the condition $\alpha^{-1} \leq 1$ is necessary but may not be sufficient for gap formation, as $e^\pm$ pairs may be supplied to the polar cap (and screen $E_\parallel$) from an accelerator located elsewhere in the magnetosphere. Simulations of the plasma-filled magnetosphere developed in recent years (Philippov and Spitkovsky 2014; Chen and Beloborodov 2014; Philippov et al. 2015a,b; Cerutti et al. 2015, 2016b; Belyaev 2015a,b) provided an opportunity to study this problem from first principles.

In the full global problem, it is not known in advance where the $e^\pm$ discharge will occur, and the problem requires a self-consistent calculation of the electromagnetic field, plasma dynamics and $e^\pm$ creation in a region that extends from the stellar surface to $r \gg R_{\rm LC}$. Self-consistent simulations of $e^\pm$ discharge showing how the magnetosphere is filled with plasma were first performed for an aligned rotator (Chen and Beloborodov 2014) and then for an inclined rotator (Philippov et al. 2015a). They demonstrated that the main particle accelerator in pulsars is the Y-shaped current sheet.

The basic technique for such simulations is the particle-in-cell (PIC) method (Birdsall and Langdon 1991). The technique was originally developed in the 1960s to study kinetic problems in plasma physics (e.g. Dawson 1962). It consists of describing a collisionless plasma from its fundamental constituents—individual charged particles ("macroparticles") and an electromagnetic field. Each macroparticle represents a large number of real particles which would be following the exact same path in phase space. From the positions and velocities of each macroparticle, the charge and current densities are deposited on the numerical grid. Then, the full time-dependent Maxwell's equations are integrated on a staggered mesh to ensure second order accuracy in both space and time (Yee 1966). All these steps are done within each time step, so that the motion of particles and the evolution of the electromagnetic fields are captured self-consistently.

In contrast to other applications (shocks, reconnection), PIC simulations of pulsars must be global, as the problem is intrinsically non-local—what is happening at the light-cylinder affects the activity at the polar cap and vice-versa. Such simulations are challenging because the separation between the macroscopic ($r_\star$, $R_{\rm LC}$) and microscopic (plasma skindepth) scales is huge in real pulsars. In practice, simulations are limited to a few orders of magnitude in scale separation, which is sufficient to capture basic physics of pulsars, such as the mechanism of gap formation and production of gamma-rays.

Apart from the numerical PIC aspects, a major challenge in pulsar simulations is the self-consistent implementation of $e^\pm$ creation. It was first implemented in an axisymmetric PIC simulation by Chen and Beloborodov (2014). Since it was not known in advance where pairs will be created, the simulation started with a "clean" initial condition—a non-rotating vacuum dipole, which was gradually spun up to a desired Ω. It was observed how the rotation-induced electric field extracted particles from the stellar surface and accelerated them to high Lorentz factors leading to curvature emission, and how the curvature photons converted to $e^\pm$ pairs. This numerical experiment demonstrated the mechanism of filling the magnetosphere with plasma, with the following results:

- There are two types of pulsar magnetospheres, depending on the physical process available to create $e^\pm$ pairs. If $e^\pm$ are created only in a strong magnetic field (due to gamma-ray absorption by the field, Erber 1966) then pair production is limited to $r \ll R_{\rm LC}$. Chen and Beloborodov (2014) called it "Type II" rotator. In this case, the aligned rotator was found to evolve to a nearly "dead" state, resembling the electrosphere described in Sect. 3. It is qualitatively different from Type I rotators where $e^\pm$ pairs can be created at $r \sim R_{\rm LC}$ through photon-photon collisions.
- Type I rotator *self-organizes to produce copious pairs inside and around the Y-shaped current sheet.* It evolves into a quasi-steady magnetic configuration close to the force-free solution (Fig. 7).
- The current sheet inside the light cylinder (the separatrix) develops a time-dependent gap stretched along the closed zone boundary: *the separatrix gap.* This accelerator enables pair creation required to sustain the current sheet and the open magnetic flux. The gap is qualitatively different from the "slot gap" and "outer gap" proposed in earlier works. The earlier models assumed that the charge density "desired" by the magnetosphere is $\rho_{\rm GJ}$. In contrast, the charge density in the current sheet greatly exceeds $\rho_{\rm GJ}$ (in the ideal force-free model it would be infinite). The accelerating voltage develops because the large ρ and J (with $\alpha < 0$) cannot be sustained without a high rate of pair creation.

The global PIC simulations also show how the plasma is accelerated and ejected through the Y-point into the equatorial current sheet (Fig. 7). Plasma outflows along the equatorial plane outside $R_{\rm LC}$ and the Y-point resembles a nozzle formed by the open magnetic fluxes of opposite polarity. Two plasma streams come to the Y-point along the boundary of the closed zone and exchange their opposite θ-momenta. Their collimation outside the light

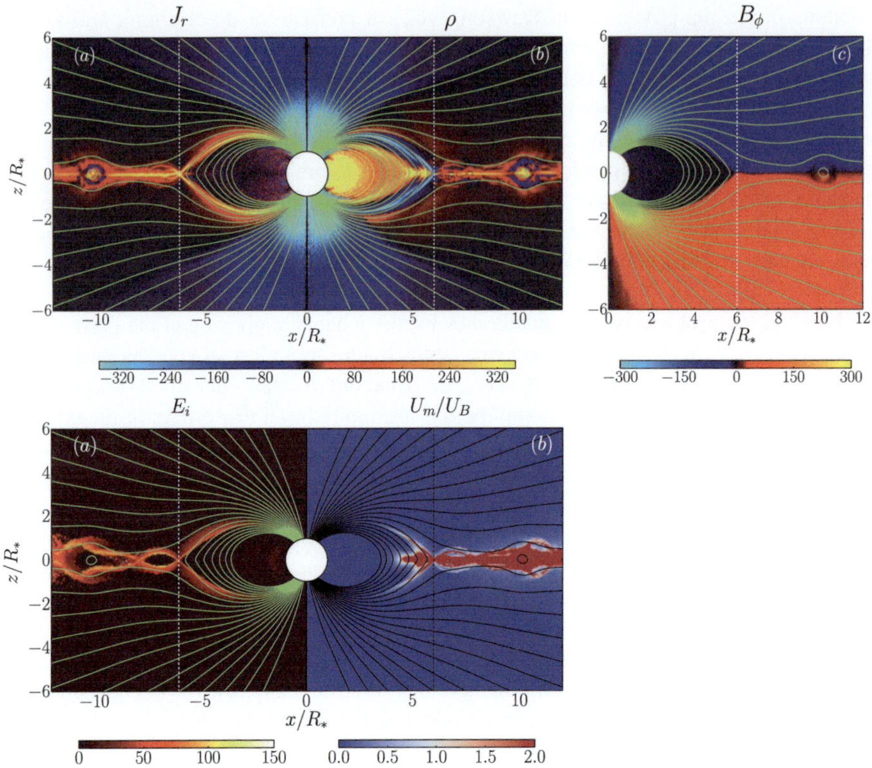

Fig. 7 Poloidal cross section of the magnetosphere of the aligned rotator (from Chen and Beloborodov 2014). *Vertical dashed line* shows the light cylinder $R_{\rm LC} = c/\Omega$. *Green curves* show the poloidal magnetic field lines. *Top*: (**a**) Radial component of the electric current density J_r. (**b**) Net charge density ρ. (**c**) Toroidal component of the magnetic field B_ϕ. Units: distance is measured in $r_\star$, charge density in $m_e c^2/4\pi e r_\star^2$, current in $m_e c^3/4\pi e r_\star^2$, and field in $m_e c^2/e r_\star$. *Bottom*: (**a**) Average ion energy in units of $m_e c^2$ (the ion rest mass was re-scaled to $5m_e$ in the simulation). One can see the acceleration of ions in the gap, their ejection through the Y-point, and gyration in the equatorial current sheet. (**b**) Ratio of matter energy density U_m to magnetic energy density $U_B = B^2/8\pi$

cylinder is achieved through gyration in the (predominantly toroidal) magnetic field, which communicates the θ-momentum from one stream to the other.

The 3D simulations of inclined rotators (Philippov et al. 2015a) show a similar mechanism of sustaining the wobbling current sheet predicted by the 3D force-free simulations of Spitkovsky (2006). A snapshot of the inclined rotator with $\chi = 60°$ is shown in Fig. 8.

The pulsar spindown reported by PIC simulations is in agreement with the ideal MHD expectations in the limit where abundant plasma is present in the magnetosphere (Fig. 6). However, all PIC studies also show significant dissipation of the spindown power $L_\star$ inside and around the light cylinder. Figure 9 (top panel) shows the fraction of $L_\star$ that is dissipated within a sphere of radius $r = 2R_{\rm LC}$ found in the literature, as a function of the pulsar inclination. The aligned rotator has the maximum dissipated fraction of about 20 %. The dissipation rate then decreases monotonically with increasing inclination to about 2.5 % for $\chi = 90°$ (Philippov et al. 2015a). The dissipated Poynting flux is not lost numerically, but instead it is self-consistently channelled into particle kinetic energy and non-thermal radiation.

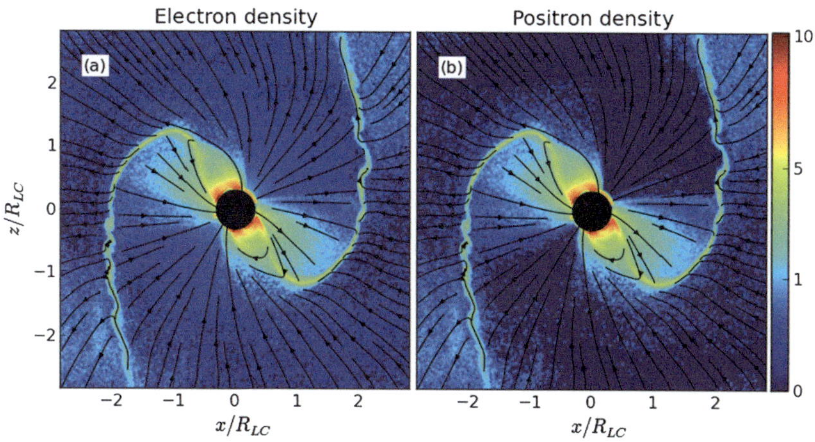

Fig. 8 Electron (*left*) and positron (*right*) densities in a 3D PIC simulation of the oblique rotator with $\chi = 60^\circ$ (from Philippov et al. 2015a). The magnetosphere is viewed in the $\mathbf{\Omega} - \boldsymbol{\mu}$ plane; *solid curves* show the magnetic field lines. Particle acceleration and $e^{\pm}$ creation occurs in part of the polar cap as well as in the current sheet

The scatter in $L/L_\star$ at $\chi = 0^\circ$ in Fig. 9 is most likely due to different prescriptions for particle injection. Cerutti et al. (2015) did not implement $e^{\pm}$ creation in the magnetosphere and instead injected a mildly relativistic flow of $e^{\pm}$ from the stellar surface. This prescription tends to screen $E_\parallel$ and somewhat reduces dissipation. Belyaev (2015a) implemented $e^{\pm}$ creation proportionally to the local $E_\parallel$. Philippov et al. (2015a) assumed instantaneous local pair creation by any particle reaching a fixed threshold Lorentz factor $\gamma_{\rm thr}$. Chen and Beloborodov (2014) performed the most detailed simulation that followed the propagation of curvature gamma-rays and their conversion to pairs. All the simulations, however, show a qualitatively similar structure of the magnetosphere, in particular the formation of the Y-shaped current sheet with significant dissipation.

The dissipated fraction must increase as the pulsar slows down and approaches its "deathline," when the pair creation rate declines so that it becomes increasingly difficult to screen $E_\parallel$. When Cerutti et al. (2015) reduced plasma supply in their simulations, they observed that the magnetosphere becomes highly charge-separated with a low spindown power ($L_\star < L_0$, some field lines remain closed outside the light-cylinder) and the dissipation fraction can reach $> 50\ \%$. In this dissipative solution, the separatrix current layers and the Y-point disappear, and the equatorial current sheet is electrostatically supported (see also Contopoulos et al. 2014 solution).

5.3 General Relativistic Effects

In agreement with the criterion in Eq. (27), Chen and Beloborodov (2014) did not observe a polar-cap accelerator in the standard aligned-rotator problem—the negative electric current flowing from the polar cap was carried by the mildly relativistic electron flow extracted from the star. Philippov et al. (2015b) observed similar behavior of moderately inclined rotators, with inclinations $\chi < 40^\circ$. For larger inclinations, the polar cap became "activated," in agreement with the analysis of the α parameter in force-free models: strongly oblique rotators have large variations of α across their polar caps (e.g., see Fig. 1 in Timokhin and Arons 2013).

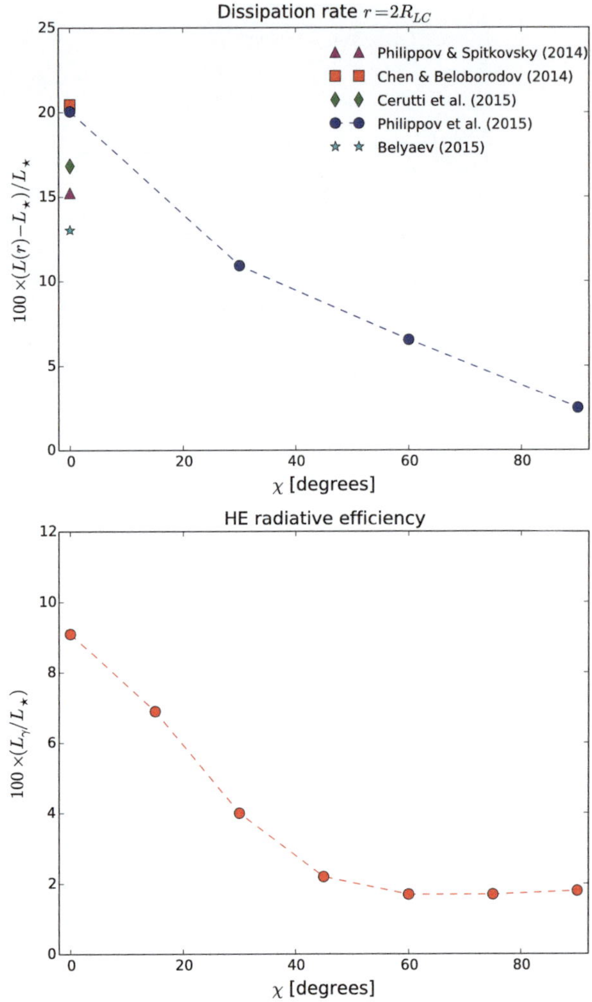

Fig. 9 *Top*: Dissipated fraction of the total radial Poynting flux $L_\star$ (converted to non-thermal particles and radiation) within a sphere of radius $r = 2R_{\mathrm{LC}}$ as a function of the pulsar inclination, as reported by global PIC studies (Philippov and Spitkovsky 2014; Chen and Beloborodov 2014; Cerutti et al. 2015; Philippov et al. 2015a; Belyaev 2015a). The values from Cerutti et al. (2015) and Belyaev (2015a) are for their highest particle injection solutions (closest to force-free); lower particle injection leads to more dissipation (20–50 %). *Bottom*: High-energy radiation efficiency reported by Cerutti et al. (2016b) as a function of the pulsar inclination. The radiative efficiency is defined as the curvature and synchrotron radiative power summed over the numerical box and over all directions, L_γ, divided by the total radial Poynting flux, $L_\star$

The lack of polar-cap activity at moderate inclinations may be problematic for explaining observed radio emission, in particular its "core" component that is usually associated with $e^\pm$ production inside the open field-line bundle. This motivated recent work on simulating general relativistic effects near pulsars, which change the parameter α by tens of percent (Philippov et al. 2015b; Gralla et al. 2016; Belyaev and Parfrey 2016). The most important effect in this context is the frame-dragging of space-time by the stellar rotation (Beskin 1990; Muslimov and Tsygan 1992; Sakai and Shibata 2003). The effective angular velocity of the magnetosphere is reduced by the Lense-Thirring angular velocity, $\omega_{\mathrm{LT}} = (2/5)\Omega(r_{\mathrm{g}}/r_\star)(r_\star/r)^3$, where $r_{\mathrm{g}} = 2GM_\star/c^2$ is the Schwarzschild radius. This leads to an increase of the α parameter,

$$\alpha_{\mathrm{GR}} \approx \frac{\alpha}{1-\omega_{\mathrm{LT}}/\Omega}, \tag{29}$$

where α on the right-hand side corresponds to the standard dipole rotator neglecting general relativistic effects. The effect is strongest at the stellar surface, where one finds $\alpha_{\mathrm{GR}} \approx 1.2\alpha$.

Philippov et al. (2015b) included the Lense-Thirring effect in the field solver (in a leading order) in a 2D axisymmetric PIC simulation, and observed the ignition of $e^{\pm}$ discharge above the polar cap for a sufficiently high compactness of the neutron star $M/r_{\star}$. They also observed that the discharge operates in a quasi-periodic manner, in agreement with Beloborodov (2008) and Timokhin and Arons (2013), and may generate radio waves.

5.4 Radiation from the Equatorial Current Sheet

The equatorial current sheet outside the light cylinder is a prominent dissipation site, especially if the magnetosphere itself is close to force-free (the pulsar is far from the deathline). The strong energy dissipation and particle acceleration in the equatorial sheet accompanies magnetic reconnection of the opposite open magnetic fluxes, which produces and ejects relativistic plasmoids (Philippov and Spitkovsky 2014; Chen and Beloborodov 2014; Cerutti et al. 2015), as predicted by Coroniti (1990) and Michel (1994). This dissipation occurs in the plasma outflow outside the light cylinder and may be studied in detail by PIC simulations even without an explicit implementation of pair creation. Instead, one can form a plasma-filled magnetosphere e.g. by sprinkling $e^{\pm}$ everywhere (Philippov and Spitkovsky 2014) or by ejecting a flow of copious $e^{\pm}$ pairs from the stellar surface (Cerutti et al. 2015, 2016b). This leads to the magnetic configuration close to the force-free solution with the Y-shaped current sheet, and to the strong dissipation in the equatorial plane outside the light cylinder.

An important parameter is the plasma magnetization defined as

$$\sigma = \frac{B^2}{4\pi \Gamma n m_e c^2}, \tag{30}$$

where n is the number density of the $e^{\pm}$ plasma and Γ is its average bulk Lorentz factor. In pulsars, $\sigma \gg 1$ everywhere in the magnetosphere except the interior of the equatorial current sheet, and dissipation of a non-negligible fraction of the spindown power can significantly increase the energy per particle. Simulations show that most of the equatorial dissipation occurs near the light cylinder, around the Y-point up to a few R_{LC}. The high-energy particles tend to pile up at an energy given by the magnetization of the plasma at the light cylinder, i.e. $\gamma \approx \sigma_{\mathrm{LC}}$ (Philippov and Spitkovsky 2014; Cerutti et al. 2015). This energy should be compared with the maximum energy that a particle can get from the vacuum potential drop across the polar cap $\phi_{\mathrm{pc}} = e\Phi_{\mathrm{pc}}/m_e c^2$, where (using Eqs. (1), (23))

$$\Phi_{\mathrm{pc}} = \int_0^{\theta_{\mathrm{pc}}} E_\theta(R_\star) R_\star \, d\theta = \frac{\mu \Omega^2}{c^2}, \tag{31}$$

so that (Cerutti et al. 2015)

$$\gamma \approx \sigma_{\mathrm{LC}} \sim \frac{\phi_{\mathrm{pc}}}{\kappa \Gamma}, \tag{32}$$

where κ is the multiplicity of $e^{\pm}$ pairs ejected in the current sheet, defined relative to the minimum particle number required to sustain the electric current. Hence, the average particle ejected by a high-multiplicity pulsar earns only a small fraction of the vacuum potential drop. Cerutti et al. (2015) tracked the motion of accelerated particles in their simulations of the aligned rotator and studied where the particles gain their energies. They found that positive charges are accelerated outward as they cross the Y-point region and get ejected along the equator. In contrast, electrons are mainly accelerated as they precipitate back towards the star along the current sheet (see Fig. 10).

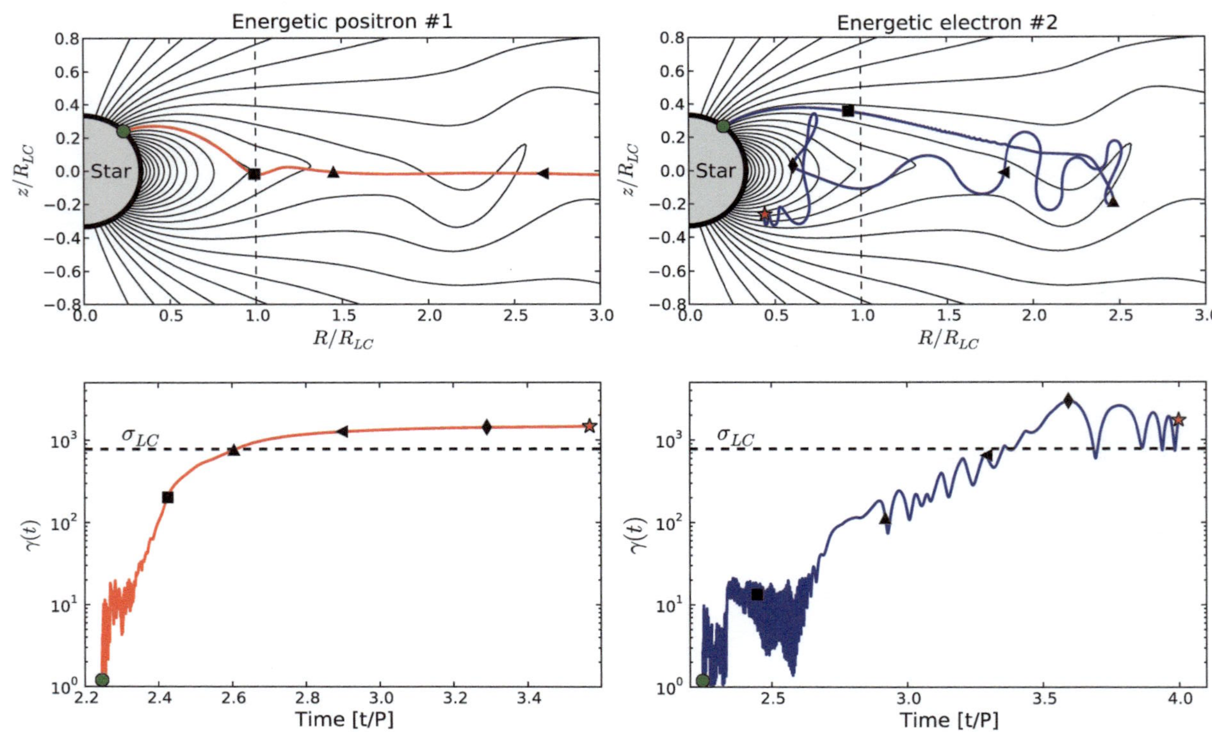

Fig. 10 *Top*: Sample trajectories of a high-energy positron (*left*) and electron (*right*) accelerated in the current sheet of an aligned rotator, shown projected on the poloidal plane. *Bottom*: The particle Lorentz factor as a function of time. *Black symbols* on the curves indicate the correspondence between the time (*bottom panels*) and position on the trajectory (*top panels*). Figure adapted from Cerutti et al. (2015)

Efficient particle acceleration leads to strong non-thermal high-energy emission. The radiation-reaction force due to the curvature and synchrotron emission can be of the same order as the Lorentz force, and particle momentum perpendicular to the field lines may be quickly radiated away. These effects are self-consistently included in the PIC simulations. In classical electrodynamics, the radiation reaction force is given by the Landau-Lifshitz formula (Tamburini et al. 2010; Cerutti et al. 2016b)

$$\mathbf{g} = \frac{2}{3} r_{\mathrm{e}}^2 \big[(\mathbf{E} + \boldsymbol{\beta} \times \mathbf{B}) \times \mathbf{B} + (\boldsymbol{\beta} \cdot \mathbf{E}) \mathbf{E} \big] - \frac{2}{3} r_{\mathrm{e}}^2 \gamma^2 \big[(\mathbf{E} + \boldsymbol{\beta} \times \mathbf{B})^2 - (\boldsymbol{\beta} \cdot \mathbf{E})^2 \big] \boldsymbol{\beta}, \tag{33}$$

where r_{e} is the classical radius of the electron. The 3D radiative PIC simulations of Cerutti et al. (2016b) showed that the strong curvature and synchrotron cooling had little impact on the overall structure of the oblique rotator and its spindown rate (Fig. 6). Radiative cooling also has little impact on the maximum energy reached by the particles accelerated in the current sheet, because the particles are accelerated while being focused deep inside the layer where the perpendicular magnetic field virtually vanishes (Kirk 2004; Uzdensky et al. 2011; Cerutti et al. 2013). However, when the particles leave or re-enter the layer they radiate efficiently. The total isotropic power radiated away in the form of synchrotron radiation reaches almost 10 % of the spindown for the aligned pulsar and decreases down to a few percent at high inclinations (bottom panel in Fig. 9), because the dissipation rate decreases with inclination. Such radiative efficiencies are consistent with observed high-energy gamma-ray efficiencies reported by the *Fermi*-LAT (Abdo et al. 2010, 2013).

The expected radiative signatures—pulse profiles and spectra—can then be calculated from the instantaneous synchrotron and curvature radiation spectrum emitted by each simulation particle,

$$F_\nu(\nu) = \frac{\sqrt{3} e^3 \tilde{B}_\perp}{m_{\mathrm{e}} c^2} \left(\frac{\nu}{\nu_{\mathrm{c}}} \right) \int_{\nu/\nu_{\mathrm{c}}}^{+\infty} K_{5/3}(x)\, dx \tag{34}$$

$$\nu_{\mathrm{c}} = \frac{3 e \tilde{B}_\perp \gamma^2}{4\pi m_{\mathrm{e}} c} \tag{35}$$

$$\tilde{B}_\perp = \sqrt{(\mathbf{E} + \boldsymbol{\beta} \times \mathbf{B})^2 - (\boldsymbol{\beta} \cdot \mathbf{E})^2}. \tag{36}$$

This is the usual formula (e.g. Blumenthal and Gould 1970) except that the effective perpendicular magnetic field, $\tilde{B}_\perp$ here includes the electric field, because it is of the same order as B in pulsar magnetospheres. Photons are then collected on a spherical screen located at infinity and they are folded over the pulsar phase to reconstruct the pulse profiles.

The simulations show that the strongest high-energy emission is produced in the equatorial current sheet in the form of synchrotron photons (Cerutti et al. 2016b), as envisioned by Lyubarskii (1996). Additional high-energy radiation is produced in the separatrix current sheets inside the light cylinder (Chen and Beloborodov 2014). The simulated high-energy pulse profiles typically have two peaks, in agreement with gamma-ray observations (Abdo et al. 2010, 2013). Each peak appears when the equatorial current sheet crosses the observer's line of sight (Fig. 11, Cerutti et al. 2016b). The form of the pulse profile (i.e., phase-separation, amplitude and width of pulses) is entirely determined by the geometrical shape of the current sheet.

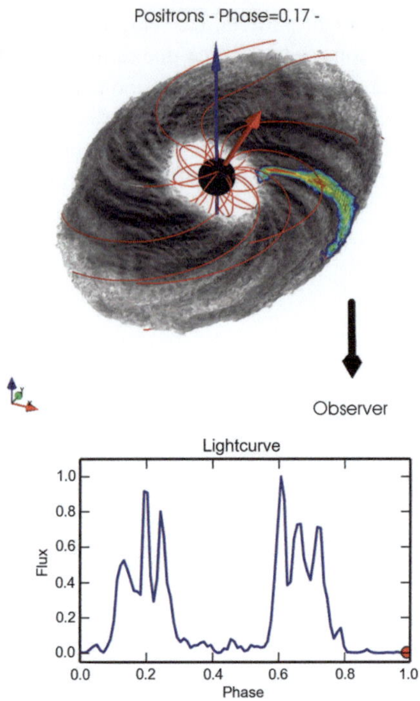

Fig. 11 *Top*: Spatial distribution of the high-energy synchrotron radiation from an oblique rotator obtained with a 3D PIC simulation. The *grey scale* shows the isotropically integrated flux, while the *color scale* shows the emitting regions at the pulsar phase 0.17 as seen by an observer looking along the equator. The angle between the rotation axis (*blue arrow*) and the magnetic axis (*red arrow*) is $\chi = 30°$. *Red curves* are the magnetic field lines. *Bottom*: Reconstructed high-energy pulse profile of radiation received by the observer. Figure adapted from Cerutti et al. (2016b)

These studies prelude the beginning of direct comparison between PIC simulations and observations. As a proof of principle, Cerutti et al. (2016a) showed that polarization modelling along with pulse profile fitting can provide independent constraints on the pulsar inclination and viewing angles. Applied to the Crab pulsar, they find an obliquity angle $\chi = 60°$ and viewing angle $\alpha = 130°$ which are in good agreement with values inferred from the morphology of the pulsar wind nebula in X-rays (Ng and Romani 2004; Weisskopf et al. 2012). A more detailed and systematic analysis extended to other pulsars should be performed in the future.

6 Summary

The theory of pulsar electrodynamics is complex. Although the physics problem is well posed, it entangles several processes such as particle acceleration, non-thermal emission, pair creation, and relativistic reconnection, which make the problem hard. The force-free electrodynamics and MHD models demonstrate some of the essential elements of the pulsar magnetosphere—the spindown power, the magnetic topology, and the pattern of electric currents—and highlight the presence of a prominent Y-shaped current sheet. The recent PIC numerical experiments, which attempt to solve the full problem from first principles, have been a game changer in pulsar theory. They show how the magnetosphere is filled with plasma and how a fraction of the spindown power is dissipated and deposited into high-energy particles that produce observed radiation. For the first time, numerical simulations became capable of predicting observables from first principles, opening new perspectives for our understanding of pulsars.

Acknowledgements The authors thank the International Space Science Institute in Bern for the invitation and hospitality during the Workshop on Jets and Winds in Pulsar Wind Nebulae and Gamma-ray Bursts in November 2015. BC acknowledges support from CNES and Labex OSUG@2020 (ANR10 LABX56). AMB pulsar research is supported by NASA grants NNX13AI34G and NNX15AU71G.

References

A.A. Abdo, M. Ackermann, M. Ajello, W.B. Atwood, M. Axelsson, L. Baldini, J. Ballet, G. Barbiellini, M.G. Baring, D. Bastieri, et al., The first Fermi Large Area Telescope catalog of gamma-ray pulsars. Astrophys. J. Suppl. Ser. **187**, 460–494 (2010). doi:10.1088/0067-0049/187/2/460

A.A. Abdo, M. Ajello, A. Allafort, L. Baldini, J. Ballet, G. Barbiellini, M.G. Baring, D. Bastieri, A. Belfiore, R. Bellazzini, et al., The second Fermi Large Area Telescope catalog of gamma-ray pulsars. Astrophys. J. Suppl. Ser. **208**, 17 (2013). doi:10.1088/0067-0049/208/2/17

J. Arons, Some problems of pulsar physics. Space Sci. Rev. **24**, 437–510 (1979). doi:10.1007/BF00172212

J. Arons, Pair creation above pulsar polar caps—geometrical structure and energetics of slot gaps. Astrophys. J. **266**, 215–241 (1983). doi:10.1086/160771

J. Arons, Pulsar emission: where to go, in *Astrophysics and Space Science Library*, ed. by W. Becker. Astrophysics and Space Science Library, vol. 357, 2009, p. 373. doi:10.1007/978-3-540-76965-1_15

J. Arons, Pulsar wind nebulae as cosmic pevatrons: a current sheet's tale. Space Sci. Rev. **173**, 341–367 (2012). doi:10.1007/s11214-012-9885-1

A.M. Beloborodov, Polar-cap accelerator and radio emission from pulsars. Astrophys. J. Lett. **683**, 41 (2008). doi:10.1086/590079

A.M. Beloborodov, C. Thompson, Corona of magnetars. Astrophys. J. **657**, 967–993 (2007). doi:10.1086/508917

M.A. Belyaev, Dissipation, energy transfer, and spin-down luminosity in 2.5D PIC simulations of the pulsar magnetosphere. Mon. Not. R. Astron. Soc. **449**, 2759–2767 (2015a). doi:10.1093/mnras/stv468

M.A. Belyaev, PICsar: a 2.5D axisymmetric, relativistic, electromagnetic, particle in cell code with a radiation absorbing boundary. New Astron. **36**, 37–49 (2015b). doi:10.1016/j.newast.2014.09.006

M.A. Belyaev, K. Parfrey, Spatial distribution of pair production over the pulsar polar cap. ArXiv e-prints (2016)

V.S. Beskin, General relativity effects on electrodynamic processes in radio pulsars. Sov. Astron. Lett. **16**, 286 (1990)

V.S. Beskin, S.V. Chernov, C.R. Gwinn, A.A. Tchekhovskoy, Radio pulsars. Space Sci. Rev. **191**, 207–237 (2015). doi:10.1007/s11214-015-0173-8

C.K. Birdsall, A.B. Langdon, *Plasma Physics via Computer Simulation* (1991)

R.D. Blandford, To the lighthouse, in *Lighthouses of the Universe: The Most Luminous Celestial Objects and Their Use for Cosmology*, ed. by M. Gilfanov, R. Sunyeav, E. Churazov (2002), p. 381. doi:10.1007/10856495_59

G.R. Blumenthal, R.J. Gould, Bremsstrahlung, synchrotron radiation, and Compton scattering of high-energy electrons traversing dilute gases. Rev. Mod. Phys. **42**, 237–271 (1970). doi:10.1103/RevModPhys.42.237

S.V. Bogovalov, On the physics of cold MHD winds from oblique rotators. Astron. Astrophys. **349**, 1017–1026 (1999)

S. Bonazzola, F. Mottez, J. Heyvaerts, General solution for the vacuum electromagnetic field in the surroundings of a rotating star. Astron. Astrophys. **573**, 51 (2015). doi:10.1051/0004-6361/201424659

R. Buckley, Pulsar magnetospheres with arbitrary geometry in the force-free approximation. Mon. Not. R. Astron. Soc. **180**, 125–140 (1977). doi:10.1093/mnras/180.2.125

R. Bühler, R. Blandford, The surprising Crab pulsar and its nebula: a review. Rep. Prog. Phys. **77**(6), 066901 (2014). doi:10.1088/0034-4885/77/6/066901

B. Cerutti, J. Mortier, A.A. Philippov, Polarized synchrotron emission from the equatorial current sheet in gamma-ray pulsars. Mon. Not. R. Astron. Soc. **463**, 89–93 (2016a). doi:10.1093/mnrasl/slw162

B. Cerutti, A.A. Philippov, A. Spitkovsky, Modelling high-energy pulsar light curves from first principles. Mon. Not. R. Astron. Soc. **457**, 2401–2414 (2016b). doi:10.1093/mnras/stw124

B. Cerutti, G.R. Werner, D.A. Uzdensky, M.C. Begelman, Simulations of particle acceleration beyond the classical synchrotron burnoff limit in magnetic reconnection: an explanation of the Crab flares. Astrophys. J. **770**, 147 (2013). doi:10.1088/0004-637X/770/2/147

B. Cerutti, A. Philippov, K. Parfrey, A. Spitkovsky, Particle acceleration in axisymmetric pulsar current sheets. Mon. Not. R. Astron. Soc. **448**, 606–619 (2015). doi:10.1093/mnras/stv042

A.Y. Chen, A.M. Beloborodov, Dead zone in the polar-cap accelerator of pulsars. Astrophys. J. **762**, 76 (2013). doi:10.1088/0004-637X/762/2/76

A.Y. Chen, A.M. Beloborodov, Electrodynamics of axisymmetric pulsar magnetosphere with electron-positron discharge: a numerical experiment. Astrophys. J. Lett. **795**, 22 (2014). doi:10.1088/2041-8205/795/1/L22

K.S. Cheng, C. Ho, M. Ruderman, Energetic radiation from rapidly spinning pulsars. I—Outer magnetosphere gaps. II—VELA and Crab. Astrophys. J. **300**, 500–539 (1986). doi:10.1086/163829

I. Contopoulos, The coughing pulsar magnetosphere. Astron. Astrophys. **442**, 579–586 (2005). doi:10.1051/0004-6361:20053143

I. Contopoulos, D. Kazanas, Toward resolving the Crab σ-problem: a linear accelerator? Astrophys. J. **566**, 336–342 (2002). doi:10.1086/324778

I. Contopoulos, C. Kalapotharakos, D. Kazanas, A new standard pulsar magnetosphere. Astrophys. J. **781**, 46 (2014). doi:10.1088/0004-637X/781/1/46

I. Contopoulos, D. Kazanas, C. Fendt, The axisymmetric pulsar magnetosphere. Astrophys. J. **511**, 351–358 (1999). doi:10.1086/306652

F.V. Coroniti, Magnetically striped relativistic magnetohydrodynamic winds—the Crab nebula revisited. Astrophys. J. **349**, 538–545 (1990). doi:10.1086/168340

L. Davis, Stellar electromagnetic fields. Phys. Rev. **72**, 632–633 (1947). doi:10.1103/PhysRev.72.632

J. Dawson, One-dimensional plasma model. Phys. Fluids **5**, 445–459 (1962). doi:10.1063/1.1706638

A.J. Deutsch, The electromagnetic field of an idealized star in rigid rotation in vacuo. Ann. Astrophys. **18**, 1 (1955)

T. Erber, High-energy electromagnetic conversion processes in intense magnetic fields. Rev. Mod. Phys. **38**, 626–659 (1966). doi:10.1103/RevModPhys.38.626

P. Goldreich, W.H. Julian, Pulsar electrodynamics. Astrophys. J. **157**, 869 (1969). doi:10.1086/150119

S.P. Goodwin, J. Mestel, L. Mestel, G.A.E. Wright, An idealized pulsar magnetosphere: the relativistic force-free approximation. Mon. Not. R. Astron. Soc. **349**, 213–224 (2004). doi:10.1111/j.1365-2966.2004.07481.x

S.E. Gralla, A. Lupsasca, A. Philippov, Pulsar magnetospheres: beyond the flat spacetime dipole. ArXiv e-prints (2016)

I.A. Grenier, A.K. Harding, Gamma-ray pulsars: a gold mine. C. R. Phys. **16**, 641–660 (2015). doi:10.1016/j.crhy.2015.08.013

A. Gruzinov, Power of an axisymmetric pulsar. Phys. Rev. Lett. **94**(2), 021101 (2005). doi:10.1103/PhysRevLett.94.021101

A.K. Harding, D. Lai, Physics of strongly magnetized neutron stars. Rep. Prog. Phys. **69**, 2631–2708 (2006). doi:10.1088/0034-4885/69/9/R03

A. Hewish, S.J. Bell, J.D.H. Pilkington, P.F. Scott, R.A. Collins, Observation of a rapidly pulsating radio source. Nature **217**, 709–713 (1968). doi:10.1038/217709a0

E.W. Hones Jr., J.E. Bergeson, Electric field generated by a rotating magnetized sphere. J. Geophys. Res. **70**, 4951–4958 (1965). doi:10.1029/JZ070i019p04951

F. Hoyle, J.V. Narlikar, J.A. Wheeler, Electromagnetic waves from very dense stars. Nature **203**, 914–916 (1964). doi:10.1038/203914a0

E.A. Jackson, Minimum energy state of the constrained pulsar atmosphere. Astrophys. J. **227**, 266–274 (1979). doi:10.1086/156725

P.B. Jones, Properties of condensed matter in very strong magnetic fields. Mon. Not. R. Astron. Soc. **218**, 477–485 (1986). doi:10.1093/mnras/218.3.477

C. Kalapotharakos, I. Contopoulos, Three-dimensional numerical simulations of the pulsar magnetosphere: preliminary results. Astron. Astrophys. **496**, 495–502 (2009). doi:10.1051/0004-6361:200810281

C. Kalapotharakos, D. Kazanas, A. Harding, I. Contopoulos, Toward a realistic pulsar magnetosphere. Astrophys. J. **749**, 2 (2012). doi:10.1088/0004-637X/749/1/2

J.G. Kirk, Particle acceleration in relativistic current sheets. Phys. Rev. Lett. **92**(18), 181101 (2004). doi:10.1103/PhysRevLett.92.181101

J.G. Kirk, Y. Lyubarsky, J. Petri, The theory of pulsar winds and nebulae, in *Astrophysics and Space Science Library*, ed. by W. Becker. Astrophysics and Space Science Library, vol. 357 (2009), p. 421. doi:10.1007/978-3-540-76965-1_16

S.S. Komissarov, Simulations of the axisymmetric magnetospheres of neutron stars. Mon. Not. R. Astron. Soc. **367**, 19–31 (2006). doi:10.1111/j.1365-2966.2005.09932.x

J. Krause-Polstorff, F.C. Michel, Electrosphere of an aligned magnetized neutron star. Mon. Not. R. Astron. Soc. **213**, 43–49 (1985a). doi:10.1093/mnras/213.1.43P

J. Krause-Polstorff, F.C. Michel, Pulsar space charging. Astron. Astrophys. **144**, 72–80 (1985b)

J. Li, A. Spitkovsky, A. Tchekhovskoy, Resistive solutions for pulsar magnetospheres. Astrophys. J. **746**, 60 (2012). doi:10.1088/0004-637X/746/1/60

Y.E. Lyubarskii, Equilibrium of the return current sheet and the structure of the pulsar magnetosphere. Sov. Astron. Lett. **16**, 16 (1990)

Y.E. Lyubarskii, A model for the energetic emission from pulsars. Astron. Astrophys. **311**, 172–178 (1996)

J. McDonald, A. Shearer, Investigations of the magnetospheric plasma distribution in the vicinity of a pulsar. I. Basic formulation. Astrophys. J. **690**, 13–19 (2009). doi:10.1088/0004-637X/690/1/13

J.C. McKinney, Relativistic force-free electrodynamic simulations of neutron star magnetospheres. Mon. Not. R. Astron. Soc. **368**, 30–34 (2006). doi:10.1111/j.1745-3933.2006.00150.x

Z. Medin, D. Lai, Condensed surfaces of magnetic neutron stars, thermal surface emission, and particle acceleration above pulsar polar caps. Mon. Not. R. Astron. Soc. **382**, 1833–1852 (2007). doi:10.1111/j.1365-2966.2007.12492.x

L. Mestel, S. Shibata, The axisymmetric pulsar magnetosphere—a new model. Mon. Not. R. Astron. Soc. **271**, 621 (1994). doi:10.1093/mnras/271.3.621

L. Mestel, J.A. Robertson, Y.-M. Wang, K.C. Westfold, The axisymmetric pulsar magnetosphere. Mon. Not. R. Astron. Soc. **217**, 443–484 (1985). doi:10.1093/mnras/217.3.443

F.C. Michel, Rotating magnetosphere: a simple relativistic model. Astrophys. J. **180**, 207–226 (1973a). doi:10.1086/151956

F.C. Michel, Rotating magnetospheres: an exact 3-D solution. Astrophys. J. Lett. **180**, 133 (1973b). doi:10.1086/181169

F.C. Michel, Rotating magnetosphere: acceleration of plasma from the surface. Astrophys. J. **192**, 713–718 (1974). doi:10.1086/153109

F.C. Michel, Magnetic structure of pulsar winds. Astrophys. J. **431**, 397–401 (1994). doi:10.1086/174493

F.C. Michel, The state of pulsar theory. Adv. Space Res. **33**, 542–551 (2004). doi:10.1016/j.asr.2003.06.019

F.C. Michel, H. Li, Electrodynamics of neutron stars. Phys. Rep. **318**, 227–297 (1999). doi:10.1016/S0370-1573(99)00002-2

A.G. Muslimov, A.K. Harding, High-altitude particle acceleration and radiation in pulsar slot gaps. Astrophys. J. **606**, 1143–1153 (2004). doi:10.1086/383079

A.G. Muslimov, A.I. Tsygan, General relativistic electric potential drops above pulsar polar caps. Mon. Not. R. Astron. Soc. **255**, 61–70 (1992). doi:10.1093/mnras/255.1.61

T. Neukirch, Equilibria of charge-separated rigidly rotating relativistic magnetospheres. Astron. Astrophys. **274**, 319 (1993)

C.-Y. Ng, R.W. Romani, Fitting pulsar wind tori. Astrophys. J. **601**, 479–484 (2004). doi:10.1086/380486

J.P. Ostriker, J.E. Gunn, On the nature of pulsars. I. Theory. Astrophys. J. **157**, 1395 (1969). doi:10.1086/150160

F. Pacini, Energy emission from a neutron star. Nature **216**, 567–568 (1967). doi:10.1038/216567a0

F. Pacini, Rotating neutron stars, pulsars and supernova remnants. Nature **219**, 145–146 (1968). doi:10.1038/219145a0

K. Parfrey, A.M. Beloborodov, L. Hui, Introducing PHAEDRA: a new spectral code for simulations of relativistic magnetospheres. Mon. Not. R. Astron. Soc. **423**, 1416–1436 (2012). doi:10.1111/j.1365-2966.2012.20969.x

J. Pétri, Relativistic stabilisation of the diocotron instability in a pulsar "cylindrical" electrosphere. Astron. Astrophys. **469**, 843–855 (2007). doi:10.1051/0004-6361:20066985

J. Pétri, The pulsar force-free magnetosphere linked to its striped wind: time-dependent pseudo-spectral simulations. Mon. Not. R. Astron. Soc. **424**, 605–619 (2012). doi:10.1111/j.1365-2966.2012.21238.x

J. Pétri, Multipolar electromagnetic fields around neutron stars: exact vacuum solutions and related properties. Mon. Not. R. Astron. Soc. **450**, 714–742 (2015). doi:10.1093/mnras/stv598

J. Pétri, Theory of pulsar magnetosphere and wind. ArXiv e-prints (2016)

J. Pétri, J. Heyvaerts, S. Bonazzola, Diocotron instability in pulsar electrospheres. I. Linear analysis. Astron. Astrophys. **387**, 520–530 (2002a). doi:10.1051/0004-6361:20020442

J. Pétri, J. Heyvaerts, S. Bonazzola, Global static electrospheres of charged pulsars. Astron. Astrophys. **384**, 414–432 (2002b). doi:10.1051/0004-6361:20020044

A.A. Philippov, A. Spitkovsky, Ab initio pulsar magnetosphere: three-dimensional particle-in-cell simulations of axisymmetric pulsars. Astrophys. J. Lett. **785**, 33 (2014). doi:10.1088/2041-8205/785/2/L33

A.A. Philippov, A. Spitkovsky, B. Cerutti, Ab initio pulsar magnetosphere: three-dimensional particle-in-cell simulations of oblique pulsars. Astrophys. J. Lett. **801**, 19 (2015a). doi:10.1088/2041-8205/801/1/L19

A.A. Philippov, B. Cerutti, A. Tchekhovskoy, A. Spitkovsky, Ab initio pulsar magnetosphere: the role of general relativity. Astrophys. J. Lett. **815**, 19 (2015b). doi:10.1088/2041-8205/815/2/L19

M. Ruderman, Matter in superstrong magnetic fields, in *Physics of Dense Matter*, ed. by C.J. Hansen. IAU Symposium, vol. 53 (1974), p. 117

M.A. Ruderman, P.G. Sutherland, Theory of pulsars—polar caps, sparks, and coherent microwave radiation. Astrophys. J. **196**, 51–72 (1975). doi:10.1086/153393

N. Sakai, S. Shibata, General relativistic electromagnetism and particle acceleration in a pulsar polar cap. Astrophys. J. **584**, 427–432 (2003). doi:10.1086/345616
E.T. Scharlemann, R.V. Wagoner, Aligned rotating magnetospheres. General analysis. Astrophys. J. **182**, 951–960 (1973). doi:10.1086/152195
S. Shibata, Confirmation of quiet pulsar model by the boundary element method. Astrophys. Space Sci. **161**, 187–194 (1989). doi:10.1007/BF00653564
I.A. Smith, F.C. Michel, P.D. Thacker, Numerical simulations of aligned neutron star magnetospheres. Mon. Not. R. Astron. Soc. **322**, 209–217 (2001). doi:10.1046/j.1365-8711.2001.04045.x
A. Spitkovsky, Time-dependent force-free pulsar magnetospheres: axisymmetric and oblique rotators. Astrophys. J. Lett. **648**, 51–54 (2006). doi:10.1086/507518
A. Spitkovsky, *Current Models of Pulsar Magnetospheres*. Astrophysics and Space Science Proceedings, vol. 21 (2011), p. 139. doi:10.1007/978-3-642-17251-9_11
A. Spitkovsky, J. Arons, Simulations of pulsar wind formation, in *Neutron Stars in Supernova Remnants*, ed. by P.O. Slane, B.M. Gaensler. Astronomical Society of the Pacific Conference Series, vol. 271 (2002), p. 81
P.A. Sturrock, A model of pulsars. Astrophys. J. **164**, 529 (1971). doi:10.1086/150865
M. Tamburini, F. Pegoraro, A. Di Piazza, C.H. Keitel, A. Macchi, Radiation reaction effects on radiation pressure acceleration. New J. Phys. **12**(12), 123005 (2010). doi:10.1088/1367-2630/12/12/123005
A. Tchekhovskoy, A. Philippov, A. Spitkovsky, Three-dimensional analytical description of magnetized winds from oblique pulsars. Mon. Not. R. Astron. Soc. **457**, 3384–3395 (2016). doi:10.1093/mnras/stv2869
A. Tchekhovskoy, A. Spitkovsky, J.G. Li, Time-dependent 3D magnetohydrodynamic pulsar magnetospheres: oblique rotators. Mon. Not. R. Astron. Soc. **435**, 1–5 (2013). doi:10.1093/mnrasl/slt076
K.O. Thielheim, H. Wolfsteller, Stable space charge configurations near the force-free surface of rotating magnets. Astrophys. J. **431**, 718–723 (1994). doi:10.1086/174522
A.N. Timokhin, On the force-free magnetosphere of an aligned rotator. Mon. Not. R. Astron. Soc. **368**, 1055–1072 (2006). doi:10.1111/j.1365-2966.2006.10192.x
A.N. Timokhin, J. Arons, Current flow and pair creation at low altitude in rotation-powered pulsars' force-free magnetospheres: space charge limited flow. Mon. Not. R. Astron. Soc. **429**, 20–54 (2013). doi:10.1093/mnras/sts298
D.A. Uzdensky, B. Cerutti, M.C. Begelman, Reconnection-powered linear accelerator and gamma-ray flares in the Crab nebula. Astrophys. J. Lett. **737**, 40 (2011). doi:10.1088/2041-8205/737/2/L40
T. Wada, S. Shibata, A particle simulation for the global pulsar magnetosphere—II. The case of dipole field. Mon. Not. R. Astron. Soc. **418**, 612–624 (2011). doi:10.1111/j.1365-2966.2011.19510.x
M.C. Weisskopf, R.F. Elsner, J.J. Kolodziejczak, S.L. O'Dell, A.F. Tennant, Unraveling the geometry of the Crab nebula's "inner ring". Astrophys. J. **746**, 41 (2012). doi:10.1088/0004-637X/746/1/41
K. Yee, Numerical solution of inital boundary value problems involving Maxwell's equations in isotropic media. IEEE Trans. Antennas Propag. **14**, 302–307 (1966). doi:10.1109/TAP.1966.1138693

Space Sci Rev (2017) 207:137–174
DOI 10.1007/s11214-017-0344-x

Modelling Jets, Tori and Flares in Pulsar Wind Nebulae

Oliver Porth[1] · Rolf Buehler[2] · Barbara Olmi[3,4,5] · Serguei Komissarov[6] · Astrid Lamberts[7] · Elena Amato[3,5] · Yajie Yuan[8,9] · Alexander Rudy[10]

Received: 12 December 2016 / Accepted: 8 March 2017 / Published online: 22 March 2017

Abstract In this contribution we review the recent progress in the modelling of Pulsar Wind Nebulae (PWN). We start with a brief overview of the relevant physical processes in the magnetosphere, the wind-zone and the inflated nebula bubble. Radiative signatures and particle transport processes obtained from 3D simulations of PWN are discussed in the context of optical and X-ray observations. We then proceed to consider particle acceleration in PWN and elaborate on what can be learned about the particle acceleration from the dynamical structures called GwispsG observed in the Crab nebula. We also discuss recent observational and theoretical results of gamma-ray flares and the inner knot of the Crab nebula, which had been proposed as the emission site of the flares. We extend the discussion to GeV flares from binary systems in which the pulsar wind interacts with the stellar wind from a companion star. The chapter concludes with a discussion of solved and unsolved problems posed by PWN.

✉ O. Porth
porth@itp.uni-frankfurt.de

1 Institute for Theoretical Physics, Max-von-Laue Str. 1, 60438 Frankfurt am Main, Germany

2 DESY, Platanenallee 6, 15738 Zeuthen, Germany

3 Dipartimento di Fisica ed Astronomia, Università degli Studi di Firenze, Via G. Sansone 1, 50019 Sesto F.no (Firenze), Italy

4 INFN—Sezione di Firenze, Via G. Sansone 1, 50019 Sesto F.no (Firenze), Italy

5 INAF Osservatorio Astrofisico di Arcetri, Largo E. Fermi 5, 50125 Firenze, Italy

6 School of Mathematics, University of Leeds, Leeds, LS29JT, UK

7 Theoretical Astrophysics, California Institute of Technology, Pasadena, CA 91125, USA

8 Department of Astrophysical Sciences, Princeton University, 4 Ivy Lane, Princeton, NJ 08544, USA

9 KIPAC, SLAC and Stanford University, Stanford, CA 94305, USA

10 Department of Astronomy & Astrophysics, University of California, Santa Cruz, CA 95064, USA

1 Introduction

At the end of their life, massive stars leave behind some of the most powerful high energy sources in the sky: neutron stars or black holes. The environment around these fascinating objects allows us to test conditions which often can not be replicated on Earth. Among these systems, pulsar wind nebulae (PWN) are of particular interest. In them, the relativistic plasma wind ejected by the rotating neutron star interacts with the ambient medium. The latter is in most cases composed of the ejecta of the original star explosion. This interaction leads to bright synchrotron and inverse Compton (IC) emission of the wind electrons, producing some of the brightest sources in the sky (electrons and positrons are referred to together as electrons throughout this text).

Modelling the emission of PWN reveals that these systems are able to accelerate particles to very high energies of up to several PeV. The ultimate energy source of this acceleration is rotation of the neutron star which induces a strong electric potential. As neutron stars are highly magnetised and conducting, they act as a unipolar inductor, creating an electric potential between poles and equator. This electromagnetic energy is carried away from the pulsar magnetosphere with the pulsar wind. Where and how it is transformed into kinetic particle energy is a matter of intense research.

In comparison to other sources of relativistic plasma, PWN can be resolved in great detail, as shown in Fig. 1. Scales close to the gyroradius of the highest energy electrons in this system can be resolved by current radio, optical and X-ray telescopes. Particularly in two systems, the Crab and Vela PWN, the plasma flow can be resolved down to scales of ≈ 10 light days. These systems provide perfect test beds to study the behaviour of relativistic magnetised plasma, which is thought to be present in other high energy sources as gamma-ray bursts or active galactic nuclei (Berger 2014; Massaro et al. 2015). The discovery of gamma-ray flares from the Crab PWN revealed very efficient and rapid acceleration of particles in this system. Strong and rapid gamma-ray flares are also observed in GRBs and AGN (see e.g. Ackermann et al. 2013 and Aharonian et al. 2009). It is likely, that common mechanisms, magnetic reconnection or shock acceleration are responsible for the acceleration in all of these systems (see Kagan et al. 2015 and Sironi et al. 2015 for recent reviews).

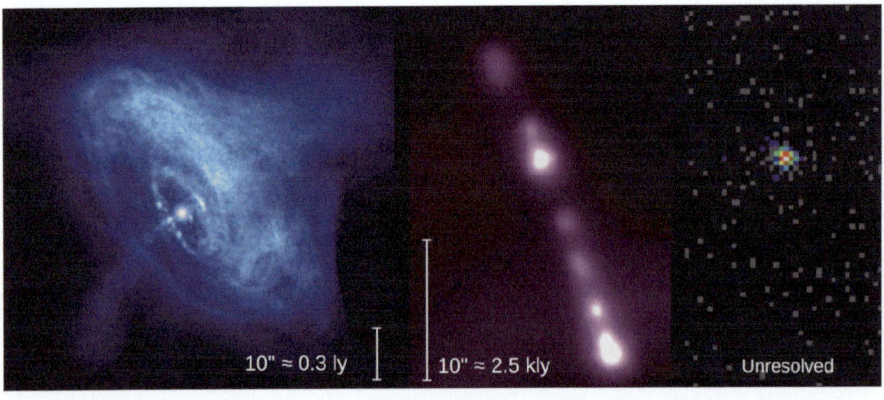

Fig. 1 Images of sources of relativistic plasma taken by the Chandra X-ray Observatory (CXO): the *right panel* shows the Crab nebula, the *middle panel* the AGN M87 and the *right panel* GRB 991216. *Each panel* shows the approximate spatial scale which can be resolved within the Chandra angular resolution CXO of ≈ 1 arcsec. Image credits: NASA/CXC/ASU/J. Hester et al./E. Perlman et al./L. Piro et al.

In contrast to the inner region of the Vela PWN, the Crab PWN has been detected across the electromagnetic spectrum. The plasma motion is resolved in space and time with unmatched resolution in this system (see Hester 2008 and Bühler and Blandford 2014 for reviews). It has therefore been the target of most theoretical studies of PWN. In particular, relativistic magnetohydrodynamic (RMHD) simulations have provided deep insights into the plasma behaviour over the past decades (see e.g. Komissarov and Lyubarsky 2004; Del Zanna et al. 2006; Porth et al. 2013). The qualitative behaviour of the global plasma flow in these systems is thought to be understood to date. The energy of the wind emitted by the pulsar is concentrated towards the equator. The wind is thought to be highly magnetised and cold (meaning that its thermal energy is much smaller than its magnetic and bulk kinetic energy). Over time, this wind has blown a bubble into the ejecta of the original star explosion. A reverse shock emerges when the pulsar wind first interacts with the surrounding medium, creating an oblate termination surface (see Fig. 4 in Sect. 2). Downstream of this termination surface the plasma flows into the equatorial region forming a torus and towards the poles. The latter flow is a result of the magnetic hoop stress and forms the jet observed perpendicular to the torus.

In this article, we will review the status of our current understanding of the plasma flow in PWN and the particle acceleration happening within it. The article is structured as follows: first, we will introduce the RMHD models of PWN in Sect. 2. We then proceed to discuss the particle acceleration in PWN. In Sect. 3 we discuss what can be learned about the particle acceleration from the dynamical structures called "wisps" observed in the Crab nebula. Then we proceed to discuss the gamma-ray flares observed in this system in Sect. 4. In this context, we will also discuss recent observational and theoretical results of the inner knot of the Crab nebula in Sect. 5, which had been proposed as the emission site of the flares. In Sect. 6 we will extend the discussion to binary systems, in which the pulsar wind interacts with the stellar wind from a companion star. The article concludes with a discussion of solved and unsolved problems of our understanding of PWN in Sect. 7.

2 RMHD Models of PWN

2.1 Simulating the Plasma Flow

PWN are customarily modelled under assumption of ideal magnetohydrodynamics (MHD). Although the validity of the fluid approximation in PWN is not without question,[1] by upholding basic conservation laws, a first insight into the physics can be attained. Let us start with a brief review of the neutron star magnetosphere and then make our way out to the termination shock and PWN contact discontinuity.

2.2 Magnetosphere

MHD can be applied if sufficient plasma is available to screen the electric field: $\mathbf{E}\cdot\mathbf{B}=\mathbf{0}$. Seminal works of Deutsch (1955) and Goldreich and Julian (1969) indicate that this is a good working assumption as an electromagnetic cascade develops that fills the "gaps" of unscreened electric fields with freshly created pair-plasma (Sturrock 1971). The number

[1] For example, on the scale of the gyroradius for X-ray emitting particles, significant wisp-like substructure is observed which prompts for a more accurate description taking into account finite gyroradius effects (MHD+PIC).

density of pairs divided by the minimal density required to screen the electric fields in the co-moving frame is known as pulsar multiplicity κ (Goldreich and Julian 1969). Due to the supply of conducting magnetospheric plasma, the early vacuum oblique dipole models (e.g. Pacini 1967; Ostriker and Gunn 1969) are not applicable, rather can the pulsar be thought of as a *unipolar inductor* whose rotation drives a current system. Torque from the $\mathbf{j} \times \mathbf{B}$ Lorentz force of the current flowing parallel to the neutron star surface serves to spin down the pulsar and extracts energy carried away in the form of Poynting flux. In the limit that EM contributions are stronger than pressure, inertial and gravitational forces, plasma can be considered force-free, that is the EM forces balance exactly:

$$\frac{1}{c}\mathbf{j} \times \mathbf{B} + \rho_e \mathbf{E} = 0. \tag{1}$$

In this regime, an important exact solution was obtained by Michel (1973). In Michel's *monopolar* solution, field lines are confined to cones in the ϖz-planes and describe perfect Archimedean spirals in all $\varpi\phi$-planes (ϖ is the cylindrical radius). Moving away from the source, the magnetic field is thus progressively winding up with $B_\phi = B_0 \varpi \Omega / c (r/r_0)^{-2}$, $B_r = B_0 (r/r_0)^{-2}$ (Ω is the angular velocity of the central object). The induced electric field is simply $E_\theta = \varpi \Omega B_r / c$ and we obtain a Poynting flux

$$\mathcal{S}_r = \frac{c}{4\pi}\mathbf{E} \times \mathbf{B} = \frac{\Omega^2 B_0^2 r_0^4 \sin^2\theta}{4\pi r^2}. \tag{2}$$

The important point to note here is that the energy flux is highly anisotropic $\propto \sin^2\theta$, thus more energy is ejected in the equatorial direction. After long struggle, a solution to the at first glance unimposing equation (1) for more realistic *dipolar* stellar field was obtained numerically by Contopoulos et al. (1999). Most notably, the dipole field rips open at the light cylinder radius $\varpi_{\rm LC} = c/\Omega$ and a radial wind streams out similar to Michel's solution i.e. with $\mathcal{S}_r \propto \sin^2\theta$. This general result for the aligned rotator has been confirmed and improved on by many groups (Goodwin et al. 2004; Gruzinov 2005; McKinney 2006; Parfrey et al. 2012; Ruiz et al. 2014). As the polarity of the dipole field reverses across hemispheres, so does the wound-up toroidal field, giving rise to an equatorial current-sheet. The ensuing dissipation violates the ideal MHD condition $\mathbf{E} \cdot \mathbf{B} = \mathbf{0}$ and also force-freeness must break down locally as magnetic dominance cannot be maintained. It was noted already by Coroniti (1990) that the current sheet can play an important role in the dynamics of the wind as a whole. In case the magnetic axis is mis-aligned with the rotation axis (oblique rotator), the current-sheet assumes a wavy or *striped*-shape. Dissipation, dynamics and particle acceleration of the striped wind is subject to active research and worthy of a review of its own (see Arons 2012). We shall return to this issue below.

The first force-free model of an *oblique* dipole magnetosphere was presented by Spitkovsky (2006) and has been confirmed by several groups and methodologies (Kalapotharakos et al. 2012; Pétri 2016; Tchekhovskoy et al. 2016). Its salient features are the complex geometry of the striped wind (as predicted by Michel 1971) and the dependence of the spin-down power on the obliquity angle α

$$L = k_1 \frac{\mu^2 \Omega^4}{c^3}\left(1 + k_2 \sin^2\alpha\right) \tag{3}$$

Here $\mu = B_p r^{\star 3}/2$ is the magnetic moment of the star (B_p measured at the poles) and $k_1 = 1 \pm 0.05$, $k_2 = 1 \pm 0.1$ were obtained by fitting to the numerical simulations. Following Eq. (3), oblique pulsars emit up to twice as much wind-power as aligned rotators.

Concerning the distribution of wind power in the orthogonal case, one obtains

$$\langle S_r \rangle_\phi^{90^\circ} \approx \frac{\Omega^2 B_0^2 r_0^4 \sin^4\theta}{8\pi r^2} \tag{4}$$

where the extra factor of $\sin^2\theta$ comes from bunching of radial field in the equatorial regions. Semi-analytic formulae for intermediate cases were presented by Tchekhovskoy et al. (2016).

Recent global particle-in-cell (PIC) simulations now confirm that given sufficient plasma supply, the magnetosphere adopts a near force free configuration consistent with the fluid models (Philippov and Spitkovsky 2014; Chen and Beloborodov 2014; Belyaev 2015; Cerutti et al. 2015) showing that the above predictions are likely robust.

2.3 Wind Zone

To estimate the plasma-parameters in the PWN, it is instructive to first study a 1D description of the wind flow. Motivated by the success of Michel's solution, we assume that the wind predominantly streams out in spherical r-direction, $\mathbf{B} = B_r \hat{\mathbf{e}}^r + B_\phi \hat{\mathbf{e}}^\phi$, $\mathbf{v} = v_r \hat{\mathbf{e}}^r + v_\phi \hat{\mathbf{e}}^\phi$, the relevant equations for the stationary flow are

$$\partial_r\left(r^2 \Gamma \rho v_r\right) = 0 \tag{5}$$

$$\partial_r\left(r^3\left(\omega v_\phi v_r \Gamma^2 - B_\phi B_r\right)\right) = 0 \tag{6}$$

$$\partial_r\left(r^2\left(\omega \Gamma^2 v_r + S_r\right)\right) = 0 \tag{7}$$

$$\partial_r(r E_\theta) = 0 \tag{8}$$

$$\partial_r\left(r^2 B_r\right) = 0, \tag{9}$$

which describe conservation of mass, angular-momentum, energy, Faraday's law of induction and the $\nabla \cdot \mathbf{B} = 0$ constraint. Adopting an ideal equation of state with adiabatic index $\hat{\gamma}$ we write the enthalpy-density $\omega = \rho c^2 + \hat{\gamma}/(\hat{\gamma} - 1)p$. In the above equations, the wind Poynting flux is given by $S = \mathbf{E} \times \mathbf{B}$ (hence setting $4\pi = c = 1$).

Let's derive some useful relations from the system (5)–(9). From the $\nabla \cdot \mathbf{B} = 0$ constraint (9), we immediately see that the radial field must decay as $B_r \propto r^{-2}$. From the induction-law (8) together with the ideal MHD electric field $E_\theta = B_\phi v_r - B_r v_\phi$ and $B_r \propto r^{-2}$, we obtain a first conservation law:

$$r\Omega \equiv v_\phi - v_r \frac{B_\phi}{B_r} \tag{10}$$

It means that the rotation of the central object Ω is conserved as "angular velocity of the field lines". Equation (10) can serve to visualise two previously mentioned aspects of the pulsar magnetosphere: 1. In Michel's solution, were the field-lines rigid radial wires sticking out of the spinning source, beyond the light-cylinder plasma would be forced to rotate faster than the speed of light. This is circumvented by induction of a toroidal component B_ϕ and winding up of the field. 2. In the closed dipolar magnetosphere, as $v_r = 0$ on the equator, again we would also obtain $v_\phi > c$, thus field lines crossing the light-cylinder must be open. Furthermore, from (10) we can estimate that at the light-cylinder $B_\phi \simeq B_r$.

Dividing the toroidal momentum-flux (6) by the particle rest-mass energy flux $r^2\Gamma\rho v_r c^2$ (5), we obtain the conserved angular momentum flux

$$l \equiv \frac{\omega\Gamma r v_\phi}{\rho c^2} - \frac{r B_\phi}{k c^2} \tag{11}$$

with the ratio of matter to magnetic flux $k \equiv \rho v_r \Gamma / B_r$ that is also constant along a streamline. For an accelerating wind close to the speed of light ($v_r \simeq c$) we now see that $B_\phi/B_r \simeq r/r_{LC}$ and the wind becomes dominated by the toroidal magnetic field. Similarly, the wind rotation follows $v_\phi \propto r^{-1}$ and far away from the light-cylinder, the flow is well described by a purely radial velocity v_r and an entirely toroidal magnetic field B_ϕ.

Dividing the energy flux (7) by the rest-mass energy flux, the conserved quantity

$$\mu \equiv \frac{\omega\Gamma}{\rho c^2} + \frac{\mathcal{S}_r}{\Gamma\rho v_r c^2} = \Gamma\left(\omega/\rho c^2 + \sigma\right) \tag{12}$$

is recovered. It is clear that the Lorentz factor of the wind cannot exceed the value of μ. In the latter equation we have introduced the magnetisation or σ-parameter:

$$\sigma \equiv \mathcal{S}_r / \left(\Gamma^2 \rho v_r c^2\right) \tag{13}$$

which compares the Poynting flux with the kinetic energy of the wind. In the cold limit $\omega \to \rho c^2$, we have $\mu = \Gamma(1+\sigma)$ which shows that accelerating the wind goes hand-in-hand with decreasing its magnetisation.

As we shall see later, the value for σ at the TS as inferred from 1D and 2D models is $\sigma \sim 10^{-3}$–10^{-2}. Quite in contrast, models for the pulsar magnetosphere predict highly magnetised plasma with $\sigma \sim 10^3$ (e.g. Arons 2012, and references therein). The conversion of magnetic energy to kinetic energy in both confined and unconfined winds has been a subject of intensive research. Although claims have been made that the ideal MHD acceleration can provide the required energy conversion (e.g. Vlahakis 2004), it is now widely accepted that relativistic MHD flows are very inefficient accelerators (e.g. Komissarov et al. 2009; Tchekhovskoy et al. 2009; Lyubarsky 2009, 2010) and can achieve $\sigma \approx 1$ at best. The discrepancy between σ obtained via MHD acceleration of the unconfined wind and the magnetisation inferred from PWN models is known as the σ-Problem.

It was noted by Coroniti (1990) that allowing for finite resistivity in the wind zone, dissipation of the striped component might present a way out of this problem. However, even when the wind-magnetisation is optimistically reduced to its non-oscillating mean value, in order to arrive at the requested $\sigma \sim 10^{-3}$–10^{-2} the pulsar needs to be very nearly perpendicular (see e.g. Komissarov 2013). A more in-depth analysis of the dissipating striped wind was presented by Lyubarsky and Kirk (2001) and Kirk and Skjæraasen (2003) who demonstrated that unless the wind multiplicity exceeds $\kappa = 10^5$, due to efficient thermal acceleration out of the dissipated magnetic energy, the growth time exceeds the crossing time from the light-cylinder to the TS in the frame of the pulsar. This somewhat high value of κ casts some doubt that dissipation can be efficient in the wind zone. As a "last exit", Lyubarsky (2003) proposed that dissipation of stripes will occur directly at the termination shock, with the shock compression leading to "driven magnetic reconnection". Interestingly, the jump conditions of this modified shock are identical to the case where stripes have been dissipated entirely in the upstream region. The PIC simulations by Sironi and Spitkovsky (2011), and two-fluid MHD models (Amano and Kirk 2013) in fact now show that the driven reconnection is a viable mechanism, but is able to reduce the magnetisation only down to the mean (non-oscillating) value.

In light of this, assuming the fraction of magnetic energy in alternating stripes $\chi_\alpha(\theta)$ is dissipated entirely, straightforward calculation for the effective magnetisation yields

$$\sigma(\theta) = \frac{\sigma_0 \chi_\alpha(\theta)}{1+\sigma_0(1-\chi_\alpha(\theta))} \tag{14}$$

where the function $\chi_\alpha(\theta)$ is given by the obliqueness angle of the pulsar α: From the equator where polarities cancel exactly $[\chi_\alpha(\pi/2)=0]$, the striped region extends by an angle $\pm\alpha$ after which only a single magnetic polarity is present and the polar regions may remain strongly magnetised. Assuming the wind is well described by the split-monopole solution far from the pulsar, further geometric considerations can yield the exact form of the $\chi_\alpha(\theta)$ function (Komissarov 2013). Equation (14) means that for low wind σ_0, the effective magnetisation increases linearly $\sigma(\theta) \propto \sigma_0\chi_\alpha(\theta)$ as naively expected, yet for very high wind magnetisation $\sigma_0 \to \infty$ the effective value is governed by the geometry alone $\sigma(\theta) \to \chi_\alpha(\theta)/(1-\chi_\alpha(\theta))$. While the stripe dissipation might not be sufficient to solve the σ-problem entirely, it can significantly reduce the effective flow magnetisation. For example, for an intermediate obliqueness angle $\alpha = 45°$, found by Harding et al. (2008) in their modelling of the high-energy emission of the Crab pulsar, the ratio of magnetic luminosity to kinetic luminosity that is injected through the shock saturates at $L_\mathrm{m}/L_\mathrm{k} \approx 0.3$ (e.g. Porth et al. 2014a). In this simplified one-dimensional view, the pulsar obliqueness angle α assumes a much more prominent role than the σ-parameter of the wind.

2.4 Termination Shock

Like the solar wind, the pulsar wind terminates when its dynamic pressure equals the thermal pressure of the surrounding shocked medium, that is the pulsar wind nebula. As the observations suggest very low magnetisation in the PWN, we will adopt a hydrodynamical model for some basic estimates. A typical scale for ram-pressure balance is readily found: $r_0 = (L/(4\pi pc))^{1/2}$ with p the pressure of the nebular plasma. If energy is supplied to the PWN at a constant rate L and its contact discontinuity evolves according to $r_\mathrm{n}(t) \propto t^{\alpha_\mathrm{r}}$ (where we will adopt $\alpha_\mathrm{r} = 6/5$; Chevalier 1977), adiabatic expansion of the ultrarelativistic shocked PWN implies $\dot{E} = L - \alpha_\mathrm{r} E/t$. Thus the energy accumulated over time in the PWN bubble is

$$E = \frac{Lt}{1+\alpha_\mathrm{r}} \tag{15}$$

with the initial condition $E(0) = 0$. Assuming a uniform distribution of the PWN pressure $p = E/(4\pi r_\mathrm{n}^3)$ we find the scale of the TS

$$r_0 = \frac{(1+\alpha_\mathrm{r})}{\alpha_\mathrm{r}} r_\mathrm{n} (v_\mathrm{n}/c)^{1/2} \tag{16}$$

Where v_n is the observed expansion speed of the nebula. In case of Crab, $v_\mathrm{n} \approx 2000\ \mathrm{km\,s^{-1}}$ and using the observed nebula radius of $r_\mathrm{n} = 1.65$ pc (Hester 2008) we find the values $r_0 = 0.1 r_\mathrm{n} = 0.17$ pc. The very good match of this estimate with the extent of the X-ray inner ring $r_\mathrm{ir} \simeq 0.14$ pc (Weisskopf et al. 2000), seems too good to be coincidental which is why the inner ring is often identified directly with the location of the TS. In terms of the light cylinder we have $r_0 = 5 \times 10^8 \varpi_\mathrm{LC}$ which means that both rotation and radial magnetic field are subdominant at the scale of the TS. We will hence proceed in the "toroidal approximation": $v_\phi/c \approx 0$, $B_r/B_\phi \approx 0$.

Given the anisotropy of the wind power (2) and (4), the TS will attain pressure equilibrium at different locations: near the poles, the wind power steeply declines and the shock retreats back towards the pulsar. In the equatorial plane where the wind power is maximal on the other hand, the shock will bulge out further. This mechanism provides a simple explanation to the torus structure observed in many PWN (Lyubarsky 2002). In Sect. 5, more detailed calculations concerning the shape of the oblique shock and radiative signatures of particles emitting close to the shock will be discussed.

Depending on their local wind magnetisation, streamlines that pass over the shock behave qualitatively very different. In high-σ regions i.e. close to the axis, the flow can remain highly relativistic with downstream Lorentz factors larger than $\sigma^{1/2}$, the value obtained for the perpendicular shock. For $\sigma \to \infty$, streamlines simply continue to flow out radially after traversing the shock. In finite-σ regions with non-vanishing shock compression, the force-free equilibrium of the wind is destroyed and flow-lines start curving towards the pole due to the pinch-force of the magnetic field. This mechanism of jet formation due to the magnetic hoop-stress has been observed in axisymmetric (e.g. Komissarov and Lyubarsky 2003) and full 3D simulations (e.g. Porth et al. 2014a). In the low-sigma, e.g. striped regions, due to the conservation (compression) of the tangential (normal) velocity fields, streamlines experience shock aberration and instead get deflected towards the equatorial plane. This leads to the occurrence of "wisps" emitted from the shock to be further discussed in Sect. 3.

2.5 Nebula

The stationary solutions of the system (5)–(8) in the toroidal approximation were exhaustively studied by Kennel and Coroniti (1984a). They showed that assuming the wind is ultra relativistic $\Gamma_1 \gg 1$ and expands adiabatically after encountering a strong shock, e.g. $\Gamma_1 \gg \Gamma_2$, $\hat{\gamma} = 4/3$, the flow can be parametrised entirely in terms of the upstream σ (here subscripts 1 and 2 indicate the upstream and downstream medium, respectively). The asymptotic flow velocity in the shocked nebula then becomes

$$v_\infty = \frac{\sigma}{1+\sigma} c \tag{17}$$

and hence it is clear that the boundary condition $v_{\mathrm{n}} \approx 2000\ \mathrm{km\,s^{-1}}$ could only be satisfied by requiring low values of $\sigma < 10^{-2}$. Going further, assuming we indeed have a weakly magnetised wind, the large sound-speed in the shocked bubble will quickly equilibrate density and pressure and thus (5) implies $v_r \propto r^{-2}$ after crossing over the termination shock (TS). Then, from (8), we find $B_\phi \propto r$. The initially sub-dominant magnetic energy increases rapidly until equipartition with the thermal energy is obtained. From then on, since $v_r(r) > v_\infty$ the velocity approaches a constant value and the magnetic energy decreases again $B_\phi \propto 1/r$. This dynamic behaviour further tightens the limit on σ in the toroidal approximation, e.g. Kennel and Coroniti (1984a) obtained a best fit to the observed boundary conditions of Crab nebula for $\sigma \simeq 0.003$.

An additional argument for low magnetisation of the injected wind was put forward by Rees and Gunn (1974): ignoring the spin-down of the pulsar for the moment, each turn adds one "loop" of magnetic field and magnetic flux conservation implies $B_\phi \sim t$. Thus for the magnetic energy accumulated in the nebula we have $\mathcal{E}_{\mathrm{mag}} \sim t^2$. At the same time, particle energy is injected at constant rate $\mathcal{E}_{\mathrm{part}} \sim t$. In order to arrive at todays approximate equipartition $\mathcal{E}_{\mathrm{mag}} \sim \mathcal{E}_{\mathrm{part}}$ (and accounting for nebula expansion and spin down of the pulsar), Rees and Gunn (1974) estimate a wind magnetisation of $\sigma \approx 0.01$.

The elegance and simplicity of the Kennel and Coroniti (1984a) model (KC) and the Rees and Gunn (1974) argument renders KC models a standard tool to recover parameters of observed PWN (e.g. Sefako and de Jager 2003; Petre et al. 2007) although there are significant problems in reproducing the X-ray spectral index maps of several PWN (Reynolds 2003; Tang and Chevalier 2012; Nynka et al. 2014; Porth et al. 2016). These deficiencies, the uncomfortably small value of σ and the obviously non-spherical *jet and torus* morphology revealed by the Chandra X-ray telescope (Kargaltsev and Pavlov 2008) make a strong case for multi-dimensional models of PWN. To date, a number of groups have carried out axisymmetric relativistic MHD simulations of pulsar wind nebulae (Komissarov and Lyubarsky 2003, 2004; Del Zanna et al. 2004; Bogovalov et al. 2005; Camus et al. 2009; Olmi et al. 2014). Although rather different computer codes were employed, all these simulations produce quite similar results: the numerical solutions reproduce well the observed jet-torus structure and suggest dynamical explanations to the wisps (Sect. 3) and curious inner knot (Sect. 5).

Despite these successes, the axisymmetric models can not significantly lift the tension on the wind magnetisation parameter: if $\sigma > 0.1$, the termination shock is pushed far back to the pulsar and excessively strong jets develop that even punch through the nebula bubble (e.g. Porth et al. 2013). Both effects are linked to the accumulating hoop stress caused by conservation of the toroidal magnetic flux. In order to accommodate wind $\sigma \approx 1$ with the observations, two additional ingredients are essential: 1. Destruction of hoop-stress via fluid instabilities (no more toroidal approximation) and 2. Magnetic dissipation in the nebula volume.

The fact that the toroidal (z-pinch) equilibria used in the axisymmetric modelling (e.g. Begelman and Li 1992) are unstable to the MHD kink instability was first discussed by Begelman (1998). The authors speculated that the disrupted configuration may be less demanding on the magnetisation of pulsar winds. Indeed, one would expect the magnetic pressure due to randomised magnetic field to dominate the mean Maxwell stress tensor, and the adiabatic compression to have the same effect on the magnetic pressure as on the thermodynamic pressure of relativistic gas. Under such conditions, the global dynamics of PWN produced by high-σ winds may not be that much different from those of PWN produced by particle-dominated winds. These expectations have received strong support from numerical studies (Mizuno et al. 2011) of the magnetic kink instability for the cylindrical magnetostatic configuration. These simulations have shown a relaxation towards a quasi-uniform total pressure distribution inside the computational domain on a dynamical time-scale.

In addition to the dissipation of magnetic stripes in the wind or at the termination shock, the magnetic dissipation could occur inside the PWN as well (Lyutikov 2010; Komissarov 2013). In fact, the development of the kink instability near the axis and Kelvin-Helmholtz instabilities operating in the equatorial region (Camus et al. 2009) are bound to facilitate such dissipation. In principle, simultaneous observations of both the synchrotron and inverse-Compton emission allow a measurement of the energy distribution between the magnetic fields and the emitting elcctrons (and positrons). From a simple "one-zone" model of the Crab nebula, it follows that its magnetic energy is only a small fraction, $\sim 1/30$, of the energy stored in the emitting particles (Meyer et al. 2010; Komissarov 2013). Thus unless the striped zone fills almost the entire wind volume, additional magnetic dissipation inside the nebula is required.

2.6 Insights from 3D Simulations of PWN

With the discussion of the previous sections, the setup of a PWN simulation becomes straight-forward: first the total energy flux of the wind $f_{\rm tot}(r,\theta)$ needs to be chosen. Typi-

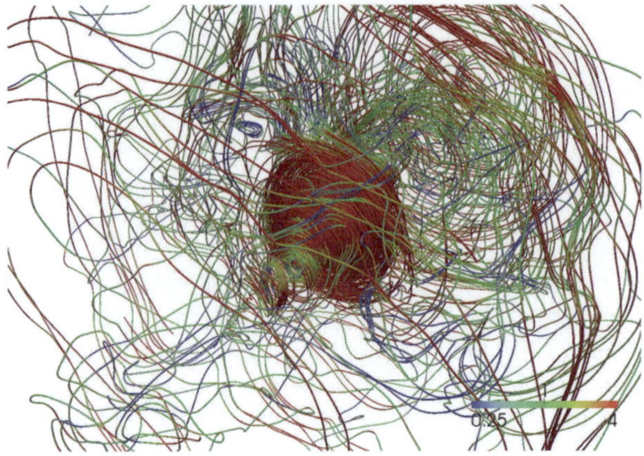

Fig. 2 Rendering of magnetic field lines in 3D simulations of PWN. *Colour* indicates B_ϕ/B_p: in the innermost regions, toroidal field (*red*) is injected which randomises further out and develops a poloidal component (*blue*). The kink instability of the leads to the corkscrew morphology of the inner jet

cally the energy flux is assumed to follow the Poynting flux of the aligned (2) or perpendicular rotator (4), however more sophisticated formulas for oblique rotators are now also being applied (see Bühler and Giomi 2016). Then the striped zone is modelled by choosing an appropriate function for $\chi_\alpha(\theta)$. Here the prescriptions of various groups differ somewhat (cf. Eq. (5) of Porth et al. 2013, and Eq. (2) of Olmi et al. 2014). With the effective wind-magnetisation given by Eq. (14), the magnetic and kinetic energy fluxes follow to

$$f_{\mathrm{m}}(r,\theta)=\sigma(\theta)\frac{f_{\mathrm{tot}}(r,\theta)}{1+\sigma(\theta)};\qquad f_{\mathrm{k}}(r,\theta)=\frac{f_{\mathrm{tot}}(r,\theta)}{1+\sigma(\theta)}. \tag{18}$$

Adopting the wind Lorentz-factor Γ, one can then solve for the density and magnetic field strength of the cold wind ($p \ll \rho$)

$$\rho(r,\theta)=f_{\mathrm{k}}(r,\theta)/\big(\Gamma^2c^2v_r\big);\qquad B_\phi(r,\theta)=\pm\sqrt{4\pi f_{\mathrm{m}}(r,\theta)/v_r} \tag{19}$$

where we have taken advantage of the toroidal approximation. Note that the magnetic field still reverses polarity from the northern to the southern hemisphere which gives rise to a magnetic null line within the nebula bubble. Due to numerical reasons, a realistic Lorentz-factor of $\sim 10^5$ cannot be realised in the simulations which typically set $\Gamma \sim 10$. In the hydrodynamic regime, it can be shown (e.g. Komissarov and Lyubarsky 2004) that the jump conditions of the ultrarelativistic shock are fairly well approximated already at this low value of Γ. Finally, the outer nebula boundary is modelled by assuming a supersonically expanding hydrodynamic shell of stellar ejecta with velocities at the inner edge $\sim 1000\ \mathrm{km\,s^{-1}}$ (e.g. Del Zanna et al. 2004).

Contrary to axisymmetric simulations, the 3D case allows the magnetic field to deform and develop a poloidal component. Since injection of fresh toroidal field competes with fluid instabilities in the downstream flow, whether the telltale jet-and-torus structure can also be established in 3D is not entirely obvious. In Fig. 2, a rendering of field lines from 3D simulations illustrates the process. Indeed, toroidally dominated field is present only in the direct vicinity of the TS. The strong jets observed for $\sigma = 1$ in axisymmetric simulations do not survive in 3D. In fact, the jet rapidly becomes unstable to the $m = 1$ kink instability as predicted by Begelman (1998). Nonetheless, outflow velocities in the polar plume reach $\approx 0.7c$, similar to the pattern speeds observed in Crab ($\sim 0.4c$) (Hester et al. 2002) and Vela ($0.3c$–$0.7c$) (Pavlov et al. 2003).

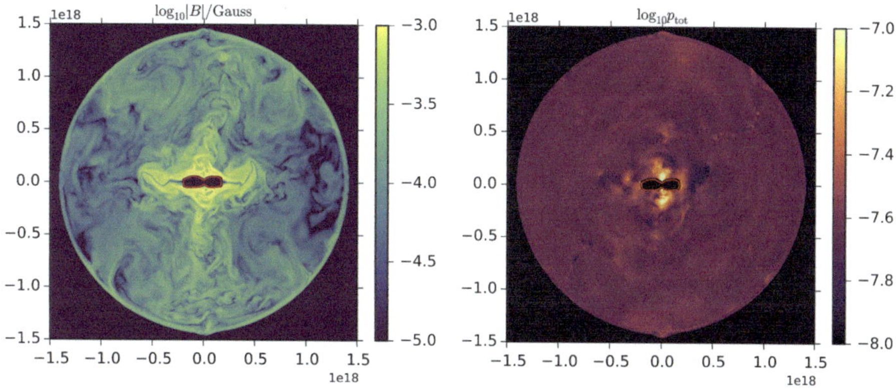

Fig. 3 Global views of the nebula showing magnetic field strength (*left*) and total pressure in the nebula (*right*). There is a significant range in magnetic field strength, with $|B| \approx 1$ mGauss near the TS and a volume-average of ≈ 100 µGauss. Apart from the hotspot in the dissipation region of the polar beam, the total pressure is close to uniform in the nebula

An overview of the large-scale dynamics in the 3D models is given in Fig. 3. As the z-pinch is rapidly destroyed, the total pressure is now more or less uniform and resembles the purely hydrodynamic models even for highly magnetised wind. Only at the base of the plumes flow compression is significantly enhanced. Because of the dramatically altered magnetic field distribution in the nebula, the termination shock does not dive back towards the pulsar and values compatible with the hydrodynamic prediction leading to Eq. (16) could be found even for $\sigma_0 = 3$, $\alpha = 45°$ (Porth et al. 2014a). The magnetic field strength exhibits a large range with maximal values in the "arch flow" just on top/below the oblique TS. As long as axisymmetry is approximately conserved, the magnetic field strength follows the scalings discussed in Sect. 2.5, hence the magnetic energy density increases until reaching equipartition with the plasma pressure. Throughout the torus region, the magnetic field remains strong and toroidally dominated, however fluid instabilities in the fast equatorial shear flow lead to turbulence starting at ≈ 2 shock radii. It is interesting to note that the shock compresses streamlines that carry magnetic field with opposing polarity towards the equator, turning the null-line into a large-scale current sheet. It is in this current sheet where the majority of magnetic dissipation takes place, a fact that was already observed in 2D simulations of Camus et al. (2009).

A closeup view of the TS is shown in Fig. 4 illustrating a typical flow field. In the vicinity of the TS, the predictions from axisymmetric calculations remain largely valid: we obtain a separation of flow lines into equatorial and polar flow. Velocities in the fast equatorial shear flow reach up to $\approx 0.5c$. Re-focussing of flow lines from the polar regions forms a plume-like vertical outflow and an inner highly unstable "polar beam". Due to the kink instability of the polar beam, the actual formation of the jet is offset from the TS—in good agreement with the observations. This dynamical behaviour also suggests an identification of the highly variable "Sprite" (see e.g. Hester et al. 2002) located at the base of the jet with the violently unstable polar flow.

It is interesting to note that (similar to the 1D KC84 model) due to expansion and deceleration of streamlines carrying a toroidal field, the magnetisation of the flow in mid-latitude regions can in fact increase over the value injected at the TS. The high σ regions in the nebula volume might present suitable locations for rapid particle acceleration in connection

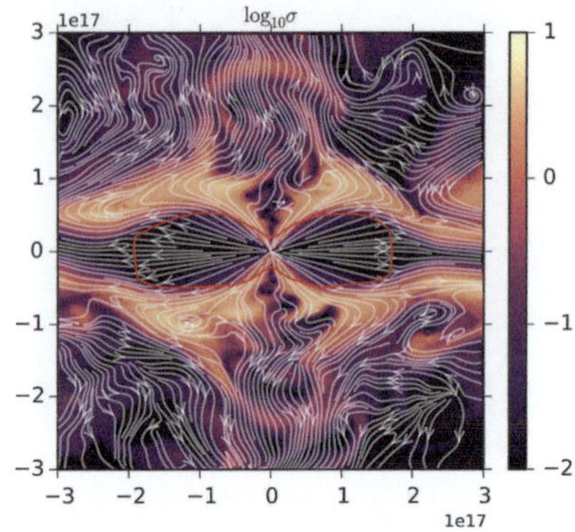

Fig. 4 Zoomed in view on the termination shock (*red contour*) showing the logarithm of the magnetisation. Parameters of the simulation are: $\sigma_0 = 1$, $\alpha = 45°$. In-plane streamlines (*white contours*) get deflected at the shock and turn towards the equator or polar axis, depending on the magnetisation. Note that the magnetisation in the nebula can be significantly higher by a factor ~ 10 than the one injected with the wind

with the Crab flares (e.g. Lyutikov et al. 2016b). Recent advances in modelling the elusive Crab flares will be discussed further in Sect. 4.

2.7 Radiative Predictions from Particle Transport Models

Radiative predictions are necessary to ultimately test our theoretical models and simulations with respect to observed systems. While the general non-thermal emission processes in PWN were rapidly identified, namely synchrotron radiation from non-thermal electrons and inverse Compton scattering of ambient radiation by the very same electrons (Shklovsky 1953; Dombrovsky 1954; Atoyan and Aharonian 1996), ever increasing spatial and temporal resolution ask for more and more refined models of the PWN emission. In their core, current emission models follow the suggestion of Kennel and Coroniti (1984b), that non-thermal electrons are accelerated at the TS and then follow streamlines in the nebula where they experience adiabatic and radiative losses. In the optical and X-ray bands where the cooling timescale is comparable to the crossing time, this promotes a "centre-filled" appearance and gradual steepening of the X-ray spectral index, observed especially in young PWN (e.g. Slane et al. 2000, 2004). Sophisticated methods to track the distribution of non-thermal particles injected at the TS via passive tracer scalars were devised (Del Zanna et al. 2006; Camus et al. 2009) and are now standard practice in MHD simulations of PWN. While acceleration at the TS appears as a good working assumption, the details of the particle acceleration process are far from understood and the large inferred acceleration efficiency appears in conflict with the superluminal nature of the shock in the framework of standard models. This will be discussed further in Sect. 3.2 where the observational signatures of various acceleration sites on the shock are compared.

On the qualitative level, synthetic maps of synchrotron emission from PWN are able to reproduce a stunning amount of features: foremost the famous jet-and-torus morphology whose dynamical origins were previously discussed, but also finer details seen primarily in the high resolution observations of the Crab nebula. We highlight the aforementioned sprite, the inner knot (Sect. 5) and the wisps (Sect. 3). As the details of energy injection into the PWN are intimately related to fundamental parameters of the rotating neutron star, one can attempt to constrain pulsar parameters with the morphology of the synchrotron emission in the nebula (e.g. Bühler and Giomi 2016). In this vein, in Fig. 5, we show synthetic emission maps of the region around the TS from two 3D simulations: one with parameters $\sigma_0 = 3$, $\alpha = 10°$ (left) and one with $\sigma_0 = 1$, $\alpha = 45°$ (right). While the right-hand panel provides a good match to the morphology of Crab, the torus structure is entirely suppressed in the left-hand panel which might find its likeness in one of the jet-dominated sources (Kargaltsev

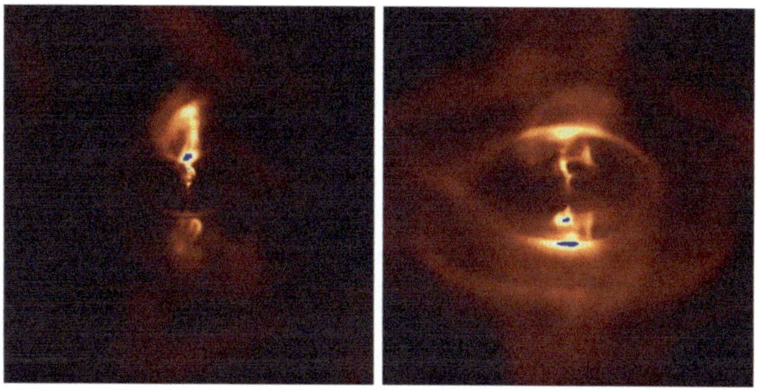

Fig. 5 Synthetic emission maps assuming energy of the non-thermal particles scales as $\epsilon \propto p$, i.e. assuming only adiabatic effects. Two simulations are compared: $\sigma_0 = 3$, $\alpha = 10°$ (*left*) and $\sigma_0 = 1$, $\alpha = 45°$ (*right*). In high-magnetisation, low obliqueness models, the torus and wisp regions are under-luminous and the emission from the jet (in fact the polar beam) dominates. Morphology and parameters of the right-hand model fit well to the optical images of the Crab PWN. *Blue colour* indicates where pixels were saturated (80% peak intensity) in order to better show faint features

and Pavlov 2008). This result is not surprising: as the extent of the striped zone is decreased, more streamlines are re-focussed into the jet (cf. Fig. 14 of Porth et al. 2014a). The recent 3D simulations presented by Olmi et al. (2016) confirm this finding. Hence the jet/torus flux ratio could yield a valuable handle on the pulsar obliqueness.

On the quantitative level, some progress could be made recently with detailed modelling of the inner knot of Crab nebula by Yuan and Blandford (2015) and Lyutikov et al. (2016a). If the knot-feature is caused by beamed emission right behind the TS, then the local magnetisation must be $\sigma < 1$, otherwise the observed properties of the knot cannot be reproduced. For an inclination of the Crab pulsar of 60° with respect to the plane of the sky and an assumed obliqueness angle of 45°, the knot falls into the striped zone and low post-shock magnetisation is in fact expected (see Sect. 5 for details).

Moving away from the TS, the uncertainties in radiative models increase: currently, the synthetic maps tend to over-predict the contrast between torus and ambient emission, as well as the optical and radio polarisation degree. For example, the unresolved polarisation degree of the Crab nebula in the optical band is $\Pi = 9.3 \pm 0.3\%$ (Oort and Walraven 1956), whereas the simulations produce an over three times higher value. Both effects are related, e.g. with decreasing emissivity in the ordered-field torus region as compared to the turbulent bulk, the average polarisation will also decrease. This points at several potential shortcomings of the current 3D models: 1. Their short duration does not allow the dense filaments due to the Rayleigh-Taylor instability to develop and penetrate deep into the nebula (e.g. Porth et al. 2014b), an effect that would likely increase the randomisation of magnetic field. 2. The finite resolution global simulations might over-predict the dissipation of magnetic field in the bulk. 3. The prescription of localised acceleration, occurring at the TS only, as in the KC84 original model, is possibly overly simplified and particle re-acceleration in the bulk might have to be taken into account. Future higher resolution and longer duration 3D simulations will need to address these questions.

Due to the inherent complexity of the data, emission maps from dynamical simulations are not well suited to yield quantitative information of PWN, and steps to reduce the dimensionality must be taken so that parameters can be obtained via model fits. In this sense, Volpi et al. (2008) fitted the spectral energy distribution (SED) of axisymmetric MHD simulations to the Crab nebula, taking into account two species of synchrotron emitting particles (radio- and optical/Xray-electrons) and realistic target photon fields for the IC emission. The result can be understood as a different flavour of the σ-problem: In order to reproduce the high-energy jet/torus morphology and due to the strong axial compression, the average field turns out to be 3–4 times smaller than predictions from KC84 models (de Jager and Harding 1992). Hence in order to arrive at the observed synchrotron flux, the electron density must be increased accordingly. This is problematic not only as it adds to the tension on the multiplicity parameter κ, but also since it leads to an exaggerated IC emission (see also the discussion in Amato 2014). With their more uniform distribution of the magnetic field even for high magnetisation, 3D models might present a way out of this dilemma and research in this direction is ongoing (Olmi et al. 2016).

An alternative technique to the advection of passive tracer scalars was followed by Porth et al. (2016): here, X-ray synchrotron electrons are treated as test-particles that experience the Lorentz force due to the MHD background fields. This allows to directly map out the advection and diffusion of particles embedded in the PWN flow. It was found that due to large velocity fluctuations, the transport exhibits a diffusive character and a typical diffusion coefficient of

$$D_{\mathrm{Ls}} \sim \frac{1}{3} v_f L_{\mathrm{s}} = 2.1 \times 10^{27} \left(\frac{v_f}{0.5c} \right) \left(\frac{L_{\mathrm{s}}}{0.42\ \mathrm{Ly}} \right) \mathrm{cm}^2\,\mathrm{s}^{-1} \tag{20}$$

is suggested following the turbulence with driving scale L_{s}—the size of the termination shock, and a typical velocity at this scale v_f. Note that the value of D_{Ls} for Crab nebula becomes very similar to the one estimated by Amato et al. (2000) and also to that obtained in the early model due to Weinberg and Silk (1976) of $1.9 \times 10^{27}\ \mathrm{cm}^2\,\mathrm{s}^{-1}$. As particles are mostly following the velocity field, e.g. the first order drift velocity in the toroidally dominated regions $\mathbf{v}_D = \mathbf{E} \times \mathbf{B}/B^2$ just corresponds to the flow velocity in the poloidal plane $\mathbf{v}_D = \mathbf{v}_p$, the diffusive transport is independent of the particle energy, as assumed already in the models of Gratton (1972), Wilson (1972). Further discussion on diffusion in PWN can be found in Tang and Chevalier (2012). With the spherically averaged MHD simulation as background, the particle transport taking into account advection, diffusion and synchrotron losses, can be fitted to X-ray observations of three young PWN confirming that diffusion makes an important contribution in PWN. In particular, the model of Porth et al. (2016) yields better fits to the gradual increase of the X-ray spectral index than a traditional KC84 model.

3 Wisps as Probes of the Particle Acceleration Mechanism in the Crab Nebula

PWN are among the most powerful accelerators in the Galaxy, with the Crab showing an acceleration efficiency of order 20% and producing particles up to PeV energies. It is a common opinion that the acceleration of those high energy particles might take place in the proximity of the wind termination shock. However the termination shock is a very hostile environment for acceleration, due to its nature of magnetised and relativistic shock. Relativistic shocks have indeed been proven to be efficient accelerators only in two cases: when the magnetic field is quasi parallel to the shock normal, in particular when it is sub-luminal (meaning that the angle between $\mathbf{B}$ and $\mathbf{v}_{\mathrm{shock}}$ is $\theta \ll \theta_c \simeq 1/\gamma_{\mathrm{shock}}$), or when the shock is poorly magnetised, with the magnetisation σ, defined as the ratio of Poynting flux to particle kinetic energy flux in the wind, below $\sim 10^{-3}$.

The observed spectrum of the Crab nebula, similar to other PWNe, seems to suggest different acceleration mechanisms for low and high energy emitting particles, since it is a broken power law $N(E) \propto E^{-\gamma_e}$, with $\gamma_e \sim 1.5$ at low energies (below ~ 100 GeV) and $\gamma_e \sim 2.2$ at higher ones (up to a few TeV).

At present, two main candidates have been invoked as possible mechanisms for particle acceleration at the Crab TS.

The slope of the radio component is compatible with driven magnetic reconnection. This mechanism was largely investigated by Sironi and Spitkovsky (2011) and Sironi et al. (2013), who have performed numerical simulations compatible with the Crab nebula's shock conditions.

They assume that the striped morphology of the pulsar wind is maintained all the way to the TS, where the compression causes the magnetic field to reconnect. Reconnection islands form in the flow and particles can be accelerated by the un-screened electric field. The properties of the resulting spectrum depend on the flow magnetisation and on the ratio between the wavelengths of the stripes and the particle Larmor radii. This ratio can be connected to the value of the pulsar pair multiplicity κ, and the result in the case of the Crab nebula is that, in order to reproduce the observed radio spectrum, a magnetisation of $\sigma \gtrsim 30$ and a multiplicity of $\kappa \gtrsim 10^8$ are needed. We remind the reader that the already mentioned pulsar multiplicity κ is related to the efficiency with which particles are generated in the pulsar magnetosphere, namely it is defined as the number of pairs produced by a single primary electron that emerges from a polar cap of the pulsar.

The main problem with this scenario is that it is very difficult to explain such a high value of the multiplicity: from observations and theoretical models we expect much lower values, with $\kappa \sim few \times 10^5$ at most (Timokhin and Harding 2015). The only possibility to reduce the required value of κ is to accelerate particles near the polar cusps of the TS, where the shock front is nearer to the pulsar, and as a consequence the particle density is much higher (given that in the wind it decreases as $\approx 1/r^2$). On the other hand, no stripes are expected to be present at these locations, and reconnection can only be obtained due to O-point dissipation.

Moreover, a wind with the required high number of pairs, $\kappa \sim 10^8$, is expected to reconnect well before the TS (Lyubarsky and Kirk 2001), causing the magnetisation to lower well below the required minimum for the mechanism to be viable (Amato 2014).

A different mechanism can be invoked to account for the much steeper high energy component of the spectrum. A photon index of $\gamma_e \simeq 2.2$ is in fact what one expects from Diffusive Shock Acceleration (DSA), i.e. Fermi I like acceleration at relativistic shocks. Contrary to driven magnetic reconnection, DSA is effective only if the magnetisation is sufficiently low, in particular $\sigma \lesssim 10^{-3}$ is required (Sironi and Spitkovsky 2009). MHD simulations taught us that, if the emission from Crab has to be reproduced, this condition can only be realised in a thin latitude strip around the pulsar equator or in the vicinities of the polar axis, while at intermediate latitudes the flow magnetisation must be substantially larger than this, likely σ of order a few (Komissarov 2013). Therefore, at least DSA can only be effective in a small sector of the shock, where magnetic dissipation is sufficiently strong so as to ensure a local condition of quasi-unmagnetised plasma (Amato 2014).

The other possibility that has received some attention in the literature is resonant absorption of ion-cyclotron waves in a ion-doped plasma (Hoshino et al. 1992; Amato and Arons 2006). This requires that most of the energy of the wind is carried by ions and its viability does not depend on the value of the magnetisation. The idea is that pairs were accelerated by resonant absorption of the cyclotron radiation emitted by ions in the wind, which are set into gyration by the enhanced magnetic field at the shock crossing (Amato 2014).

At present there are no indications for significant role of ions in the pulsar wind energetics, or that ions are present at all. The observations of high energy neutrinos provides the primary tool to assess this matter (e.g. Amato et al. 2003). While current limits, set by

6 years of observations with IceCube, are still not very stringent, upcoming neutrino facilities are promising to either detect neutrinos from a few bright PWNE or to constrain the fraction of pulsar wind energy flux carried by ions to levels below 10% (Di Palma et al. 2016). Given the current uncertainties, however, the possibility that electron acceleration occurs through resonant absorption of ion cyclotron waves will not be discussed further in this work.

3.1 Wisps from Crab

The inner region of the Crab nebula has been known to be highly variable at optical wavelengths since the late 60s, when Scargle (1969) first identified some bright, arc-shaped and variable features, that he named "wisps". They appear to be periodically produced at about the expected location of the TS, which can be seen as the first bright ring surrounding the central dark region of the pulsar wind, and then move outwards with mildly relativistic velocities, as one expects for a post-shock hydrodynamic flow. Their typical period between appearance close to the TS and disappearance in the bulk of the nebula can be of the order of days, months or even years, depending on the energy band in which they are observed (X-rays, optical or radio respectively). They also appear to be more prominent in the north-west sector than in the south-east.

Few years after the first wisps identification, the properties of such features were extensively studied at different wavelengths, in particular focussing on radio and optical bands, and more recently also on the X-ray band. The most important conclusion was that wisps at different wavelengths are not coincident and that they are seen to propagate at various speeds (Bietenholz et al. 2004; Schweizer et al. 2013). The observed discrepancies between radio and optical wisps led (Bietenholz et al. 2004) to conclude that the two populations must have a different acceleration mechanism and/or site, and possibly the same conclusion might be drawn for X-ray wisps.

3.2 Wisps as Probes of the Particle Acceleration

As we discussed previously, the different acceleration mechanisms proposed require very different physical conditions to be viable, which are not realised everywhere along the shock surface. Identifying the location at which particles are accelerated can then put strong constraints on the acceleration mechanism at work.

One way to test different scenarios of acceleration is by comparison with observations of the variability they entail in the inner nebula at various frequencies: since wisps are seen to start so close to the TS, at least at optical and X-ray frequencies, where radiative losses are important, they trace freshly injected particles, and in the simulated maps, their appearance and motion depends on the location at which the emitting particles are injected in the nebula.

In Olmi et al. (2014) it was shown that wisps' properties at radio wavelengths are well reproduced without invoking any *ad hoc* mechanisms, with the bulk flow of the nebula acting as the main driver for the observed wisps appearance and motions. In MHD models, wisps appearance is thus totally due to the combined effects of the locally enhanced magnetic field, just downstream of the TS, and the Doppler boosting effect, since channels with significant v/c form along oblique sectors of the shock surface. In particular, Doppler boosting is responsible for the angular profile of the wisps, as well as for the enhanced brightness of the front side of the nebula with respect to the back side, that appears to be very faint. The intensity contrast between distinct wisps is, on the contrary, strongly connected to the local magnetic field strength.

Within an MHD description, the fact that wisps arise at different wavelengths with different properties (different locations and outward velocities) suggests a difference in the acceleration sites of the particles responsible of such emission. In fact, assuming that the emitting particles are all accelerated at the TS, including the radio component, the discrepancies can only be explained by choosing different acceleration regions along the TS for distinct distributions. If particles with different energies are injected at different latitudes along the TS, the paths induced by the post-shock flow structures, and the adiabatic and synchrotron losses, will also be different, and features at different energies are expected to be not coincident, as observed.

In Olmi et al. (2015) axisymmetric relativistic MHD simulations are used to constrain the acceleration sites of particles responsible for the observed wisps at the different frequencies in the Crab nebula.

Considering several different scenarios, with particles of different energies being injected in different sectors of the TS, with polar and equatorial cones defined with various angular extents, wisp properties are extracted on top of the simulated emission maps at various wavelengths.

The entire Crab nebula is simulated up to its present age ($\sim$ 1000 yr), with outputs sampled with monthly frequency during the last 10 years of evolution. In analogy with the observation based study carried out by Schweizer et al. (2013), wisp intensity profiles are extracted from radio, optical and X-ray emission maps from a $3''$ wide slice centred on the polar axis in the upper hemisphere of the nebula, where wisps are seen to be more prominent. For ease of comparison with observational data, profiles are also convolved with the appropriate instrumental PSF (relative to the instrument used for the data considered for comparison in each band) and only intensity peaks with $I \geq I_{\max}/3$ are taken into account, where $I_{\max}$ is the maximum value of the intensity in each map. This cut off is applied in order to remove the background of weaker variations, that are not useful for the comparison with data.

This procedure is then repeated for each output in the considered interval of 10 years, and for each one of the defined cases for particles injection, namely:

1. particles are injected uniformly along the entire shock front, for all the three families;
2. particles are injected in a wide equatorial sector along the shock ($\theta \in [20°, 160°]$) or in the opposite narrow polar one ($\theta \in [0°, 20°] \cup [160°, 180°]$);
3. particles are injected in a narrow equatorial sector ($\theta \in [70°, 110°]$) or in the opposite wide polar one ($\theta \in [0°, 70°] \cup [110°, 180°]$).

The polar and equatorial angular sectors of case (2) and case (3) are represented, respectively, by the green and red cones of Fig. 6, where the structure of the velocity field is shown in the proximity of the TS.

An explanation for the choice of these injection sectors is in order. The equatorial sector of case (3) roughly represents the wind striped region in the considered wind model (for a complete description see Olmi et al. 2014), while the polar sector of case (2) mimics a scenario in which particles can be accelerated in the low-σ region around the polar axis thanks to Fermi I acceleration or O-point dissipation.

3.3 Wisp Profiles in the Different Scenarios

The three hypotheses are tested for radio, optical and X-ray particle families. In particular two different distribution functions are defined to account for radio and X-ray emission, while the optical one is obtained as a mixed contribution of the other two. Since radio and

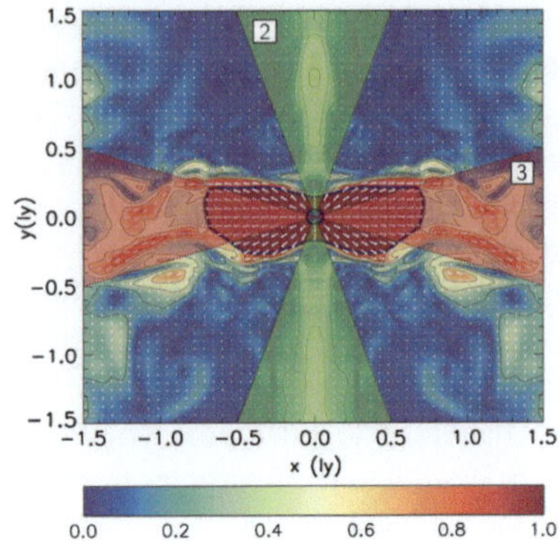

Fig. 6 Plot of the flow structure, with colours indicating the velocity magnitude in terms of c and arrows the velocity field direction. The TS front is highlighted by the *blue contour line* of $0.99c$. Sectors corresponding to case (2) (polar) and case (3) (equatorial) are represented, respectively, by the *green* and *red cones*

X-ray distribution functions are normalised in order to fit the complete integrated spectrum of Crab, the optical spectrum is naturally determined as the superposition of radio and X-ray contributions.

Radio particles are injected with the following spectrum:

$$f_{0\mathrm{R}}(\epsilon_0) \propto \begin{cases} 0 & \text{if } \epsilon_0 < \epsilon_{\mathrm{minR}}, \\ \epsilon_0^{-p_\mathrm{R}} \exp(-\epsilon_0/\epsilon_\mathrm{R}^*) & \text{if } \epsilon_0 > \epsilon_{\mathrm{minR}}, \end{cases} \tag{21}$$

and X-ray ones with:

$$f_{0\mathrm{X}}(\epsilon_0) \propto \begin{cases} 0 & \text{if } \epsilon_0 < \epsilon_{\mathrm{minX}}, \\ \epsilon_0^{-p_\mathrm{X}} \exp(-\epsilon_0/\epsilon_\mathrm{X}^*) & \text{if } \epsilon_0 > \epsilon_{\mathrm{minX}}, \end{cases} \tag{22}$$

where ϵ_0 is the Lorentz factor of the particle at the injection site.

Power-law indices and cut-off energies that appear in the previous formulas, are all determined based on the comparison of the simulated emission with data. The best set of those parameters results to be: $p_\mathrm{R} = 1.6$, $\epsilon_{\mathrm{minR}} = 10^3$, $\epsilon_\mathrm{R}^* = 2 \times 10^6$, $p_\mathrm{X} = 2.8$, $\epsilon_{\mathrm{minX}} = 1.5 \times 10^6$ and $\epsilon_\mathrm{X}^* = 10^{10}$.

The first interesting point to look at is the difference between wisps profile under case (1) assumptions (i.e. uniform injection at TS) and one of the other two cases. In Fig. 7 wisp profiles are shown as plots of the radial position of the intensity maxima as a function of time, with case (1) on the left and case (3) (X-ray particles injected in the equatorial sector and radio in the polar one) on the right. As expected, wisps appear to be coincident at the different wavelengths in the hypothesis of case (1). This is no more true as soon as particles of different families are injected in different sectors of the shock, as can be easily seen in the plot on the right.

Excluding the case of uniform injection, which clearly does not reproduce the expected behaviour of wisps, the remaining possibilities are case (2) and (3). All the different scenarios of these two cases are shown and compared in Fig. 8. Here particles responsible for radio and X-ray wisps are injected either in the polar sector (stars and crosses) or in the equatorial region (diamonds and circles).

When particles are injected in the wide equatorial region of case (2), wisps appear to be almost identical of those of the uniform injection case. On the contrary, particles injected in the narrow polar sector are not seen to produce wisp-like features, with the most prominent structure being a quasi-stationary feature at a distance of $5''$ from the pulsar, which is something very different from what observations show.

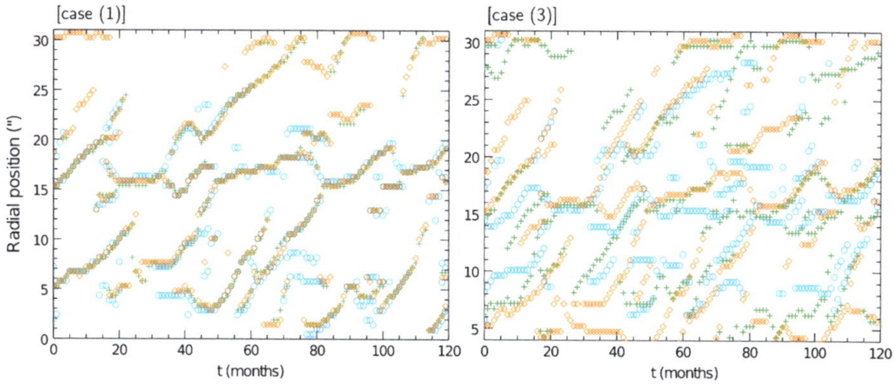

Fig. 7 Radial positions of the local intensity maxima (in arcseconds) as a function of time (in months) with orange diamonds identifying radio wisps ($\nu_r = 5$ GHz), *green crosses* optical ones ($\nu_o = 3.75 \times 10^{14}$ Hz) and *light-blue circles* for X-rays (1 keV). *On the left* case (1) is shown. *On the right* case (3) is shown, with X-ray particles injected in the equatorial zone and radio ones injected in the complementary sector

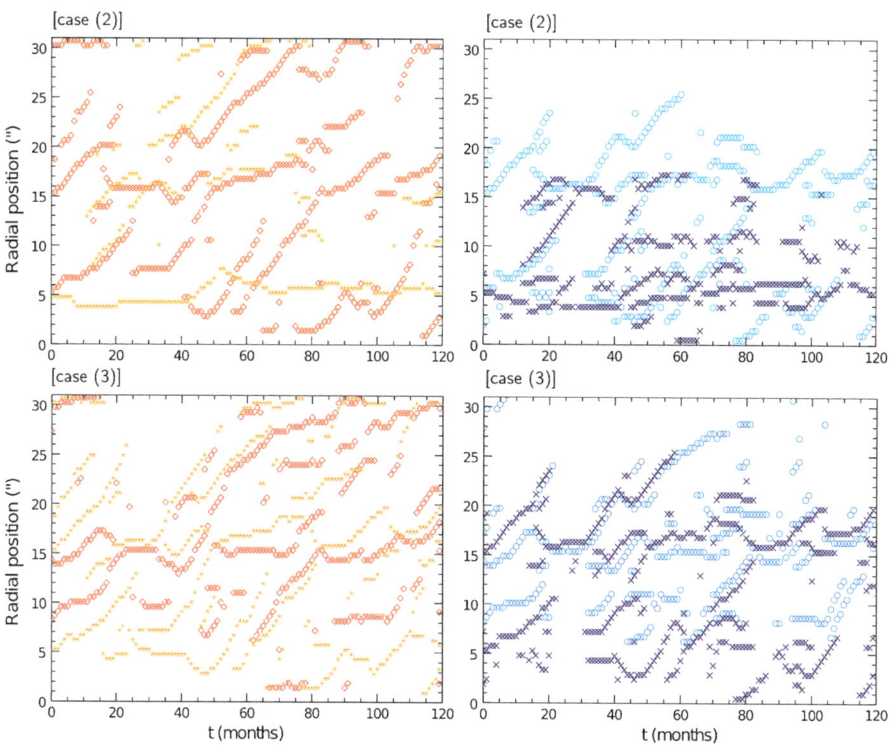

Fig. 8 Radial position of the local intensity maxima as a function of time for case (2), in the *upper row*, and case (3), in the *bottom* one. *On the left*, different possibilities for the injection of radio particles are shown. *Stars and diamonds* represent, respectively, polar and equatorial injection. On the *left-hand side*, the same cases are shown for X-ray particles: *violet crosses* are used for polar injection and *light blue circles* for equatorial one

Under hypotheses of case (2) it is thus impossible to define two non coincident sectors in which radio and X-ray particles can be injected, so as to give rise to different wisps at the different energies.

Finally, in the bottom row of the same Fig. 8, the alternative injection scenarios considered in case (3) are shown. Here wisps appear in any case, no matter whether particles are injected in the wide polar sector or they are injected in the narrow equatorial region. The strongest constraint for discriminating between the possibilities comes from Schweizer et al. (2013), where X-ray wisps were shown to be present only beyond $\sim 6''$ from the pulsar. The only case in which this behaviour is correctly reproduced is when X-ray particles are injected in the equatorial region, which approximatively corresponds to the striped zone of the wind, where dissipation of magnetic field is most efficient. This may indicate that X-ray particles are produced via Fermi I acceleration in the striped zone, where magnetisation can be low enough to make this mechanism viable.

Outward apparent velocities of wisps at the different frequencies have also been investigated for case (3), leading to a range of $0.08c \lesssim v \lesssim 0.38c$, which is in good agreement with what observed.

Strong constraints on radio emission are on the other hand very difficult to draw, and the only case that can be effectively excluded is the one in which radio particles are injected in a narrow polar cone. But, in order to have non coincident wisps at the different wavelengths, radio particles must be accelerated elsewhere than the equatorial region of case (3). The remaining possibilities are thus that acceleration happens in a wider equatorial region or in the complementary polar zone with respect to X-ray particles, where conditions for driven magnetic reconnection to be at work might be locally satisfied.

4 Gamma-Ray Flares of the Crab Nebula

4.1 The Flare Observations

Particle acceleration was observed in real time for the first time in the Crab nebula with the discovery of the gamma-ray flares in 2010. The nebula emission was expected to be stable over time scales of years. It therefore came as a huge surprise when the AGILE and Fermi-LAT satellites observed strong high energy (HE, > 100 MeV) gamma-ray flares (Tavani et al. 2011; Abdo et al. 2011a). When observed in gamma-ray band with energies above 100 MeV, the nebula emission is highly variable. In this energy band one observes the high energy end of the synchrotron component, and the onset of the inverse Compton component in the SED of the nebula, as shown in Fig. 9. The monthly flux variations observed in this frequency band in the first 8 years of the Fermi-LAT mission are shown in Fig. 10. A statistical analysis of the flux variations shows that the nebula emission in HE gamma-rays varies on all time scales that can be resolved by current instruments. The power density distribution (PDS) of the frequency of flux variation can be described by a red noise process with an index of $PDS \propto \nu^{-0.9}$ from yearly to daily time scales (Buehler et al. 2012).

Extreme gamma-ray outbursts are observed approximately once per year. During these flares the photon flux above 100 MeV can increase up to a factor ≈ 30. During the two brightest flares in April 2011 and March 2013 flux doubling time scales of $t_{var} \approx 6$ hours were observed (Buehler et al. 2012; Mayer et al. 2013). The isotropic luminosity release at the peak of these flares was $L_{\gamma,iso} \approx 2 \times 10^{36}\ \mathrm{erg\,s^{-1}}$, about 1% of the spin-down power of the pulsar. The temporal structure of the flares is generally different between them, as are their spectral properties. Their photon flux, f_{ph}, as a function of energy is typically well

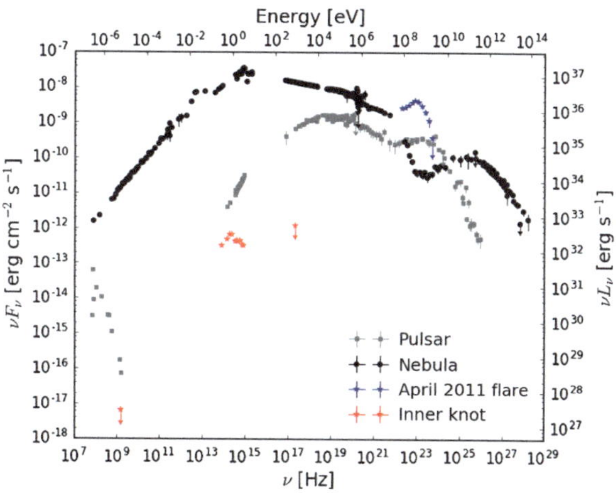

Fig. 9 SED of the Crab nebula and pulsar. See Bühler and Blandford (2014) for the references to the data from the average nebula and pulsar SEDs. The Crab knot spectrum is also shown, its data was taken from Lobanov et al. (2011), Sandberg and Sollerman (2009), Rudy et al. (2015). Also shown is the SED at the maximum of the April 2011 gamma-ray flare from Buehler et al. (2012)

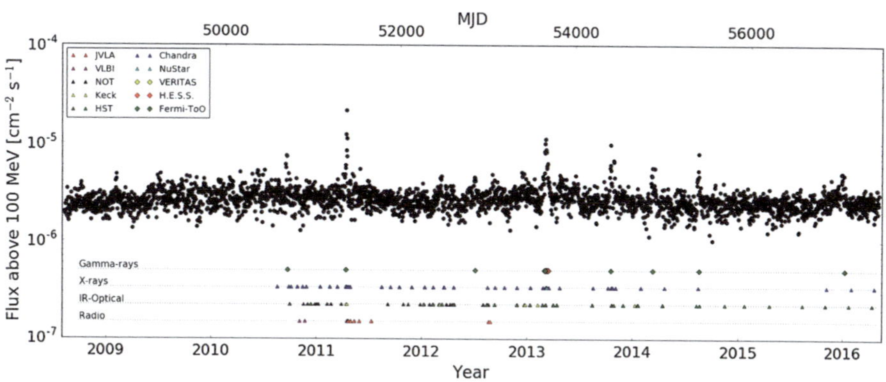

Fig. 10 The *black markers* show the daily averaged gamma-ray flux above 100 MeV from the direction of the Crab nebula measured by the Fermi-LAT. This data is routinely updated and publicly available at http://fermi.gsfc.nasa.gov/ssc/data/access/lat/msl_lc/. *Coloured markers* indicate the times at which instruments at other wavebands observed the source (see text)

described by a power law $f_{ph} \propto E^{-\gamma}$, with wide range of photon indices ranging between $\gamma \approx 1.2 - 3.7$. For the flares of April 2011 and March 2013, a high energy cutoff of the spectrum could be statistically resolved (see Fig. 9). The cutoff energy is approx $E_{cut} \lesssim 1$ GeV. This appears to be a general feature of the flare spectra, for no significant emission was detected beyond this energy. Interestingly, the time evolution of the cutoff could be resolved in time during the April 2011 flare. It was observed that the total energy flux of the gamma-ray emission, f_e, scales as a function of the cutoff energy as $f_e \propto E_{cut}^{3.4\pm0.9}$ (Buehler et al. 2012).

Unfortunately, the angular resolution in gamma-rays is not sufficient to pinpoint the site of the emission region within the nebula. However, observations at lower frequencies can resolve the nebula morphology in great detail. Therefore, extensive multi-wavelength campaigns have been in place since 2010 to find correlated emission to the gamma-ray flares at lower frequencies. Some of the most sensitive instruments in the world participated in these multi-wavelength campaigns, as the Hubble Space Telescope (HST), the Keck Observatory, Chandra X-ray Observatory (Weisskopf 2013; Rudy et al. 2015), the Very Long Baseline

Interferometer (Lobanov et al. 2011) the Jansky Very Large Array (Bietenholz et al. 2014), the H.E.S.S., VERITAS and MAGIC Cherenkov telescopes (Abramowski et al. 2014; Aliu et al. 2014; Aleksić et al. 2015) and NuSTAR (Madsen et al. 2015). The dates of these observations are shown in Fig. 10. In total more than 1 Msec of observations were taken by these instruments since 2010. Simultaneous observations on monthly intervals were carried out and additional observations were taken in intervals of a few days during times in which the HE gamma-ray flux was high. Surprisingly, to date, these observations have not revealed any emission correlated to the gamma-ray flares (Weisskopf 2013; Rudy et al. 2015).

4.2 Theoretical Models of the Flares

The Crab nebula flares pose severe challenges to models of particle acceleration. It is usually assumed that the flare emission originates from synchrotron radiation of freshly accelerated electrons. The reason is that the high gamma-ray energy of the flares (≈ 1 GeV) implies the presence of electrons with multi PeV energies for typical magnetic fields of $B \approx 200$ μG expected in the nebula. The cooling time of these particles is $\tau_{cool} \lesssim 20$ days. Therefore the emission region cannot be far from the acceleration site (however, other views have also been proposed, see e.g. Bykov et al. 2012 and Zrake 2015). Assuming the acceleration and radiation region are the same, the high luminosity of the flares poses severe constraints on the acceleration efficiency during these events: the total isotropic fluence during the brightest flares is $\varepsilon \approx L_{\gamma,iso} \times t_{var} \approx 4 \times 10^{40}$ erg (again adopting $t_{var} \approx 6$ hours). Causality implies that the emission region has a volume $\approx (ct_{var})^3$ if there is no strong Doppler beaming. The magnetic energy in such a region is $\varepsilon_B \approx \frac{c^3}{8\pi} t_{var}^3 B^2$. The radiation efficiency can then be estimated as $\epsilon \equiv \varepsilon/\varepsilon_B \approx 3700 \times B_{-3}^{-2}$, where B_{-3} is the magnetic field measures in mG.

The RMHD simulations discussed in Sect. 2 show that magnetic fields reach a maximum of a few mG within the body of the nebula. This implies a radiation efficiency $\epsilon \gg 1$ for the flare emission. This contradiction shows that flare emission cannot be isotropic and that relativistic beaming is likely playing an important role. However, it is not easy to explain highly relativistic plasma motions in the nebula: these are not found in RMHD simulations and observations show only mildly relativistic speeds, of $\approx 0.5c$ on the resolvable spatial scales. It is therefore likely, that a combination of boosting and a high magnetic field in the emission region is required. Furthermore, the field energy in this macroscopic region must be transferred very rapidly to kinetic particle energy (macroscopic compared to the kinetic scales as the plasma skin depth).

The general concept of catastrophic dissipation of magnetic energy to non-thermal particles on macroscopic scales has been named *magnetoluminescence* by Blandford et al. (2014). The process might be triggered by ideal MHD instabilities on large scales and produce regions with non-ideal conditions where rapid particle acceleration takes place. Recently, several works have studied this concept using force-free/MHD (East et al. 2015; Lyutikov et al. 2016b; Zrake and East 2016) and PIC simulations (Lyutikov et al. 2016b; Nalewajko et al. 2016; Yuan et al. 2016). For example, East et al. (2015) found that highly magnetised plasma configurations, in the form of multiply packed flux ropes in a 3D periodic box, can be unstable to ideal modes if they possess "free energy"—the amount of energy dispensable while preserving the overall topology. These instabilities evolve rapidly and release the magnetic free energy over a single dynamic time scale. PIC simulations further showed that the large scale instability forces current sheets to form self-consistently over dynamic time scales—these are the main sites of particle acceleration and electromagnetic dissipation (Lyutikov et al. 2016b; Nalewajko et al. 2016; Yuan et al. 2016). Similar to the planar current sheet reconnection case (Cerutti et al. 2013, 2014), particles accelerated in

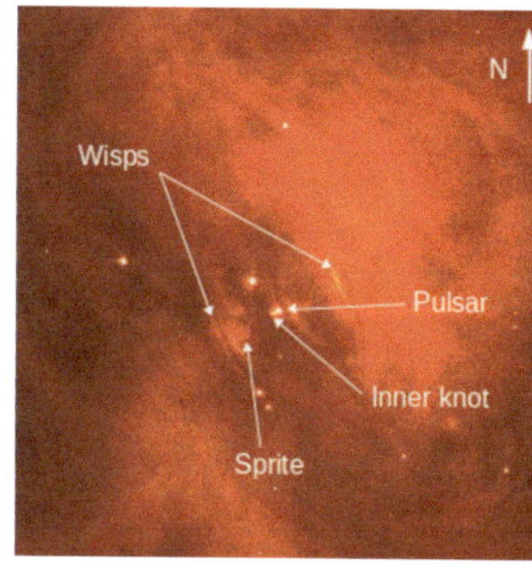

Fig. 11 Optical image of the Crab nebula taken by the Hubble Space Telescope. The image is 0.02° on each side. Different structures of the nebula are indicated (e.g. Hester et al. 2002). Particularly, the "inner knot" can be seen, it is located ≈ 0.6 arcsec away from the pulsar

the current sheets are beamed and at the same time bunched by the tearing modes, resulting in rapid variability of observed synchrotron radiation (Yuan et al. 2016). While it is unlikely that these highly stylised flux rope configurations used in simulations correspond to realistic field structures in the Crab nebula, they provide some insights into rapid dissipation of electromagnetic energy. Lyutikov et al. (2016b) also studied similar processes in a more natural configuration, where two adjacent flux tubes with a zero total current merge and accelerate particles efficiently. It still remains to be seen if the aforementioned schemes produce all the features of the Crab flares, but the progress so far makes them promising candidates.

Independent of the actual particle acceleration process, it appears clear from the discussion above that regions of high magnetisation and large relativistic motion are preferred sites for the gamma-ray flares. The intermediate latitude region just downstream of the reverse shock is one of the most promising regions (Lyutikov et al. 2016b). Interestingly, the shock in this region might correspond to the inner knot, observed very close to the pulsar, as will be discussed in Sect. 5.

4.3 Do the Flares Originate from the Inner Knot?

The inner knot is the closest feature observed around the pulsar, located ≈0.6 arcsec south east of it (see Fig. 11). The knot is detected in the infrared and optical bands. In radio and X-rays it has so far not be detected, the obtained upper limit on the flux are shown in Fig. 9. The inner knot is a dynamical feature, its distance to the pulsar can vary by ≈20% and its brightness by ≈50% (Sandberg and Sollerman 2009). Its emission has a high degree of linear polarisation (Moran et al. 2013). Recent observations suggest that the polarisation angle might change over time (Moran et al. 2015).

Its variability, compactness, high-level of polarisation and proximity to the pulsar made the inner knot one of the prime candidates for the site of origin of the gamma-ray flares. In fact, it was proposed by Komissarov and Lyutikov (2011) that almost all of the emission above 100 MeV could come from this feature. Therefore, particularly dense observations targeting the inner knot where performed by the Keck Observatory and HST. Surprisingly however, none of the knot's properties was found to correlate with the gamma-ray emission (Rudy et al. 2015). One example is shown in Fig. 12, where the gamma-ray flux is plotted as a function of the distance of the knot from the pulsar. The inner knot is therefore likely not the site of origin of the gamma-ray flares. However, the observations revealed the dynamical behaviour of the inner knot with unprecedented detail. Its size was found to increase with the distance of the pulsar. At the same time its luminosity decreases with increasing distance, as shown in Fig. 13. As will be presented in detail in Sect. 5, modelling of the knot shows that

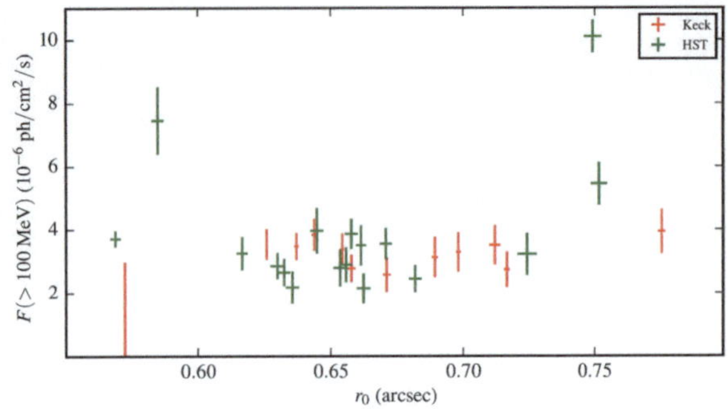

Fig. 12 Gamma-ray flux above 100 MeV as a function of the distance of the inner knot to the Crab pulsar. The figure is reproduced from Rudy et al. (2015)

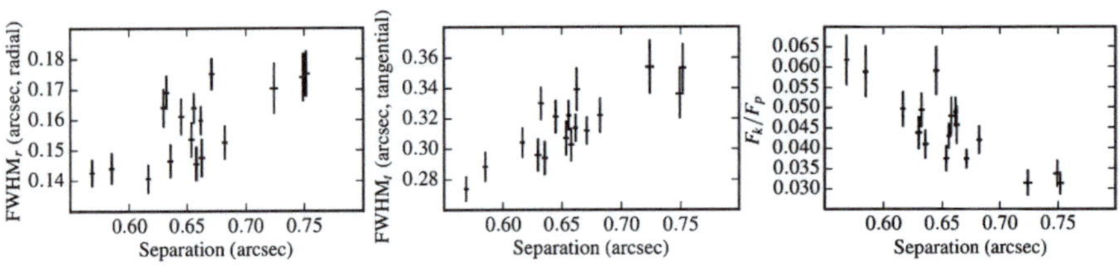

Fig. 13 Properties of the inner knot as a function of its distance from the Crab pulsar. *Left*: size of the knot perpendicular to the pulsar-knot axis, *Middle*: size of the knot along the pulsar-knot axis, *Right*: *k*-band flux of the knot normalised to the pulsar flux. The figures are reproduced from Rudy et al. (2015)

it is likely the high-latitude shock of the pulsar wind, giving us the first radiative signature of the wind particles.

5 Inner Knot of the Crab Nebula

The discovery of the jet-torus structure of the inner Crab nebula coincided with the emergence of computer codes for relativistic MHD capable of dealing with highly relativistic magnetised flows (Komissarov 1999). This was most handy, as theoretical MHD models of the Crab nebula allowed for reasonable ideas on the origin of the structure but could not address the problem rigorously due to the complicated nature of non-spherically symmetric flows. The key properties of pulsar winds giving rise to the jet-torus appearance of the nebula in these models are (1) their anisotropy, with the wind power increasing towards the equatorial plane of the pulsar rotation (2) their magnetisation, with purely azimuthal magnetic fields aligned with the pulsar's rotational axis. The first property naturally leads to the torus component, whereas the second opens the possibility of magnetic collimation of the flow toward the polar axis. In fact, this collimation mechanism fails for the wind itself, with possible exception for a tiny polar section, due to its highly super-magnetosonic motion. However, downstream of the wind termination shock the corresponding Mach number drops, the causal connectivity across the flow is restored and the magnetic hoop stress regains its collimating potential. Thus, in the MHD model the Crab's jet is not present in the un-shocked pulsar wind but forms in the shocked part of the wind and thus already inside the nebula.

The first computer-generated models of the Crab nebula focussed on the case of weakly magnetised pulsar wind, thus adopting the conclusions of the Rees-Gunn-Kennel-Coroniti theory (Rees and Gunn 1974; Kennel and Coroniti 1984a,b). To the delight of theorists,

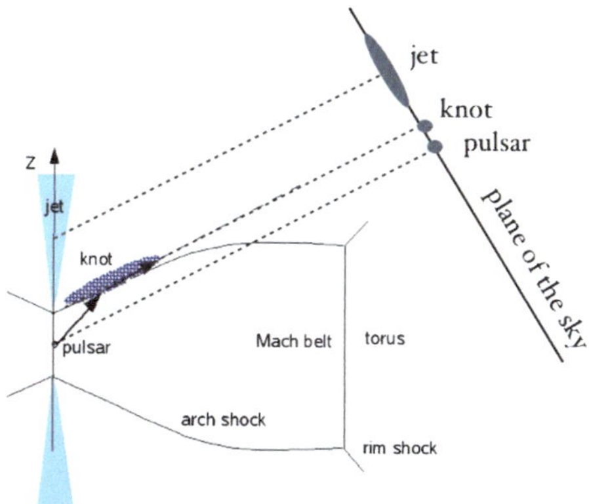

Fig. 14 In this sketch, we show the plane defined by the rotational axis of the pulsar and the location of fiducial observer. The region around the pulsar corresponds to the un-shocked pulsar wind, the continuous lines show the wind termination shock, indicating its topological components, the Mach belt, arch and rim shocks. The *blue shaded region* shows the region where the shocked wind plasma flows towards the observer with moderately relativistic speed. The *dashed lines* are the lines of sight, which show the order in which the jet the knot and the pulsar appear in the plane of the sky of the observer

the simulations confirmed the possibility of the separation of the post-shock flow into the equatorial (torus) and polar (jets) components, provided the wind magnetisation parameter σ exceeded the critical value of few $\times 10^{-3}$ (Komissarov and Lyubarsky 2003, 2004; Del Zanna et al. 2004; Bogovalov et al. 2005). This could be concluded immediately from the analysis of the velocity field and the distribution of other fluid parameters but in order to compare the results with the observations a synthetic emission map is highly desirable. Some elements required for the synchrotron emission calculations were readily provided by the numerical models. These are the magnetic field strength and the orientation, as well as the velocity field, which is important for the relativistic beaming and Doppler effects. The missing part, concerning the spectrum of the ultra-relativistic electrons, had to be recovered indirectly, via a crude model connecting the spectrum to the fluid pressure of the MHD solution.[2] The un-shocked wind zone was not expected to make a significant contribution and hence was explicitly excluded from the emission calculations.

One of the most prominent and yet unexpected features of the synthetic synchrotron images of the simulated PWN was a very bright feature located right in the centre, where the image of the model's pulsar would be if its emission were included (but it was not: Komissarov and Lyubarsky 2003, 2004). A closer look revealed that the feature was a slightly off-centred and extended knot. In order to identify the feature of the synthetic images with a particular feature of the numerical solution, a detailed inspection of the data was carried out and it revealed that the emission originated close to the termination shock, where the shocked wind plasma was still flowing with (moderately) relativistic speed towards the fiducial observer, and for this reason was subject to significant Doppler-boosting. This is illustrated in Fig. 14. Subsequent studies by other groups (Del Zanna et al. 2006) and the recent 3D simulations (Porth et al. 2013, 2014a) confirmed the result.

Once we had understood the nature of the feature, it became clear that a counterpart must be present in the real images of the Crab nebula, unless something is very wrong with the MHD model. The oblate geometry of the termination shock ensures that there always be

[2] A more sophisticated technique was used in more recent simulations.

a shock section where the post-shock plasma is streaming towards the observer with relativistic speed and hence its emission is Doppler-boosted. Provided the Crab's synchrotron electrons are indeed accelerated at the termination shock, as proposed in the Kennel-Coroniti model, this ensures the existence of a bright knot inside the area occupied in the plane of the sky by the un-shocked pulsar wind zone. This knot must be a very bright permanent feature, located on the projected symmetry axis. Since the polar jets are also subject to relativistic beaming, the knot must be on the jet side of the pulsar (see Fig. 14) and have no counterpart on the counter-jet side. A comparison with the high-resolution HST images revealed that such a feature is indeed present in the Crab nebula—it is called the inner knot (or knot 1; Hester et al. 1995) and is located approximately 0.65 arcsec away from the Crab pulsar (see Fig. 11). So far it has been detected only in the optical-IR range (the highest energy detection was in NUV band by Melatos et al. 2005). The emission is strongly polarised with the electric polarisation vector aligned with the rotational axis in the plane of the sky. This is exactly what is expected, as in the close vicinity of the termination shock the magnetic field should still preserve the highly regular azimuthal structure it has in the wind. In a way, the inner knot was predicted by the simulations and then confirmed by the observations. Indeed, even if the observational discovery of the Crab's knot preceded the simulations, it was not immediately connected to the termination shock and its emergence in the synthetic maps was a complete surprise.

The identification of the inner knot with the termination shock is already very interesting as it implies a unique opportunity for studying properties of highly relativistic shocks in magnetised plasma. The discovery of gamma-ray flares in the Crab nebula added interest as the short life-time of the electrons emitting at such energies suggests that the gamma-ray emission originates from the termination shock, provided it is the main acceleration site for the synchrotron electrons of all energies. Moreover, the beamed nature of the emission in this region ensures the domination of the inner knot contribution to the observed gamma-ray flux (Komissarov and Lyutikov 2011). Unfortunately, the angular resolution of gamma-ray telescopes is not sufficient to test this conclusion directly. Under such circumstances, the only way to localise the flares is via their identification with structural variability in the radio, optical-IR and X-ray windows where the resolution of modern instruments is much higher. Such observations have been carried out but they have not led to a positive identification so far. Moreover, they seem to have ruled out the inner knot as the source of gamma-ray flares as its variability did not show any correlation with the flares (Rudy et al. 2015). On the other hand, these studies significantly increased the available information on the knot properties, allowing to test its theoretical connection with termination shock in greater detail.

The knot does not seem to have much of an internal structure, with brightness gradually decreasing from the peak value in the centre to the background one at the periphery. It is elongated in the direction normal to the rotational axis and a little bit bowed away from the pulsar. The knot is clearly separated from the pulsar and the separation varies with a 30% amplitude on the time scale of few months, which is similar to the wisp production time scale. The knot's flux anti-correlates with the separation. The variability implies that in the vicinity of the termination shock the flow of shocked plasma is not steady but varies on the time-scale of the shock light-crossing time. In fact, such a variability, accompanied by a strong variation of the shock geometry, was discovered in numerical simulations before the observations. In the high-resolution 2D simulations by Camus et al. (2009), it was found that this variability led naturally to emergence of wisps in synthetic synchrotron maps. These wisps were identified with inhomogeneities created by the variable shock in the equatorial outflow and advected with the outflow speed. Thus, the similar time scales of the inner knot dynamics and the wisp production are easily explained by the fact that both are traced back

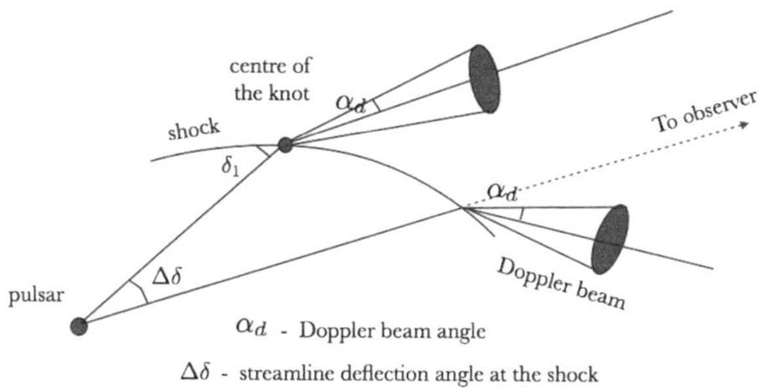

Fig. 15 In this sketch, we show Doppler beams at two locations on the termination shock, selected by the lines of sight originated at the pulsar and at the brightness peak of the knot, where the post-shock velocity is parallel to the line of sight. In order for the knot not to extend all the way to the pulsar in the plane of the sky, the Doppler beam at the first location should point away from the observer and hence $\alpha_d < \Delta\delta$

to the termination shock variability. Moreover, the recent 3D simulations of the Crab nebula are consistent with the observed knot's flux-separation anticorrelation (Porth et al. 2014a).

A couple of attempts have been made recently to build a simple semi-analytical model of Crab's inner knot (Yuan and Blandford 2015; Lyutikov et al. 2016a). The main motivation behind these attempts is based on the fact that the structure of the upstream wind is relatively simple and one of the key factors determining the knot appearance, namely the Doppler beaming, is relatively easy to predict based on the properties of relativistic transverse MHD shock. The most important conclusion of the studies is that the observed clear separation of the knot from the pulsar can only be achieved in models with low wind magnetisation ($\sigma \ll 1$). Indeed, suppose that the brightness peak of the knot corresponds to the point on the shock where the shocked plasma flows directly towards the observer. Then the observed angular distance between the peak and the pulsar is determined by flow deflection angle $\Delta\delta$ at the shock (see Fig. 15). The knot will not engulf the pulsar provided the half-opening angle of the Doppler beam, α_d, is smaller than the deflection angle along the line of sight passing through the pulsar (see Fig. 15). Given the ultra-relativistic nature of the wind, both these angles depend mostly on the upstream σ and hence the condition $\alpha_d < \Delta\delta$ translates into the condition on σ.

The deduced low σ is in contrast with the theoretical expectation of high σ in pulsar winds. This conflict can be settled if the knot is produced by the striped equatorial component of the wind. In the striped section, σ can be reduced via magnetic dissipation of the stripes either in the wind or inside the hybrid MHD shock which terminates it, with the post-shock flow dynamics being identical in both these cases.

Using the Doppler-beamed post-shock emissivity as a proxy for the knot brightness, one can estimate a number of observed knot parameters such as its elongation, the distance from the pulsar in the units of the equatorial radius of the termination shock, and its polarisation degree. The results are somewhat dependent on the utilised model for the termination shock shape, but generally agree with the observational measurements quite well provided the knot is formed by the low-sigma part of the wind. The total flux polarisation degree is strongly effected by the relativistic aberration of light, which leads to a noticeable rotation of the polarisation vector across the knot (see Fig. 16). Yuan and Blandford (2015) concluded that the rotation imposes an upper limit of $\simeq 50\%$ on the overall polarisation degree of the knot, which is significantly lower than the observed $\simeq 60\%$ (Moran et al. 2013). However, in the observations the integration was carried out only over the central part of the knot, thus excluding outer regions where the rotation is strong. Lyutikov et al. (2016a) have demon-

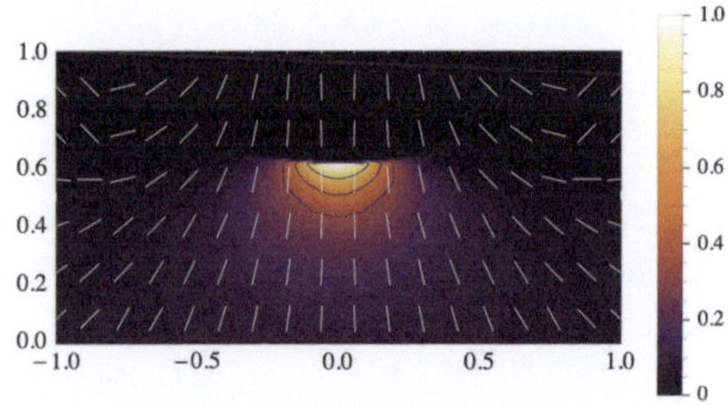

Fig. 16 The emissivity distribution over the shock surface projected onto the plane of the sky and the polarisation vector of the shock emission

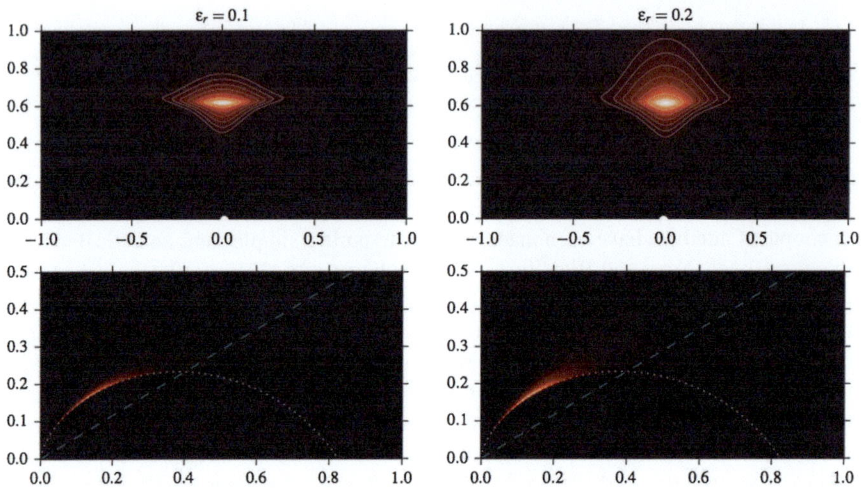

Fig. 17 The *top panels* show the knot images resulting from emitting region of thickness equal to 0.1 (*left*) and 0.2 (*right*) of the local shock radius. The *bottom panels* illustrate the emissivity models used for these calculations. The *dotted line* shows the termination shock surface, the *dashed line* shows the line of sight and the mono-coloured image shows the corresponding volume emissivity distribution

strated that the polarisation degree decreases with the size of the integration area and this can explain the disagreement between the observations and the upper limit found by Yuan and Blandford (2015). Moreover, the polarisation degree of the synthetic knot reproduced in the 3D simulation (Porth et al. 2014a) agrees with the observed values exceptionally well.

The shape of the knot based on the emissivity isophotes was another matter of concern (Yuan and Blandford 2015). Instead of being bowed away from the pulsar the knot looked bowed towards it, reflecting the shape of the termination shock "shadow" in the plane of the sky (see Fig. 16). In order to understand the significance of this result one has to study the role of other factors influencing the knot appearance. For example, the effective geometric thickness of the emitting layer may have a strong impact as well as the variations of the velocity field in this layer (Lyutikov et al. 2016a; Yuan and Blandford 2015). Figure 17 illustrates the role of finite thickness in a model where the emissivity above the termination shock is set via an extrapolation of that at the shock surface (Lyutikov et al. 2016a). One can see that the effect is rather strong.

In these calculations, it was assumed that the spectrum of the knot emission is a power law. This implies that the termination shock is an efficient accelerator of non-thermal particles. Although this assumption is a corner stone of the popular and successful model for the Crab nebula emission by Kennel and Coroniti (1984b), its validity has been questioned by the PIC simulations of shocks in striped flow (Sironi and Spitkovsky 2011). These suggest that the non-thermal particle acceleration can only be efficient in a small equatorial sector where the downstream plasma becomes almost unmagnetised. Otherwise, the plasma is simply heated and its synchrotron emission has a thermal spectrum. Interestingly, the current failure to detect both the radio and X-ray emission from the knot is consistent with such a spectrum.

6 Gamma-Ray Binaries: Pulsar Winds Interacting with a Massive Companion

6.1 Introduction

The last decade revealed a new group of gamma-ray emitters, composed of a fast-rotating pulsar and a massive star. The emission, which peaks in the MeV band, arises from the shocked region between the stellar wind and the pulsar wind. Particle acceleration at the relativistic shock creates emission up to several TeV, the highest emission observed for binary systems. Figure 18 presents a schematic view of interaction region. A handful of these systems have been discovered so far: PSR B1259-63 (H.E.S.S. Collaboration 2005), LS 5039 (Aharonian et al. 2005), LSI+61° 303 (Albert et al. 2006), HESS J0632+057 (Bongiorno et al. 2011), 1FGL J1018.6-5856 (Abdo et al. 2010), and more recently HESS J1832-093 (Eger et al. 2016) and LMC P3 in the LMC (Corbet et al. 2016). While pulsed emission has been observed for PSR B1259 (thanks to its wide orbit), and LSI+61° 303 has shown magnetar type flares (Torres et al. 2012), the nature of the compact object is not firmly established in the other systems. However, given the very similar emission patterns, the colliding wind scenario is now firmly established. As such, the major aspects of the colliding wind structure and the high-energy emission mechanisms have been assessed, and while many questions remain, the study of gamma-ray binaries has matured enough to become a new window on pulsar wind physics. These systems provide a unique opportunity to study otherwise very elusive pulsar winds. The binary interaction typically takes place around a fraction of AU from the pulsar (or about 10^4 times the light cylinder), about 5 orders of magnitude closer than for pulsars interacting with the ISM or supernova remnants. Modelling strongly

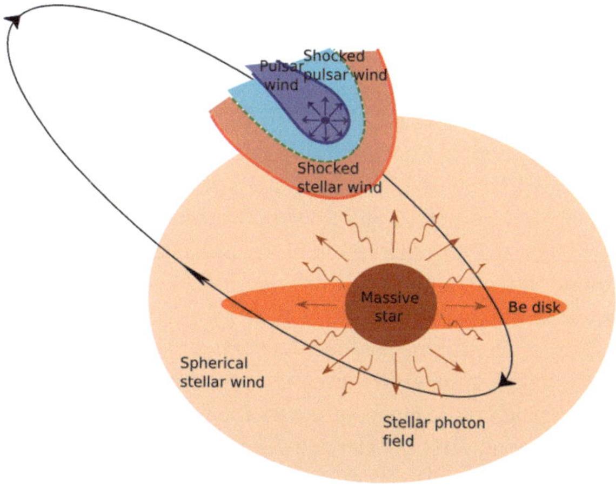

Fig. 18 Schematic view of a gamma-ray binary. The pulsar wind interacts with the stellar wind, photon field and circumstellar disk if the companion is a Be star

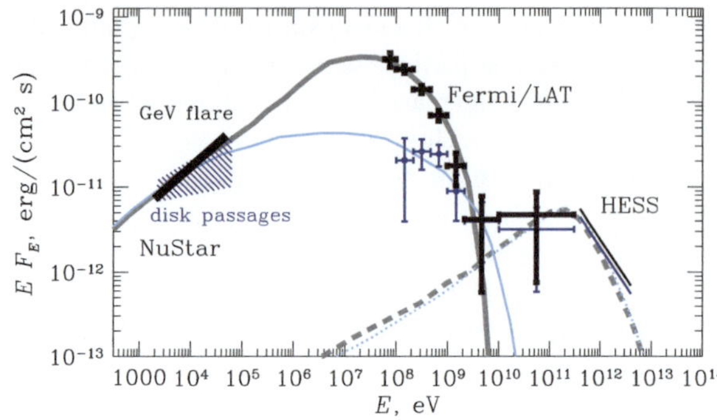

Fig. 19 High energy spectral energy distribution of PSR B1259-63, during the disk passages (*blue*) and flares (*grey*) (from Chernyakova et al. 2015)

benefits from information provided by orbital variability and the well constrained environment created by the companion star, both in terms of density and photons fields (at least compared to a typical region of the ISM or a supernova remnant). While an excellent review can be found in Dubus (2013), this section provides some updates and focuses on how to determine the pulsar wind properties from the emission of gamma-ray binaries.

6.2 γ-Ray Binaries: Puzzling Observations

While most of the energy is emitted in MeV photons (hence the name), gamma-ray binaries emit all the way from radio to a few TeV. Figure 19 shows the high energy spectral energy distribution of PSR B1259-63 with emission resulting from the electrons accelerated at the shocks between the winds. Acceleration at the relativistic shock produces synchrotron emission (up to a few GeV) and Inverse Compton emission on seed photons of the massive star (up to a few TeV).

In theses systems, most of the emission (with the exception of the low frequency radio band (Marcote et al. 2015)) shows orbital variability. Figure 20 shows lightcurves of LS 5039, a binary with a 3.9 day period: X-ray and TeV emission show similar orbital variability, in opposition with the variability observed with *Fermi/LAT*. The MeV emission in LS 5039, with an exponential cutoff at a few GeV before the harder but fainter TeV emission cannot be reconciled while assuming emission from a single population (Dubus et al. 2008). The different origin of the GeV and TeV emission is also suggested by the absence of emission below 200 GeV in HESS J0632+057 (Malyshev and Chernyakova 2016) and around periastron in PSR B1259-63, while both show strong TeV emission.

LSI+61303 shows additional variability on superorbital timescales (Chernyakova et al. 2012), attributed to long-term variations in the Be disk surrounding the companion star (Paredes-Fortuny et al. 2015). Interactions with the Be disk are likely responsible for the suborbital optical line variability in HESS J0632+057 (Moritani et al. 2015).

In PSR B1259-63, which has a 3.4 year period, X-rays, TeV and radio flares are observed about 15 days before and after periastron. Optical spectroscopy indicates these flares are consistent with the disruption of the Be disk as the pulsar crosses it (van Soelen et al. 2016). About 60 days after periastron, *Fermi/LAT* detects strong emission with no counterpart at any other wavelength (Abdo et al. 2011b). This flare, which contains almost all the spindown power of the pulsar, has been confirmed at the 2014 periastron passage (Caliandro et al. 2015) with a similar flux level and spectral shape but slightly different temporal evolution.

Extended radio emission traces the contours of the shocked region (Dhawan et al. 2006; Moldón et al. 2011). Extended X-ray emission has also been found for most (if not all) of these systems but its origin remains unclear (Kargaltsev et al. 2014).

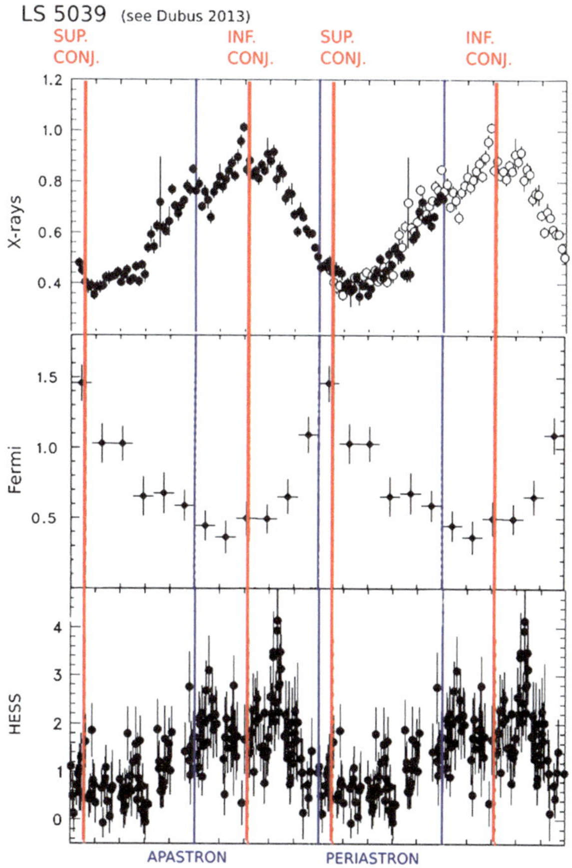

Fig. 20 X-ray, GeV and TeV modulation for LS 5039 (adapted from Kishishita et al. 2009; Fermi-LAT Collaboration 2009 and H.E.S.S. Collaboration 2006)

6.3 Ingredients for Successful Models

The wide variety of observational constraints is both a challenge and a tremendous opportunity to understand these systems. Until recently, studies have either focussed on sophisticated one-zone emission models or aimed at reproducing the hydrodynamical structure of the colliding wind region, with little connection to the non-thermal emission. While both approaches have been fruitful, they have not been able to consistently reproduce the orbital variability and spectral features from radio up to TeV. The next sections describe both approaches and how they can be combined to yield a better understanding of pulsar wind physics.

The first ingredients are the non-thermal emission processes, dominated by leptonic processes (Dubus 2006), causing a synchrotron bump and inverse Compton bump. Models have included refined representations of the ambient photon field, including self-Compton on the nebula (Cerutti et al. 2010) and the infrared photons of the Be disk when present (van Soelen et al. 2012). Reproducing the TeV lightcurve of LS 5039 requires anisotropic inverse Compton emission (Dubus et al. 2008) and an extended emission region. The latter is necessary to prevent complete absorption by pair production at superior conjunction in LS 5039 (when the compact object is behind the massive star). Inverse-Compton pair cascades could extend the emission region (Bednarek 2006; Cerutti et al. 2010). Or, the emission could also result from one or more distant emission regions (Zabalza et al. 2013). Similarly, the absence of occultation features in the X-ray lightcurves suggests an extended emission region (Bosch-Ramon et al. 2007; Szostek and Dubus 2011). As the emission originates from the relativistic shocked pulsar wind, Doppler boosting is also at work at inferior conjunction (when the pulsar wind points toward the observer) and explains some of the GeV modulations (Dubus et al. 2010).

Single zone models have so far failed to consistently reproduce the full variability of these systems, even when cascade emission is allowed. The exponential cutoff around a few GeV is hard to reconcile with the hard TeV emission. A single particle distribution with IC, synchrotron and adiabatic cooling fails to reproduce this spectral feature (Khangulyan et al. 2008; Zabalza et al. 2013). The presence of an additional emission component has been suggested. While the GeV emission is analogous to pulsar magnetospheric emission observed by *Fermi*, its orbital variations are hard to explain (Hadasch et al. 2012). The emission of the unshocked pulsar wind (Khangulyan et al. 2007) has been considered but is insufficient (Sierpowska-Bartosik and Torres 2008). A shocked mono-energetic component has also been considered (Dubus and Cerutti 2013). An alternative path is the presence of a different acceleration region, with different properties in terms of magnetic field, photon field and velocity structure (Zabalza et al. 2013).

Over the years, it has become clearer that the only path towards fully understanding the emission in gamma-ray binaries would require an elaborate model for the geometry of the system, coupled with a refined model for the non-thermal emission.

The flow dynamics in gamma-ray binaries is dominated by the double shock structure. It shares many similarities with colliding stellar winds, which have been studied for decades. Simulations of colliding wind binaries (Pittard 2009; Lamberts et al. 2011) show strong instabilities which may yield to mixing in the winds (Zdziarski et al. 2010) and affect the spiral structure expected at large scales (Lamberts et al. 2012; Bosch-Ramon et al. 2012).

However, the relativistic nature of the pulsar wind affects the structure of the interaction region. Relativistic hydrodynamics induces more complex shock patterns, where parallel velocities (to the shock normal) play a role. Multidimensional simulations in the ultrarelativistic regime of pulsar winds (Lorentz factor $\simeq 10^3$–10^5) are far beyond current computational and numerical capabilities. Some rescaling may be necessary even when focussing on the shocked winds,which have $\Gamma \lesssim 10$ close to the pulsar. Simulations suggest a narrower opening angle for the pulsar wind (Lamberts et al. 2013), and a reacceleration of the pulsar wind up to its initial velocity (Bogovalov et al. 2008). Keeping in mind these intrinsic difficulties, relativistic simulations are crucial in order to determine Doppler boosting, which is a key ingredient to orbital modulation and maybe to explain the GeV flares in PSR B1259 (Kong et al. 2012). Simulations suggest the presence of a back shock behind the pulsar, which can provide an additional site for particle acceleration, further away from the binary. They also indicate the presence of a large scale spiral (Bosch-Ramon et al. 2012; Barkov and Bosch-Ramon 2016)

Dubus et al. (2015) developed the first high energy emission model based on a fully three-dimensional relativistic simulation of LS 5039. As described in Sect. 2.7, the non-thermal emission is determined during post-processing, with a particle distribution injected at the relativistic shock and followed along the streamlines in the shocked pulsar wind. The model self-consistently represents adiabatic, inverse Compton and synchrotron cooling. The resulting emission takes into account pair creation, anisotropic inverse Compton and directly benefits from the simulation to model Doopler boosting. Figures 21 and 22 show the resulting emission maps and lightcurves. Comparison with observations suggested a

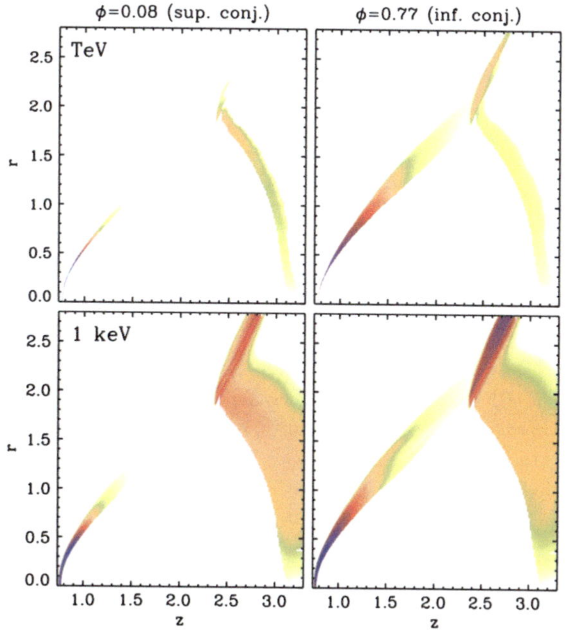

Fig. 21 TeV and X-ray emission maps for LS 5039 at superior and inferior conjunctions. Image taken from Dubus et al. (2015)

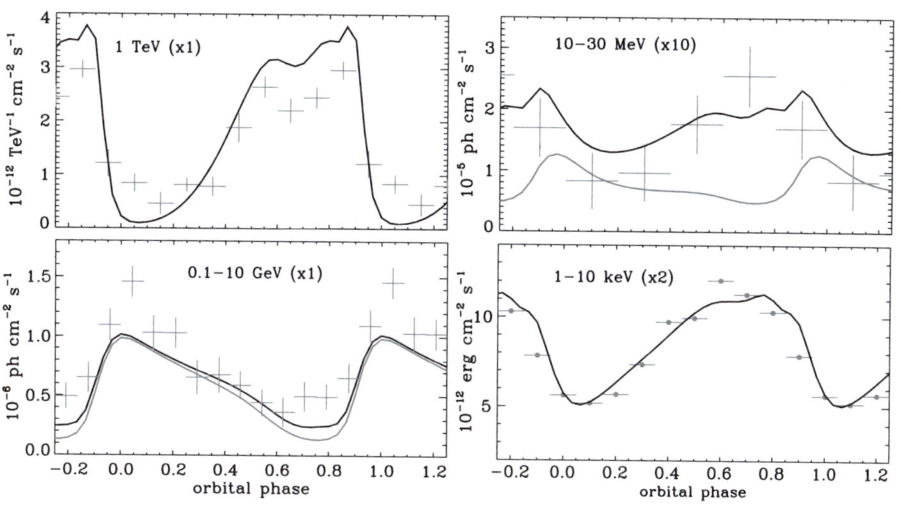

Fig. 22 Simulated lightcurves for LS 5039 with a powerlaw particle distribution (*black*) and an additional mono-energetic component (*grey*). Image taken from Dubus et al. (2015)

strongly magnetised, rather slow pulsar wind ($\Gamma \simeq 10^3$, $\sigma \simeq 1$). While small discrepancies exist with observations, and radio emission will not be modelled properly without taking into account the magnetic field structure, the clear success of the combination of relativistic hydrodynamics with refined emission models shows the path forward in this field.

7 Discussion

Over the past two decades, modelling of PWN has lead to a good understanding of the plasma flow in these systems. Predominantly, most studies have focussed on the Crab nebula. The global plasma flow and dynamics in this system are well described by the simulations. The origin of its jet, the torus, the inner knot and the wisps are understood. However, other features as the knots of the inner ring or the "thin wisp" (Hester et al. 1995) are not present in the simulations. Also, the morphology of other PWN appears to be more complex. In the Vela PWN for instance, the receding jet appears to be brighter than the one pointed to the observer (Durant et al. 2013), at odds with the expectation of Doppler beaming in plasma flow models.

Recently, an important step forward was done by simulating the Crab nebula in 3D. The 3D randomisation of the magnetic field leads to magnetic dissipation inside the PWN. This removes the constraint of low magnetisation of the pulsar wind imposed by 1D and 2D axisymmetric models and is likely the solution to the longstanding "σ-Problem". As always, a deeper understanding comes with many new questions. Also, several older questions remain unanswered. We will summarise the ones we consider most important in the following:

1. *Where and how does non-thermal particle acceleration take place? How does the particle spectrum evolve in the nebula, particularly along the termination shock?*
   Magnetic reconnection appears to be the main process by which particles are accelerated in the nebula. It remains unclear where and how this happens. In particular it is unclear if different acceleration mechanisms take place at different latitudes along the termination shock and what the contribution of second order Fermi acceleration within the body of the nebula is.
2. *What is the effect of the feedback of the acceleration on the plasma flow?*
   As a significant fraction of the total energy goes into accelerating particles, their back reaction on the plasma could be important. To date, only test particle simulations have been performed, neglecting the feedback of the accelerated particles on the global plasma flow. The simple ideal MHD approximations will need to be extended to a more realistic plasma description.
3. *Where within the nebulae do the gamma-ray flares originate? Are these exceptional events with little impact on the overall evolution/emission of PWN or a high-energy tail of the main dissipative processes?*
   Significant progress has been made in understanding the processes that can lead to explosive gamma-ray flares. However, if and where within the nebula these processes happen remains unclear. It is puzzling that no gamma-ray flares have been discovered from any other isolated PWN than the Crab nebula. It is not clear if similar processes are responsible for Crab nebula flares and the GeV flares of the binary PSR B1259-63.
4. *What is the origin of radio electrons? Are they supplied by the pulsar wind and if not then why does the spectrum of emission not show a discontinuity between the radio and optical parts?*
   The hard energy spectrum of the radio emitting electrons suggests that they are accelerated via magnetic reconnection. However, where these electrons come from remains an open question. If the electrons are provided by the wind their large number is in conflict with magnetospheric models. An alternative is that electrons are stripped from the gas and dust in the nebula's filaments. In this case, however, spectral continuity with the higher energy particle population would be a mystery, and just as mysterious would be

the similarity between the radio spectra of PWN and those of bow-shock PWN, where there are no filaments from which low energy particles could come.

5. *What is the structure of termination shock in the polar region?*
   In the polar region the pulsar wind has no stripes and hence a very high magnetisation is expected both upstream and downstream of the termination shock. It is unclear what the fate of the highly magnetised flow injected into the nebula in this region is and if its magnetic energy can be dissipated on a light-crossing time.
6. *How many electrons/positrons escape into the ISM, and with which energies? Can PWN potentially explain the positron excess?*
   The interaction of PWN with the surrounding medium is complex, as they evolve within the remnant of their progenitors explosions. It is unclear how many electrons can escape the magnetic confinement of these systems and if their number is sufficient to explain the positron excess observed at Earth. Of course, if the radio emitting particles, which are the relevant ones for this matter, come from filaments, and are therefore all electrons (no positrons), then PWN cannot contribute to the positron excess.
7. *Is an energetically significant amount of non-thermal ions present in PWN?*
   To date, no observational signature of ions has been found in PWN. However, due to their low radiative efficiency, there could still be an energetically significant number of ions present in the pulsar wind and it is interesting to notice that these would be ions with energies in the PeV range.

The search for answers of these questions is ongoing. Fortunately, deeper observations and increasingly realistic simulations are powerful tools at our disposal in this quest.

Acknowledgements OP is supported by the ERC synergy grant "BlackHoleCam: imaging the Event Horizon of Black Holes" (Grant No. 610058). BO acknowledges support from the INFN—TEONGRAV initiative (local PI: Luca Del Zanna) and from the University of Florence grant "Fisica dei plasmi relativistici: toeria e applicazioni moderne". Support for AL was provided by an Alfred P. Sloan Research Fellowship, NASA ATP Grant NNX14AH35G and NSF Collaborative Research Grant 1411920 and CAREER grant 1455342.

References

A.A. Abdo et al., Astrophys. J. Suppl. Ser. **188**, 405 (2010)
A.A. Abdo et al., Science **331**, 739 (2011a)
A.A. Abdo et al., Astrophys. J. **736**, L11 (2011b)
A. Abramowski et al., Astron. Astrophys. **562**, L4 (2014)
M. Ackermann et al., Science **343**, 42 (2013)
F. Aharonian et al., Science **309**, 746 (2005)
F. Aharonian et al., Astron. Astrophys. **502**, 749 (2009)
J. Albert et al., Science **312**, 1771 (2006)
J. Aleksić et al., J. High Energy Astrophys. **5–6**, 30 (2015)
E. Aliu et al., Astrophys. J. **781**, L11 (2014)
T. Amano, J.G. Kirk, Astrophys. J. **770**, 18 (2013)
E. Amato, Int. J. Mod. Phys. Conf. Ser. **28**, 60160 (2014)
E. Amato, J. Arons, Astrophys. J. **653**, 325 (2006)
E. Amato, M. Salvati, R. Bandiera, F. Pacini, L. Woltjer, Astron. Astrophys. **359**, 1107 (2000)
E. Amato, D. Guetta, P. Blasi, Astron. Astrophys. **402**, 827 (2003)
J. Arons, Space Sci. Rev. **173**, 341 (2012)
A.M. Atoyan, F.A. Aharonian, Mon. Not. R. Astron. Soc. **278**, 525 (1996)
M.V. Barkov, V. Bosch-Ramon, Mon. Not. R. Astron. Soc. **456**, L64 (2016)
W. Bednarek, Mon. Not. R. Astron. Soc. **371**, 1737 (2006)
M.C. Begelman, Astrophys. J. **493**, 291 (1998)
M.C. Begelman, Z.-Y. Li, Astrophys. J. **397**, 187 (1992)
M.A. Belyaev, Mon. Not. R. Astron. Soc. **449**, 2759 (2015)

E. Berger, Annu. Rev. Astron. Astrophys. **52**, 43 (2014)
M.F. Bietenholz, J.J. Hester, D.A. Frail, N. Bartel, Astrophys. J. **615**, 794 (2004)
M.F. Bietenholz, Y. Yuan, R. Buehler, A.P. Lobanov, R. Blandford, Mon. Not. R. Astron. Soc. **446**, 205 (2014)
R. Blandford, P. Simeon, Y. Yuan, Nucl. Phys. B, Proc. Suppl. **256–257**, 9 (2014)
S.V. Bogovalov, V.M. Chechetkin, A.V. Koldoba, G.V. Ustyugova, Mon. Not. R. Astron. Soc. **358**, 705 (2005)
S.V. Bogovalov, D.V. Khangulyan, A.V. Koldoba, G.V. Ustyugova, F.A. Aharonian, Mon. Not. R. Astron. Soc. **387**, 63 (2008)
S.D. Bongiorno, A.D. Falcone, M. Stroh, J. Holder, J.L. Skilton, J.A. Hinton, N. Gehrels, J. Grube, Astrophys. J. **737**, L11 (2011)
V. Bosch-Ramon, C. Motch, M. Ribó, R. Lopes de Oliveira, E. Janot-Pacheco, I. Negueruela, J.M. Paredes, A. Martocchia, Astron. Astrophys. **473**, 545 (2007)
V. Bosch-Ramon, M.V. Barkov, D. Khangulyan, M. Perucho, Astron. Astrophys. **544**, A59 (2012)
R. Buehler et al., Astrophys. J. **749**, 26 (2012)
R. Bühler, R. Blandford, Rep. Prog. Phys. **77**, 066901 (2014)
R. Bühler, M. Giomi, Mon. Not. R. Astron. Soc. **462**, 2762 (2016)
A.M. Bykov, G.G. Pavlov, A.V. Artemyev, Y.A. Uvarov, Mon. Not. R. Astron. Soc. Lett. **421**, L67 (2012)
G.A. Caliandro, C.C. Cheung, J. Li, J.D. Scargle, D.F. Torres, K.S. Wood, M. Chernyakova, Astrophys. J. **811**, 68 (2015)
N.F. Camus, S.S. Komissarov, N. Bucciantini, P.A. Hughes, Mon. Not. R. Astron. Soc. **400**, 1241 (2009)
B. Cerutti, J. Malzac, G. Dubus, G. Henri, Astron. Astrophys. **519**, A81 (2010)
B. Cerutti, G.R. Werner, D.A. Uzdensky, M.C. Begelman, Astrophys. J. **770**, 147 (2013)
B. Cerutti, G.R. Werner, D.A. Uzdensky, M.C. Begelman, Astrophys. J. **782**, 104 (2014)
B. Cerutti, A. Philippov, K. Parfrey, A. Spitkovsky, Mon. Not. R. Astron. Soc. **448**, 606 (2015)
A.Y. Chen, A.M. Beloborodov, Astrophys. J. **795**, L22 (2014)
M. Chernyakova, A. Neronov, S. Molkov, D. Malyshev, A. Lutovinov, G. Pooley, Astrophys. J. **747**, L29 (2012)
M. Chernyakova et al., Mon. Not. R. Astron. Soc. **454**, 1358 (2015)
R.A. Chevalier, in *Supernovae*, ed. by N. Schramm. Astrophysics and Space Science Library, vol. 66 (1977), p. 53. doi:10.1007/978-94-010-1229-4_5
I. Contopoulos, D. Kazanas, C. Fendt, Astrophys. J. **511**, 351 (1999)
R.H.D. Corbet et al., Astrophys. J. **829**, 105 (2016)
F.V. Coroniti, Astrophys. J. **349**, 538 (1990)
O.C. de Jager, A.K. Harding, Astrophys. J. **396**, 161 (1992)
L. Del Zanna, E. Amato, N. Bucciantini, Astron. Astrophys. **421**, 1063 (2004)
L. Del Zanna, D. Volpi, E. Amato, N. Bucciantini, Astron. Astrophys. **453**, 621 (2006)
A.J. Deutsch, Ann. Astrophys. **18**, 1 (1955)
V. Dhawan, A. Mioduszewski, M. Rupen, in *VI Microquasar Workshop: Microquasars and Beyond* (2006)
I. Di Palma, D. Guetta, E. Amato, Preprint (2016), arXiv:1605.01205
V. Dombrovsky, Dokl. Akad. Nauk SSSR **94**, 1021 (1954)
G. Dubus, Astron. Astrophys. **456**, 801 (2006)
G. Dubus, Astron. Astrophys. Rev. **21**, 64 (2013)
G. Dubus, B. Cerutti, Astron. Astrophys. **557**, A127 (2013)
G. Dubus, B. Cerutti, G. Henri, Astron. Astrophys. **477**, 691 (2008)
G. Dubus, B. Cerutti, G. Henri, Astron. Astrophys. **516**, A18 (2010)
G. Dubus, A. Lamberts, S. Fromang, Astron. Astrophys. **581**, A27 (2015)
M. Durant, O. Kargaltsev, G.G. Pavlov, J. Kropotina, K. Levenfish, Astrophys. J. **763**, 72 (2013)
W.E. East, J. Zrake, Y. Yuan, R.D. Blandford, Phys. Rev. Lett. **115**, 095002 (2015)
P. Eger, H. Laffon, P. Bordas, E. de Oña Whilhelmi, J. Hinton, G. Pühlhofer, Mon. Not. R. Astron. Soc. **457**, 1753 (2016)
Fermi-LAT Collaboration, Astrophys. J. **706**, L56 (2009)
P. Goldreich, W.H. Julian, Astrophys. J. **157**, 869 (1969)
S.P. Goodwin, J. Mestel, L. Mestel, G.A.E. Wright, Mon. Not. R. Astron. Soc. **349**, 213 (2004)
L. Gratton, Astrophys. Space Sci. **16**, 81 (1972)
A. Gruzinov, Phys. Rev. Lett. **94**, 021101 (2005)
H.E.S.S. Collaboration, Astron. Astrophys. **442**, 1 (2005)
H.E.S.S. Collaboration, Astron. Astrophys. **460**, 743 (2006)
D. Hadasch et al., Astrophys. J. **749**, 54 (2012)
A.K. Harding, J.V. Stern, J. Dyks, M. Frackowiak, Astrophys. J. **680**, 1378 (2008)
J.J. Hester, Annu. Rev. Astron. Astrophys. **46**, 127 (2008)
J.J. Hester et al., Astrophys. J. **448**, 240 (1995)

J.J. Hester et al., Astrophys. J. **577**, L49 (2002)
M. Hoshino, J. Arons, Y.A. Gallant, A.B. Langdon, Astrophys. J. **390**, 454 (1992)
D. Kagan, L. Sironi, B. Cerutti, D. Giannios, Space Sci. Rev. **191**, 545 (2015)
C. Kalapotharakos, I. Contopoulos, D. Kazanas, Mon. Not. R. Astron. Soc. **420**, 2793 (2012)
O. Kargaltsev, G.G. Pavlov, in *40 Years of Pulsars: Millisecond Pulsars, Magnetars and More*, ed. by C. Bassa, Z. Wang, A. Cumming, V.M. Kaspi. American Institute of Physics Conference Series, vol. 983 (2008), pp. 171–185. arXiv:0801.2602. doi:10.1063/1.2900138
O. Kargaltsev, B. Rangelov, J. Hare, G.G. Pavlov, Astron. Nachr. **335**, 301 (2014)
C.F. Kennel, F.V. Coroniti, Astrophys. J. **283**, 694 (1984a)
C.F. Kennel, F.V. Coroniti, Astrophys. J. **283**, 710 (1984b)
D. Khangulyan, S. Hnatic, F. Aharonian, S. Bogovalov, Mon. Not. R. Astron. Soc. **380**, 320 (2007)
D. Khangulyan, F. Aharonian, V. Bosch-Ramon, Mon. Not. R. Astron. Soc. **383**, 467 (2008)
J.G. Kirk, O. Skjæraasen, Astrophys. J. **591**, 366 (2003)
T. Kishishita, T. Tanaka, Y. Uchiyama, T. Takahashi, Astrophys. J. **697**, L1 (2009)
S.S. Komissarov, Mon. Not. R. Astron. Soc. **303**, 343 (1999)
S.S. Komissarov, Mon. Not. R. Astron. Soc. **428**, 2459 (2013)
S.S. Komissarov, Y.E. Lyubarsky, Mon. Not. R. Astron. Soc. **344**, L93 (2003)
S.S. Komissarov, Y.E. Lyubarsky, Mon. Not. R. Astron. Soc. **349**, 779 (2004)
S.S. Komissarov, M. Lyutikov, Mon. Not. R. Astron. Soc. **414**, 2017 (2011)
S.S. Komissarov, N. Vlahakis, A. Königl, M.V. Barkov, Mon. Not. R. Astron. Soc. **394**, 1182 (2009)
S.W. Kong, K.S. Cheng, Y.F. Huang, Astrophys. J. **753**, 127 (2012)
A. Lamberts, S. Fromang, G. Dubus, Mon. Not. R. Astron. Soc. **418**, 2618 (2011)
A. Lamberts, G. Dubus, G. Lesur, S. Fromang, Astron. Astrophys. **546**, A60 (2012)
A. Lamberts, S. Fromang, G. Dubus, R. Teyssier, Astron. Astrophys. **560**, A79 (2013)
A.P. Lobanov, D. Horns, T.W.B. Muxlow, Astron. Astrophys. **533**, 1 (2011)
Y.E. Lyubarsky, Mon. Not. R. Astron. Soc. **329**, L34 (2002)
Y.E. Lyubarsky, Mon. Not. R. Astron. Soc. **345**, 153 (2003)
Y. Lyubarsky, Astrophys. J. **698**, 1570 (2009)
Y.E. Lyubarsky, Mon. Not. R. Astron. Soc. **402**, 353 (2010)
Y. Lyubarsky, J.G. Kirk, Astrophys. J. **547**, 437 (2001)
M. Lyutikov, Mon. Not. R. Astron. Soc. **405**, 1809 (2010)
M. Lyutikov, S.S. Komissarov, O. Porth, Mon. Not. R. Astron. Soc. **456**, 286 (2016a)
M. Lyutikov, L. Sironi, S. Komissarov, O. Porth, Preprint (2016b), arXiv:1603.05731
K.K. Madsen et al., Astrophys. J. **801**, 66 (2015)
D. Malyshev, M. Chernyakova, Preprint (2016), arXiv:1601.08216
B. Marcote, M. Ribó, J.M. Paredes, C.H. Ishwara-Chandra, Mon. Not. R. Astron. Soc. **451**, 59 (2015)
F. Massaro, D.J. Thompson, E.C. Ferrara, Astron. Astrophys. Rev. **24**, 2 (2015)
M. Mayer, R. Buehler, E. Hays, C.C. Cheung, M.S. Dutka, J.E. Grove, M. Kerr, R. Ojha, Astrophys. J. **775**, L37 (2013)
J.C. McKinney, Mon. Not. R. Astron. Soc. **368**, L30 (2006)
A. Melatos et al., Astrophys. J. **633**, 931 (2005)
M. Meyer, D. Horns, H.-S. Zechlin, Astron. Astrophys. **523**, A2 (2010)
F.C. Michel, Comments Astrophys. Space Phys. **3**, 80 (1971)
F.C. Michel, Astrophys. J. **180**, L133 (1973)
Y. Mizuno, Y. Lyubarsky, K.-I. Nishikawa, P.E. Hardee, Astrophys. J. **728**, 90 (2011)
J. Moldón, S. Johnston, M. Ribó, J.M. Paredes, A.T. Deller, Astrophys. J. **732**, L10 (2011)
P. Moran, A. Shearer, R.P. Mignani, A. Slowikowska, A.D. Luca, C. Gouiffes, P. Laurent, Mon. Not. R. Astron. Soc. **433**, 2564 (2013)
P. Moran, G. Kyne, C. Gouiffès, P. Laurent, G. Hallinan, R.M. Redfern, A. Shearer, Mon. Not. R. Astron. Soc. **456**, 2974 (2015)
Y. Moritani et al., Astrophys. J. **804**, L32 (2015)
K. Nalewajko, J. Zrake, Y. Yuan, W.E. East, R.D. Blandford, Astrophys. J. **826**, 115 (2016)
M. Nynka et al., Astrophys. J. **789**, 72 (2014)
B. Olmi, L. Del Zanna, E. Amato, R. Bandiera, N. Bucciantini, Mon. Not. R. Astron. Soc. **438**, 1518 (2014)
B. Olmi, L. Del Zanna, E. Amato, N. Bucciantini, Mon. Not. R. Astron. Soc. **449**, 3149 (2015)
B. Olmi, L. Del Zanna, E. Amato, N. Bucciantini, A. Mignone, Preprint (2016), arXiv:1610.07956
J.H. Oort, T. Walraven, Bull. Astron. Inst. Neth. **12**, 285 (1956)
J.P. Ostriker, J.E. Gunn, Nature **223**, 813 (1969)
F. Pacini, Nature **216**, 567 (1967)
X. Paredes-Fortuny, M. Ribó, V. Bosch-Ramon, J. Casares, O. Fors, J. Núñez, Astron. Astrophys. **575**, L6 (2015)

K. Parfrey, A.M. Beloborodov, L. Hui, Mon. Not. R. Astron. Soc. **423**, 1416 (2012)
G.G. Pavlov, M.A. Teter, O. Kargaltsev, D. Sanwal, Astrophys. J. **591**, 1157 (2003)
R. Petre, U. Hwang, S.S. Holt, S. Safi-Harb, R.M. Williams, Astrophys. J. **662**, 988 (2007)
J. Pétri, Mon. Not. R. Astron. Soc. **455**, 3779 (2016)
A.A. Philippov, A. Spitkovsky, Astrophys. J. **785**, L33 (2014)
J.M. Pittard, Mon. Not. R. Astron. Soc. **396**, 1743 (2009)
O. Porth, S.S. Komissarov, R. Keppens, Mon. Not. R. Astron. Soc. **431**, L48 (2013)
O. Porth, S.S. Komissarov, R. Keppens, Mon. Not. R. Astron. Soc. **438**, 278 (2014a)
O. Porth, S.S. Komissarov, R. Keppens, Mon. Not. R. Astron. Soc. **443**, 547 (2014b)
O. Porth, M.J. Vorster, M. Lyutikov, N.E. Engelbrecht, Mon. Not. R. Astron. Soc. **460**, 4135 (2016)
M.J. Rees, J.E. Gunn, Mon. Not. R. Astron. Soc. **167**, 1 (1974)
S.P. Reynolds, ArXiv Astrophysics e-prints (2003)
A. Rudy et al., Astrophys. J. **811**, 24 (2015)
M. Ruiz, V. Paschalidis, S.L. Shapiro, Phys. Rev. D **89**, 084045 (2014)
A. Sandberg, J. Sollerman, Astron. Astrophys. **504**, 525 (2009)
J.D. Scargle, Astrophys. J. **156**, 401 (1969)
T. Schweizer, N. Bucciantini, W. Idec, K. Nilsson, A. Tennant, M.C. Weisskopf, R. Zanin, Mon. Not. R. Astron. Soc. **433**, 3325 (2013)
R.R. Sefako, O.C. de Jager, Astrophys. J. **593**, 1013 (2003)
I. Shklovsky, Dokl. Akad. Nauk SSSR **90**, 983 (1953)
A. Sierpowska-Bartosik, D.F. Torres, Astropart. Phys. **30**, 239 (2008)
L. Sironi, A. Spitkovsky, Astrophys. J. **707**, L92 (2009)
L. Sironi, A. Spitkovsky, Astrophys. J. **741**, 39 (2011)
L. Sironi, A. Spitkovsky, J. Arons, Astrophys. J. **771**, 54 (2013)
L. Sironi, U. Keshet, M. Lemoine, Space Sci. Rev. **191**, 519 (2015)
P. Slane, Y. Chen, N.S. Schulz, F.D. Seward, J.P. Hughes, B.M. Gaensler, Astrophys. J. **533**, L29 (2000)
P. Slane, D.J. Helfand, E. van der Swaluw, S.S. Murray, Astrophys. J. **616**, 403 (2004)
A. Spitkovsky, Astrophys. J. **648**, L51 (2006)
P.A. Sturrock, Astrophys. J. **164**, 529 (1971)
A. Szostek, G. Dubus, Mon. Not. R. Astron. Soc. **411**, 193 (2011)
X. Tang, R.A. Chevalier, Astrophys. J. **752**, 83 (2012)
M. Tavani et al., Science **331**, 736 (2011)
A. Tchekhovskoy, J.C. McKinney, R. Narayan, Astrophys. J. **699**, 1789 (2009)
A. Tchekhovskoy, A. Philippov, A. Spitkovsky, Mon. Not. R. Astron. Soc. **457**, 3384 (2016)
A.N. Timokhin, A.K. Harding, Astrophys. J. **810**, 144 (2015)
D.F. Torres, N. Rea, P. Esposito, J. Li, Y. Chen, S. Zhang, Astrophys. J. **744**, 106 (2012)
B. van Soelen, P.J. Meintjes, A. Odendaal, L.J. Townsend, Mon. Not. R. Astron. Soc. **426**, 3135 (2012)
B. van Soelen, P. Väisänen, A. Odendaal, L. Klindt, I. Sushch, P.J. Meintjes, Mon. Not. R. Astron. Soc. **455**, 3674 (2016)
N. Vlahakis, Astrophys. J. **600**, 324 (2004)
D. Volpi, L. Del Zanna, E. Amato, N. Bucciantini, Astron. Astrophys. **485**, 337 (2008)
S.L. Weinberg, J. Silk, Astrophys. J. **205**, 563 (1976)
M.C. Weisskopf, Mem. Soc. Astron. Ital. **84**, 582 (2013)
M.C. Weisskopf et al., Astrophys. J. **536**, L81 (2000)
A.S. Wilson, Mon. Not. R. Astron. Soc. **157**, 229 (1972)
Y. Yuan, R.D. Blandford, Mon. Not. R. Astron. Soc. **454**, 2754 (2015)
Y. Yuan, K. Nalewajko, J. Zrake, W.E. East, R.D. Blandford, Astrophys. J. **828**, 92 (2016)
V. Zabalza, V. Bosch-Ramon, F. Aharonian, D. Khangulyan, Astron. Astrophys. **551**, A17 (2013)
A.A. Zdziarski, A. Neronov, M. Chernyakova, Mon. Not. R. Astron. Soc. **403**, 1873 (2010)
J. Zrake, Preprint (2015), arXiv:1512.05426
J. Zrake, W.E. East, Astrophys. J. **817**, 89 (2016)

Space Sci Rev (2017) 207:175–234
DOI 10.1007/s11214-017-0356-6

Pulsar-Wind Nebulae and Magnetar Outflows: Observations at Radio, X-Ray, and Gamma-Ray Wavelengths

Stephen P. Reynolds[1] · **George G. Pavlov** · **Oleg Kargaltsev** · **Noel Klingler** · **Matthieu Renaud** · **Sandro Mereghetti**

Received: 23 January 2017 / Accepted: 22 March 2017 / Published online: 30 March 2017

Abstract We review observations of several classes of neutron-star-powered outflows: pulsar-wind nebulae (PWNe) inside shell supernova remnants (SNRs), PWNe interacting directly with interstellar medium (ISM), and magnetar-powered outflows. We describe radio, X-ray, and gamma-ray observations of PWNe, focusing first on integrated spectral-energy distributions (SEDs) and global spectral properties. High-resolution X-ray imaging of PWNe shows a bewildering array of morphologies, with jets, trails, and other structures. Several of the 23 so far identified magnetars show evidence for continuous or sporadic emission of material, sometimes associated with giant flares, and a few "magnetar-wind nebula" have been recently identified.

Keywords First keyword · Second keyword · More

1 Introduction

Pulsars emit relativistic winds in a variety of forms, not well understood at this time. The outflows consist of some combination of highly relativistic leptons (e^+/e^- pairs) and perhaps ions as well, and magnetic field. The winds initially appear to be "dark" because they are cold in the comoving frame, but become thermalized somehow at a wind termination shock, which may or may not resemble a traditional perpendicular shock. Its location is fixed by pressure balance between the outgoing wind and the local ambient medium, which is either a shell supernova-remnant (SNR) interior (see Fig. 1), or for older pulsars that have outlived their SNR, undisturbed interstellar medium (ISM). Beyond this point, radiation from the outflows is apparent, and the observed object is known as a pulsar-wind nebula (PWN). [See Gaensler and Slane (2006) and Kargaltsev and Pavlov (2008) for general reviews of PWNe.] The broadband spatially integrated spectral-energy distribution (SED) appears to consist of two parts: a lower-energy spectrum of synchrotron emission, responsible for emission from

✉ S.P. Reynolds
reynolds@ncsu.edu

1 Physics Department, North Carolina State University, Raleigh, USA

Fig. 1 Cartoon of PWN inside a shell SNR. The SNR blast wave is the outermost circle

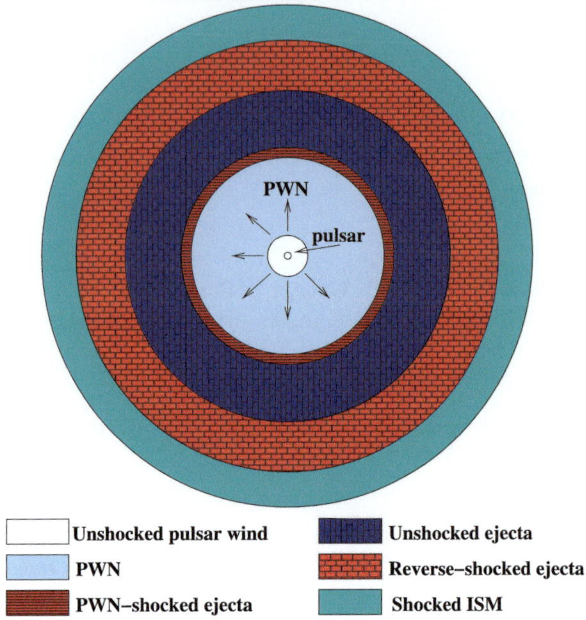

low radio frequencies to X-rays and into the MeV range (in the few cases where it is detectable at those energies), and a higher-energy spectrum in the GeV–TeV range, in most cases attributed to inverse-Compton upscattering of any of various possible photon fields: the PWN's own synchrotron photons ("synchrotron self-Compton," SSC), the cosmic microwave background (ICCMB), or local optical/IR radiation fields. In a few cases, it may be that this emission is produced by relativistic ions colliding inelastically with thermal gas and producing both charged and neutral pions. The π^0 mesons decay to photons, producing a continuum above the kinematic threshhold of about 150 MeV. But this hadronic process is not thought to be a major contributor to most PWN GeV–TeV spectra.

We structure this review mainly from young to older objects. Pulsars live far longer than their natal supernova remnant, so a relatively small fraction of PWNe should still be found within a SNR. However, this subclass offers an excellent chance to understand pulsar outflows, as information from the SNR can add to what can be deduced from the PWN alone. Other PWNe without clear evidence of a surrounding shell still have unmistakable signs of youth. The first part of this review will focus on those two subclasses, on their integrated SEDs and on their spatially resolved spectral properties. Pulsars older than the typical lifetime of an SNR continue to emit winds, which interact directly with the surrounding medium. The second part of this review deals with radio and X-ray emission from these PWNe. Some of both young and older X-ray PWNe emit gamma-rays as well, but the gamma-ray class is predominantly made up of much older objects which may be unprepossessing or undetectable at longer wavelengths. The third part of the review covers the gamma-ray properties of PWNe of all ages. Finally, magnetars have unique characteristics in terms of outflows; these characteristics are described in the last part of this chapter.

While the gross properties of PWNe (mean sizes, integrated spectra) provide one kind of information on the basic physics, a great deal can also be learned from detailed morphological investigations. The *Chandra* X-ray Observatory has provided a rich collection of imaging information down to sub-arcsecond scales, and we review that collection as well. Many of these objects are also older, with the pulsar wind interacting directly with ISM, sometimes in a well-defined bow shock, but in other cases in unusual and perplexing morphologies. Finally, we describe evidence for outflows from magnetars, neutron stars with very strong magnetic fields and a propensity to emit giant flares. Evidence for ejection of material seems strong. While clear cases of steady outflows ("magnetar-wind nebulae") have not yet been firmly confirmed, candidates have been identified. We also mention gamma-ray binaries, in which the relativistic pulsar wind interacts with a wind from the binary companion, but these systems are described more fully elsewhere.

2 Synchrotron Emission from Young Pulsar-Wind Nebulae: Radio to X-Rays

The synchrotron spectrum of PWNe contains the most specific information about the particle spectrum injected into the nebula at the wind termination shock (WTS), though the unknown magnetic-field structure can complicate the extraction of that information. Furthermore, the injected spectrum can change with time and location, due to evolutionary effects and particle propagation, e.g., advection or diffusion. Disentangling these effects is essential to understand the nature of particle energization at the WTS.

Figure 2 illustrates four PWN/SNR combinations, in which the pulsars are well studied. The PWNe are asymmetric and structured, and for G292.0+1.8, not centered on the pulsar. However, these systems are amenable to detailed interpretation based on analysis of both the PWN nonthermal radiation and the properties of the surrounding SNR. In both B0540-693 (Williams et al. 2008) and G11.2-0.3 (Borkowski et al. 2016), the PWN's interaction with the SNR inner ejecta is important in the overall characterization of the object.

Young objects like the Crab or 3C 58 (unlikely to be the remnant of an event in 1181 AD, but still only of order 2000 yr old; Chevalier 2005), without clear evidence of a shell, also show properties consistent with expansion into a low-density medium which is probably itself expanding. Simple 1-D models show that as long as the pulsar maintains its original energy-loss rate, the PWN expands into uniformly expanding ejecta with radius $R \propto t^{6/5}$ (Reynolds and Chevalier 1984; van der Swaluw et al. 2001), producing a shock wave into the inner ejecta that strengthens with time. (See Fig. 1).

These two classes of young PWNe, with and without SNR shells, will be the focus of this section.

2.1 General Properties of the PWN Synchrotron Spectrum

Pulsar-wind nebulae were originally defined by radio properties: "flat" radio spectrum (that is, energy spectral index $\alpha \sim 0$–0.4 where $S_\nu \propto \nu^{-\alpha}$ is the energy flux), center-brightened morphology, and high radio polarization (e.g., Weiler and Panagia 1978). The catalog remained small, however, until the high spatial resolution of *Chandra* and XMM-*Newton* allowed the identification of PWNe in shells and in confused regions, at which point a large increase in identified PWNe began which continues to this day. The catalog by Kargaltsev et al. (2013) lists 70 X-ray PWNe with known pulsars, and 6 more objects identified at gamma-ray energies without currently known X-ray counterparts. Many of the PWNe discovered at X-ray or gamma-ray energies have weak, poorly known, or undetectable radio emission. In Green's catalog of SNRs and PWNe (Green 2009), 26 PWNe in shells (formerly known as "composite" SNRs) and another 13 isolated PWNe (once called "plerions")

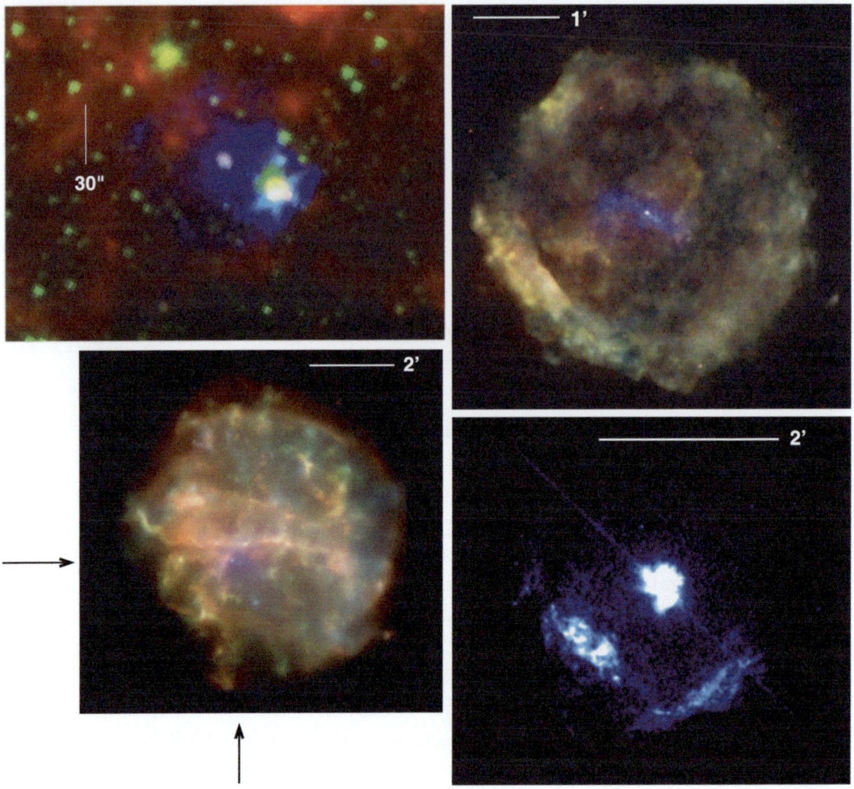

Fig. 2 *Upper left*: B0540-693 in the Large Magellanic Cloud (Williams et al. 2008). *Red*, *Spitzer* 8 μm; *green*, *Spitzer* 3.6 μm; *blue*, *Chandra* 0.5–8 keV. The pulsar and surrounding PWN is barely resolved at the center of the frame, just above and to the left of the bright foreground star. *Upper right*: G11.2-0.3 with *Chandra* (Borkowski et al. 2016). *Red*: 0.6–1.2 keV; *green*, 1.2–3.3 keV; *blue*, 3.3–8 keV. The PWN is visible in *blue*. *Lower left*: G292.0+1.8 with *Chandra* (Park et al. 2007). *Red*: 0.3–0.8 keV; *green*, 0.8–1.7 keV; *blue*, 1.7–8 keV. The PWN is the purplish region slightly SE of the center, and the pulsar (indicated by *arrows*) is at its NE edge. *Lower right*: Kes 75 with *Chandra* (Gavriil et al. 2008). The pulsar is highly piled up, as the readout streak indicates

Fig. 3 Distribution of radio spectral indices α ($S_\nu \propto \nu^{-\alpha}$) of 39 PWNe in shells (*blue*) and isolated (*red*). Many of the former have poorly known radio spectral indices. Typical uncertainties on α are ± 0.1. The three steep-spectrum PWNe ($\alpha \geq 0.6$) are all anomalous in other ways as well

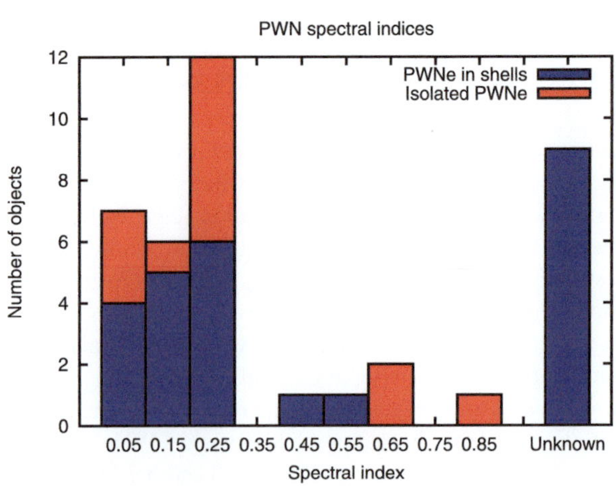

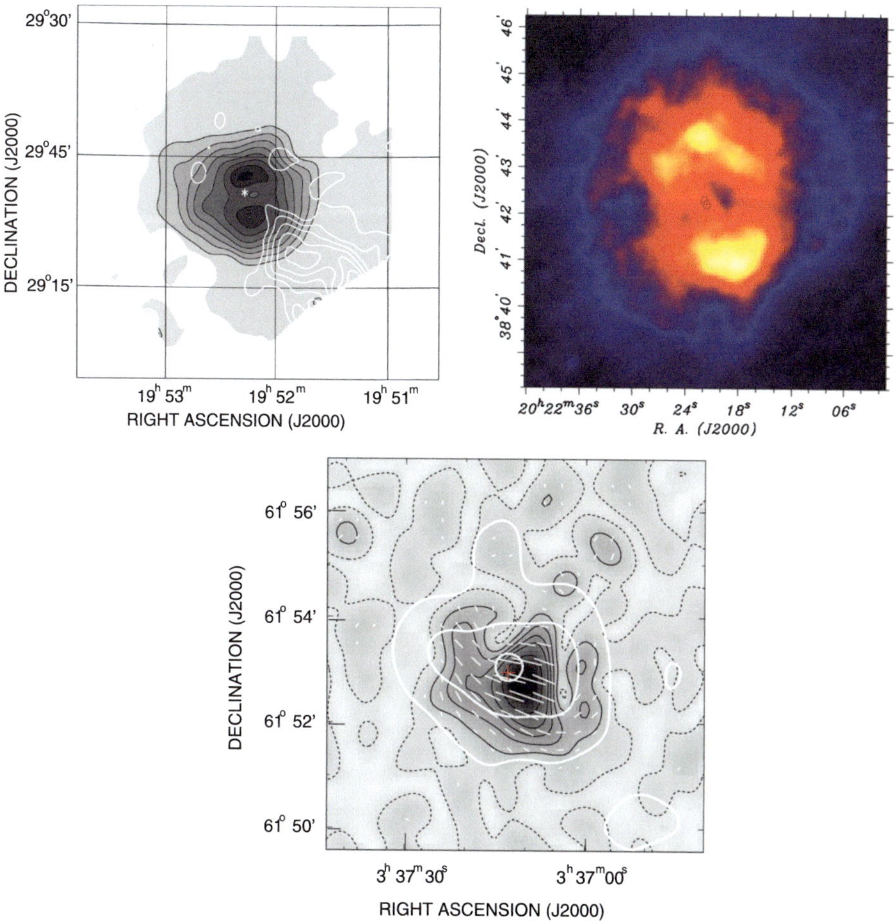

Fig. 4 Three PWNe with steep radio spectra. *Upper left*: DA 495 (G65.7+1.2) at 1.4 GHz. *White contours*: a background H II region. *Asterisk*: a compact X-ray source (Kothes et al. 2008). *Upper right*: G76.9+1.0 (Arzoumanian et al. 2011) at 1.4 GHz. Almost invisible *contours* of extent about 40″ near the center show the X-ray nebula. *Bottom*: G141.2+5.0 at 1.4 GHz in polarized intensity (*greyscale*) with contours indicating total intensity (Kothes et al. 2014), and the *red cross* indicating a pointlike X-ray source (Reynolds and Borkowski 2016)

are listed. This catalog began as a listing of radio SNRs, and does not include pulsar bow-shock nebulae or other manifestations of pulsar outflows without clear radio counterparts. With the addition of two more recent discoveries, Fig. 3 shows the distribution of PWN radio spectral indices. Since there are a few good cases of steeper-spectrum PWNe ($\alpha > 0.4$), it is possible that more such objects exist but have been selected against. However, the three steep-spectrum PWNe ($\alpha \geq 0.6$) are all anomalous in other ways as well. It is also possible that some very flat-spectrum objects ($\alpha \sim 0.1$) have been mistaken for H II regions, since strong radio polarization, the usual discriminator, is not found in all objects.

The three anomalous steep-spectrum PWNe are shown in Fig. 4. The first two, DA 495 and G76.9+1.0, with $\alpha = 0.87 \pm 0.1$ (Kothes et al. 2008) and 0.62 ± 0.04 (Landecker et al. 1993), have strikingly similar double-lobed morphology, with tiny extended X-ray sources in the center of each. G141.2+5.0 (Reynolds and Borkowski 2016) shows a simple center-

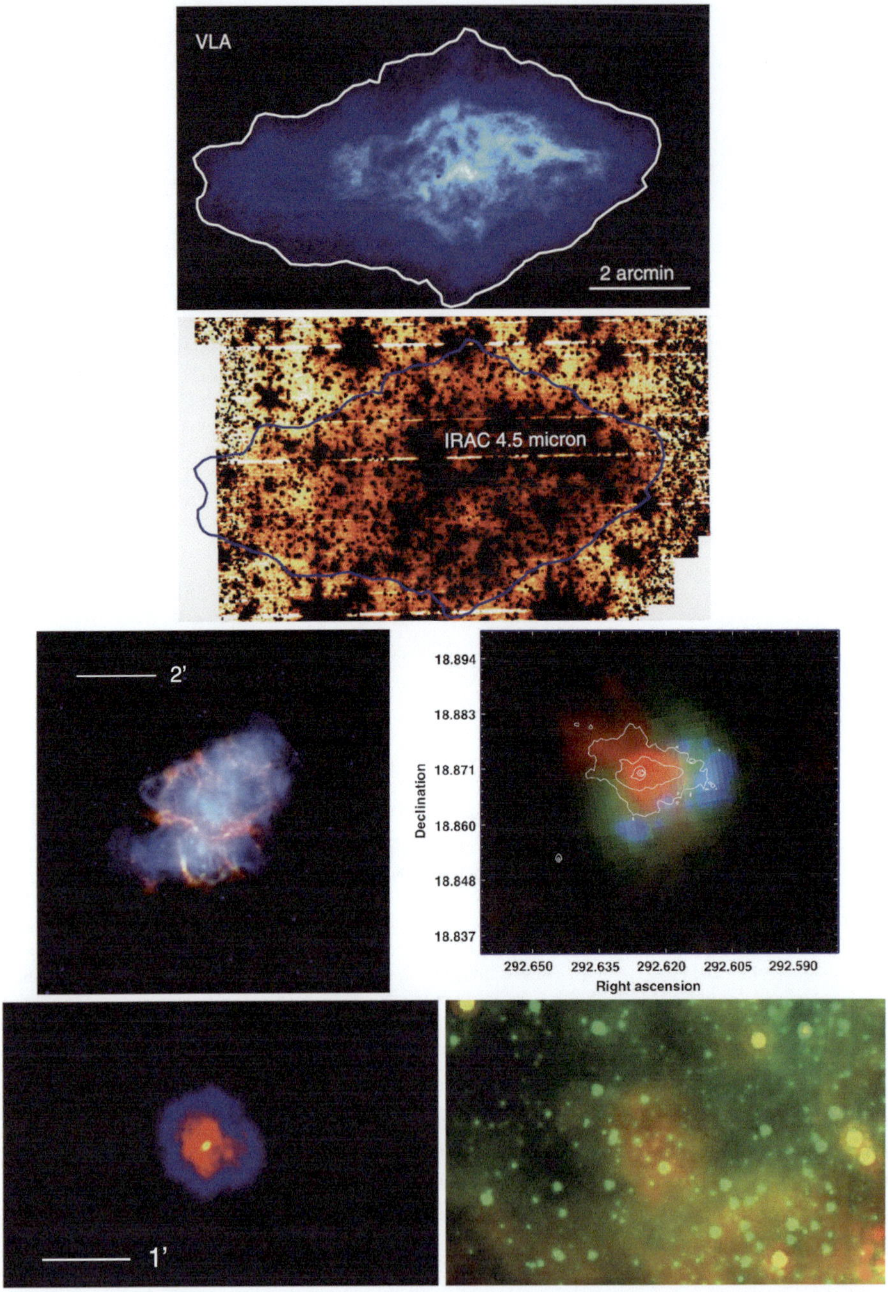

Fig. 5 *Top*: 3C 58 (Slane et al. 2008). (*Upper*: radio, VLA at 1.4 GHz, Reynolds and Aller 1988; *lower*, IR, *Spitzer* at 4.5 μm.) *Middle left*: Crab Nebula (Temim et al. 2006). *Red*, 24 μm (mainly [O IV] at 26 μm); *green*, 8 μm (mainly [Ar II] at 7 μm); *blue*, 3.6 μm (mainly synchrotron continuum). *Middle right*: G54.1+0.3 (Temim et al. 2010). *Contours*, X-ray. *Red*, radio. *Green*, 70 μm. *Blue*, 24 μm. *Lower*: G21.5-0.9 (Zajczyk et al. 2012). *Left*: *Chandra* image. *Right*: *Red*, 24 μm; *green*, 8 μm

Table 1 *Spitzer* observations of PWNe

Object	Type	Detectors[a]	Result	Reference
Crab	PWN	I, M, S	Images, spectrum	Temim et al. (2006)
G21.5-0.9	PWN	I, M	Images	Zajczyk et al. (2012)
3C58	PWN	I, M	Images	Slane et al. (2008)
G54.1+0.3	PWN, IR shell	I, M, S	Images, spectrum	Temim et al. (2010)
B0540-693	PWN, shell	I, M, S	Images, spectrum	Williams et al. (2008)

[a]I: IRAC (3.6, 4.5, 5.8, & 8.0 μm). M: MIPS (24, 70, 160 μm). S: IR Spectrograph (5–35 μm)

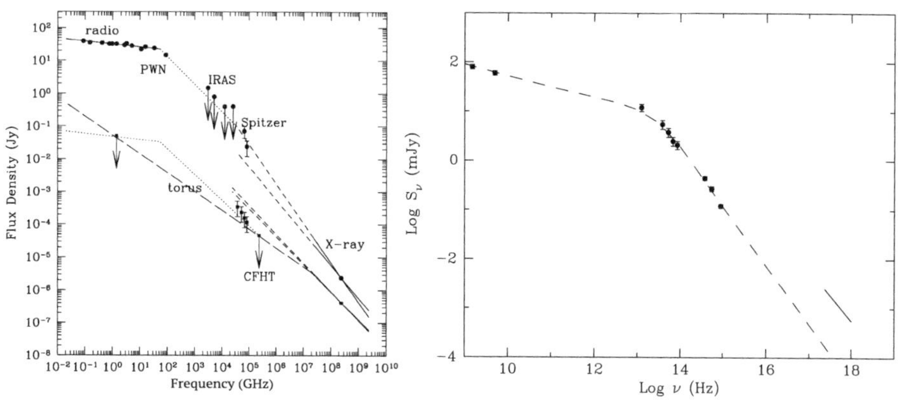

Fig. 6 *Left*: SED of 3C 58 (Slane et al. 2008). The spatially integrated spectrum is above, and is roughly describable by two power-laws of slopes α of 0.1 and $\sim$1, with a break near 100 GHz. (The small-scale torus requires a considerably more complex spectrum.) *Right*: SED of B0540-693 in the Large Magellanic Cloud (Williams et al. 2008). *Triangles* are *Spitzer* observations. The *solid line* is the *Chandra* spectrum (Kaaret et al. 2001), which is affected by pileup in the pulsar. The PWN X-ray spectrum does not appear to be consistent with the power-law extrapolation from the optical-IR (Serafimovich et al. 2004)

brightened morphology, but with a central X-ray source unresolved by *Chandra*. It is not known how these objects fit into the overall PWN scheme; no data are available for any at other wavelengths.

For many PWNe, no other observations are available at frequencies below the X-ray regime. But a few have been imaged with *Spitzer* (see Fig. 5 and Table 1). Much of the emission is thermal radiation from dust grains, or fine-structure spectral lines, but some synchrotron continuum is evident in the Crab, 3C 58, and G21.5-0.9.

PWNe are often observed to be smaller at X-ray than at longer wavelengths, as is clearly the case with the Crab Nebula. However, several other prominent PWNe show X-rays extending to the edges of the radio contours, though with greater center-to-edge brightness contrast. Figures 7 and 8 show that both 3C 58 and G54.1+0.3 have X-ray extents comparable to their radio extents.

2.2 Imaging PWNe Above 10 keV with *NuSTAR*

Spectral inhomogeneity in PWNe is commonly observed at X-ray energies, where energy losses are becoming important. Spectra steepen with distance from the central pulsar, pre-

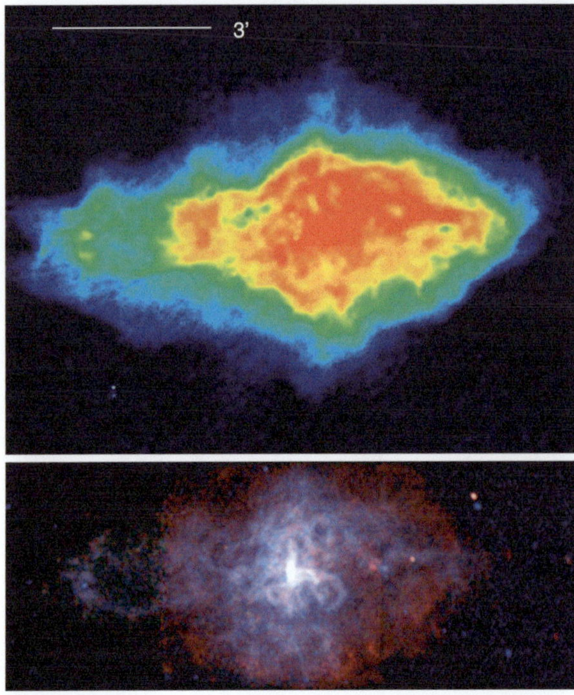

Fig. 7 *Top*: VLA image of 3C 58 at 1.4 GHz (Reynolds and Aller 1988). *Bottom*: *Chandra* X-rays (Slane et al. 2008). Note the close correspondence between detailed features, and the similar maximum extent in radio and X-rays

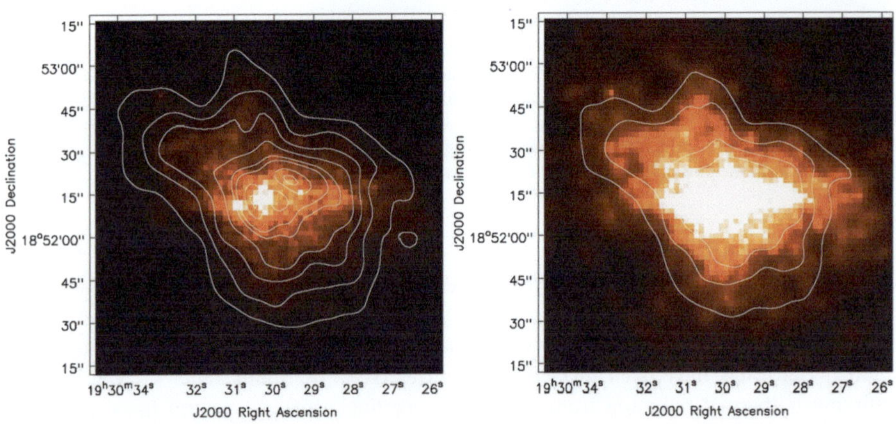

Fig. 8 Images of G54.1+0.3 in radio (*contours*; VLA at 4.7 GHz) and X-ray (*Chandra*), with two different stretches to emphasize the extent of faint X-ray emission (Lang et al. 2010)

sumably as higher-energy electrons are depleted. The steepening trend tends to begin immediately, as shown in Fig. 9 for three PWNe (Bocchino and Bykov 2001). This behavior is explored more fully below. This means, however, that integrated SEDs may hide important spectral variations. Unfortunately, until recently true imaging at X-ray energies above about 10 keV was not available. However, the *NuSTAR* mission, launched in 2012 (Harrison et al. 2013), has brought that capability to the study of PWNe. Results of observations of three PWNe with *NuSTAR* are summarized below; details are in the primary publications listed.

Fig. 9 Steepening of the X-ray spectrum with scaled distance from the center, for three PWNe (Bocchino and Bykov 2001). The distance is in units of the distance at which the surface brightness has dropped a factor of 2 below its peak. The *line* is a straight-line fit to the IC 443 data only

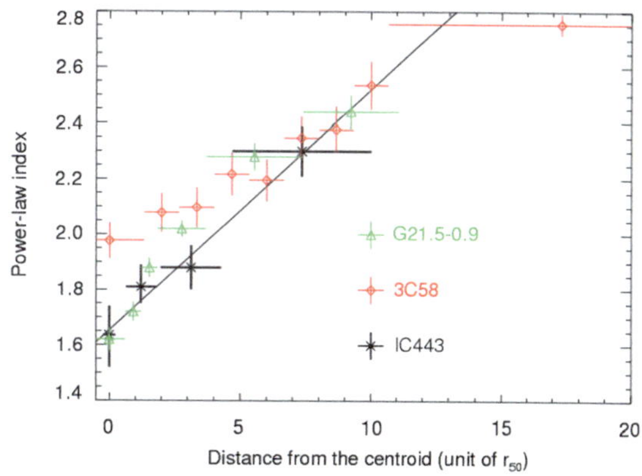

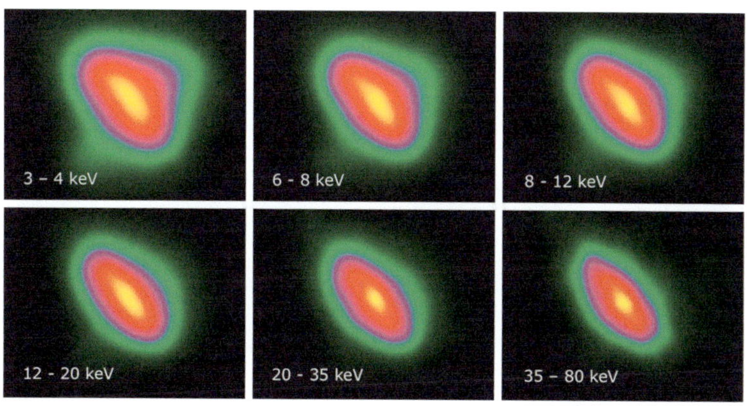

Fig. 10 The Crab Nebula with *NuSTAR* (Madsen et al. 2015), after maximum-entropy deconvolution. The nebula shrinks with increasing photon energy, at different rates in different directions

2.2.1 Crab Nebula

For PWNe with a bright pulsar, disentangling the PWN emission from the pulsar, and then searching for spectral variations, requires good resolution in both time and space. Madsen et al. (2015) describe this process in detail for the Crab. The observations were divided into 13 phase intervals of the pulsar rotation period, and only bins 10–12, where pulsar emission was negligible, were used for the PWN spatial/spectral analysis (see Figs. 10 and 11). The spatially and temporally integrated signal was calibrated to the accepted power-law shape with $\Gamma = 2.1$ [$F(E) \propto E^{-\Gamma}$ photons $\mathrm{cm^{-2}\,s^{-1}\,keV^{-1}}$]. But the spectrum is substantially harder at smaller radii. This was known previously through *Chandra* observations (Mori et al. 2004). However, *NuSTAR* has revealed another important feature: the spectrum of the inner nebula (primarily the torus) appears to steepen by a substantial amount, $\Delta\Gamma \cong 0.25$, above a break energy of about 9 keV. Figure 12 shows the low and high-energy Γ values, based on the *NuSTAR* data.

The original spherical, steady-state MHD model of Kennel and Coroniti (1984a,b) predicted how the size of the Crab Nebula in the X-ray range should scale with photon energy E: $R \propto E^m$ with $m = -1/9$. This prediction was consistent with results found by several sounding-rocket measurements with lunar occultations in the 1970's (Kestenbaum et al.

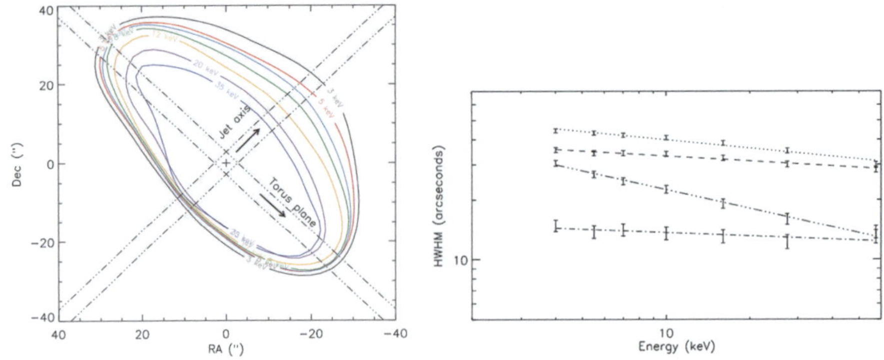

Fig. 11 *Left*: Crab 50% peak intensity contours at different energies (Madsen et al. 2015). From outside in, 3–5, 5–6, 6–8, 8–12, 12–20, 20–35, and 35–78 keV. *Right*: Fits to the HWHM along different directions. From *top* to *bottom*: NE torus plane, SW torus plane, NW jet axis, and SE jet axis

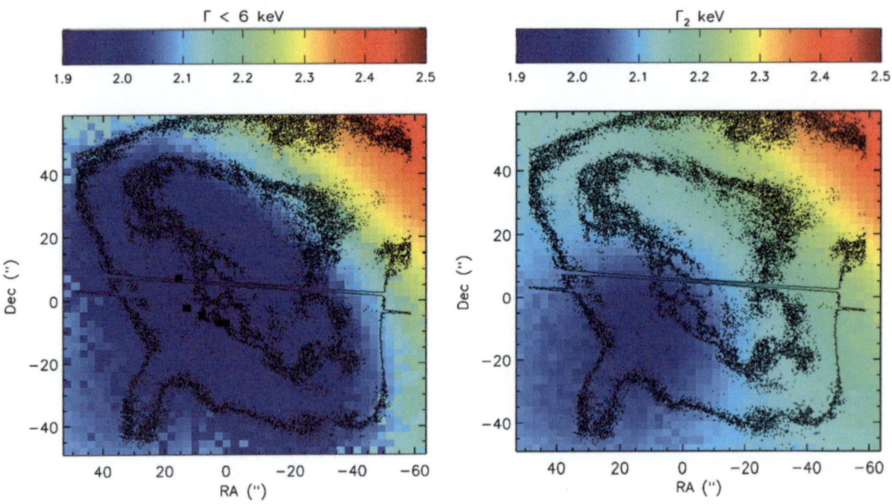

Fig. 12 *Left*: Crab photon index Γ below 6 keV; *right*: Γ above 10 keV (Madsen et al. 2015). *Contours* are from *Chandra* (Mori et al. 2004)

1975; Ku et al. 1976) which obtained FWHM $\propto E^{-0.148\pm0.012}$, fitting Gaussians to the data. This result was regarded as adequate agreement. The *NuSTAR* data show that the shrinkage is spatially varying (Figs. 10 and 11). The fitted values of m are -0.086 ± 0.025 along the NE direction and -0.073 ± 0.028 along the SW, or about -0.08 ± 0.03 along the torus plane. For the SE, m is consistent with zero, but along the NW ("counter-jet" direction), the rate is substantially larger: $m = -0.218 \pm 0.040$. The torus rate is consistent with the Kennel and Coroniti (1984a,b) prediction, but the counter-jet clearly shrinks much more rapidly.

2.2.2 G21.5-0.9

This bright PWN was used for calibration by *NuSTAR* and other missions. It has been reported as a TeV source (Djannati-Ataï et al. 2008, see Fig. 15); while those data appear to lie roughly on an extrapolation from the measurements of *NuSTAR* and INTEGRAL, the

TeV emission must be due to a separate process as normal synchrotron emission is limited to photon energies below a few hudred MeV. However, the radio–X-ray spectrum appears to be reasonably well described by two power-laws, requiring a steepening in mid-IR by $\Delta \equiv \alpha(\text{high}) - \alpha(\text{low}) \sim 1.0$, where $\alpha(\text{high})$ is the X-ray energy index ($\alpha_X \equiv \Gamma - 1$).

The fairly symmetric appearance of G21.5-0.9, and its similar size and morphology in radio, IR, and X-rays, are unusual for PWNe. Figure 13 shows that the PWN sits in the center of an apparent shell of diffuse X-rays. This diffuse emission shows some apparent limb-brightening in the SE and irregular structures to the N which have thermal spectra and appear to be the SNR shell (Bocchino 2005), but there is also a substantial X-ray halo due to scattering by interstellar dust providing the symmetric component of diffuse emission that drops with radius (Bandiera and Bocchino 2004).

NuSTAR observations (Nynka et al. 2014) show substantial shrinkage with X-ray energy (Fig. 14, though the FWHM shrinkage is less dramatic than the drop in total flux). Figure 15 shows the FWHM shrinkage, well described by a power-law with $m = -0.21 \pm 0.01$, similar to the value found for the Crab counter-jet. The northern part of the shell is detectable in the images to at least 20 keV, suggesting that part of the emission may be nonthermal. The *NuSTAR* spectrum shows, as for the Crab torus, a spectral softening that can be fit with a broken power-law, with Γ steepening from 1.996 ± 0.013 below a break energy $E_b = 9.7^{+1.2}_{-1.4}$ keV to 2.093 ± 0.013 above that energy. (Dust scattering efficiency drops with increasing photon energy and should not affect the spectrum in the *NuSTAR* energy range above 3 keV.)

2.2.3 MSH 15-52

This complex object has almost no radio counterpart (Gaensler et al. 2002), but is bright at X-ray wavelengths (Fig. 16). It contains a well-known 150 ms pulsar, PSR B1509-58, with a 1600-yr spindown timescale (Seward and Harnden 1982). A bright, curved jet extends to the SE, while long, straighter "fingers" reach to the NW to an H II region, RCW89, which contains small knots of radio and X-ray emission. A TeV detection has been reported (Aharonian et al. 2005). The full SED is shown in Fig. 17.

NuSTAR observations (An et al. 2014) show the same progressive steepening evident in other PWNe. Combined fitting with *Chandra* and *NuSTAR* gives a rate that appears to slow with distance, and can be described by two power-laws (Fig. 18). This steepening is reflected in the broader measure of source extent as a function of photon energy (Fig. 19). Profiles were created by summing transversely in a rectangular box of width $100''$ extended along the jet axis. Again, combined data from *Chandra* and *NuSTAR* were used. Power-laws describe the data fairly well, with $m \sim -0.2$ (Fig. 19), a value similar to that found for G21.5-0.9 and for the counterjet of the Crab.

2.3 Spectral Breaks in PWNe

The integrated synchrotron spectrum of all known PWNe steepens sharply between radio and X-rays. Steepening due to radiative losses is expected, but steepening can also be caused by evolutionary effects, such as the dropoff of pulsar luminosity after a few hundred years (e.g., Pacini and Salvati 1973), or by intrinsic structure in the spectrum of particles injected into the nebula at the wind termination shock. It is important to be able to characterize the injected spectrum to understand the physics of particle acceleration, so determining the cause of spectral structure is a necessary part of using PWNe as laboratories for the study of astrophysical particle energization.

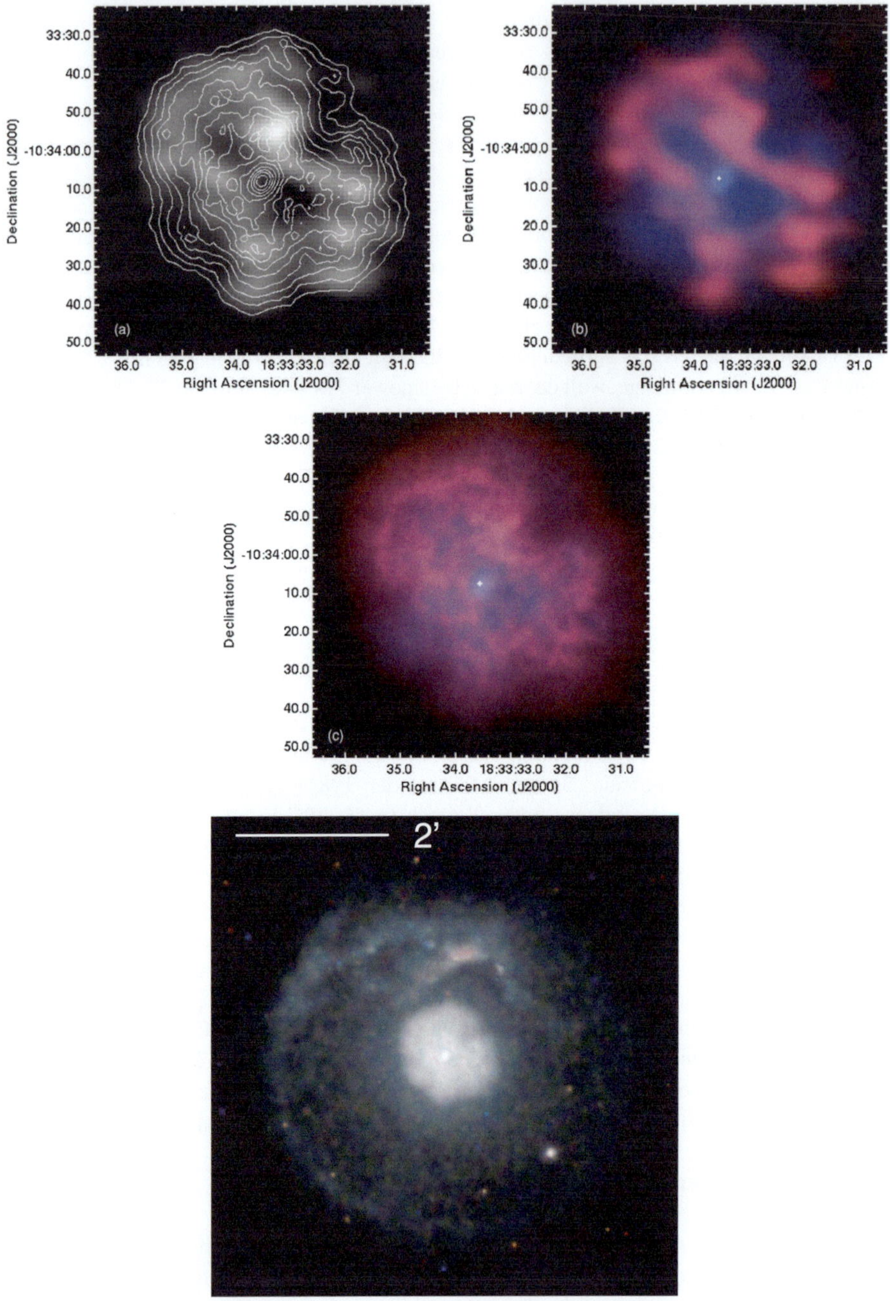

Fig. 13 G21.5-0.9 in radio and X-rays (Matheson and Safi-Harb 2010). *Upper left*: radio (*greyscale*; 22.3 GHz from the Nobeyama Millimeter-Wave Array, resolution 8″ (Fuerst et al. 1988) and X-rays (*Chandra* contours; Matheson and Safi-Harb 2010). *Upper right*: 22.3 GHz data in *red*, *Chandra* in *blue*. *Center*: 4.75 GHz radio (VLA; Bietenholz and Bartel 2008) in *red*, same X-rays in *blue*. *Bottom*: *Chandra* (Matheson and Safi-Harb 2010)

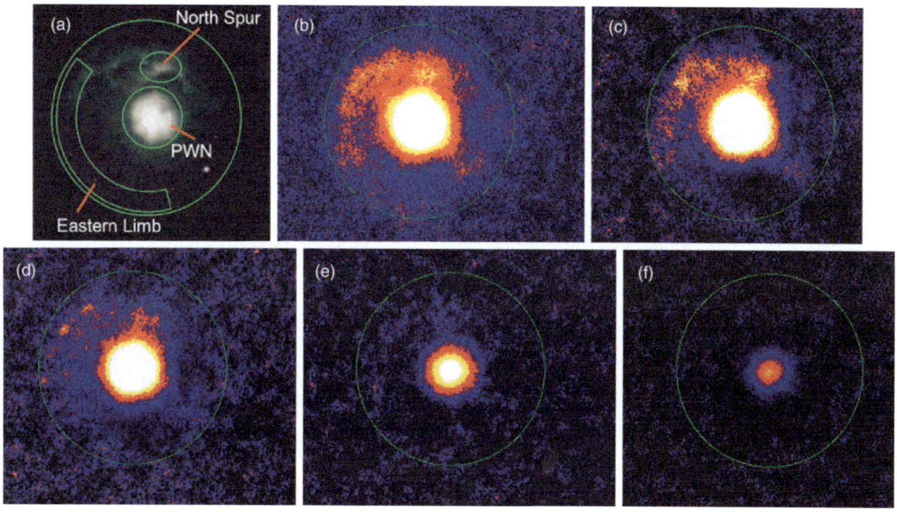

Fig. 14 (**a**): *Chandra* image of G21.5-0.9 between 3 and 6 keV (Matheson and Safi-Harb 2005). (**b**)–(**f**): Deconvolved *NuSTAR* images in various bands: (**b**) 3–6 keV; (**c**) 6–10 keV; (**d**) 10–15 keV; (**e**) 15–20 keV; and (**f**) 20–25 keV. The *green circles* in panels (**b**) through (**f**) have radii of 165″

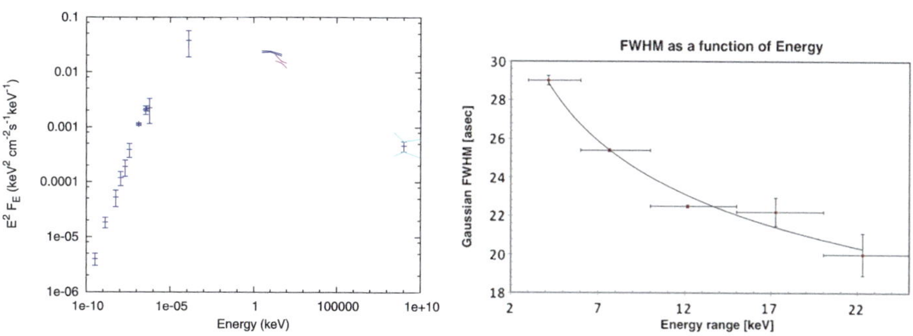

Fig. 15 *Left*: G21.5-0.9 SED (PWN only). Radio–mm points from Salter et al. (1989); IR point from Zajczyk et al. (2012); *blue* bowtie, *NuSTAR* (Nynka et al. 2014); *magenta* bowtie, INTEGRAL (de Rosa et al. 2009); TeV bowtie in *cyan*, H.E.S.S. (Djannati-Ataï et al. 2008). *Right*: Shrinkage of G21.5-0.9 with photon energy (Nynka et al. 2014)

While the small-scale torus of a PWN often has a complex, non-monotonic SED (see, for example, Figs. 6 and 20 for the cases of 3C 58 and G292.0+1.8), the larger-scale PWN often has simpler spectral behavior. The spectrum of an initially straight power-law distribution of electrons subject to synchrotron losses steepens at an energy at which the source age equals the synchrotron-loss time (Kardashev 1962). The amount of spectral steepening $\Delta \equiv \alpha(\mathrm{high}) - \alpha(\mathrm{low})$ is 0.5 in the case of a uniform source with constant particle injection, but for inhomogeneous sources, naturally arising from outflow, Δ can differ from 0.5 (Kennel and Coroniti 1984b; Reynolds 2009). Since $\Delta > 0.5$ is far more common than $\Delta = 0.5$ (see Table 2, and Fig. 6 for the examples of 3C58 and 0540-693, both with $\Delta \sim 1$), either these breaks are not due to losses (in which case the absence of loss breaks poses serious modeling problems), or PWNe are significantly inhomogeneous. Outflow models, however, predict spatial dependence of steepening which is at odds with observations (uniform spectra until

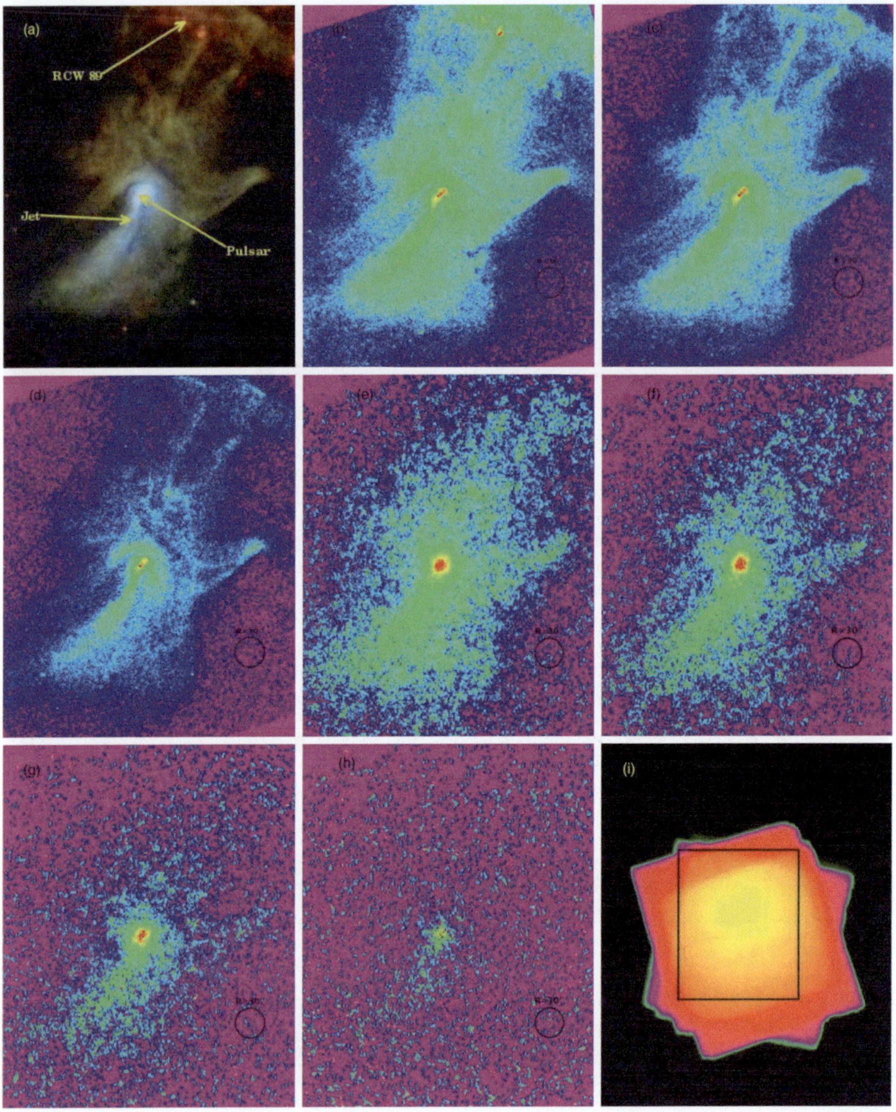

Fig. 16 MSH 15-52 with *Chandra* and *NuSTAR* (An et al. 2014). From *left* to *right*, *top* to *bottom*: *Chandra+NuSTAR*, 0.5–40 keV; *Chandra*, 0.5–2 keV; *Chandra*, 2–4 keV; *Chandra*, 4–7 keV; *NuSTAR*, 3–7 keV; *NuSTAR*, 7–12 keV; *NuSTAR*, 12–25 keV; *NuSTAR*, 25–40 keV; *NuSTAR* exposure map. All frames except the last are $10' \times 12'$

sudden steepening near the edge, clearly at variance with observed behaviors, e.g., Figs. 9 and 18).

Simple PWN models offer two possibilities for spectral steepening, both based on radiative losses of electrons, but invoking different pictures of particle transport from the wind shock through the nebula. In pure-advection models like those of Kennel and Coroniti (1984a) and Reynolds (2009), steepening can be thought of as due to the shrinkage of effective source size with increasing photon energy, as energy losses deplete electrons with

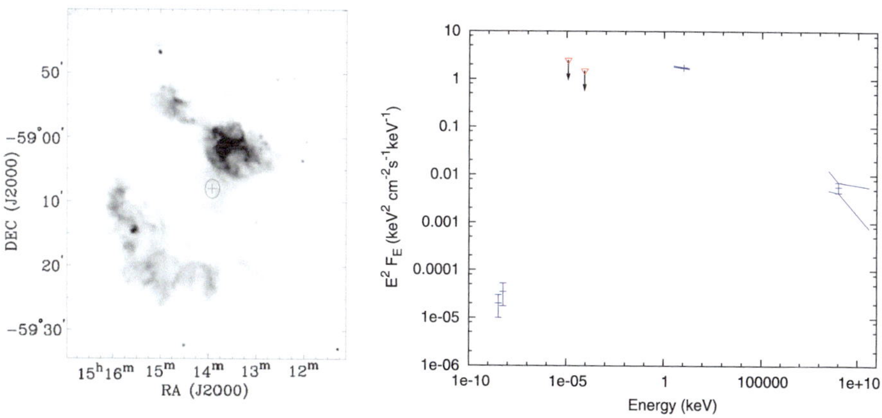

Fig. 17 *Left*: Radio image of MSH 15-52 at 1.3 GHz with ATCA (Gaensler et al. 1999). The bright horseshoe to the N is RCW 89. There is little evidence of the PWN; the *cross* marks the pulsar location. *Right*: SED for MSH 15-52. Radio: Gaensler et al. (2002). IR upper limits are from Koo et al. (2011). X-rays: An et al. (2014). TeV: Aharonian et al. (2005)

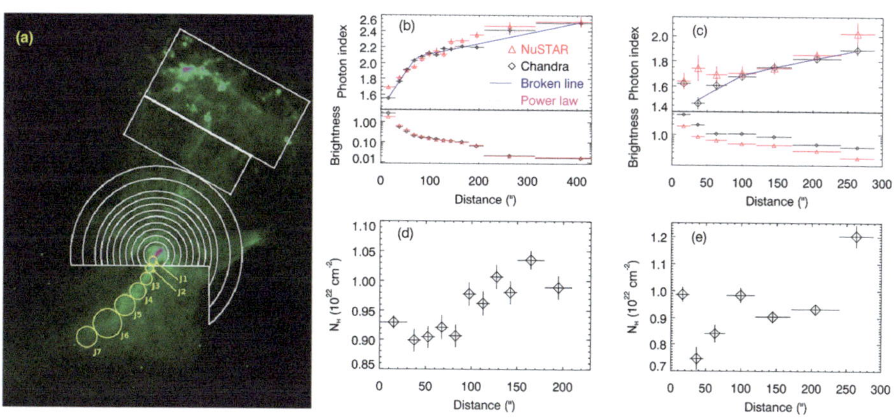

Fig. 18 *Left* (panel (**a**)): Regions of MSH 15-52 used for spatially resolved spectroscopy. Panel (**b**): Photon index as a function of distance to the NW (in the partial annuli). Panel (**c**): Photon index along the jet to the SE. Panels (**d**) and (**e**) give fitted column densities N_H from *Chandra* data as a function of distance from the pulsar, for the northern nebula and the jet, respectively. Figures from An et al. (2014)

higher energies at smaller radii. However, such models all share one important failing: they predict that the PWN spectral index should be constant throughout the source until a sudden steepening at the (energy-dependent) edge radius (Fig. 21).

However, advection models in inhomogeneous sources can produce the frequently observed steepenings by values Δ greater than 0.5. The models can be tested because they achieve such values of Δ by source shrinkage at related rates; that is, the source size R obeys $R \propto E^m$ with m and Δ related by the model. The Kennel–Coroniti model predicts $\Delta = (4+\alpha)/9$, where $\alpha = 0.6$ characterizes the assumed injected spectrum, explaining optical and higher-energy emission. A triumph of this model was the successful explanation $(m = -1/9)$ of the rocket observations of the Crab in the 1970's, as described above.

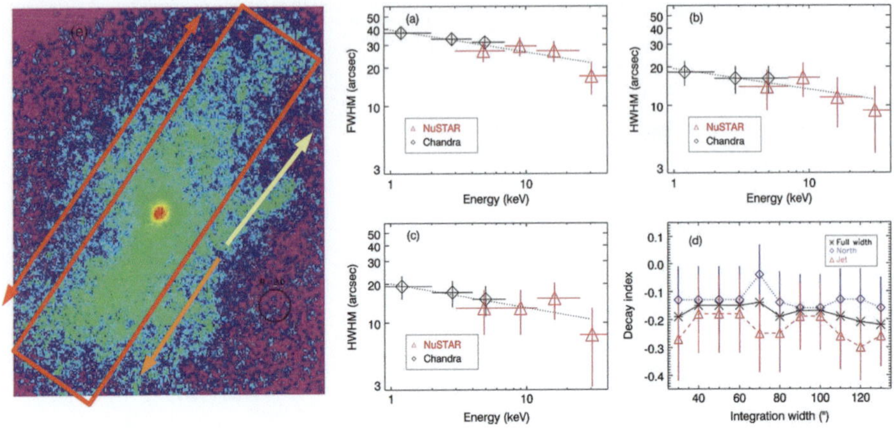

Fig. 19 *Left*: *Box* used to extract profiles from *NuSTAR* image of MSH 15-52. Profiles were summed along the minor dimension (100″), and extents as a function of energy displayed in panels to the *right*. Panel (**a**): FWHM of profiles centered on the pulsar extending in both directions. (**b**): Extending from the pulsar to the NW. (**c**) Extending from the pulsar to the SE (along the jet). (**d**) Energy shrinkage ("decay") indices m (with extents $\propto E^m$). Plots from An et al. (2014)

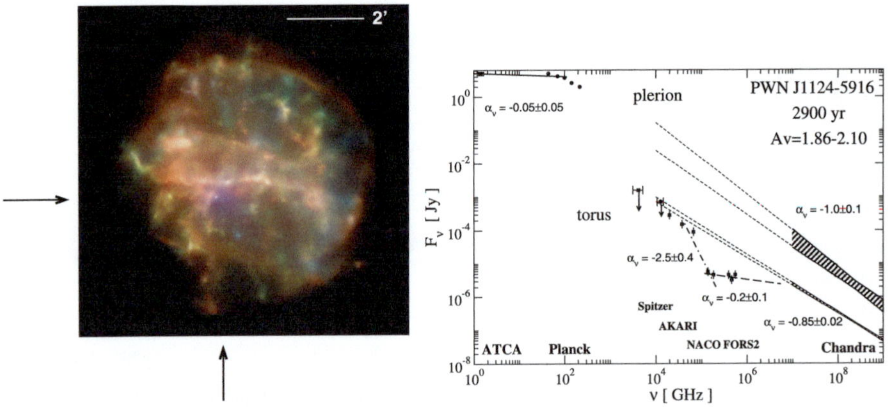

Fig. 20 *Left*: The shell SNR G292.0+1.8, with the pulsar indicated and the PWN in *blue* (Park et al. 2007). *Right*: SEDs of the entire PWN and of the torus [seen in VLT images at H and K_s bands; Zharikov et al. (2013)]. The torus spectrum is complex, but the PWN could be described by a double power-law with a single spectral break around 100 GHz. However, the steepening Δ is almost 1

More extreme source gradients are required to produce Δ significantly in excess of 0.5, however. Reynolds (2009) considered ad-hoc source gradients in flow-tube radius, magnetic field, density, and velocity, and connected them to predicted values of Δ. The large parameter space is significantly constrained if the source shrinkage with size can actually be measured. If magnetic-field strength varies as r^{m_B} and density as r^{m_ρ}, conical flows (jets or spherical winds) with $\alpha = 0$ (for simplicity) satisfy

$$\frac{\Delta}{-m} = 3 + m_B + m_\rho. \tag{1}$$

Table 2 Spectral breaks in PWNe

Object	α_{radio}	$\alpha_{\text{X-ray}}$	Δ	Reference
Crab	0.3	1.1	0.8	Manchester et al. (1993)
G11.2-0.3	$\sim$0	0.9	0.9	Roberts et al. (2003)
Kes 75	0.2	0.9	0.7	Blanton and Helfand (1996)
G54.1+0.3	0.28	0.8	0.5	Lu et al. (2002)
3C 58	0.1	1.1	1.0	Green and Scheuer (1992)
B0540-693	0.25	1.2	0.95	Manchester et al. (1993)
G21.5-0.9	$\sim$0	1.0	1.0	Salter et al. (1989), Nynka et al. (2014)
MSH 15-52	0.2	1.1	0.9	Gaensler et al. (2002)
G292.0+1.8	0.05	0.9	0.85	Gaensler and Wallace (2003)

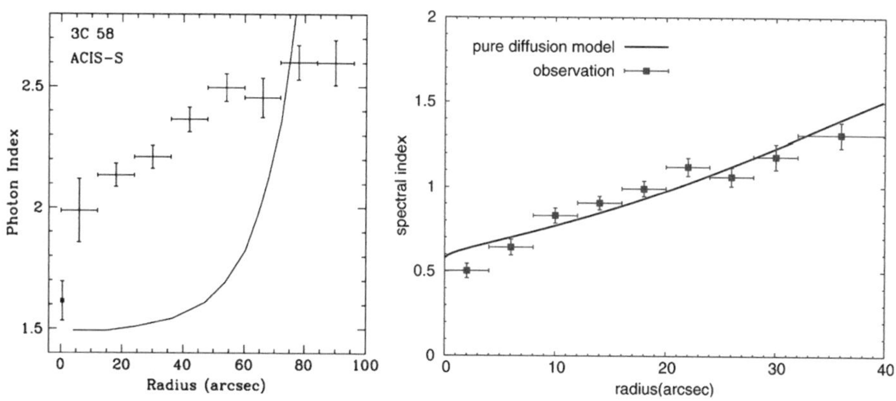

Fig. 21 *Left*: Observed photon index Γ as a function of radius in 3C 58 (Slane et al. 2004). *Curve*: Prediction from Kennel–Coroniti pure-MHD spherical advection model (Reynolds 2003). *Right*: Diffusion model for G21.5-0.9 (Tang and Chevalier 2012, data from Slane et al. (2000))

For G21.5-0.9, $\Delta/(-m) = 4.29$ so $m_B + m_\rho = 1.3$, requiring that either magnetic field B or gas density ρ, or both, *rise* with radius. This could conceivably happen due to turbulent amplification of magnetic field and/or mass loading of the PWN outflow (Lyutikov 2003), but seems unlikely. A similar result holds for MSH 15-52, requiring $m_B + m_\rho \sim 1$. While a value for m can be determined for the Crab's counter-jet, an independent value of Δ for that region alone is difficult to obtain, since that feature is embedded in the radio nebula and is not morphologically distinct.

Alternatively, particle transport may occur through diffusion. Purely diffusive models (Gratton 1972; Reynolds and Jones 1991; Tang and Chevalier 2012) can produce fairly linear increases in spectal index with radius (Fig. 21). However, they do not predict a change of source size with photon energy, and the spectral breaks must be intrinsic. Combination models of diffusion and advection (Tang and Chevalier 2012) offer more flexibility, but are not amenable to analytic investigation and have not yet been explored in great detail. These models all remain quite simple, with spherical or simple one-dimensional outflow geometries. The gradual increase of spectral index with radius probably requires a mix of particles of different ages at each radius; such a mix can be achieved by diffusion, but also by more complicated flow geometries, such as the backflows found in relativistic-MHD simulations such as those of Komissarov and Lyubarsky (2004) or Del Zanna et al. (2004).

The question of the origin of the full SED of the synchrotron spectrum from PWNe is thus still unsettled, so the observations cannot yet be brought to bear directly on theories of particle energization in PWNe. Even the fundamental origin of the radio-emitting particles is still mysterious. Relativistic-MHD simulations still lack the ability to make detailed spectral predictions, as these require propagating the full particle spectrum with each fluid element, and are computationally prohibitive at this time. It is most likely, however, that further progress will require the development of this capability. In the meantime, the observational characterization of PWN SEDs and spatially resolved spectra remain important tools for constraining models, and may ultimately contribute essential clues to solutions of these basic problems.

All three PWNe observed with *NuSTAR* have been detected at TeV wavelengths. Gamma-ray observations hold out the promise of further constraining the particle distributions of PWNe known at other wavelengths, but also of discovering new types of object that are predominantly TeV emitters. Both cases are discussed in Sect. 4.

3 X-Ray Pulsar Wind Nebulae Outside Supernova Remnants

3.1 Expected General Properties of PWNe of Supersonically Moving Pulsars

For the first 500–1000 years after the SN explosion, the SNR's radius $R_{\rm snr}$ almost linearly increases with time with a typical speed of $\sim$10,000 km s^{-1}, much faster than the typical puslar speed of a few hundred km s^{-1}. At larger ages the SNR expansion slows down ($R_{\rm snr} \propto t^{2/5}$ and $t^{3/10}$ in the Sedov and pressure-driven snowplow stages, respectively), while the pulsar keeps moving with about the same velocity and eventually leaves the SNR at an age of $\sim$20–200 kyr (see Fig. 4 in Arzoumanian et al. 2002). This means that most of the known pulsars are moving in the ISM with the speed $V_{\rm psr}$ considerably exceeding the speed of sound in the ambient medium, $c_s = (\gamma_{\rm ad} kT/\mu m_{\rm H})^{1/2} \sim$ 3–30 km s^{-1}, where $\gamma_{\rm ad}$ is the adiabatic index ($\gamma_{\rm ad} = 5/3$ for monoatomic gases), T is the temperature, and μ is the molecular weight. The supersonic motion of the pulsar drastically changes the PWN morphology (see van der Swaluw et al. 2004, Gaensler and Slane 2006, and references therein). Since the ram pressure $p_{\rm ram} = \rho_{\rm amb} V_{\rm psr}^2 = 1.5 \times 10^{-9} n_b (V_{\rm psr}/300\,{\rm km\,s^{-1}})^2$ dyn cm^{-2} (where n_b is the ambient baryon number density in units of cm^{-3}) exceeds the ambient pressure $p_{\rm amb}$,

$$p_{\rm ram}/p_{\rm amb} = \gamma_{\rm ad}\mathcal{M}^2 \gg 1\,, \quad (2)$$

($\mathcal{M} = V_{\rm psr}/c_s$ is the Mach number), the PWN acquires a cometary shape with a compact head around the pulsar and a long tail behind it (Fig. 22). In an idealized picture, the interaction of the pulsar wind (PW) with the ambient (circumpulsar) medium creates three distinct regions. The bullet-shaped cavity around the pulsar is filled with the *unshocked relativistic PW* confined within the termination shock (TS). The *shocked PW*, which is the main source of synchrotron radiation in X-rays, is flowing away between the TS and the contact discontinuity (CD) that separates the shocked PW from the shocked ambient medium. Finally, the compressed and heated *shocked ambient medium* between the CD and the forward bow shock (FS) is expected to emit IR-optical-UV radiation in spectral lines and continuum.

For an isotropic pulsar wind, a characteristic distance from the pulsar at which the ram pressure of the unshocked PW, $\dot{E}_w/(4\pi c r^2)$, balances the ram pressure of the ambient

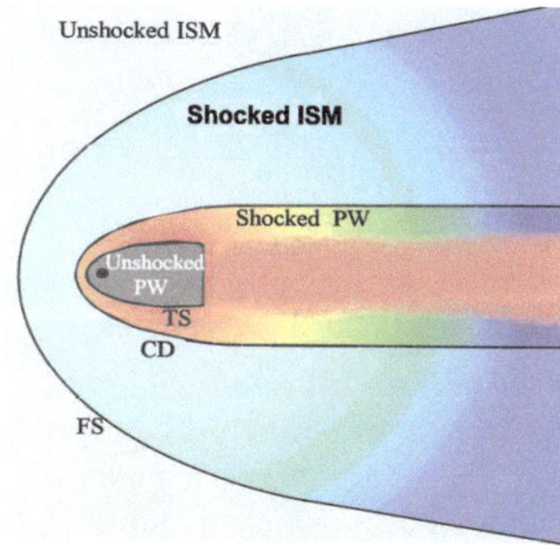

Fig. 22 Cartoon of a head-tail PWN created by a supersonically moving pulsar. A synchrotron (e.g., X-ray) PWN is produced by the shocked PW flowing between the TS and CD surfaces, while the shocked circumstellar medium between the FS and CD surfaces is expected to be the source of IR-optical-UV radiation

medium (the so-called stagnation point) is

$$R_0 = \left(\frac{\dot{E}_w}{4\pi c p_{\rm ram}}\right)^{1/2} = 1.3\times 10^{16}\dot{E}_{w,35}^{1/2} n_b^{-1/2}\left(V_{\rm psr}/300\,{\rm km\,s^{-1}}\right)^{-1}\,{\rm cm}, \tag{3}$$

where $\dot{E}_w = \xi_w \dot{E} = 10^{35}\dot{E}_{w,35}$ erg s^{-1}, $\dot{E}$ is the pulsar's spin-down power, and $\xi_w < 1$ is the fraction of $\dot{E}$ that powers the PW. For an unmagnetized PW, R_0 is approximately equal to the distance of the CD apex ahead of the pulsar, $R_{\rm cd} \approx R_0$ (van der Swaluw et al. 2004). The characteristic angular separation between the pulsar and the sky projection of the CD surface is

$$\theta_0 \approx R_0/d = 0''.89 d_{1.0}^{-1}\dot{E}_{w,35}^{1/2} n_b^{-1/2}\left(V_\perp/300\,{\rm km\,s^{-1}}\right)^{-1}\sin i, \tag{4}$$

where $d = 1.0 d_{1.0}$ kpc is the distance,[1] i is the angle between the line of sight and the pulsar velocity,[2] and $V_\perp = V_{\rm psr}\sin i$. The small value of θ_0 implies that subarcsecond resolution (provided only by *Chandra* among the currently active X-ray observatories) is required to resolve the PWN head from the pulsar even for nearby pulsars.

The unshocked PW consists of relativistic particles (likely electrons and positrons) and a magnetic field. The magnetization parameter σ, defined as the ratio of the Poynting flux, $B^2 c/(4\pi)$, to the particle enthalpy flux, is unknown. Pulsar models predict $\sigma \gg 1$ immediately outside the pulsar magnetosphere, while PWN models require $\sigma \lesssim 1$ (or even $\sigma \ll 1$; Kennel and Coroniti 1984a) just upstream of the TS. The decrease of σ with distance from the pulsar could be due to transfer of the magnetic field energy to the particles, e.g., by magnetic field reconnection in the striped PW (see Kirk et al. 2009, and references therein). For a given magnetization, the magnetic field upstream of the TS at the stagnation point can be estimated as

$$B \sim \left[\frac{\dot{E}_w \sigma}{c R_0^2(\sigma+1)}\right]^{1/2} = \left(\frac{4\pi p_{\rm ram}\sigma}{\sigma+1}\right)^{1/2} \approx 140\left(\frac{n_b \sigma}{\sigma+1}\right)^{1/2}\frac{V_{\rm psr}}{300\,{\rm km\,s^{-1}}}\,\mu{\rm G}. \tag{5}$$

The magnetic field can be somewhat higher in the shocked PW, up to a factor of 3 immediately downstream of TS, at $\sigma \ll 1$ (Kennel and Coroniti 1984a). Thus, characteristic PWN magnetic field values are expected to be of the order of 10–100 μG.

[1] In this Section the subscript of d means the scaling distance in kpc.

[2] Equation (4) is applicable at not too small $\sin i$; see Romani et al. (2010).

Typical energies of synchrotron photons emitted in such fields can be estimated as

$$E = \zeta \frac{heB_\perp \gamma^2}{2\pi m_e c} = 1.16\zeta B_{-5}\gamma_8^2 \text{ keV} = 4.43\eta B_{-5}(E_e/100 \text{ TeV})^2 \text{ keV} \quad (6)$$

where $B_\perp = 10^{-5} B_{-5}$ G is the magnetic field component perpendicular to the electron velocity, $\gamma = E_e/(m_e c^2) = 10^8 \gamma_8$ is the electron Lorentz factor, and $\zeta \sim 1$ is a numerical factor. The synchrotron emission spectum is determined by the electron spectrum, which depends on the still poorly understood acceleration mechanism. The commonly considered Fermi acceleration mechanism at fronts of relativistic shocks (e.g, the TS) gives a power-law (PL) electron spectrum, $dN_e/d\gamma \propto \gamma^{-p}$ in the range $\gamma_{\rm min} < \gamma < \gamma_{\rm max}$, with $p \gtrsim 2$ (see, e.g., Chap. 6 of the review by Bykov et al. 2012). Such an electron spectrum produces a PL photon spectrum, $dN/dE \propto E^{-\Gamma}$ in the $E_{\rm min} < E < E_{\rm max}$ range, with the photon index $\Gamma = (p+1)/2 \gtrsim 1.5$. The maximum Lorentz factor of accelerated electrons,[3] $\gamma_{\rm max} \lesssim (e/m_e c^2)[\dot{E}\sigma/c(\sigma+1)]^{1/2} \approx 1.1 \times 10^9 \dot{E}_{35}^{1/2}[\sigma/(\sigma+1)]^{1/2}$, can be estimated from the condition $R_g < R_0$, where $R_g = \gamma m_e c^2/(eB) = 1.7 \times 10^{16} \gamma_8 B_{-5}^{-1}$ cm is the gyration radius. This corresponds to the maximum synchrotron photon energy

$$E_{\rm max} \lesssim 130\zeta \dot{E}_{35} B_{-5}\sigma/(\sigma+1) \text{ keV}. \quad (7)$$

This equation shows that one should not expect X-ray PWNe from very old, low-power pulsars, but head-tail PWNe could be expected at UV-optical-IR wavelengths.

The qualitative head-tail PWN picture is generally confirmed by analytical estimates (e.g., Romanova et al. 2005) and numerical simulations. For instance, Bucciantini et al. (2005) presented relativistic MHD axisymmetric simulations for $\mathcal{M} = 30$ and three values of the PW magnetization σ upstream of the TS. Figure 1 of that paper shows that for an isotropic PW with a toroidal magnetic field the TS, CD and FS apices are at distances of $\approx R_0$, $1.3R_0$ and $1.7R_0$, respectively, at $\sigma = 0.002$, while the radius of the cylindrical tail (confined by the CD surface behind the pulsar) is $r_{\rm cd} \approx 4R_0$, almost independent of magnetization. The bulk flow velocity in the tail reaches 0.8–0.9c at its periphery (closer to the CD), being 0.1–0.3c in the central channel (behind the TS bullet). Simulated maps of synchrotron brightness (Figs. 4 and 5 in Bucciantini et al. 2005) show that the brightness is maximal at the PWN head, gradually decreasing with distance from the pulsar in the PWN tail.

The inner and outer channels of the tail flow should mix with each other at larger distances from the pulsar due to shear instability. At even larger distances the flow should slow down due to mass-loading of the ambient matter, which leads to additional tail broadening (e.g., Morlino et al. 2015).

As the outflowing electrons lose their energy to synchrotron radiation, we can expect spectral softening with increasing distance from the pulsar, which means the observed tail's length should increase with decreasing photon energy (e.g., it should be larger in the radio than in the X-rays). The length scale of the tail at photon energy E (in keV) can be estimated as

$$l_{\rm tail} \sim V_{\rm flow}\tau_{\rm syn} \sim 18\big(V_{\rm flow}/10{,}000 \,\text{km s}^{-1}\big)\zeta^{1/2} E^{-1/2} B_{-5}^{-3/2} \text{ pc} \quad (8)$$

[3] Note that $\gamma_{\rm max} m_e c^2 \sim e\Phi\sigma/(\sigma+1)$, where $\Phi = B_{\rm LC} R_{\rm LC} \sim B_{\rm NS} R_{\rm NS}^3 \Omega^2/c^2$ is the potential drop across the pulsar's polar cap, $R_{\rm LC} = c/\Omega$ is the light cylinder radius, $\Omega = 2\pi/P$, $B_{\rm LC}$ and $B_{\rm NS}$ are the magnetic field values at the light cylinder and the neutron star surface. The maximum electron energies in other acceleration models (e.g., Romanova et al. 2005) are also fractions of $e\Phi$.

where $V_{\rm flow}$ is a typical flow velocity and $\tau_{\rm syn} = 1.6\gamma_8^{-1} B_{-5}^{-2}$ kyr $\sim 1.7\zeta^{1/2} E^{-1/2} B_{-5}^{-3/2}$ kyr is the synchrotron cooling time. We should note that such an estimate is very crude because both the flow velocity and the magnetic field vary with the distance from the pulsar.

The simulations by Bucciantini et al. (2005) were done for an isotropic PW. We, however, know that at least in some young pulsars the PW is mostly concentrated around the equatorial plane, perpendicular to the spin axis, and the PWN has equatorial and polar components (e.g., the torus and jets in the Crab pulsar; Weisskopf et al. 2000). In this case the ram pressure of the unshocked PW becomes anisotropic, which changes not only the distance to the stagnation point but also the overall PWN appearance. In particular, the PWN shape strongly depends on the angle between the pulsar's velocity and spin vectors, as well as on the angle between the spin vector and the line of sight. Three-dimensional simulations for several cases of anisotropic PW were presented by Vigelius et al. (2007), assuming nonrelativistic, unmagnetized flows (see also Wilkin 2000 and Romani et al. 2010 for analytical approximations). The PW anisotropy should strongly affect the shape of the PWN head, which may become substantially different from the bullet-like one. On the other hand, the shape of tail should not be so strongly affected, especially at large distances behind the pulsar.

In addition to synchrotron (and inverse Compton—see Sect. 4) emission from the shocked PW, one can expect emission from the shocked ISM between the FS and the CD. While passing through the FS, it is compressed and heated up to temperatures

$$T \approx (3/16)(\mu m_p/k) V_{\rm psr}^2 = 1.2\times 10^6 (\mu/0.6)\big(V_{\rm psr}/300\,{\rm km\,s^{-1}}\big)^2\ {\rm K}, \tag{9}$$

where μm_p is the mean mass per particle (including electrons; see Bykov et al. 2008 for details). The shocked ISM emits in both continuum and spectral lines in the optical, UV, and even soft X-ray ranges. If there are neutral hydrogen atoms in the ambient medium ahead of the pulsar, they can be excited at the FS and emit spectral lines in the course of radiative de-excitation.

Thus, a rotation-powered pulsar that has left its parent SNR should be accompanied by a nebula that consists of a head-tail synchrotron component, emitting in a broad energy range, perhaps from the radio to soft γ-rays, enveloped by a (forward) bow-shock component observable in the optical-UV (particularly, in the Balmer lines). Below we will see, however, that observational results do not always coincide with the predictions of the current simple models.

3.2 Observational Results

Although most of the known rotation-powered pulsars have left their SNRs and are moving with supersonic velocities, their spin-down powers (and the PWN luminosities) have significantly decreased with age. Therefore, the number of detected head-tail (bow-shock) PWNe is relatively small. Among over 70 PWNe and PWN candidates detected in X-rays (Kargaltsev et al. 2013), only about 15 PWNe are certainly created by old (and/or fast) enough pulsars that have left their parent SNRs. These have spin-down powers $0.01 \lesssim \dot{E}_{35} \lesssim 30$ and characteristic (spin-down) pulsar ages $\tau_{\rm sd} \gtrsim 20$ kyr. Of the remainder, about 25–30 are still inside SNRs, and the rest are ambiguous, or not yet confirmed as PWNe. An even smaller number of head-tail PWNe were observed with exposures deep enough to accurately measure the X-ray PWN properties. Below we present a few examples of such observations.

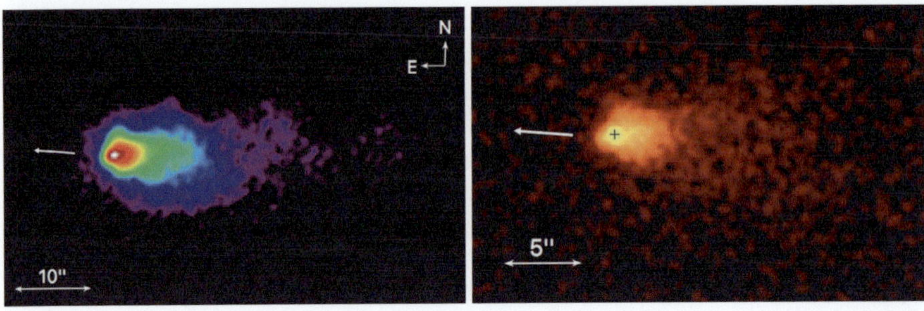

Fig. 23 *Left*: Merged image from 5 *Chandra* ACIS observations of the Mouse (0.5–8 keV, 154 ks exposure). *Right*: *Chandra* HRC image of the Mouse (58 ks exposure). The *cross* in the *right* image marks the pulsar position; the *arrows* show the direction of proper motion

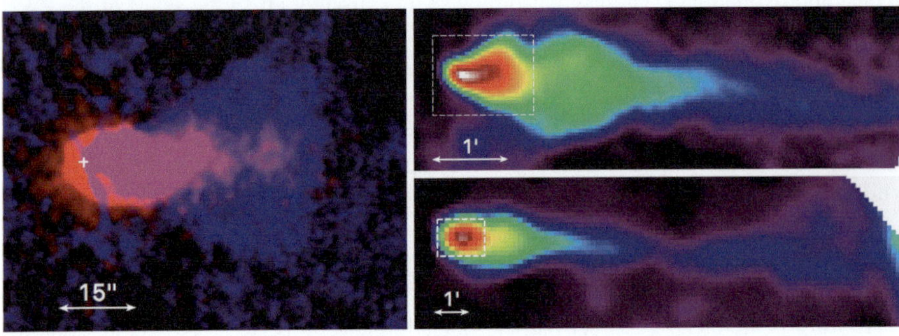

Fig. 24 *Left*: Composite X-ray (*red*; ACIS) and radio (*blue*; VLA, $1''.07$ beam) image of the Mouse PWN. The pulsar position is shown by the *white cross*. *Right*: VLA radio images (*top*: $11''$ beam, *bottom*: $32''$ beam) showing the extended tail of the Mouse. The field of view of the *left* image is shown by the *dashed white box* in the *lower right* panel. The radio images were obtained from the NRAO VLA Archive

3.2.1 The Mouse: A Textbook Example?

The Mouse PWN was discovered in a radio survey of the Galactic center region (Yusef-Zadeh and Bally 1987). The VLA image showed a bright compact head ("snout"), a bulbous $\sim 2'$ long "body", and a remarkably long, $\sim 12'$, narrow "tail" (see the right panels in Fig. 24). Camilo et al. (2002) discovered a 98 ms pulsar, J1747-2958, within the Mouse's head, with a spin-down power $\dot{E} = 2.5 \times 10^{36}$ erg s^{-1} and a characteristic age $\tau_{sd} = 25$ kyr. Hales et al. (2009) measured the proper motion of the radio PWN head,[4] $\mu = 12.9 \pm 1.8$ mas yr^{-1}, which corresponds to a transverse velocity $V_\perp = (306 \pm 43) d_5$ km s^{-1}, where $d_5 = d/5$ kpc. Based on the projected tail length, $\sim 17 d_5$ pc, and the lack of an SNR that could possibly be associated with the pulsar (see, however, Yusef-Zadeh and Gaensler 2005), Hales et al. (2009) argue that the true age of the pulsar is $\gtrsim 160$ kyr.

The Mouse PWN was observed with *Chandra* by Gaensler et al. (2004) and Klingler et al. (2017, in preparation), with 36 ks and 120 ks exposures, respectively. These observations have shown an X-ray nebula with a compact bright head and a tail of $\sim 45''$ length, a factor of 16 shorter than the radio tail (see Fig. 23). The X-ray luminosity of the PWN, $L_{0.5\text{–}8\,\mathrm{keV}} \approx$

[4]It can differ from the pulsar's proper motion if the pulsar moves in a non-uniform ambient medium, but we will neglect this difference here.

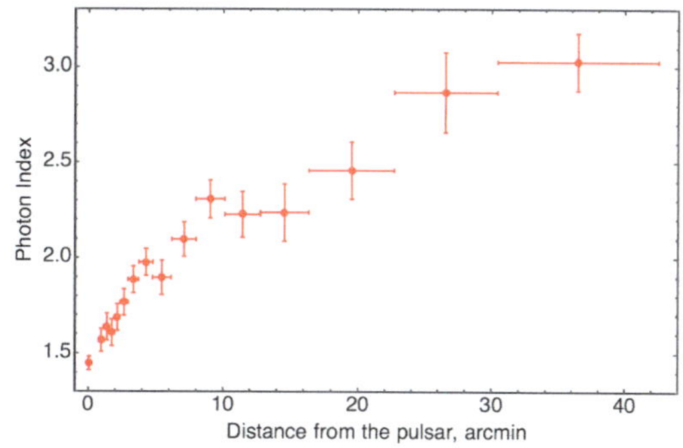

Fig. 25 Photon index Γ as a function of distance from the pulsar along the tail of the Mouse (Klingler et al. 2017, in preparation)

$3.3 \times 10^{34} d_5^2$ erg s^{-1}, is about $0.013 d_5^2$ of the pulsar's spin-down power, unusually high compared to other PWNe (Kargaltsev and Pavlov 2008). The distance from the pulsar to the projected leading edge of the head, $\theta_0 \approx 1''$, corresponds to $n_b \sim 0.8\xi_w d_5^{-4} \sin^2 i$ cm^{-3} (see Eq. (4)), typical for a warm phase of the ISM (e.g., Bykov et al. 2008). For a sound speed of ~10 km s^{-1}, expected for this phase, the pulsar's Mach number can be estimated as $\mathcal{M} \sim 30 d_5/\sin i$. The spatially-resolved X-ray spectrum showed a significant increase of the photon index with increasing distance from the pulsar, from $\Gamma \approx 1.6$ in the immediate vicinity of the pulsar to $\Gamma \approx 3.0$ at ~40″ from the pulsar (see Fig. 25). The spectral softening could be caused by synchrotron cooling. Assuming equipartition between the magnetic field energy and the kinetic energy of relativistic electrons, the magnetic field strength estimated from the observed X-ray emission is a few hundred μG, substantially higher than in other head-tail PWNe.[5] For such a magnetic field the projected length of the X-ray tail, $l_{\rm tail} \sim 1\, d_5$ kpc, corresponds to a flow velocity $V_{\rm flow} \sim 20{,}000\, B_{-4}^{3/2}\, d_5/\sin i$ km s^{-1}, much higher than $V_{\rm psr}$ but significantly lower than the mildly relativistic speeds predicted by Bucciantini et al. (2005). Moreover, the comparison of the X-ray and high-resolution radio images (see the left panel of Fig. 24) suggests that the flow is faster in the middle of the tail, contrary to the model predictions.

The radio tail of the Mouse is longer than in any other known PWN. Just behind the pulsar, the radio image looks like a cone with an ≈25° half-opening angle (much broader than the Mach cone at the above-estimated $\mathcal{M} \sim 30$) and a vertex at the pulsar position. The cone abruptly narrows at ~1′ from the pulsar. Such behavior is not explained by the current PWN models. The Mouse is one of the few PWNe with a mapped radio polarization. Polarization mesurements by Yusef-Zadeh and Gaensler (2005) suggest that the magnetic field wraps around the bow shock structure near the apex of the system, but runs parallel to the direction of the pulsar's motion in the tail behind the pulsar. Such a magnetic field distribution is different from the toroidal one assumed in the models by Bucciantini et al. (2005). Thus, the Mouse has a few features consistent with the model predictions, but the models do not fully agree with the observations, particularly in the radio.

3.2.2 Geminga: An Odd "Three-Tail" PWN

The X-ray PWN created by the radio-quiet γ-ray pulsar Geminga ($P = 237$ ms, $\dot{E} = 3.3 \times 10^{34}$ erg s^{-1}, $\tau_{\rm sd} = 340$ kyr, $d = 0.25^{+0.23}_{-0.08}$ kpc) looks quite different from the

[5] We should note, however, that such estimates assume that the magnetic field strength is about the same throughout the emitting volume, which may be far from reality.

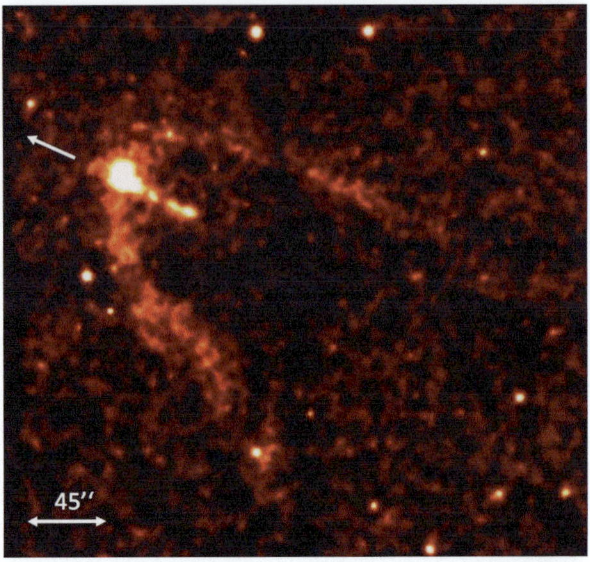

Fig. 26 Combined *Chandra* ACIS image of the Geminga PWN (0.5–8 keV, 540 ks). The *arrow* shows the direction of the pulsar's proper motion

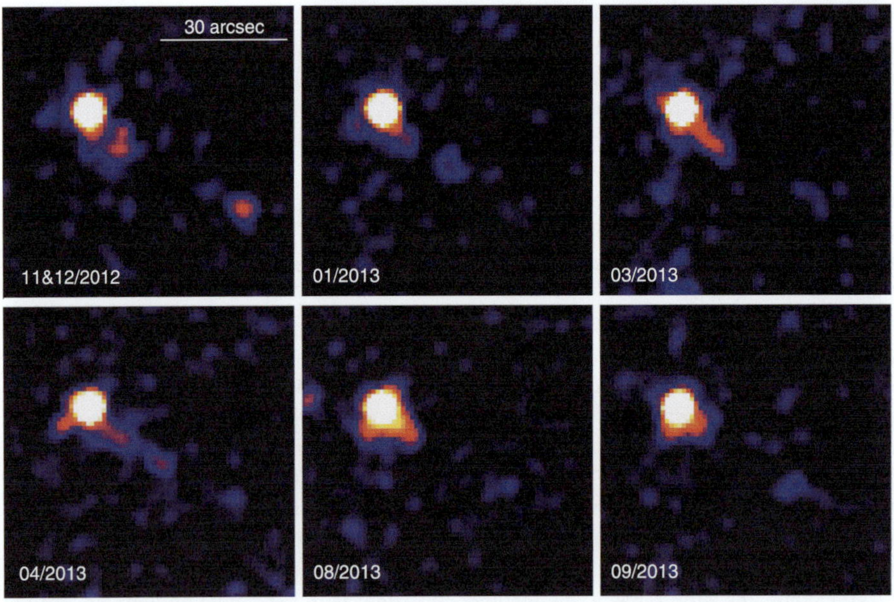

Fig. 27 Images of the central tail of the Geminga PWN in separate *Chandra* observations, in the 0.5–8 keV band

Mouse[6] and from the predictions of PWN models. The proper motion of this pulsar, $\mu = 178.2 \pm 1.8$ mas yr^{-1} (Faherty et al. 2007), corresponds to the transverse velocity $V_\perp = (211 \pm 2) d_{0.25}$ km s^{-1}. Observations with *XMM-Newton* revealed two bent "tails" be-

[6]One should bear in mind, however, that much smaller spatial scales can be probed in the nearby Geminga PWN than in the Mouse.

hind the pulsar, on both sides of its sky trajectory (Caraveo et al. 2003), while *Chandra* observations with higher spatial resolution showed a shorter third tail between the two lateral tails (Pavlov et al. 2006, 2010). The most detailed data on the Geminga PWN were provided by a series of 12 *Chandra* observations carried out in 2012–2013, with a total exposure of about 580 ks (Posselt et al. 2016). Figure 26 shows a summed image from these observations, where we see two lateral tails of $\sim 3'$ ($0.2 d_{0.25}$ pc) length and one $\sim 0.45''$ ($0.05 d_{0.25}$ pc) long central tail. Surprisingly, there is only a hint of bow-like emission ahead of the pulsar and no bright, filled 'bullet' predicted by the PWN models assuming an isotropic PW. The 0.3–8 keV luminosities of the northern and southern lateral tails, and the central tail are 1.6, 2.6, and $0.9 \times 10^{29} d_{0.25}^2$ erg s^{-1}, respectively, i.e., the total PWN luminosity is a fraction of $1.5 \times 10^{-5} d_{0.25}^2$ of the pulsar's spin-down power, three orders of magnitude smaller than for the Mouse. Images from separate exposures show that the central tail is formed by isolated "blobs" seen at different distances from the pulsar in different observations (see Fig. 27). However, there is no evidence of constant or decelerated motion of the blobs. The spectra of the lateral tails are very hard, $\Gamma \approx 0.7$–1.0, much harder than the spectrum of the central tail, $\Gamma \approx 1.6$, and they do not show significant changes with increasing distance from the pulsar.

The nature of the three tails is not certain yet. One could assume that the lateral tails represent a limb-brightened paraboloid shell of shocked PW downstream of the TS and their unusually hard spectrum is emitted by electrons accelerated by the Fermi mechanism at the shocks that form in two colliding flows (in the reference frame of CD)—the PW and the oncoming ambient medium. However, a lack of diffuse emission in between the lateral tails strongly suggests that the shell emissivity is not azimuthally symmetric with respect to the shell axis (i.e., the direction of motion). Such an asymmetry could be caused by a strong asymmetry of the PW (e.g., because the equatorial plane around the pulsar spin axis, where the PW is presumbly concentrated, is strongly misaligned with the direction of motion) or an azumuthally asymmetric magnetic field in the shell (see Posselt et al. 2016 for details). Alternatively, the lateral tails could be interpreted as strongly collimated polar outflows (jets) bent by the ram pressure of the oncoming ambient medium. The blobs in the central tail could be short-lived plasmoids formed by magnetic field reconnection in the relativistic plasma behind the moving pulsar, resembling the magnetotails of the Solar system planets. Observations at different wavelengths could clarify the PWN nature, but the Geminga PWN was not detected in the radio or Hα. For any interpretation, we can conclude that the Geminga's PW is strongly anisotropic, and new models are required to explain the morphologies and spectra of such PWNe.

3.2.3 The Guitar: First Example of a Misaligned Outflow

The Guitar nebula is produced by the relatively old, low-power pulsar B2224+65 ($P =$ 683 ms, $\dot{E} = 1.2 \times 10^{33}$ erg s^{-1}, $\tau_{\rm sd} = 1.12$ Myr). The pulsar is among the highest velocity neutron stars known; its proper motion, $\mu = 182 \pm 3$ mas yr^{-1}, corresponds to the transverse velocity $V_\perp = 860$–1730 km s^{-1} (the uncertainty is caused by the uncertain distance, $d = 1$–2 kpc). The guitar-shaped Hα nebula was discovered by Cordes et al. (1993) and further studied in several papers (see Dolch et al. 2016 and references therein). One could expect a head-tail X-ray PWN within the Hα bow shock, but a high-resolution observation with *Chandra* showed instead a straight $2'$ ($0.3 d_1$ pc) long feature inclined by $118°$ to the direction of the pulsar's proper motion (Wong et al. 2003; Hui and Becker 2007; see Figs. 28 and 29). A second *Chandra* observation 6 years later showed that the sharp leading edge of the jet-like feature had the same proper motion as the pulsar, and it provided evidence for

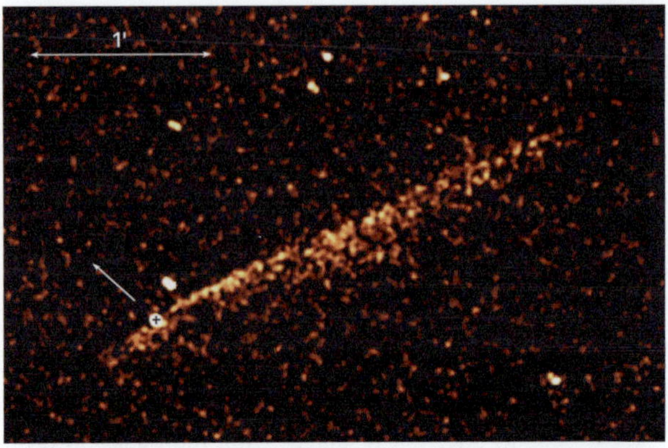

Fig. 28 Combined *Chandra* ACIS image (0.5–8 keV, 195 ks total exposure) of PSR B2224+65 and its misaligned outflow, in a coordinate frame moving with the pulsar. The image is a combination of two images separated by an interval of 6 years, during which the pulsar moved 1″ on the sky

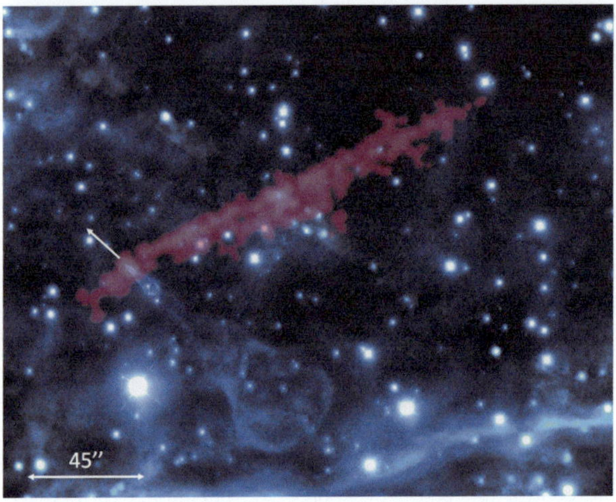

Fig. 29 Smoothed X-ray image from Fig. 28 overlaid onto the Hα image showing the Guitar. The Hα image is taken from http://chandra.harvard.edu/photo/2015/archives/

the presence of a counter-feature, albeit substantially shorter and fainter than the main one (Johnson and Wang 2010). The feature shows a power-law spectrum with $\Gamma \approx 1.6$, comparable to that of the point-like source (the pulsar plus an unresolved PWN?). The luminosity of the feature, $L_{0.3\text{–}7\,\mathrm{keV}} \sim 7 \times 10^{30} d_1^2$ erg s^{-1}, exceeds that of the point-like source by a factor of 3–4, and is a fraction of $\sim 6 \times 10^{-3} d_1^2$ of the pulsar's spin-down power.

The lack of a resolved X-ray head-tail PWN could be explained by the very high pulsar velocity and low spin-down power. Indeed, according to Eq. (3), the characteristic PWN size ahead of the pulsar, $R_0 \sim 5 \times 10^{14} \xi_w^{1/2} n_b^{-1/2} d_1^{-1}$ cm, correponds to the angular distance[7] as small as $\theta_0 \sim 0.''03 (\xi_w/n_b)^{1/2} d_1^{-2}$, much smaller than the angular resolution of *Chandra*. The lack (shortness) of the X-ray tail could be due to a high magnetic field (hence fast synchrotron cooling) in the shocked PW associated with the small stand-off distance (see Eq. (5)). Another reason could be a low maximum energy of accelerated electrons at the low spin-down power of B2224+65 (see Eq. (7)).

The nature of the elongated feature remains unclear. It might be a pulsar jet, but such a jet should be bent by the ram pressure of the oncoming ambient medium while no bending

[7] Chatterjee and Cordes (2002) estimated $\theta_0 = 0.''06 \pm 0.''02$ from modeling an Hα image obtained with the *Hubble Space Telecope*.

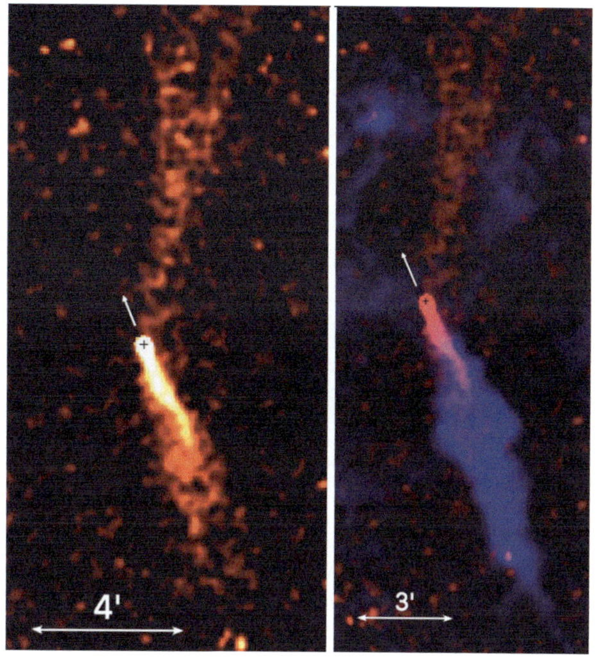

Fig. 30 *Left*: *Chandra* ACIS image of the J1509-5850 PWN (0.5–8 keV, 374 ks) showing the southwest tail and the misaligned outflow toward the north. *Right*: Combined *Chandra* ACIS (*red*) and VLA (*blue*) image of the same PWN. The *arrows* show an assumed direction of proper motion

is observed. Bandiera (2008) suggested that the feature is produced by synchrotron radiation of highest energy electrons ($\gamma \sim 10^8$) accelerated at the TS and leaked into the ISM along its magnetic field. This scenario, however, requires a rather high ambient magnetic field ($\sim$45 μG, according to Bandiera 2008) and it remains unclear why the counter-feature is so much fainter than the main one. Since the true nature of the feature is not certain yet, we will call it simply a *misaligned outflow*. It should be emphasized that, most likely, the misaligned outflow is not a (magneto)hydrodynamical flow but rather a stream of high-energy particles not interacting with each other and with the ISM gas.

3.2.4 J1509-5850: Another Misaligned Outflow, a "Three-Tail" Compact Nebula, and a Long Tail

PSR J1509-5850 is a middle-aged ($\tau_{sd} = 154$ kyr) pulsar with $P = 89$ ms, $\dot{E} = 5.1 \times 10^{35}$ erg s^{-1}, and a dispersion-measure distance $d \approx 3.8$ kpc. Its X-ray PWN, consisting of a compact "head" and a long "tail" southwest of the pulsar, was discovered in a *Chandra* observation by Kargaltsev et al. (2008). Deep follow-up *Chandra* observations (374 ks total exposure) are described by Klingler et al. (2016a). In addition to the previously detected southwest tail extending up to 7′ ($7.7d_{3.8}$ pc), these observations revealed similarly long (but fainter) diffuse emission stretched toward the north and the fine structure of the PWN "head" (see Fig. 30). The "head" (dubbed the Compact Nebula [CN] by Klingler et al. 2016a) is resolved into two lateral tails and one short central tail (Fig. 31), remarkably similar to the Geminga PWN. Although the pulsar's proper motion has not been measured, the overall CN and southwest tail morphology provides strong evidence that the pulsar is moving northeast. In this case the northern structure is another example of a misaligned outflow.

Klingler et al. (2016a) estimated upper and lower limits for the transverse velocity, $V_\perp \lesssim 640 d_{3.8}$ km s^{-1} and $V_\perp \gtrsim 160 n_b^{-1/2} d_{3.8}^{-1}$ km s^{-1}, using upper limits on the pulsar's

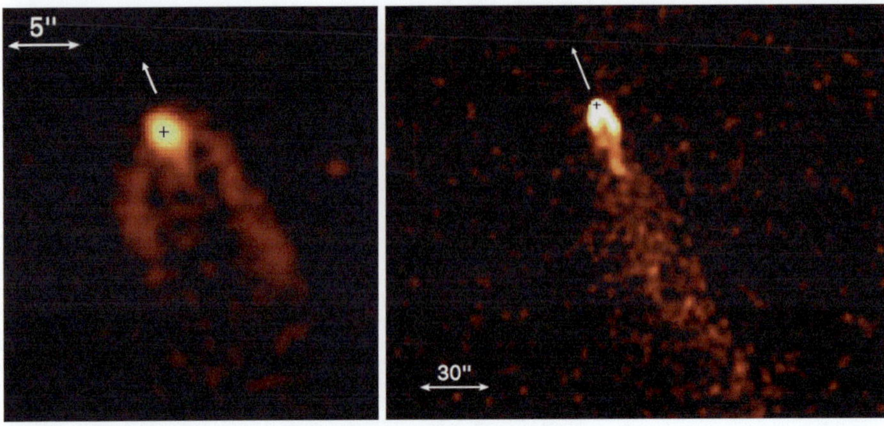

Fig. 31 *Left*: Compact X-ray nebula in the vicinty of PSR J1509-5850. *Right*: ACIS image demonstrating the transition from the compact nebula to the southwest tail

proper motion and stand-off distance (the latter estimate assumes an isotropic PW). Being morphologically similar to the Geminga PWN, the CN of J1509-5850 is a factor of $\sim$200 more luminous (e.g., $L_{0.5\text{–}8\,\mathrm{keV}} \approx 7.5 \times 10^{31} d_{3.8}^2$ erg s^{-1} for the CN lateral tails) and a factor of $\sim$10 more X-ray efficient ($\eta_{0.5\text{–}8\,\mathrm{keV}} \equiv L_{0.5\text{–}8\,\mathrm{keV}}/\dot{E} \approx 1.5 \times 10^{-4} d_{3.8}^2$ vs. $1.5 \times 10^{-5} d_{0.25}^2$ for Geminga). In addition, the spectra of the lateral tails are much softer in the CN than in the Geminga PWN ($\Gamma \approx 1.8$ vs. $\Gamma \approx 1$, respectively). The reason of these differences is currently unclear.

Being aligned with the CN symmetry axis, the extended tail southwest of the pulsar is obviously composed of a shocked PW collimated by ram pressure. Its luminosity is $L_{0.5\text{–}8\,\mathrm{keV}} \approx 1 \times 10^{33} d_{3.8}^2$ erg s^{-1}, and its spectral slope, $\Gamma \approx 1.9$, does not show any increase (rather a hint of decrease) with increasing distance from the pulsar. The lack of spectral softening suggests a very high speed of the outflowing matter. Alternatively, there could be some "reheating" due to in situ conversion of magnetic field energy into particle energy, e.g., via turbulent processes and accompanying reconnection, which might explain the hint of spectral hardening at large distances from the pulsar. The tail is also seen in radio up to about 10′ from the pulsar (Ng et al. 2010). Surprisingly, the radio emission brightens with distance from the pulsar (contrary to the Mouse tail), becomes broader than the X-ray emission beyond $\approx$3′, and then narrows again beyond $\approx$5′ (see Fig. 30). Another difference from the Mouse tail is the predominant magnetic field orientation, stretched along the tail in the Mouse and helical, with the helix axis parallel to the pulsar's direction of motion, in the J1509 tail. The different magnetic field geometries possibly reflect different spin-velocity alignments of the parent pulsars.

The median of the 7′ long wedge-like northern structure is inclined to the CN symmetry axis (alleged direction of proper motion) by $\approx$33°. Its luminosity, $L_{0.5\text{–}8\,\mathrm{keV}} \approx 4 \times 10^{32} d_{3.8}^2$, is a factor of 2.5 lower than that of the southwest tail, while the spectral slope is about the same, with a slight hint of softening with increasing distance from the pulsar.

3.2.5 The Complex PWN Created by PSR B0355+54

PSR B0355+54, located at a parallax distance of $d = 1.0 \pm 0.2$ kpc, is a middle-aged radio pulsar ($\tau_{\mathrm{sd}} = 560$ kyr) with a spin-down power $\dot{E} = 4.5 \times 10^{34}$ erg s^{-1} and a period

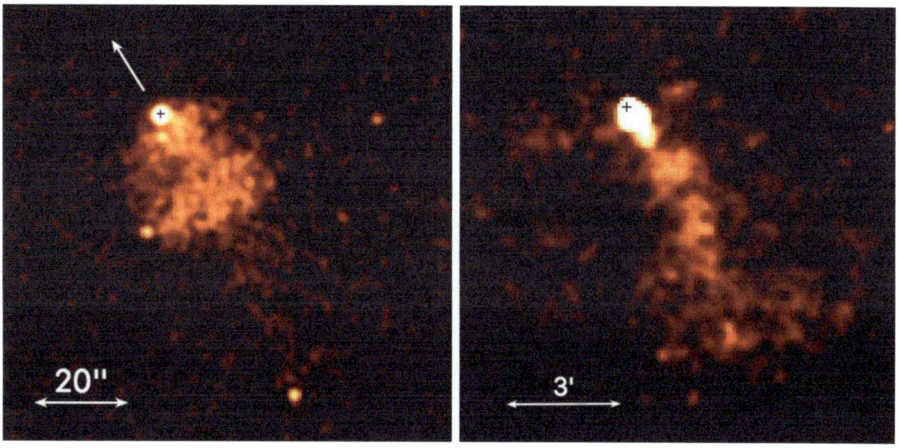

Fig. 32 *Chandra* ACIS images of the B0355+54 PWN (0.5–8 keV, 395 ks). *Left*: The compact PWN in the pulsar vicinity (the Mushroom). *Right*: The large-scale PWN, including the diffuse tail and the "whiskers". The *arrows* show the direcion of the proper motion

$P = 156$ ms. Its transverse velocity, $V_\perp = 61^{+12}_{-9} d_1$ km$'$, s^{-1} towards the northeast, is among the lowest observed. Observations with *Chandra* and *XMM-Newton* revealed the presence of a PWN (dubbed the Mushroom by Kargaltsev and Pavlov 2008) consisting of a compact "cap" and a dimmer "stem", with a hint of extended emission visible up to $\sim 7'$ (2 pc) southwest of the pulsar (McGowan et al. 2006). A series of 8 *Chandra* observations, performed over an 8-month period in 2012–2013 (total exposure of 395 ks) revealed the detailed structure of the B0355+54 PWN (see Fig. 32) and allowed us to measure the spectra of its elements (Klingler et al. 2016b). In particular, they showed a "filled" morphology of the cap, in contrast with the "hollow" morphologies of the Geminga PWN and the CN of the J1509-5850 PWN. The cap has a sharp trailing edge behind the pulsar and is brightened along the axis; its spectral slope is $\Gamma \approx 1.5$, a typical value for a PWN head. The stem is split into two structures that apparently originate from the pulsar and slightly diverge from each other further away. Klingler et al. (2016b) speculate that these structures could be pulsar's jets swept back by the ram pressure, which could also explain the brightening along the cap's axis. Overall, the "mushroom" morphology suggests a small angle between the pulsar's spin axis and our line of sight, in agreement with the lack of γ-ray pulsations. The cap and stem luminosities are 1.8×10^{31} and 2.6×10^{30} erg s^{-1}, respectively.

A long diffuse tail behind the "mushroom" is likely due to synchrotron emission of the shocked PW behind the pulsar. Its luminosity is about 3.8×10^{31} erg s^{-1}, more than a half of the total PWN luminosity, $L_{0.5\text{–}8\,\mathrm{keV}} \approx 6.4 \times 10^{31}$ erg s^{-1} that corresponds to a total PWN efficiency $\eta_{0.5\text{–}8\,\mathrm{keV}} \approx 1.4 \times 10^{-3}$. The spectrum of the tail, with a slope $\Gamma \sim 1.7$–1.8, shows only a slight hint of cooling with increasing distance from the pulsar. This implies either a fast flow speed (or a very low magnetic field), or particle re-acceleration within the tail.

The deep observation also allowed Klingler et al. (2016b) to detect two additional very faint, extended features (dubbed "whiskers") on either side of the compact nebula, likely another example of a misaligned outflow.

Thus, the B0355+54 PWN shows a particularly rich structure, which remains to be explained by PWN models. Radio and Hα observations could shed light on its nature, but the PWN has not been detected at these wavelengths.

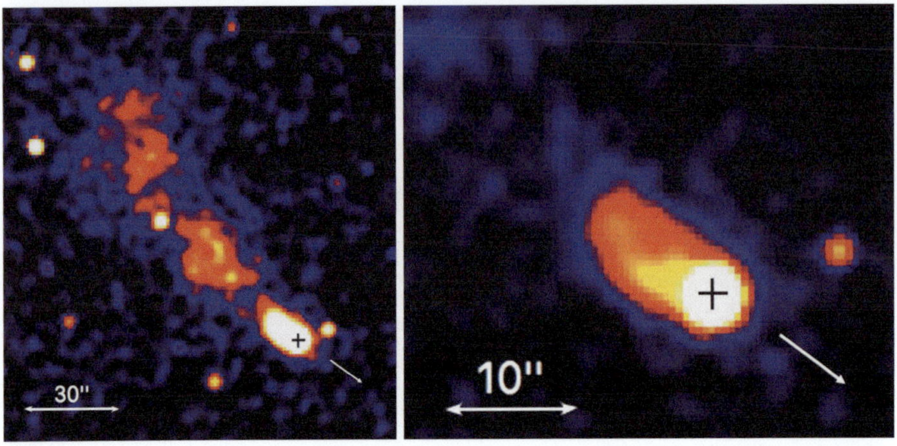

Fig. 33 Summed 0.3–6 keV *Chandra* ACIS images of the entire J1741-2054 PWN (*left*) and the leading bright component (*right*) from 5 observations in 2013 (282 ks total exposure). *Black crosses* show the pulsar position, *white arrows* show the direction of proper motion

3.2.6 J1741-2054: Another Tail Behind a Nearby Middle-aged Pulsar

Another example of a tail-like structure behind a moving pulsar is shown in Fig. 33. This PWN was discovered by Romani et al. (2010) and investigated in detail by Auchettl et al. (2015) using results from 6 *Chandra* observations carried out in 2013 (282 ks total exposure). It is created by the nearby ($d \sim 0.38$ kpc) middle-aged ($\tau_{sd} = 390$ kyr) pulsar J1741-2044 ($P = 413$ ms, $\dot{E} = 9.5 \times 10^{33}$). The pulsar's proper motion $\mu = 109 \pm 10$ mas yr^{-1}, measured by Auchettl et al. (2015) from the X-ray images, corresponds to the transverse velocity $V_\perp = (196 \pm 18) d_{0.38}$ km s^{-1}. In Fig. 33 we can see a tail-like structure in the direction opposite to that of the proper motion. The structure consists of a compact elongated nebula of $\sim 15''$ length and a fainter diffuse tail seen up to $\sim 1'.7$ ($\sim 0.2 d_{0.38}$ pc) from the pulsar. The tail is slightly bent and apparently consists of two "lobes". No small-scale structure (PWN head) is resolved around the pulsar. The 0.5–10 keV luminosities of the compact nebula and the entire PWN, about 5×10^{29} and 3×10^{30} erg s^{-1} at $d = 0.38$ kpc, are $\sim 5 \times 10^{-5}$ and $\sim 3 \times 10^{-4}$ of the pulsar's spin-down power. The spectra of the PWN elements are described by a PL model with $\Gamma \approx 1.5$–1.7, with only a hint of spectral softening with increasing distance from the pulsar.

For an isotropic PW, one could expect a PWN head with a leading edge at an angular distance $\theta_0 \approx 1'' d_{0.38}^{-2} n_b^{-1/2} \sin i$ ahead of the pulsar, too small to resolve separately from the pulsar's PSF. The actual distance to the stagnation point is likely even smaller, as seen from the Hα image (see Fig. 34). The flat front of the Hα bow shock allows one to assume that the wind of this pulsar is originally concentrated in the plane perpendicular to the pulsar's velocity, presumably the equatorial plane, which implies that the pulsar's rotational axis is parallel to the velocity vector. Being deflected by the ram pressure, the shocked PW forms the tail seen as a brighter compact component (Fig. 33, right). The matter flowing in the tail is likely decelerated by the ISM entrainment, which leads to the broadening seen in the diffuse longer tail. To confirm this interpretation, flow velocities in the compact and extended tail components should be measured, but it was not possible with the data available.

Fig. 34 The Hα nebula (*red*) and the leading tail component of the X-ray PWN created by PSR J1741-2051 (Auchettl et al. 2015)

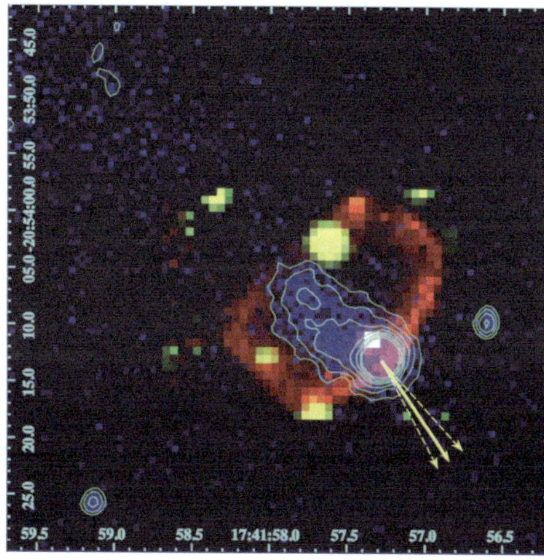

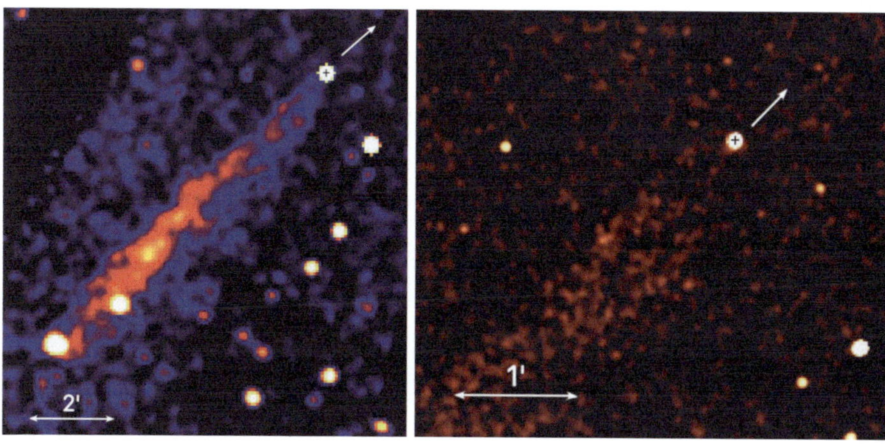

Fig. 35 *Left*: Merged *Chandra* ACIS image of the tail behind PSR J0357+3205 (0.5–6 keV, 4 observations, 136 ks total exposure). *Right*: Zoomed-in part of the same image showing the lack of tail emission in the pulsar vicinity

3.2.7 J0357+3205: A Tail Detached from the Pulsar

An interesting X-ray nebula created by a radio-quiet γ-ray pulsar J0357+3205 ($P = 444$ ms, $\tau_{\rm sd} = 540$ kyr, $\dot{E} = 5.9 \times 10^{33}$ erg s^{-1}) is shown in Fig. 35 (De Luca et al. 2011). The pulsar's proper motion is $\mu = 164 \pm 20$ mas yr^{-1} (De Luca et al. 2013), but the distance is unknown. For an assumed $d = 0.5$ kpc, its transverse velocity is $V_\perp = (389 \pm 47) d_{0.5}$ km s^{-1}. The *Chandra* and *XMM-Newton* images show a 9′ ($1.3 d_{0.5}$ pc) long, relatively straight tail behind the pulsar, but no PWN head is seen. Moreover, the tail is detached from the pulsar (not seen up to 50″), and its brightness increases with increasing distance from the pulsar, reaching a maximum at about 4′. Another unusual property of the tail is the asymmetric brightness profile across the tail, with a sharp northeastern edge, resembling the "misaligned outflow" in the Gutar nebula. The tail's luminosity, $L_{0.5\text{–}10\,\rm keV} \approx 8.8 \times 10^{30} d_{0.5}^2$ erg s^{-1}, is a fraction of $\sim 1.5 \times 10^{-3}$ of the pulsar's spin-down power.

The tail's spectrum fits a PL model with a photon index $\Gamma \approx 2$, without a significant dependence on the distance from the pulsar. Such a spectrum is consistent with the synchrotron emission from relativistic electrons in a shocked PW, but the lack of a PWN head is challenging for such an interpretation. Another potential problem is the relatively low value for the maximum synchrotron photon energy, $E_{\max} \lesssim 30\eta\sigma B_{-5}$ keV (see Eq. (7)), which is below 1 keV if $\eta\sigma B_{-5} \lesssim 0.03$. To circumvent these problems, Marelli et al. (2013) suggest that the tail's emission is in fact thermal bremsstrahlung from the shocked ISM material with a temperature of about 4 keV. To heat the ISM up to such a high temperature, a very high pulsar velocity, $V_{\rm psr} \sim 1900\ \mathrm{km\,s^{-1}}$, is required (see Eq. (9)), larger than observed for any other pulsar, which would also imply a small angle, $i < 20°$, between the velocity vector and the line of sight. The lack of diffuse emission at small distances from the pulsar could then be caused by the considerable time required for the energy transfer from ions, heated by the shock, to radiating electrons. If this scenario is confirmed by future observations, the J0357+3205 nebula would be the first example of a new class of thermally emitting nebulae associated with high-velocity pulsars.

3.2.8 Unexpectedly Faint X-Ray PWNe

From the above examples one could expect that any sufficiently powerful pulsar that has left its parent SNR produces a head-tail PWN, possibly with some misaligned outflows. However, observations of several nearby pulsars show either very faint extended emission around the pulsar or no extended emission at all. The most convincing examples of very faint (or undetected) PWNe were provided by observations of nearby pulsars B1055-52 and B0656+14.

PSR B1055-52 ($P = 197$ ms, $\dot{E} = 3.0 \times 10^{34}\ \mathrm{erg\,s^{-1}}$, $\tau_{\rm sd} = 535$ kyr) is a bright γ-ray pulsar at an estimated distance of $\sim$350 pc (Mignani et al. 2010). Its proper motion, $\mu = 42 \pm 5\ \mathrm{mas\,yr^{-1}}$, corresponds to the tranverse pulsar velocity $V_\perp \approx 70 d_{0.35}\ \mathrm{km\,s^{-1}}$. A dedicated 56 ks *Chandra* ACIS observation by Posselt et al. (2015) showed some enhancement (with respect to the model PSF) in radial count distribution from $2''$ to $20''$, better seen in the soft X-ray band (0.3–1 keV), corresponding to the luminosity of $1\text{–}2 \times 10^{29} d_{0.35}^2\ \mathrm{erg\,s^{-1}}$, which is $(3\text{–}6) \times 10^{-6} d_{0.35}^3$ of the pulsar's spindown power. The alleged extended emission showed only a hint of azimuthal asymmetry (an excess in the quadrant that includes the proper motion direction), at a 3σ level. This extended emission (if real) could be, at least partly, a dust scattering halo, but a very faint X-ray PWN cannot be excluded. Posselt et al. (2015) speculate that such a faint, nearly round PWN could be produced if the pulsar is moving away from us almost along the line of sight, i.e., $V_{\rm psr} \gg V_\perp$. It, however, remains unclear whether this interpretation is correct.

Very similar results were obtained by Bîrzan et al. (2016) for PSR B0656+14 ($P = 385$ ms, $\dot{E} = 3.8 \times 10^{34}\ \mathrm{erg\,s^{-1}}$, $\tau_{\rm sd} = 111$ kyr, and $d = 0.29 \pm 0.03$ kpc from parallax measurements). This pulsar also has a low transverse velocity, $V_\perp = (60 \pm 7) d_{0.29}\ \mathrm{km\,s^{-1}}$. From the analysis of archival *Chandra* ACIS and HRC data, Bîrzan et al. (2016) found a slight enhancement over the model PSF in an annulus of about $3''\text{–}15''$ around the pulsar, with a luminosity of $\sim 8 \times 10^{28} d_{0.29}^2\ \mathrm{erg\,s^{-1}}$. This luminosity is $\sim 2 \times 10^{-6} d_{0.29}^2$ of the pulsar's spindown power, a factor of $\sim$7 lower than the X-ray efficiency of the PWN of Geminga that has a similar (slightly higher) spindown power and is a factor of 3 older than PSR B0656+14. No azimuthal asymmetry was detected in the images. The spectrum of the enhancement is apparently very soft, $\Gamma \sim 8$, but its uncertainty is very large because the imaging ACIS observation was very short, about 5 ks. As in the case of B1055-52, the extended emission (if real) could be a combination of a dust scattering halo and a PWN created by the pulsar moving almost along the line of sight. The PWN and halo contributions could be disentangled from a longer ACIS observation.

Thus, a plausible explanation for the lack of the expected head-tail morphology and a very low PWN luminosity might be due to smallness of the angle between the pulsar velocity direction and the line of sight, which is also indicated by the small values of $V_\perp$. However, the transverse velocity of PSR B0355+54 is similarly low, but that pulsar is accompanied by a PWN with a rich structure (see Fig. 32). Moreover, the spindown power of PSR B0355+54 is similar to those of B1055-52 and B0656+14, but its PWN luminosity is at least a factor of 300 higher. Obviously, there must be some other factors that affect the X-ray efficiency and appearance of PWNe created by pulsars moving in the ISM. A possible reason for these differences could be different orientations of the pulsar rotational axes (hence the equatorial planes) with respect to their velocities. Another parameter on which the PWN properties should depend is the angle between the spin and magnetic axes, which affects the conversion of the PW magnetic energy into kinetic energy and particle acceleration. In particular, the very low efficiencies of some PWNe could be associated with nearly aligned spin and magnetic axes. To check these hypotheses, it would be useful to look for a correlation between the PWN properties and multiwavelength pulsations.

3.2.9 General Overview of the X-Ray PWNe Created by Supersonically Moving Pulsars

About 15 X-ray PWNe created by pulsars moving through the ISM have been detected. The spindown powers $\dot{E}$ of these pulsars are in the range from 1.2×10^{33} erg s^{-1} (PSR B2224+65, the Guitar PWN) to 2.5×10^{36} erg s^{-1} (PSR J1747-2958, the Mouse PWN). Electrons/positrons of PWs of less powerful pulsars, which consitute the majority of rotation-powered pulsars, apparently cannot be accelerated to energies high enough to emit X-ray synchrotron radiation, and even if the energy is sufficient, the PWN luminosity may be too low to detect it, even from nearby sources.

The examples presented here show that most supersonically moving pulsars are accompanied by tails, with typical lengths of a few parsecs. However, the appearances of PWN heads vary considerably in different sources. Some of the well-resolved PWNe have a filled PWN head morphology (e.g., the Mouse, B0355+54, J1741-2044) while others show "hollow" morphologies (Geminga, J1509-5850). Moreover, there is at least one example, J0357+3205, which shows a long tail but no resolved PWN head around the pulsar. The diversity of PWN heads suggests that PWs of old pulsars are anisotropic, perhaps concentrated around the equatorial plane (as in the Crab and some other young pulsars) in many cases. The different appearances of the compact PWN components could be due to different orientations of the pulsar's spin axis with respect to the velocity direction and the line of sight. A particularly puzzling morphology is seen in the Geminga PWN, with its three "tails", which can be considered as a hollow-morphology compact PWN component observed from a close distance (it might have a much longer tail that is perhaps too faint to be detected by the current instruments).

Quite unexpected was the discovery of "misaligned outflows" in X-ray observations of several pulsars: B2224+65 (the Guitar), J1509-5850, B0355+54, described above, and likely the spectacular Lighthouse nebula created by PSR J1101-6101 (Pavan et al. 2016 and references therein; see Fig. 36). Their nature still remains puzzling. A hypothesis was suggested by Bandiera (2008) that these features are produced by synchrotron radiation of very

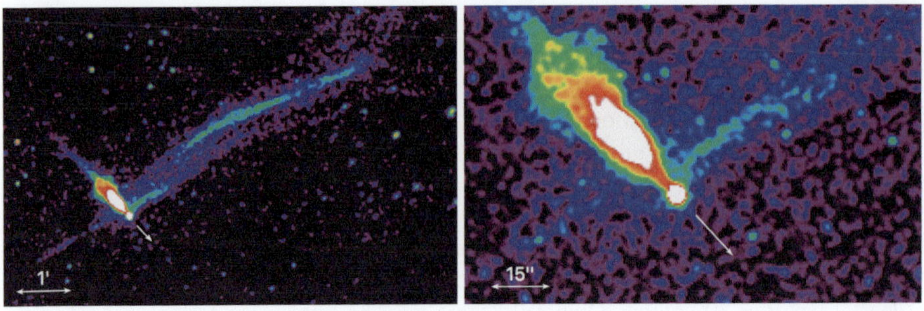

Fig. 36 *Chandra* ACIS image of the Lighthouse PWN (0.5–8 keV, 300 ks) created by PSR J1101-6101 ($\dot{E} = 1.4 \times 10^{36}$ erg s^{-1}, $\tau_{sd} = 116$ kyr) moving from the SNR MSH 11-61A (Pavan et al. 2016). The image shows a bright tail behind the fast-moving pulsar ($V_\perp \sim 1000 d_7$ km s^{-1}) and an $11 d_7$ pc long jet-like feature, possibly another example of a "misaligned outflow"

high energy particles leaked from the TSs into the ambient ISM along the ISM magnetic field, but this interpretation remains to be confirmed by quantitative modeling.

Although the flow speeds along the tails or misaligned outflows have not been directly measured, they have been crudely estimated for some tails based on circumstantial arguments. These estimates show that flow speeds considerably exceed the pulsar speeds, but they are well below the mildly relativistic speeds predicted by the numerical simulations.

The X-ray efficiencies, $\eta_X = L_X/\dot{E}$, of the observed PWNe vary between $\lesssim 2 \times 10^{-6}$ (the alleged PWN around PSR B0656+14) and $\sim 10^{-2}$ (the Mouse). The reason for such a huge scatter remains unclear. At least partly, it can be due to different orientations of the pulsar equatorial planes, where the PWs are presumably concentrated, with respect to the pulsar velocities. It also may be that the fraction ξ_w of the spindown power lost to the PW is different in different pulsars (because the fraction of $\dot{E}$ radiated from pulsar magnetospheres depends on pulsar parameters). Another likely reason for different PWN efficiencies is associated with conversion of the magnetic PW energy into the kinetic energy of particles, which should increase with increasing angle between the magnetic and spin axes.

The X-ray spectra of PWNe created by supersonically moving pulsars are usually well described by a PL model, which supports their synchrotron interpretation. Typical spectral slopes Γ in the compact nebula components (PWN heads) are in the range of 1.5–2.0, but the lateral tails of the Geminga PWN are much harder, $\Gamma \sim 0.7$–1.0. Some of the PWN tails (e.g., in the Mouse and Lighthouse) show a substantial softening with increasing distance from the pulsar, up to $\Delta\Gamma \approx 1.0$–1.5, while others show no softening at all, sometimes even a hint of hardening. The fast softening is an indication of a relatively high magnetic field (e.g., up to a few hundreds of μG in the Mouse, the highest value found so far, which is a factor of 10 higher than typical magnetic fields).

A few head-tail PWNe have been detected in the radio. The radio tails are usually longer than the X-ray ones, as expected from the synchrotron cooling. The additional radio data allow one to examine a broad-band PWN spectrum, which is usually harder in the radio than in the X-rays, and get an idea about the spectrum of emitting particles. Measuring the spatially resolved radio polarization makes it possible to map the directions of the magnetic field within the PWN, but the two PWNe for which such mapping was done (the Mouse and J1509-5850) show very different distributions.

The X-ray spectral slopes $\Gamma < 2$ suggest that the main contribution to the total synchrotron PWN luminosities are provided by photons with energies above the observed X-ray range, i.e., above ~10 keV. However, none of the head-tail PWNe has been detected at hard X-rays or γ-rays, perhaps because the current detectors are not sensitive enough.

3.3 Open Questions

Although many of the observed properties of PWNe produced by supesonically moving pulsars are qualitatively understood, there remain several problems that require further investigations, both observational and theoretical.

First of all, we should understand the reason(s) for the the *great diversity of PWN shapes*. Although elongated X-ray tails behind the moving pulsars have been observed in many of them, it is not quite clear which parameters determine the tail properties (shape, length, collimation and divergence, separation from the pulsar in some cases). Even less clear is the origin of the divesrity of PWN heads (e.g., filled vs. hollow morphology). One of such parameters is obviously the angle between the velocity vector and the line of sight, but other parameters, such as the angle between the spin and magnetic axes, and between the spin axis and the pulsar velocity, can play an important role. It is also possible that the direction and strength of the ambient ISM magnetic field can affect the observed surface brightness distribution. To assess the contribution of the different factors, a study of correlation of the PWN shape with the shapes and phases of pulsar pulses at different wavelengths (e.g., radio and γ-rays), supplemented by PWN modeling in the case of anisotropic PW, would be particularly useful.

The most puzzling features among the recently discovered PWN components are the "*misaligned outflows*", whose directions are strongly misaligned with respect to the pulsar velocities. It is tempting to interpret them as jets along the pulsar spin axes, similar to those observed in PWNe of young pulsars, but such jets are expected to be bent by the ram pressure on much smaller scales than the observed lengths of these nearly straight, elongated features. Only qualitative interpretations of such features have been suggested so far, which remain to be confirmed by quantitative modeling.

It remains unclear whether "ordinary" *jets along the spin axes* have been detected in old PWNe outside SNRs. May it be that the outflows along the spin axis are less powerful (at least, less luminous) in old pulsars than in young ones? To answer this question, we should obtain independent information on the spin axis directions, which could be done with the aid of multiwavelength polarimetry, in addition to the pulse analysis.

Since the X-ray PWN emission is synchrotron radiation from relativistic electrons and/or positrons, one should expect softening of the PWN spectrum with increasing distance from the pulsar due to synchrotron cooling. Such softening has indeed been observed in some of the tails, but other tails, as well as the misaligned outflows, show *no spectral softening* at all, but hints of hardening. What is the reason for such behavior? Is it an indication of an additional (re)acceleration (heating) along the tails? What is the acceleration mechanism? Is it the same mechanism that is responsible for thc unusually hard spectra of the Geminga's lateral tail? Why does it operate only in some tails? To answer these questions, deep high-resolution X-ray observations are required, which would allow spatially resolved spectral analysis to accurately measure the spectral changes. If the lack of softening (or even spectral hardening) is confirmed, possible acceleration mechanisms (e.g., the Fermi acceleration at fronts of oblique shocks or magnetic turbulence) should be studied.

Another puzzling problem is the *very low X-ray efficiency (or even the absence) of PWNe around some nearby pulsars* that are powerful enough to create an observable X-ray PWN. While it might be partly expained by closeness of the pulsar velocity direction to the line of sight, it is certainly not a full explanation. Are these PWNe so faint because the conversion of the PW magnetic energy into the particle energy (e.g., via magnetic field reconnection in the striped wind zone) is inefficient, as expected for pulsars with nearly aligned magnetic and spin axes? Are there other mechanisms that suppress the production and acceleration of the particle component of PWs? To understand the true reason, more nearby pulsars should be observed, and the absence or presence of PWNe should be confronted with the observed pulsar properties.

To conclude, significant progress in our understanding of PWNe of supersonically moving pulsars has been achieved, thanks to the high resolution and sensitivity of the *Chandra* and *XMM-Newton* observatories, but there remain a number of open problems that could be resolved with further deep X-ray observations and more realistic modeling.

4 Gamma-Ray Observations of Pulsar-Wind Nebulae

Gamma-ray astronomy has entered a golden age during the last decade thanks to the latest generation of space telescopes in the High-Energy (HE; $0.1 < E < 100$ GeV) domain (*Fermi*-LAT, *AGILE*), and ground-based instruments in the very-high energy (VHE; $0.1 < E < 100$ TeV) domain, in particular the Imaging Atmospheric Cherenkov Telescopes (IACTs) such as H.E.S.S., VERITAS and MAGIC, featuring unprecedented performance.[8] *Fermi*-LAT, with a large field-of-view of 2.4 sr, an angular resolution varying from $\sim 5^\circ$ at 100 MeV to $\lesssim 0.1^\circ$ above 50 GeV, and a point-source sensitivity at the level of $\sim 10^{-12}$ erg cm^{-2} s^{-1} in the GeV domain, has revealed more than 3000 (*3FGL*) sources in the 0.1–300 GeV band (Acero et al. 2015) and 360 (*2FHL*) sources above 50 GeV (Ackermann et al. 2016). IACTs have so far detected more than 170 VHE sources, two third of which being located in the Galaxy .[9] In particular, the H.E.S.S. experiment, exhibiting a 5°-wide field-of-view, an angular resolution of $\sim 0.06^\circ$ and a point-source sensitivity of a few 10^{-12} erg cm^{-2} s^{-1} in the TeV domain, has surveyed the inner Galactic Plane during ten years, resulting in the detection of more than 70 sources.[10]

These GeV–TeV instruments offer for the first time the possibility to spatially resolve large enough sources such as middle-aged pulsar wind-nebulae (PWNe) and shell-type supernova remnants (SNRs), and to study source spectra over more than five decades in energy. Spectro-morphological studies are of prime importance as they allow one to pinpoint the origin of γ-ray emission from SNR systems in which the pulsar (PSR), the associated wind-nebula, and the host SNR shell, can contribute to the observed emission. In this regard, while *Fermi*-LAT has detected more than 200 PSRs in the HE domain,[11] IACTs, and H.E.S.S. in particular, have revealed more than 30 TeV PWNe and PWN candidates associated with energetic PSRs (Klepser et al. 2013; Kargaltsev et al. 2013), a dozen of which being also detected at (multi-)GeV energies (Acero et al. 2013; Ackermann et al. 2016). Nevertheless, a large fraction of the HE/VHE sources in the Galaxy still remain unassociated due to the

[8]In what follows, the angular resolutions are provided as the 68% containment radii. The *Fermi*-LAT sensitivity is given at TS = 25 in 10 y, for a source at $(\ell, b) = (0^\circ, 30^\circ)$ with Pass 8 data (https://www.slac.stanford.edu/exp/glast/groups/canda/lat_Performance.htm). The H.E.S.S. sensitivity is provided at 5 σ in 25 h, for a source near zenith (https://www.mpi-hd.mpg.de/hfm/HESS/pages/home/proposals/). See also the dashed lines in Fig. 38.

[9]See http://tevcat.uchicago.edu/.

[10]From Abramowski et al. (H.E.S.S. Collaboration) 2016, in prep. See also https://www.mpi-hd.mpg.de/hfm/HESS/pages/home/som/2016/01/.

[11]https://confluence.slac.stanford.edu/display/GLAMCOG/Public+List+of+LAT-Detected+Gamma-Ray+Pulsars.

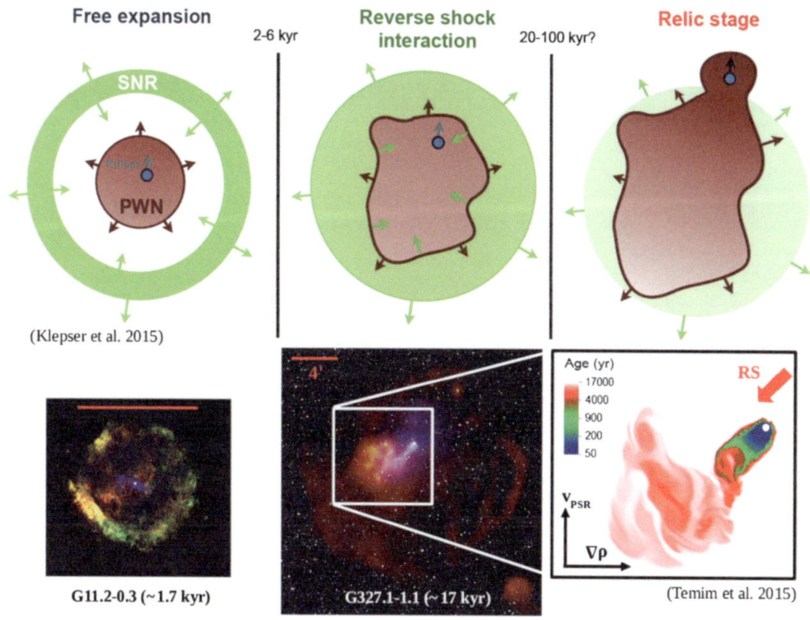

Fig. 37 PWN evolutionary phases. *Left:* young systems where the nebula lies at the center of the host shell–type SNR, as examplified by G11.2-0.3 seen in X-rays. *Middle:* evolved nebulae, after the so-called crushing phase caused by the interaction with the asymmetric reverse shock (RS) due to an inhomogeneous ISM and/or a high PSR's proper motion, as illustrated by G327.1-1.1 observed in radio (*red*) and X-rays (*blue*). The *inset image* shows the age distribution of particles injected by the PSR some 17000 years ago, based on the hydrodynamical simulation of Temim et al. (2015) with the assumed ISM density gradient and PSR's velocity depicted by the two *arrows*. *Right:* bow-shock nebulae, when the pulsar motion through the SNR or the ISM becomes supersonic

limitations of gamma-ray instruments to precisely characterize the source morphologies in most cases and to the difficulty for radio/X-ray telescopes to reveal structures at scales of the order of the typical VHE source sizes ($\sigma \sim 0.1°$–$0.3°$).

Gamma-ray emission from PWNe is usually interpreted in the so-called leptonic scenario,[12] where the accelerated electron–positron pairs emit through inverse Compton (IC) scattering off the ambient low-energy photons, from the CMB and the interstellar radiation fields (ISRFs) made of infrared emission from dust and optical/UV starlight. In this framework, HE/VHE observations allow one to derive the spectral distribution and energy content of the high-energy particles (provided the ISRFs are known), to reveal their spatial distribution, and to set important constraints on the particle acceleration and transport mechanisms. Furthermore, these measurements, when combined with radio/X-ray observations of the synchrotron (SC) component, provide estimates of the magnetic field strength and distribution in these sources, as we shall see below.

Two classes of γ-ray PWNe can be distinguished according to their evolutionary stages, as shown in Fig. 37 (no clear evidence of γ-ray emission associated with a bow-shock nebula [see Sect. 3] has been reported so far). First, *young* PWNe, such as the Crab neb-

[12]The possibility that a hadronic component could carry a significant fraction of the energy content in PWNe has been investigated in several works (see Di Palma et al. 2016 and references therein), but the observational evidence of the presence of accelerated ions in PWNe, through e.g. pion production in hadronic interactions and the subsequent emission of γ-rays and neutrinos, has been so far elusive.

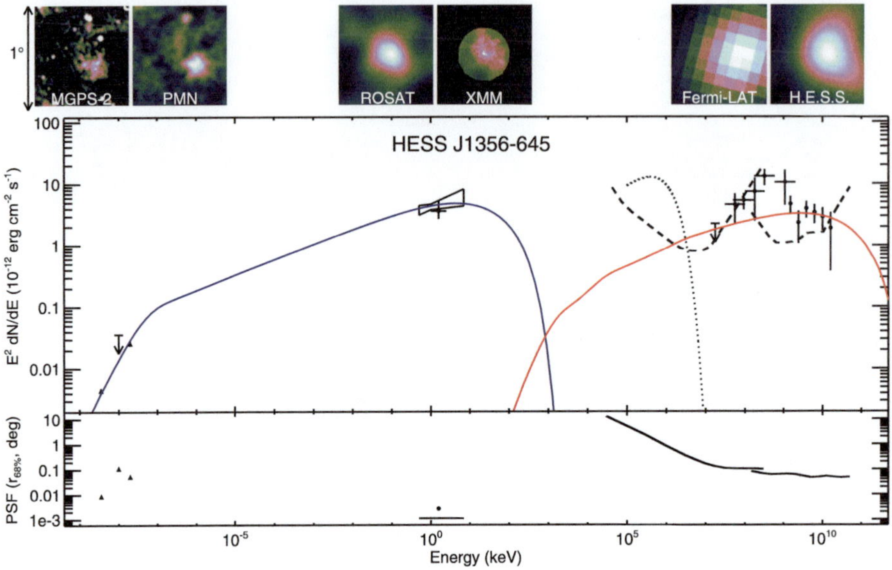

Fig. 38 Broadband spectrum of the PWN HESS J1356-645 (H.E.S.S. Collaboration et al. 2011; Ackermann et al. 2016). The *blue* and *red solid curves* give the SC and IC emission from a one-zone, time-independent, modeling performed in H.E.S.S. Collaboration et al. (2011). The *dotted line* represents the best-fit phase-averaged spectrum of PSR J1357-6429 as obtained in Lemoine-Goumard et al. (2011). The two *dashed lines* show the *Fermi*-LAT and H.E.S.S. point-source sensitivities (see text for more details). The 1°-wide multi--wavelength images, as obtained in H.E.S.S. Collaboration et al. (2011), Ackermann et al. (2016), centered on the PSR position, are also shown. The *lower panel* presents the angular resolutions of the considered radio, X-ray and gamma-ray instruments

ula, MSH 15-52, G0.9+0.1, G21.5-0.9 and Kes 75 (discussed in Sect. 4.1), are usually found at the center of their host shell-type SNRs and are seen as unresolved or compact γ-ray sources. HESS J1818-154, a compact TeV source located at the center of the radio SNR G15.4+0.1, is worth to be mentioned as being the first PWN discovered by TeV observations in a composite SNR (H.E.S.S. Collaboration et al. 2014). Second, *middle-aged* PWNe (with PSR characteristic ages of $\gtrsim 10^4$ yr), such as HESS J1356-645 (see Fig. 38), HESS J1825-137 (Sect. 4.2.1), and the peculiar Vela PWN (Sect. 4.2.2), are usually resolved in the HE/VHE domains and are found to be significantly offset from the current position of the associated PSR, with large size ratios between the X-ray and γ-ray emission regions. The evolution of the SNR into an inhomogeneous ISM, through the interaction with an asymmetric reverse shock (RS), and/or the high PSR's velocity (Blondin et al. 2001; van der Swaluw et al. 2004; Temim et al. 2015) can lead to a displacement of the crushed PWN from the SNR center. Also, mean magnetic field strengths within these γ-ray PWNe as low as ~3–5 μG are required in order to prevent the TeV-emitting electrons from suffering from severe radiative losses, enabling the majority of them to survive from (and hence probe) early epochs of the PWN evolution (de Jager and Djannati-Ataï 2009). All these effects can thus explain the large-offset X-ray-faint long-lived γ-ray sources as being the relic nebulae from the past history of the pulsar wind inside its host SNR.

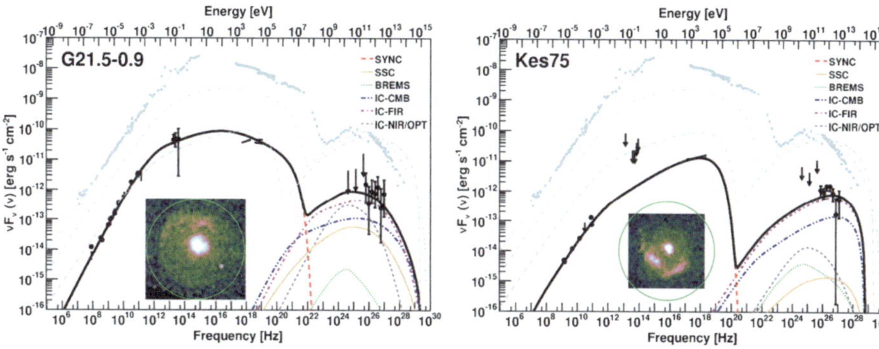

Fig. 39 Broadband spectra of G21.5-0.9 (*left*) and Kes 75 (*right*) with the best-fit models from Torres et al. (2014) shown as *solid lines*. The *grey dots* and *lines* correspond to the multi-wavelength data and the associated models (at different ages) for the Crab nebula. *Chandra* X-ray images of these two PWNe with the typical H.E.S.S. PSF depicted as a *green circle* are also shown

4.1 Young PWNe

Several time-dependent models of the radiative (sometimes coupled with the dynamical) evolution of PWNe have been developed in order to estimate, from the observed broadband non-thermal emission, the energy distribution and content of particles and the wind magnetization, and to constrain the particle acceleration mechanisms at work in these sources (Gelfand et al. 2009; Fang and Zhang 2010; Tanaka and Takahara 2010, 2011; Bucciantini et al. 2011; Martín et al. 2012; Vorster et al. 2013; Torres et al. 2014). Figure 39 shows the broadband spectra from G21.5-0.9 and Kes 75, two young PWNe respectively associated with the energetic pulsars PSR J1833-1034 and PSR J1846-0258 and detected in the VHE domain (Djannati-Ataï et al. 2008), together with the best-fit models from Torres et al. (2014). The common findings of all the above-mentioned modelings are as follows: the particle energy distribution is well described by a broken power law with an intrinsic break[13] (i.e. assumed to be of non-radiative origin) at a Lorentz factor of $\sim 10^{5-6}$, and almost all of the young PWNe considered in these studies are particle-dominated. This is translated into large pair multiplicities ($\kappa \gtrsim 10^4$) and magnetic fractions of $\lesssim$ a few percent (i.e. with magnetic fields lower than the equipartition values), with TeV emission dominated by IC scattering off (far-)infrared photons (with energy densities generally larger than the Galactic large-scale ISRFs provided by Porter and Strong (2005)). Although these constraints are valuable to pinpoint the particle acceleration mechanisms (Amato 2015), it should be noted that these modelings assume a one-zone emission region (in the 1D approximation, known to face several issues such as the so-called σ problem, cf. a brief discussion in Sect. 3.1), whereas the unresolved TeV emission from e.g. G21.5-0.9 and Kes 75 (see Fig. 39) and the debated origin of the radio emission from these young PWNe could question this assumption. Also, relying on (or rescaling) the commonly used Galactic ISRFs as an estimate of the *local* fields from dust and stars is another major source of uncertainty which directly affects the predicted IC emission.

[13]The modeling presented in Vorster et al. (2013) assumes that the particle spectrum has a discontinuity at the transition between the low- and high-energy components.

4.2 Middle-aged PWNe

As discussed above, middle-aged PWNe are observed in the HE/VHE domain as extended sources, offset from the PSR position. These two characteristics make them difficult to identify as such from γ-ray observations, and many of the so-called *dark* VHE sources, with no obvious radio/X-ray counterpart, could actually be such ancient nebulae (de Jager et al. 2009). Besides dedicated pulsars search (with *Fermi*-LAT in particular) within the VHE source extent, multi-wavelength investigation, as shown in Fig. 38 in the case of HESS J1356-645, and energy-dependent γ-ray morphological analysis (as commonly done in X-rays) are the two means to unveil their nature.

4.2.1 HESS J1825-137: Energy-Dependent Morphology and Particle Transport Mechanisms

HESS J1825-137 is the archetypal middle-aged PWN, discovered by H.E.S.S. (Aharonian et al. 2006a) as a bright, $\sim 1°$-large (i.e. ~ 70 pc at 4 kpc) source offset from the energetic radio pulsar PSR J1826-1334, and later detected with *Fermi*-LAT (Grondin et al. 2011). A detailed spectro-morphological analysis has revealed for the first time in the VHE domain a steepening of the energy spectrum with increasing distance from the pulsar, likely due to the cooling losses of electrons during their transport in the nebula. This is illustrated by the three-colour image in Fig. 40 (left) revealing the shrinking of the nebula with increasing energy and hence the physical connection between the TeV PWN with PSR J1826-1334, also responsible for the compact hard-index X-ray nebula of size $\sim 30''$ embedded in a $\sim 10'$-large softer structure (Pavlov et al. 2008; Van Etten and Romani 2011). These γ-ray measurements, when combined with X-ray observations, provide important constraints on the particle injection, transport and cooling within the nebula: a detailed 3D time-dependent multi-zone spectro-morphological modeling (Van Etten and Romani 2011) has shown a good agreement with the data by including radially decreasing advection velocity and magnetic field profiles and substantial particle diffusion in order to explain the presence of multi-TeV particles ~ 80 pc away from the pulsar. Such a rapid diffusion is at odds with the toroidal magnetic field structure effective at smaller scales in many PWNe leading to strong magnetic confinement of particles.

4.2.2 Vela X: Multi-wavelength Picture and Particle Escape

The energetic Vela pulsar (PSR B0833-45, $\tau_c = 11$ kyr, P = 89 ms, $\dot{E} = 7 \times 10^{36}$ erg s^{-1}), embedded in the nearby 8°-large Vela SNR located at d $\sim$ 290 pc, is known to power several manifestations of wind nebulae seen at different spatial scales: a compact X-ray nebula composed of two toroidal arcs at sub-arcmin scales and a $4'$ long collimated jet (Helfand et al. 2001; Pavlov et al. 2003), an extended hard X-ray emission (E > 18 keV) north of the pulsar (Mattana et al. 2011), a $45'$ elongated X-ray structure dubbed the *Vela cocoon*, also detected at TeV energies with H.E.S.S. (Aharonian et al. 2006b) and partially coincident with a bright radio filament (LaMassa et al. 2008), and the Vela X nebula (referred to as the *Vela halo*, encompassing the cocoon area), a large-scale $2° \times 3°$ non-thermal radio region offset by $\sim 40'$ from the pulsar that is also detected at GeV energies with *Fermi*-LAT (Abdo et al. 2010). Two distinct lepton populations have been suggested to explain the different broadband spectra of the latter two emission regions: a young population producing the X-ray/VHE cocoon and a relic one responsible for the radio/HE halo (de Jager et al. 2009), with similar magnetic field strengths of $\sim$3–5 μG (Abdo et al. 2010). However,

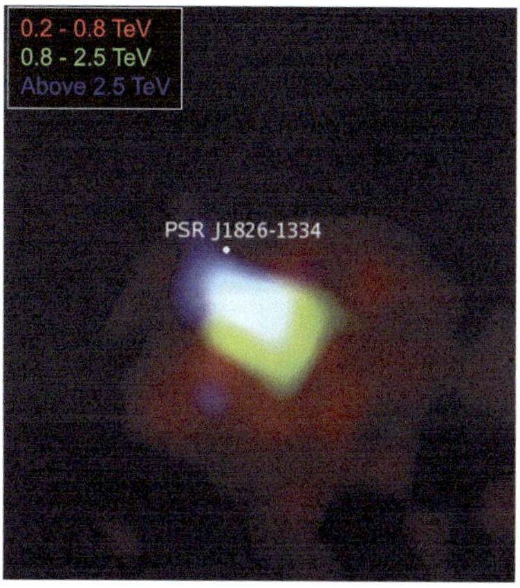

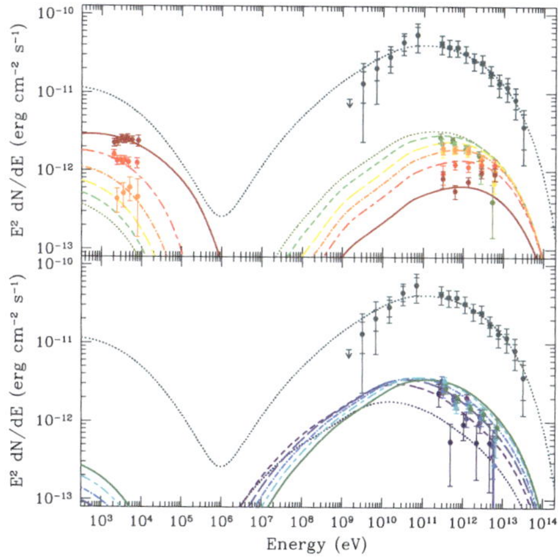

Fig. 40 *Left:* Three-color 3°-large image of HESS J1825-137 as observed with H.E.S.S. in different energy bands (Funk et al. 2008) showing the shrinking of the nebula with increasing energy. *Right:* Broadband spectra from the whole PWN (slate data *points* and *dotted line*) and from different wedges in HESS J1825-137 with a color scale evolving from *dark red* (inner region) to *green*, *blue* and *violet* (outer region), as defined in Aharonian et al. (2006a). *Lines* represent the best-fit model presented in Van Etten and Romani (2011) (see text for more details)

such an interpretation has been challenged by several new observational evidences: TeV emission beyond the cocoon, extended over much of the halo and featuring a similar VHE spectrum, has been detected with H.E.S.S. (Abramowski et al. 2012), and a detailed spectro-morphological *Fermi*-LAT data analysis has revealed two new spatial HE features matching the so-called northern and southern wings of Vela X as seen with *WMAP* and *Planck*, with marginally different spectra (Grondin et al. 2013). Such a complex morphology, with several emission components at different scales, is highlighted in the multi-wavelength image shown in Fig. 41 (left).

These new γ-ray measurements provide direct evidence that high-energy leptons are present in the extended halo, ~10 pc away from the pulsar. In order to explain the steep spectra measured with *Fermi*-LAT, diffusive escape of particles from the radio nebula has been invoked by Hinton et al. (2011) (see Fig. 41, right), as in the case of HESS J1825-137 discussed in Sect. 4.2.1. While particle confinement is thought to be important during the early PWN evolution, the interaction with the SNR reverse shock could thus make the diffusion of particles out of the nebula possible. Such an energy-dependent escape in this nearby PWN should produce a clear signature in the local CR lepton spectrum (Hinton et al. 2011; Della Torre et al. 2015). Latest measurements with AMS-02 (Aguilar et al. 2014a,b) have shown an increase of the positron fraction with increasing energy, which could be explained by the Galactic PSR population and in particular by a few nearby PSRs (Hooper et al. 2009; Di Mauro et al. 2014; Boudaud et al. 2015). Therefore, continuing HE/VHE observations of middle-aged and bow-shock PWNe are crucial to estimate the total energy content of high-energy particles residing inside these sources in order to assess the importance of escape mechanisms throughout their evolution, and hence their contribution to the Galactic CR lepton spectra observed at Earth.

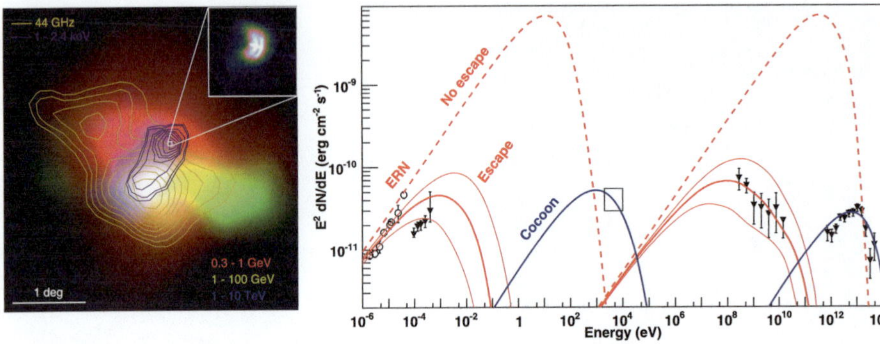

Fig. 41 *Left*: multi-wavelength picture of Vela PWN. The 4°-wide three-color image (in Galactic coordinates) is composed of the 0.3–1 GeV (*red*) and 1–100 GeV (*green*) emission measured with *Fermi*-LAT (Grondin et al. 2013), and of the 1–10 TeV (*blue*) emission detected with H.E.S.S. (Abramowski et al. 2012). *Contours* represent the 44 GHz radio emission measured by *Planck* (Planck Collaboration et al. 2016) and the X-ray emission above 1 keV observed with *ROSAT* (in *blue*). The well-known X-ray compact nebula around the Vela PSR is shown in the inset on the *upper right corner* (*Chandra*/ACIS 0.3–10 keV image). *Right*: broadband spectrum of Vela PWN (ERN: Extended Radio Nebula) with (*solid lines*) and without (*dashed lines*) the assumption of particle escape throughout its evolution, according to Hinton et al. (2011)

5 Outflows in Magnetars

Magnetars differ from the rotation-powered NS discussed in the previous sections because a strong magnetic field is the main energy source which ultimately powers their persistent and bursting/flaring emission (Mereghetti et al. 2015). Unlike ordinary pulsars, which show no or very little variability, magnetars are characterized by a variety of transient phenomena on timescales from a few milliseconds to years and involving flux changes as large as several orders of magnitude. There are many indications that their magnetosphere is highly dynamic and characterized by a complex non-dipolar geometry. It is thus expected that magnetars might also be able to accelerate charged particles and produce outflows which, in principle, can lead to the formation of diffuse nebulae.

Owing to their long spin periods (>2 s), magnetars have rotational energy losses much smaller than those of the energetic pulsars typically associated with PWNe. The magnetar with the highest spin-down rate, 1E 1547.0-5408, has $\dot{E}_{ROT} = 2.1 \times 10^{35}$ erg s^{-1}, but more typical values are in the range $\sim 10^{32}$–10^{34} erg s^{-1}. While the ratio $L_{PWN}/\dot{E}_{ROT}$ is a useful description of the efficiency of classical PWNe, the analogous quantity for magnetars, $L_{MWN}/\dot{E}_{ROT}$ is less relevant, because their nebulae might be powered, at least in part, by magnetic energy. In fact, the observation of time-variable radio and hard X-ray emission after giant flares, as well as the presence of long-lived nebulae around some magnetars, provide evidence that these neutron stars can emit relativistic particle outflows.

In this section we discuss two kinds of observational results which give evidence for relativistic outflows from magnetars: (a) the detection of time-variable emission associated to the occurrence of giant flares, and (b) the presence of persistent diffuse emission at radio and/or X-ray energies around some magnetars.

5.1 Outflows During Giant Flares

Giant flares are the most extreme manifestations of magnetars, involving radiated energies up to about 10^{46} erg (assuming isotropic emission). They are rare events: only three have

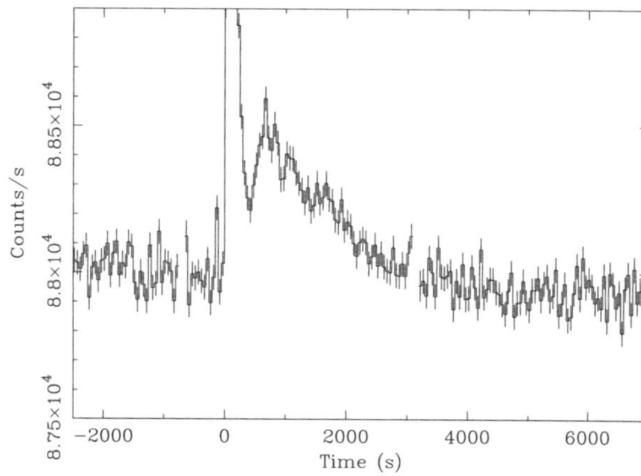

Fig. 42 High-energy (>80 keV) light curve of the 27 December 2004 giant flare of SGR 1806-20 measured with the INTEGRAL satellite (from Mereghetti et al. 2005). After the bright flare truncated in the plot (reaching $> 2 \times 10^6$ counts s^{-1}) a long tail of hard X-ray emission, which peaks at $t \sim 500$ s and lasts more than one hour is clearly visible

been observed in more than 40 years (each one from a different source, in a sample now totalling at least two dozen magnetars (Olausen and Kaspi 2014)). The famous giant flare of 1979 March 5, from the Large Magellanic Cloud source SGR 0526-66, was crucial for our understanding of magnetars, but due to its large distance and unexpected occurrence, it could not be studied in much detail.

More data could be obtained for the giant flare emitted on 1998 August 27 from the Galactic magnetar SGR 1900+14. Observations carried out with the VLA about one week after the outburst revealed a faint (~0.3 mJy), unresolved radio source (angular diameter $\theta < 0.8''$), which became undetectable after a few days (Frail et al. 1999). A power law index $\alpha = 0.74 \pm 0.15$ was derived from the fluxes at 1.43 and 4.86 GHz. This radio emission was interpreted as a cloud of synchrotron emitting relativistic particles ejected during the giant flare (or in the phase of intense bursting activity which preceded it). Simple equipartition arguments led to an estimate for the nebula minimun energy of $1.6 \times 10^{43}\ (d/10\ \mathrm{kpc})^{17/7}\ (\theta/0.4'')^{9/7}$ erg.

The event of 2004 December 27 from SGR 1806-20 was by far the most energetic and best studied giant flare, with an isotropic energy release in hard X-rays and γ-rays of more than $5 \times 10^{45}\ (d/10\ \mathrm{kpc})^2$ erg. The expanding radio nebula detected after this flare could be observed for more than one year (Gaensler et al. 2005; Cameron et al. 2005; Granot et al. 2006). It had a peak flux of ~170 mJy at 1.4 GHz in the first observation, carried out about one week after the giant flare. The flux then decreased as a steep power law of time, $F(t) \propto t^{-\delta}$ with $\delta \sim 3$, and, after a brief rebrightening at ~25 days after the flare, it followed a shallower power law decay with $\delta \sim 1.1$. The radio source was spatially resolved and featured an elliptical shape; it expanded from ~60 mas to ~400 mas in a couple of months, while at the same time its centroid moved by about 200 mas, along the direction of the elongation. The power-law spectrum and linear polarization indicate that the radio emission is synchrotron radiation. The minimum energy in the radio nebula was of the order of a few 10^{43} erg. This is much larger than the energy available in the electron/positron pairs escaping the initial fireball (Nakar et al. 2005), implying that that the relativistic flow powering the nebula was loaded by baryons or Poynting flux. Indeed, the observed properties of the expanding nebula are well explained by an asymmetric ejection of few 10^{24} g of mildly relativistic baryons, $v/c \sim 0.3\ (d/10\ \mathrm{kpc})$, which interact with a pre-existing shell of matter surrounding the magnetar (Gelfand et al. 2005; Granot et al. 2006).

High-energy data obtained immediately after the SGR 1806-20 giant flare provided additional evidence for the ejection of relativistic matter. The *INTEGRAL* satellite revealed

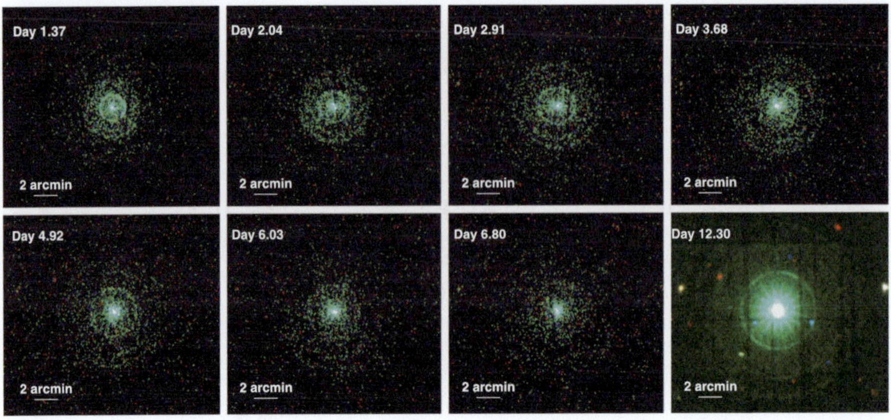

Fig. 43 Images of the expanding X-ray rings caused by interstellar dust scattering around the transient magnetar 1E 1547.0-5408 (from Tiengo et al. 2010b). The seven *Swift* observations and the *XMM-Newton* observation (*bottom right panel*) were obtained from one to 12 days after a very bright burst that occurred on 2009 January 22. The apparent angular expansion of the three rings is due to the longer pathlength of the burst radiation scattered by three layers of dust along the line of sight

emission at energy above ~80 keV, which peaked about 11 minutes after the start of the flare and then decreased approximately as $F(t) \propto t^{-0.85}$ (Fig. 42). Evidence for this long-lasting high-energy emission was later found also in the *Konus/WIND* and *RHESSI* data (Frederiks et al. 2007; Boggs et al. 2007), which indicated a duration of at least 10^4 s and a hard spectrum (power-law photon index $\Gamma \sim 1.6$). No pulsations at the NS rotation period of 7.56 s were seen in this component, consistent with an origin far from the star surface and/or magnetosphere. This long-lasting emission can be interpreted as a hard X-ray afterglow produced by the fireball ejected in the initial spike of the giant flare (Mereghetti et al. 2005).

5.2 Magnetar Wind Nebulae

The identification and study of "magnetar wind nebulae" (MWNe) is complicated by the fact that most magnetars lie in crowded and highly absorbed regions of the Galactic plane and can be surrounded by diffuse emission of different origins, such as supernova remnants, molecular clouds, and H II regions. For example, SGR 1806-20 was initially associated with a variable radio nebula (Frail et al. 1997), but, when a better localization of this magnetar was obtained, it became apparent that a different object[14] is powering the radio emission. Other magnetars are located inside radio-emitting supernova remnants, but no signs of enhanced radio emission directly connected with the neutron star has been found. Therefore, there is no evidence up to now for diffuse emission produced by magnetars in the radio band, besides that of the transient nebulae associated to giant flares described above.

In the X-ray range, where diffuse emission has been seen around several magnetars, a further complication results from the effect of interstellar dust scattering. Remarkable evidence for the importance of this effect was demonstrated by the expanding X-ray rings (Fig. 43) seen around 1E 1547.0-5408 after its January 2009 outburst (Tiengo et al. 2010a).

[14]The luminous blue variable star LBV 1806-20 (Figer et al. 2004).

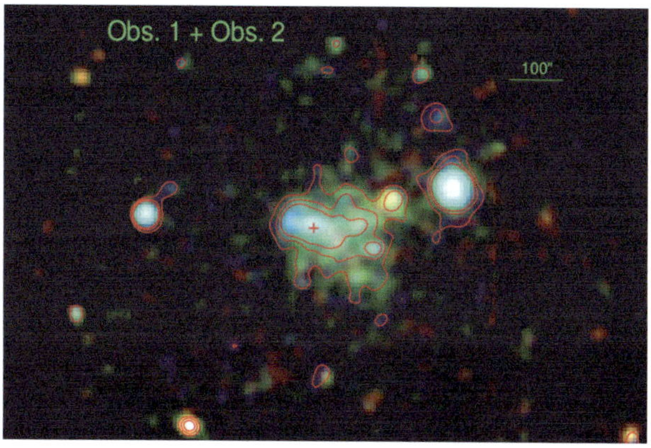

Fig. 44 X-ray image of the region of Swift J1834.9-0846 obtained with the *XMM-Newton* satellite (from Younes et al. 2016). The *cross* indicates the position of the magnetar. The *colors* code the X-ray energy (*red* 2–3 keV, *green* 3–4.5 keV, 4.5–10 keV *blue*). The *contours* are at the 2.5, 3.0, and 3.5 σ level. For a distance of 4 kpc, 100$''$ correspond to ~2 pc (*horizontal bar*)

Excess X-ray emission over the *XMM-Newton* PSF was detected at radii from ~10$''$ up to ~2$'$ around the transient SGR 1833-0832 (Esposito et al. 2011). Its spectrum was softer than that of the central source, as expected for a halo caused by interstellar dust, due to the E^{-2} dependence of the scattering cross section. Given the high absorption of SGR 1833-0832 ($N_H = 10^{23}$ cm^{-2}), a large amount of dust is likely present along its line of sight. Thus, the diffuse X-ray emission around this source, as well as that reported for other highly absorbed sources like SGR 1900+14 (Kouveliotou et al. 2001) and SGR 1806-20 (Kaplan et al. 2002; Viganò et al. 2014), is probably due to dust scattering. Recently, diffuse X-rays with a steep power-law spectrum have been detected around SGR J1935+2154, on an angular scale of about one arcmin with *XMM-Newton* (Israel et al. 2016; Younes et al. 2017). This could be a dust scattering halo, but a possible contribution from a wind nebula cannot be excluded.

To date, the best evidence for a MWN is provided by the case of Swift J1834.9-0846. This magnetar is surrounded by extended X-ray features with different spatial scales (Fig. 44). The X-ray emission within a radius of ~40$''$, resolved in *Chandra* images (Kargaltsev et al. 2012), is most likely due to dust scattering of the outburst emission of this transient. Its flux varied in correlation with that of the central source, as expected for such small angles in case of dust scattering (Esposito et al. 2013). The more extended X-ray emission, with an elongated shape (~2$'$ × 1$'$) and a constant flux detected with *XMM-Newton* in 2005, 2011 and 2014, is instead best explained as a MWN (Younes et al. 2012, 2016). In favour of this interpretation are the relatively hard spectrum of the nebula, which is well fit by a power law with photon index $\Gamma = 2.2$, and the apparent lack of variability. For a distance of 4 kpc, suggested by the possible association of Swift J1834.9-0846 with the supernova remnant W41, the nebula has an X-ray luminosity of $\sim 3 \times 10^{33}$ erg s^{-1}. This luminosity, which might be underestimated if the source is at a larger distance, represents a considerable fraction of the the spin-down power, $\dot{E}_{ROT} = 2.1 \times 10^{34}$ erg s^{-1}, of Swift J1834.9-0846.

We presented above the evidence for relativistic ejections during the giant flares, but, according to the magnetar model, the acceleration of particles in the magnetosphere is not restricted to these extremely energetic events. The normal bursting activity is expected to produce particle outflows, and Alfvén waves can drive a steady wind also during "quiescent" periods (Thompson and Duncan 1996; Harding et al. 1999). In this respect, it is more meaningful to compare L_{MWN} with the total reservoir of magnetic energy. For example, if we assume for Swift J1834.9-0846 the age of $\sim 10^5$ yr estimated for the W41 supernova remnant (Tian et al. 2007), the total energy radiated in X-rays in the nebula (a few 10^{38} erg) is a very small fraction of the magnetic energy $\sim R^2 \Delta R B^2 \sim 10^{45} B_{14}^2$ erg, if the field is in a crustal depth of $\Delta R \sim 1$ km (although probably only a small part of such energy may be available for powering a nebula).

Indeed, several authors recently proposed models in which the nebula of Swift J1834.9-0846 is magnetically-powered (Tong 2016; Granot et al. 2017). Considering the few observational data available up to now for this alleged MWN, there is considerable degeneracy in the numerous involved parameters and it is not surprising that such models led to rather different estimates for the PWN and magnetar properties. For example, the wind braking scenario of Tong (2016) invokes a flow of particles with luminosity between 10^{36} and 10^{38} erg s^{-1} and requires a rather high magnetic field in the nebula, $\sim 10^{-4}$ G, while Granot et al. (2017) estimate a likely upper limit of $\sim$30 μG. An alternative interpretation, in which the Swift J1834.9-0846 nebula is powered by rotational energy, like normal PWNe, has been proposed by Torres (2016). This is energetically possible if one considers the rotationally-powered wind injected over the whole lifetime of the magnetar and the reverberation effect due to the location in a particularly dense ambient medium.

6 Gamma-Ray Binaries: Pulsar Winds Interacting with a Massive Companion

The last decade has revealed a new group of gamma-ray emitters, composed of a fast-rotating pulsar and a massive star. The emission, which peaks in the MeV band, arises from the shocked region between the stellar wind and the pulsar wind. The binary interaction typically takes place around one AU from the pulsar, about 5 orders of magnitude closer than for pulsars interacting with the ISM. Although only a handful of these systems have been discovered, theoretical work benefits from the well-constrained environment created by the binary companion and the wealth of information provided by orbital variability. As such, gamma-ray binaries have opened a new window on pulsar wind physics. This subject is reviewed fully elsewhere.

7 Conclusions

The phenomenology of pulsar winds and their impact on their environments is a rich one, documented from radio to TeV photon energies. These objects pose various problems of particle acceleration and propagation, magnetic-field evolution, and neutron-star physics. Here we have summarized the spectra and morphology (primarily at X-ray and gamma-ray energies) of PWNe still in their natal shell supernova remnants, young PWNe which for some reason lack shells, PWNe from much older pulsars, interacting directly with the interstellar medium; and magnetars, with tantalizing but not yet definitive evidence for magnetar wind nebulae (MWNe).

Young PWNe exhibit their pulsars at near-birth properties, generally with high spin-down luminosities. The PWNe interact with expanding ejecta in SNR interiors, tending to produce fairly symmetric objects, at least compared to morphologies observed for much older objects. These objects radiate synchrotron radiation from radio through hard X-rays, whose SED contains important information on particle acceleration and propagation. The flat radio spectrum of these PWNe is still unexplained; the few anomalous PWNe with steep radio spectra have very unusual properties, such as very large ratios of radio to X-ray sizes. The spectra steepen in the mm–IR region of the spectrum, generally by larger amounts than can be accounted for by the simplest models.

NuSTAR observations above 10 keV have provided important new information on the X-ray properties of young PWNe. PWN sizes decrease with increasing photon energy; that energy-dependence encodes information on electron propagation (advective, diffusive, or a combination). It is still not clear if intrinsic spectral structure is required to explain the observations, or whether synchrotron losses in inhomogeneous sources with some combination of advection and diffusion can explain the observed size decrease and spectral steepening with increasing distance from the pulsar. *NuSTAR* observations also show unexpected slight steepening of X-ray spectra in the interiors of the Crab Nebula and G21.5-0.9. This effect may be a clue to some feature of particle energization near the pulsar-wind shock not yet explored.

The spatial resolution of *NuSTAR* allows the determination of anisotropic morphological changes with increasing photon energy that support strong asymmetry in the pulsar outflows beyond the termination shock. In the Crab Nebula, the energy-dependence of the torus radius is about what is predicted in the Kennel–Coroniti spherically symmetric model, while the NW counterjet length drops much more steeply with increasing energy, at a rate comparable to that seen in both MSH 15-52 and G21.5-0.9. This resemblance suggests a commonality of origin not yet explained. MHD models of young PWNe will need to have more accurate treatment of the spatial and temporal evolution of relativistic-particle distributions in order to confront these data effectively.

Pulsars that have left their host SNRs move in the ISM with supersonic velocities, which drastically changes PWN morphologies. In particular, long tails, with a typical length of a few parsecs, form behind the fast-moving pulsars. These tails, collimated by the ram pressure of the oncoming ISM, are fast outflows of the shocked PW emitting synchrotron radiation from radio to hard X-rays. In addition, PWN "heads" in the pulsar vicinity are seen in deep high-resolution X-ray images. The appearance of PWN heads, very different in different objects, depends on the direction of the pulsar velocity with respect to the line of sight, as well as on the inclinations of the velocity vector and the magnetic axis to the spin axis. The observed shapes of the heads differ significantly from the current PWN models. The main reason for these differences is likely the unrealistic model assumption that the unshocked PW is isotropic.

In addition to the PWN tails and heads, deep X-ray observations have revealed "misaligned outflows" in some PWNe, at large angles to the pulsar velocities. Such outflows are not predicted by the current PWN models, and their true nature remains puzzling.

The X-ray spectra of the head-tail PWNe are usually well described by power-law models, with typical photon indices $\Gamma \sim 1.5$–2.0 in the pulsar vicinity. This implies power-law spectral energy distributions of the X-ray emitting relativistic electrons/positrons, $dN_e/d\gamma \propto \gamma^{-p}$, with $p \sim 2$–3. Some PWNe show spectral softening with increasing distance from the pulsar, with $\Delta\Gamma \approx 1.0$–1.5, which can be due to synchrotron cooling. However, no softening is seen in other PWNe, which remains unexplained. Moreover, the lateral tails of the Geminga PWN have unusually hard spectra, $\Gamma \sim 0.7$–1.0, without measurable spectral softening, which might suggest an unusual acceleration mechanism for the radiating electrons.

Typical equipartition magnetic fields in head-tail PWNe are $\sim$10–100 μG. Interpreting the spectral softening observed in some tails as caused by synchrotron cooling, one can crudely estimate flow speeds $V_{\rm flow} \sim 10^4\ {\rm km\,s^{-1}}$, much faster than the pulsar speed but much slower than the mildly relativistic speeds predicted by some models.

Head-tail PWNe have been detected in X-rays only for sufficently powerful pulsars, $\dot{E} \gtrsim 10^{33}\ {\rm erg\,s^{-1}}$, likely because electrons cannot be accelerated to high enough energies in less powerful pulsars. Unexpectedly, the X-ray efficiency, $\eta_X = L_X/\dot{E}$, of head-tail PWNe shows a huge scatter, up to 4 orders of magnitude. Moreover, several nearby pulsars with $\dot{E} > 10^{34}\ \mathrm{erg\,s^{-1}}$ show either very faint PWNe or no PWN at all. This might be associated with small angles between the spin and magnetic axes, but the true reason remains unclear. To explain this and other puzzling properties of the X-ray PWNe created by supersonically moving pulsars, more such objects should be observed in X-rays and other spectral domains, with long exposures and high spatial resolution, and new, more realistic models should be developed.

Gamma-ray observations of PWNe have provided important insights on several aspects such as the spatial and spectral distributions of the high-energy particles, their total energy content and the wind magnetization. From a theoretical point of view, 3D time-dependent multi-zone spectro-morphological models [e.g. Van Etten and Romani (2011)] are required in order to grasp the processes of particle acceleration, transport and escape in these nebulae. From an observational point of view, the identification of these HE/VHE PWNe requires extensive multi-wavelength investigation and/or detailed spectro-morphological studies, only feasible for bright and resolved sources. In most cases, the association of a γ-ray source with counterparts at other wavelengths is uncertain.[15] In this regard, thanks to a factor $\sim$10 improvement in sensitivity above 100 GeV, with substantially better angular and spectral resolutions and wider field-of-view than the current IACTs, the Cherenkov Telescope Array (CTA; Acharya et al. 2013), currently in its pre-production phase, will have the potential to reveal hundreds of sources through a uniform Galactic Plane survey (Dubus et al. 2013). This will undoubtedly trigger detailed morphological and spectral investigations towards a large number of PWNe and meaningful population studies de Oña-Wilhelmi et al. (2013).

Acknowledgements SPR thanks the *NuSTAR* team for their outstanding work and permission to present their PWN results. GGP, OK, and NK are grateful to Andrei Bykov and Maxim Lyutikov for helpful discussions, and to Blagoy Rangelov and Bettina Posselt for their help in data analysis. Support for their work was provided by the National Aeronautics and Space Administration through *Chandra* Awards G03-14082 and G03-14057 issued by the *Chandra* X-ray Observatory Center, which is operated by the Smithsonian Astrophysical Observatory for and on behalf of the National Aeronautics Space Administration under contract NAS8-03060. The work was also partly supported by NASA grant NNX08AD71G. MR is grateful to Jamie Cohen and Marco Ajello for providing the *Fermi*-LAT 2FHL image towards HESS J1356-645 and to Marie-Hélène Grondin for the *Fermi*-LAT images of Vela X.

All of us warmly thank the organizers for the invitation to participate in this very stimulating ISSI workshop on *Jets and Winds in Pulsar Wind Nebulae, Gamma-ray Bursts and Blazars: Physics of Extreme Energy Release*, and the International Space Science Institute (ISSI) for support.

References

A.A. Abdo, M. Ackermann, M. Ajello, A. Allafort, L. Baldini, J. Ballet, G. Barbiellini, D. Bastieri, K. Bechtol, R. Bellazzini, B. Berenji, R.D. Blandford, E.D. Bloom, E. Bonamente, A.W. Borgland, A. Bouvier, J. Bregeon, A. Brez, M. Brigida, P. Bruel, T.H. Burnett, S. Buson, G.A. Caliandro, R.A. Cameron, P.A. Caraveo, S. Carrigan, J.M. Casandjian, C. Cecchi, Ö. Çelik, A. Chekhtman, C.C. Chung, J. Chiang, S. Ciprini, R. Claus, J. Cohen-Tanugi, J. Conrad, A. de Angelis, F. de Palma, M. Dormody, E.d.C.e. Silva, P.S. Drell, R. Dubois, D. Dumora, C. Farnier, C. Favuzzi, S.J. Fegan, W.B. Focke, P. Fortin, M. Frailis, Y. Fukazawa, S. Funk, P. Fusco, F. Gargano, N. Gehrels, S. Germani, G. Giavitto, N. Giglietto, F. Giordano, T. Glanzman, G. Godfrey, I.A. Grenier, M.-H. Grondin, J.E. Grove, L. Guillemot, S. Guiriec, A.K. Harding, M. Hayashida, E. Hays, D. Horan, R.E. Hughes, M.S. Jackson, G. Jóhannesson, A.S. Johnson, T.J. Johnson, W.N. Johnson, S. Johnston, T. Kamae, H. Katagiri, J. Kataoka, N. Kawai, M. Kerr, J. Knödlseder, M. Kuss, J. Lande, L. Latronico, S.-H. Lee, M. Lemoine-Goumard,

[15] As illustrated by HESS J1640-645 Abramowski et al. (2014) coincident with the radio SNR G338.3-0.0 hosting one compact and one extended X-ray source in its center which turned out to be an energetic pulsar (PSR J1640-4631, Gotthelf et al. 2014) and its wind nebula.

M. Llena Garde, F. Longo, F. Loparco, B. Lott, M.N. Lovellette, P. Lubrano, A. Makeev, M. Marelli, M.N. Mazziotta, J.E. McEnery, C. Meurer, P.F. Michelson, W. Mitthumsiri, T. Mizuno, A.A. Moiseev, C. Monte, M.E. Monzani, A. Morselli, I.V. Moskalenko, S. Murgia, T. Nakamori, P.L. Nolan, J.P. Norris, A. Noutsos, E. Nuss, T. Ohsugi, N. Omodei, E. Orlando, J.F. Ormes, M. Ozaki, D. Paneque, J.H. Panetta, D. Parent, V. Pelassa, M. Pepe, M. Pesce-Rollins, M. Pierbattista, F. Piron, T.A. Porter, S. Rainò, R. Rando, P.S. Ray, N. Rea, A. Reimer, O. Reimer, T. Reposeur, S. Ritz, A.Y. Rodriguez, R.W. Romani, M. Roth, F. Ryde, H.F.-W. Sadrozinski, D. Sanchez, A. Sander, P.M. Saz Parkinson, J.D. Scargle, C. Sgrò, E.J. Siskind, D.A. Smith, P.D. Smith, G. Spandre, P. Spinelli, M.S. Strickman, D.J. Suson, H. Tajima, H. Takahashi, T. Takahashi, T. Tanaka, J.B. Thayer, J.G. Thayer, D.J. Thompson, L. Tibaldo, D.F. Torres, G. Tosti, A. Tramacere, Y. Uchiyama, T.L. Usher, A. Van Etten, V. Vasileiou, C. Venter, N. Vilchez, V. Vitale, A.P. Waite, P. Wang, P. Weltevrede, B.L. Winer, K.S. Wood, T. Ylinen, M. Ziegler, Fermi Large Area Telescope observations of the Vela-X pulsar wind nebula. Astrophys. J. **713**, 146–153 (2010). doi:10.1088/0004-637X/713/1/146

A. Abramowski, F. Acero, F. Aharonian, A.G. Akhperjanian, G. Anton, S. Balenderan, A. Balzer, A. Barnacka, Y. Becherini, J. Becker Tjus, K. Bernlöhr, E. Birsin, J. Biteau, A. Bochow, C. Boisson, J. Bolmont, P. Bordas, J. Brucker, F. Brun, P. Brun, T. Bulik, S. Carrigan, S. Casanova, M. Cerruti, P.M. Chadwick, A. Charbonnier, R.C.G. Chaves, A. Cheesebrough, G. Cologna, J. Conrad, C. Couturier, M. Dalton, M.K. Daniel, I.D. Davids, B. Degrange, C. Deil, P. deWilt, H.J. Dickinson, A. Djannati-Ataï, W. Domainko, L.O.. Drury, F. Dubois, G. Dubus, K. Dutson, J. Dyks, M. Dyrda, K. Egberts, P. Eger, P. Espigat, L. Fallon, C. Farnier, S. Fegan, F. Feinstein, M.V. Fernandes, D. Fernandez, A. Fiasson, G. Fontaine, A. Förster, M. Füßling, M. Gajdus, Y.A. Gallant, T. Garrigoux, H. Gast, B. Giebels, J.F. Glicenstein, B. Glück, D. Göring, M.-H. Grondin, S. Häffner, J.D. Hague, J. Hahn, D. Hampf, J. Harris, S. Heinz, G. Heinzelmann, G. Henri, G. Hermann, A. Hillert, J.A. Hinton, W. Hofmann, P. Hofverberg, M. Holler, D. Horns, A. Jacholkowska, C. Jahn, M. Jamrozy, I. Jung, M.A. Kastendieck, K. Katarzyński, U. Katz, S. Kaufmann, B. Khélifi, D. Klochkov, W. Kluźniak, T. Kneiske, N. Komin, K. Kosack, R. Kossakowski, F. Krayzel, P.P. Krüger, H. Laffon, G. Lamanna, J.-P. Lenain, D. Lennarz, T. Lohse, A. Lopatin, C.-C. Lu, V. Marandon, A. Marcowith, J. Masbou, G. Maurin, N. Maxted, M. Mayer, T.J.L. McComb, M.C. Medina, J. Méhault, U. Menzler, R. Moderski, M. Mohamed, E. Moulin, C.L. Naumann, M. Naumann-Godo, M. de Naurois, D. Nedbal, N. Nguyen, J. Niemiec, S.J. Nolan, S. Ohm, E. de Oña Wilhelmi, B. Opitz, M. Ostrowski, I. Oya, M. Panter, D. Parsons, M. Paz Arribas, N.W. Pekeur, G. Pelletier, J. Perez, P.-O. Petrucci, B. Peyaud, S. Pita, G. Pühlhofer, M. Punch, A. Quirrenbach, M. Raue, A. Reimer, O. Reimer, M. Renaud, R. de los Reyes, F. Rieger, J. Ripken, L. Rob, S. Rosier-Lees, G. Rowell, B. Rudak, C.B. Rulten, V. Sahakian, D.A. Sanchez, A. Santangelo, R. Schlickeiser, A. Schulz, U. Schwanke, S. Schwarzburg, S. Schwemmer, F. Sheidaei, J.L. Skilton, H. Sol, G. Spengler, Ł. Stawarz, R. Steenkamp, C. Stegmann, F. Stinzing, K. Stycz, I. Sushch, A. Szostek, J.-P. Tavernet, R. Terrier, M. Tluczykont, C. Trichard, K. Valerius, C. van Eldik, G. Vasileiadis, C. Venter, A. Viana, P. Vincent, H.J. Völk, F. Volpe, S. Vorobiov, M. Vorster, S.J. Wagner, M. Ward, R. White, A. Wierzcholska, D. Wouters, M. Zacharias, A. Zajczyk, A.A. Zdziarski, A. Zech, H.-S. Zechlin, Probing the extent of the non-thermal emission from the Vela X region at TeV energies with H.E.S.S. Astron. Astrophys. **548**, 38 (2012). doi:10.1051/0004-6361/201219919

A. Abramowski, F. Aharonian, F.A. Benkhali, A.G. Akhperjanian, E. Angüner, G. Anton, S. Balenderan, A. Balzer, A. Barnacka, Y. Becherini, et al., HESS J1640-465—an exceptionally luminous TeV γ-ray supernova remnant. Mon. Not. R. Astron. Soc. **439**, 2828–2836 (2014). doi:10.1093/mnras/stu139

F. Acero, M. Ackermann, M. Ajello, A. Allafort, L. Baldini, J. Ballet, G. Barbiellini, D. Bastieri, K. Bechtol, R. Bellazzini, R.D. Blandford, E.D. Bloom, E. Bonamente, E. Bottacini, T.J. Brandt, J. Bregeon, M. Brigida, P. Bruel, R. Buehler, S. Buson, G.A. Caliandro, R.A. Cameron, P.A. Caraveo, C. Cecchi, E. Charles, R.C.G. Chaves, A. Chekhtman, J. Chiang, G. Chiaro, S. Ciprini, R. Claus, J. Cohen-Tanugi, J. Conrad, S. Cutini, M. Dalton, F. D'Ammando, F. de Palma, C.D. Dermer, L. Di Venere, E.d.C.e. Silva, P.S. Drell, A. Drlica-Wagner, L. Falletti, C. Favuzzi, S.J. Fegan, E.C. Ferrara, W.B. Focke, A. Franckowiak, Y. Fukazawa, S. Funk, P. Fusco, F. Gargano, D. Gasparrini, N. Giglietto, F. Giordano, M. Giroletti, T. Glanzman, G. Godfrey, T. Grégoire, I.A. Grenier, M.-H. Grondin, J.E. Grove, S. Guiriec, D. Hadasch, Y. Hanabata, A.K. Harding, M. Hayashida, K. Hayashi, E. Hays, J. Hewitt, A.B. Hill, D. Horan, X. Hou, R.E. Hughes, Y. Inoue, M.S. Jackson, T. Jogler, G. Jóhannesson, A.S. Johnson, T. Kamae, T. Kawano, M. Kerr, J. Knödlseder, M. Kuss, J. Lande, S. Larsson, L. Latronico, M. Lemoine-Goumard, F. Longo, F. Loparco, M.N. Lovellette, P. Lubrano, M. Marelli, F. Massaro, M. Mayer, M.N. Mazziotta, J.E. McEnery, J. Mehault, P.F. Michelson, W. Mitthumsiri, T. Mizuno, C. Monte, M.E. Monzani, A. Morselli, I.V. Moskalenko, S. Murgia, T. Nakamori, R. Nemmen, E. Nuss, T. Ohsugi, A. Okumura, M. Orienti, E. Orlando, J.F. Ormes, D. Paneque, J.H. Panetta, J.S. Perkins, M. Pesce-Rollins, F. Piron, G. Pivato, T.A. Porter, S. Rainò, R. Rando, M. Razzano, A. Reimer, O. Reimer, T. Reposeur, S. Ritz, M. Roth, R. Rousseau, P.M. Saz Parkinson, A. Schulz, C. Sgrò, E.J. Siskind, D.A. Smith, G. Spandre, P. Spinelli, D.J. Suson, H. Takahashi, Y. Takeuchi, J.G. Thayer, J.B. Thayer,

D.J. Thompson, L. Tibaldo, O. Tibolla, M. Tinivella, D.F. Torres, G. Tosti, E. Troja, Y. Uchiyama, J. Vandenbroucke, V. Vasileiou, G. Vianello, V. Vitale, M. Werner, B.L. Winer, K.S. Wood, Z. Yang, Constraints on the galactic population of TeV pulsar wind nebulae using Fermi Large Area Telescope observations. Astrophys. J. **773**, 77 (2013). doi:10.1088/0004-637X/773/1/77

F. Acero, M. Ackermann, M. Ajello, A. Albert, W.B. Atwood, M. Axelsson, L. Baldini, J. Ballet, G. Barbiellini, D. Bastieri, A. Belfiore, R. Bellazzini, E. Bissaldi, R.D. Blandford, E.D. Bloom, J.R. Bogart, R. Bonino, E. Bottacini, J. Bregeon, R.J. Britto, P. Bruel, R. Buehler, T.H. Burnett, S. Buson, G.A. Caliandro, R.A. Cameron, R. Caputo, M. Caragiulo, P.A. Caraveo, J.M. Casandjian, E. Cavazzuti, E. Charles, R.C.G. Chaves, A. Chekhtman, C.C. Cheung, J. Chiang, G. Chiaro, S. Ciprini, R. Claus, J. Cohen-Tanugi, L.R. Cominsky, J. Conrad, S. Cutini, F. D'Ammando, A. de Angelis, M. DeKlotz, F. de Palma, R. Desiante, S.W. Digel, L. Di Venere, P.S. Drell, R. Dubois, D. Dumora, C. Favuzzi, S.J. Fegan, E.C. Ferrara, J. Finke, A. Franckowiak, Y. Fukazawa, S. Funk, P. Fusco, F. Gargano, D. Gasparrini, B. Giebels, N. Giglietto, P. Giommi, F. Giordano, M. Giroletti, T. Glanzman, G. Godfrey, I.A. Grenier, M.-H. Grondin, J.E. Grove, L. Guillemot, S. Guiriec, D. Hadasch, A.K. Harding, E. Hays, J.W. Hewitt, A.B. Hill, D. Horan, G. Iafrate, T. Jogler, G. Jóhannesson, R.P. Johnson, A.S. Johnson, T.J. Johnson, W.N. Johnson, T. Kamae, J. Kataoka, J. Katsuta, M. Kuss, G. La Mura, D. Landriu, S. Larsson, L. Latronico, M. Lemoine-Goumard, J. Li, L. Li, F. Longo, F. Loparco, B. Lott, M.N. Lovellette, P. Lubrano, G.M. Madejski, F. Massaro, M. Mayer, M.N. Mazziotta, J.E. McEnery, P.F. Michelson, N. Mirabal, T. Mizuno, A.A. Moiseev, M. Mongelli, M.E. Monzani, A. Morselli, I.V. Moskalenko, S. Murgia, E. Nuss, M. Ohno, T. Ohsugi, N. Omodei, M. Orienti, E. Orlando, J.F. Ormes, D. Paneque, J.H. Panetta, J.S. Perkins, M. Pesce-Rollins, F. Piron, G. Pivato, T.A. Porter, J.L. Racusin, R. Rando, M. Razzano, S. Razzaque, A. Reimer, O. Reimer, T. Reposeur, L.S. Rochester, R.W. Romani, D. Salvetti, M. Sánchez-Conde, P.M. Saz Parkinson, A. Schulz, E.J. Siskind, D.A. Smith, F. Spada, G. Spandre, P. Spinelli, T.E. Stephens, A.W. Strong, D.J. Suson, H. Takahashi, T. Takahashi, Y. Tanaka, J.G. Thayer, J.B. Thayer, D.J. Thompson, L. Tibaldo, O. Tibolla, D.F. Torres, E. Torresi, G. Tosti, E. Troja, B. Van Klaveren, G. Vianello, B.L. Winer, K.S. Wood, M. Wood, S. Zimmer, Fermi-LAT Collaboration, Fermi Large Area Telescope third source catalog. Astrophys. J. Suppl. Ser. **218**, 23 (2015). doi:10.1088/0067-0049/218/2/23

B.S. Acharya, M. Actis, T. Aghajani, G. Agnetta, J. Aguilar, F. Aharonian, M. Ajello, A. Akhperjanian, M. Alcubierre, J. Aleksić, et al., Introducing the CTA concept. Astropart. Phys. **43**, 3–18 (2013). doi:10.1016/j.astropartphys.2013.01.007

M. Ackermann, M. Ajello, W.B. Atwood, L. Baldini, J. Ballet, G. Barbiellini, D. Bastieri, J. Becerra Gonzalez, R. Bellazzini, E. Bissaldi, R.D. Blandford, E.D. Bloom, R. Bonino, E. Bottacini, T.J. Brandt, J. Bregeon, P. Bruel, R. Buehler, S. Buson, G.A. Caliandro, R.A. Cameron, R. Caputo, M. Caragiulo, P.A. Caraveo, E. Cavazzuti, C. Cecchi, E. Charles, A. Chekhtman, C.C. Cheung, J. Chiang, G. Chiaro, S. Ciprini, J.M. Cohen, J. Cohen-Tanugi, L.R. Cominsky, J. Conrad, A. Cuoco, S. Cutini, F. D'Ammando, A. de Angelis, F. de Palma, R. Desiante, M. Di Mauro, L. Di Venere, A. Domínguez, P.S. Drell, C. Favuzzi, S.J. Fegan, E.C. Ferrara, W.B. Focke, P. Fortin, A. Franckowiak, Y. Fukazawa, S. Funk, A.K. Furniss, P. Fusco, F. Gargano, D. Gasparrini, N. Giglietto, P. Giommi, F. Giordano, M. Giroletti, T. Glanzman, G. Godfrey, I.A. Grenier, M.-H. Grondin, L. Guillemot, S. Guiriec, A.K. Harding, E. Hays, J.W. Hewitt, A.B. Hill, D. Horan, G. Iafrate, D. Hartmann, T. Jogler, G. Jóhannesson, A.S. Johnson, T. Kamae, J. Kataoka, J. Knödlseder, M. Kuss, G. La Mura, S. Larsson, L. Latronico, M. Lemoine-Goumard, J. Li, L. Li, F. Longo, F. Loparco, B. Lott, M.N. Lovellette, P. Lubrano, G.M. Madejski, S. Maldera, A. Manfreda, M. Mayer, M.N. Mazziotta, P.F. Michelson, N. Mirabal, W. Mitthumsiri, T. Mizuno, A.A. Moiseev, M.E. Monzani, A. Morselli, I.V. Moskalenko, S. Murgia, E. Nuss, T. Ohsugi, N. Omodei, M. Orienti, E. Orlando, J.F. Ormes, D. Paneque, J.S. Perkins, M. Pesce-Rollins, V. Petrosian, F. Piron, G. Pivato, T.A. Porter, S. Rainò, R. Rando, M. Razzano, S. Razzaque, A. Reimer, O. Reimer, T. Reposeur, R.W. Romani, M. Sánchez-Conde, P.M. Saz Parkinson, J. Schmid, A. Schulz, C. Sgrò, E.J. Siskind, F. Spada, G. Spandre, P. Spinelli, D.J. Suson, H. Tajima, H. Takahashi, M. Takahashi, T. Takahashi, J.B. Thayer, D.J. Thompson, L. Tibaldo, D.F. Torres, G. Tosti, E. Troja, G. Vianello, K.S. Wood, M. Wood, M. Yassine, G. Zaharijas, S. Zimmer, 2FHL: the second catalog of hard Fermi-LAT sources. Astrophys. J. Suppl. Ser. **222**, 5 (2016). doi:10.3847/0067-0049/222/1/5

M. Aguilar, D. Aisa, A. Alvino, G. Ambrosi, K. Andeen, L. Arruda, N. Attig, P. Azzarello, A. Bachlechner, F. Barao, et al., Electron and positron fluxes in primary cosmic rays measured with the Alpha Magnetic Spectrometer on the International Space Station. Phys. Rev. Lett. **113**(12), 121102 (2014a). doi:10.1103/PhysRevLett.113.121102

M. Aguilar, D. Aisa, B. Alpat, A. Alvino, G. Ambrosi, K. Andeen, L. Arruda, N. Attig, P. Azzarello, A. Bachlechner, et al., Precision measurement of the (e^{+}+e^{-}) flux in primary cosmic rays from 0.5 GeV to 1 TeV with the Alpha Magnetic Spectrometer on the International Space Station. Phys. Rev. Lett. **113**(22), 221102 (2014b). doi:10.1103/PhysRevLett.113.221102

F. Aharonian, A.G. Akhperjanian, K.-M. Aye, A.R. Bazer-Bachi, M. Beilicke, W. Benbow, D. Berge, P. Berghaus, K. Bernlöhr, C. Boisson, O. Bolz, I. Braun, F. Breitling, A.M. Brown, J. Bussons Gordo, P.M. Chadwick, L.-M. Chounet, R. Cornils, L. Costamante, B. Degrange, A. Djannati-Ataï, L. O'C. Drury, G. Dubus, D. Emmanoulopoulos, P. Espigat, F. Feinstein, P. Fleury, G. Fontaine, Y. Fuchs, S. Funk, Y.A. Gallant, B. Giebels, S. Gillessen, J.F. Glicenstein, P. Goret, C. Hadjichristidis, M. Hauser, G. Heinzelmann, G. Henri, G. Hermann, J.A. Hinton, W. Hofmann, M. Holleran, D. Horns, O.C. de Jager, B. Khélifi, N. Komin, A. Konopelko, I.J. Latham, R. Le Gallou, A. Lemière, M. Lemoine-Goumard, N. Leroy, T. Lohse, O. Martineau-Huynh, A. Marcowith, C. Masterson, T.J.L. McComb, M. de Naurois, S.J. Nolan, A. Noutsos, K.J. Orford, J.L. Osborne, M. Ouchrif, M. Panter, G. Pelletier, S. Pita, G. Pühlhofer, M. Punch, B.C. Raubenheimer, M. Raue, J. Raux, S.M. Rayner, I. Redondo, A. Reimer, O. Reimer, J. Ripken, L. Rob, L. Rolland, G. Rowell, V. Sahakian, L. Saugé, S. Schlenker, R. Schlickeiser, C. Schuster, U. Schwanke, M. Siewert, H. Sol, R. Steenkamp, C. Stegmann, J.-P. Tavernet, R. Terrier, C.G. Théoret, M. Tluczykont, G. Vasileiadis, C. Venter, P. Vincent, H.J. Völk, S.J. Wagner, Discovery of extended VHE gamma-ray emission from the asymmetric pulsar wind nebula in MSH 15–52 with HESS. Astron. Astrophys. **435**, 17–20 (2005). doi:10.1051/0004-6361:200500105

F. Aharonian, A.G. Akhperjanian, A.R. Bazer-Bachi, M. Beilicke, W. Benbow, D. Berge, K. Bernlöhr, C. Boisson, O. Bolz, V. Borrel, I. Braun, A.M. Brown, R. Bühler, I. Büsching, S. Carrigan, P.M. Chadwick, L.-M. Chounet, R. Cornils, L. Costamante, B. Degrange, H.J. Dickinson, A. Djannati-Ataï, L. O'C. Drury, G. Dubus, K. Egberts, D. Emmanoulopoulos, P. Espigat, F. Feinstein, E. Ferrero, A. Fiasson, G. Fontaine, S. Funk, S. Funk, M. Füßling, Y.A. Gallant, B. Giebels, J.F. Glicenstein, P. Goret, C. Hadjichristidis, D. Hauser, M. Hauser, G. Heinzelmann, G. Henri, G. Hermann, J.A. Hinton, A. Hoffmann, W. Hofmann, M. Holleran, D. Horns, A. Jacholkowska, O.C. de Jager, E. Kendziorra, B. Khélifi, N. Komin, A. Konopelko, K. Kosack, I.J. Latham, R. Le Gallou, A. Lemière, M. Lemoine-Goumard, T. Lohse, J.M. Martin, O. Martineau-Huynh, A. Marcowith, C. Masterson, G. Maurin, T.J.L. McComb, E. Moulin, M. de Naurois, D. Nedbal, S.J. Nolan, A. Noutsos, K.J. Orford, J.L. Osborne, M. Ouchrif, M. Panter, G. Pelletier, S. Pita, G. Pühlhofer, M. Punch, B.C. Raubenheimer, M. Raue, S.M. Rayner, A. Reimer, O. Reimer, J. Ripken, L. Rob, L. Rolland, G. Rowell, V. Sahakian, A. Santangelo, L. Saugé, S. Schlenker, R. Schlickeiser, R. Schröder, U. Schwanke, S. Schwarzburg, A. Shalchi, H. Sol, D. Spangler, F. Spanier, R. Steenkamp, C. Stegmann, G. Superina, J.-P. Tavernet, R. Terrier, C.G. Théoret, M. Tluczykont, C. van Eldik, G. Vasileiadis, C. Venter, P. Vincent, H.J. Völk, S.J. Wagner, M. Ward, Energy dependent γ-ray morphology in the pulsar wind nebula HESS J1825-137. Astron. Astrophys. **460**, 365–374 (2006a). doi:10.1051/0004-6361:20065546

F. Aharonian, A.G. Akhperjanian, A.R. Bazer-Bachi, M. Beilicke, W. Benbow, D. Berge, K. Bernlöhr, C. Boisson, O. Bolz, V. Borrel, I. Braun, F. Breitling, A.M. Brown, R. Bühler, I. Büsching, S. Carrigan, P.M. Chadwick, L.-M. Chounet, R. Cornils, L. Costamante, B. Degrange, H.J. Dickinson, A. Djannati-Ataï, L. O'C. Drury, G. Dubus, K. Egberts, D. Emmanoulopoulos, B. Epinat, P. Espigat, F. Feinstein, E. Ferrero, G. Fontaine, S. Funk, S. Funk, Y.A. Gallant, B. Giebels, J.F. Glicenstein, P. Goret, C. Hadjichristidis, D. Hauser, M. Hauser, G. Heinzelmann, G. Henri, G. Hermann, J.A. Hinton, W. Hofmann, M. Holleran, D. Horns, A. Jacholkowska, O.C. de Jager, B. Khélifi, N. Komin, A. Konopelko, I.J. Latham, R. Le Gallou, A. Lemière, M. Lemoine-Goumard, T. Lohse, J.M. Martin, O. Martineau-Huynh, A. Marcowith, C. Masterson, T.J.L. McComb, M. de Naurois, D. Nedbal, S.J. Nolan, A. Noutsos, K.J. Orford, J.L. Osborne, M. Ouchrif, M. Panter, G. Pelletier, S. Pita, G. Pühlhofer, M. Punch, B.C. Raubenheimer, M. Raue, S.M. Rayner, A. Reimer, O. Reimer, J. Ripken, L. Rob, L. Rolland, G. Rowell, V. Sahakian, L. Saugé, S. Schlenker, R. Schlickeiser, U. Schwanke, H. Sol, D. Spangler, F. Spanier, R. Steenkamp, C. Stegmann, G. Superina, J.-P. Tavernet, R. Terrier, C.G. Théoret, M. Tluczykont, C. van Eldik, G. Vasileiadis, C. Venter, P. Vincent, H.J. Völk, S.J. Wagner, M. Ward, First detection of a VHE gamma-ray spectral maximum from a cosmic source: HESS discovery of the Vela X nebula. Astron. Astrophys. **448**, 43–47 (2006b). doi:10.1051/0004-6361:200600014

E. Amato, Particle acceleration and radiation in Pulsar Wind Nebulae. ArXiv e-prints (2015)

H. An, K.K. Madsen, S.P. Reynolds, V.M. Kaspi, F.A. Harrison, S.E. Boggs, F.E. Christensen, W.W. Craig, C.L. Fryer, B.W. Grefenstette, C.J. Hailey, K. Mori, D. Stern, W.W. Zhang, High-energy X-ray imaging of the pulsar wind nebula MSH 15-52: constraints on particle acceleration and transport. Astrophys. J. **793**, 90 (2014). doi:10.1088/0004-637X/793/2/90

Z. Arzoumanian, D.F. Chernoff, J.M. Cordes, The velocity distribution of isolated radio pulsars. Astrophys. J. **568**, 289–301 (2002). doi:10.1086/338805

Z. Arzoumanian, E.V. Gotthelf, S.M. Ransom, S. Safi-Harb, R. Kothes, T.L. Landecker, Discovery of an energetic pulsar associated with SNR G76.9+1.0. Astrophys. J. **739**, 39 (2011). doi:10.1088/0004-637X/739/1/39

K. Auchettl, P. Slane, R.W. Romani, B. Posselt, G.G. Pavlov, O. Kargaltsev, C.-Y. Ng, T. Temim, M.C. Weisskopf, A. Bykov, D.A. Swartz, X-ray analysis of the proper motion and pulsar wind nebula for PSR J1741-2054. Astrophys. J. **802**, 68 (2015). doi:10.1088/0004-637X/802/1/68

R. Bandiera, On the X-ray feature associated with the Guitar Nebula. Astron. Astrophys. **490**, 3–6 (2008). doi:10.1051/0004-6361:200810666
R. Bandiera, F. Bocchino, The X-ray halo of G21.5-0.9. Adv. Space Res. **33**, 398–402 (2004). doi:10.1016/j.asr.2003.04.025
M.F. Bietenholz, N. Bartel, The expansion and radio spectral index of G21.5-0.9: is PSR J1833-1034 the youngest pulsar? Mon. Not. R. Astron. Soc. **386**, 1411–1416 (2008). doi:10.1111/j.1365-2966.2008.13058.x
L. Bîrzan, G.G. Pavlov, O. Kargaltsev, Chandra observations of the elusive pulsar wind nebula around PSR B0656+14. Astrophys. J. **817**, 129 (2016). doi:10.3847/0004-637X/817/2/129
E.L. Blanton, D.J. Helfand, ASCA observations of the composite supernova remnant G29.7-0.3. Astrophys. J. **470**, 961 (1996). doi:10.1086/177922
J.M. Blondin, R.A. Chevalier, D.M. Frierson, Pulsar wind nebulae in evolved supernova remnants. Astrophys. J. **563**, 806–815 (2001). doi:10.1086/324042
F. Bocchino, Detection of thermal X-ray emission in the halo of the plerionic supernova remnant G21.5-0.9. Adv. Space Res. **35**, 1003–1006 (2005). doi:10.1016/j.asr.2005.01.045
F. Bocchino, A.M. Bykov, The plerion nebula in IC 443: the XMM-Newton view. Astron. Astrophys. **376**, 248–253 (2001). doi:10.1051/0004-6361:20010882
S.E. Boggs, A. Zoglauer, E. Bellm, K. Hurley, R.P. Lin, D.M. Smith, C. Wigger, W. Hajdas, The giant flare of 2004 December 27 from SGR 1806-20. Astrophys. J. **661**, 458–467 (2007)
K.J. Borkowski, S.P. Reynolds, M.S.E. Roberts, G11.2-0.3: The young remnant of a stripped-envelope supernova. Astrophys. J. **819**, 160 (2016). doi:10.3847/0004-637X/819/2/160
M. Boudaud, S. Aupetit, S. Caroff, A. Putze, G. Belanger, Y. Genolini, C. Goy, V. Poireau, V. Poulin, S. Rosier, P. Salati, L. Tao, M. Vecchi, A new look at the cosmic ray positron fraction. Astron. Astrophys. **575**, 67 (2015). doi:10.1051/0004-6361/201425197
N. Bucciantini, E. Amato, L. Del Zanna, Relativistic MHD simulations of pulsar bow-shock nebulae. Astron. Astrophys. **434**, 189–199 (2005). doi:10.1051/0004-6361:20042205
N. Bucciantini, J. Arons, E. Amato, Modelling spectral evolution of pulsar wind nebulae inside supernova remnants. Mon. Not. R. Astron. Soc. **410**, 381–398 (2011). doi:10.1111/j.1365-2966.2010.17449.x
A.M. Bykov, F.B.S. Paerels, V. Petrosian, Equilibration processes in the warm–hot intergalactic medium. Space Sci. Rev. **134**, 141–153 (2008). doi:10.1007/s11214-008-9309-4
A. Bykov, N. Gehrels, H. Krawczynski, M. Lemoine, G. Pelletier, M. Pohl, Particle acceleration in relativistic outflows. Space Sci. Rev. **173**, 309–339 (2012). doi:10.1007/s11214-012-9896-y
P.B. Cameron, P. Chandra, A. Ray, S.R. Kulkarni, D.A. Frail, M.H. Wieringa, E. Nakar, E.S. Phinney, A. Miyazaki, M. Tsuboi, S. Okumura, N. Kawai, K.M. Menten, F. Bertoldi, Detection of a radio counterpart to the 27 December 2004 giant flare from SGR 1806-20. Nature **434**, 1112–1115 (2005). doi:10.1038/nature03605
F. Camilo, R.N. Manchester, B.M. Gaensler, D.R. Lorimer, Heartbeat of the mouse: a young radio pulsar associated with the axisymmetric nebula G359.23-0.82. Astrophys. J. Lett. **579**, 25–28 (2002). doi:10.1086/344832
P.A. Caraveo, G.F. Bignami, A. de Luca, S. Mereghetti, A. Pellizzoni, R. Mignani, A. Tur, W. Becker, Geminga's tails: a pulsar bow shock probing the interstellar medium. Science **301**, 1345–1348 (2003). doi:10.1126/science.1086973
S. Chatterjee, J.M. Cordes, Bow shocks from neutron stars: scaling laws and Hubble Space Telescope observations of the Guitar Nebula. Astrophys. J. **575**, 407–418 (2002). doi:10.1086/341139
R.A. Chevalier, Young core-collapse supernova remnants and their supernovae. Astrophys. J. **619**, 839–855 (2005). doi:10.1086/426584
J.M. Cordes, R.W. Romani, S.C. Lundgren, The Guitar Nebula—a bow shock from a slow-spin, high-velocity neutron star. Nature **362**, 133–135 (1993). doi:10.1038/362133a0
O.C. de Jager, A. Djannati-Ataï, Implications of HESS observations of pulsar wind nebulae, in *Astrophysics and Space Science Library*, vol. 357 (2009), p. 451 doi:10.1007/978-3-540-76965-1_17
O.C. de Jager, S.E.S. Ferreira, A. Djannati-Ataï, M. Dalton, C. Deil, K. Kosack, M. Renaud, U. Schwanke, O. Tibolla, Unidentified Gamma-Ray Sources as Ancient Pulsar Wind Nebulae. ArXiv e-prints (2009)
A. de Luca, M. Marelli, R.P. Mignani, P.A. Caraveo, W. Hummel, S. Collins, A. Shearer, P.M. Saz Parkinson, A. Belfiore, G.F. Bignami, Discovery of a faint X-ray counterpart and a parsec-long X-ray tail for the middle-aged, γ-ray-only pulsar PSR J0357+3205. Astrophys. J. **733**, 104 (2011). doi:10.1088/0004-637X/733/2/104
A. de Luca, R.P. Mignani, M. Marelli, D. Salvetti, N. Sartore, A. Belfiore, P. Saz Parkinson, P.A. Caraveo, G.F. Bignami, PSR J0357+3205: a fast-moving pulsar with a very unusual X-ray trail. Astrophys. J. Lett. **765**, 19 (2013). doi:10.1088/2041-8205/765/1/L19
E. de Oña-Wilhelmi, B. Rudak, J.A. Barrio, J.L. Contreras, Y. Gallant, D. Hadasch, T. Hassan, M. Lopez, D. Mazin, N. Mirabal, G. Pedaletti, M. Renaud, R. de los Reyes, D.F. Torres, CTA Consortium,

Prospects for observations of pulsars and pulsar wind nebulae with CTA. Astropart. Phys. **43**, 287–300 (2013). doi:10.1016/j.astropartphys.2012.08.009

A. de Rosa, P. Ubertini, R. Campana, A. Bazzano, A.J. Dean, L. Bassani, Hard X-ray observations of PSR J1833-1034 and its associated pulsar wind nebula. Mon. Not. R. Astron. Soc. **393**, 527–530 (2009). doi:10.1111/j.1365-2966.2008.14160.x

L. Del Zanna, E. Amato, N. Bucciantini, Axially symmetric relativistic MHD simulations of pulsar wind nebulae in supernova remnants. On the origin of torus and jet-like features. Astron. Astrophys. **421**, 1063–1073 (2004). doi:10.1051/0004-6361:20035936

S. Della Torre, M. Gervasi, P.G. Rancoita, D. Rozza, A. Treves, Pulsar wind nebulae as a source of the observed electron and positron excess at high energy: the case of Vela-X. J. High Energy Astrophys. **8**, 27–34 (2015). doi:10.1016/j.jheap.2015.08.001

M. Di Mauro, F. Donato, N. Fornengo, R. Lineros, A. Vittino, Interpretation of AMS-02 electrons and positrons data. J. Cosmol. Astropart. Phys. **4**, 006 (2014). doi:10.1088/1475-7516/2014/04/006

I. Di Palma, D. Guetta, E. Amato, Revised predictions of neutrino fluxes from Pulsar Wind Nebulae. ArXiv e-prints (2016)

A. Djannati-Ataï, O.C. deJager, R. Terrier, Y.A. Gallant, S. Hoppe, New companions for the lonely Crab? VHE emission from young pulsar wind nebulae revealed by H.E.S.S., in *ICRC*, vol. 2 (2008), pp. 823–826

T. Dolch, S. Chatterjee, D.P. Clemens, J.M. Cordes, L.R. Cashmen, B.W. Taylor, Recent H-alpha results on pulsar B2224+65's bow-shock nebula, the "Guitar". J. Astron. Space Sci. **33**, 167–172 (2016). doi:10.5140/JASS.2016.33.3.167

G. Dubus, J.L. Contreras, S. Funk, Y. Gallant, T. Hassan, J. Hinton, Y. Inoue, J. Knödlseder, P. Martin, N. Mirabal, M. de Naurois, M. Renaud, CTA Consortium, Surveys with the Cherenkov Telescope Array. Astropart. Phys. **43**, 317–330 (2013). doi:10.1016/j.astropartphys.2012.05.020

P. Esposito, G.L. Israel, R. Turolla, F. Mattana, A. Tiengo, A. Possenti, S. Zane, N. Rea, M. Burgay, D. Götz, S. Mereghetti, L. Stella, M.H. Wieringa, J.M. Sarkissian, T. Enoto, P. Romano, T. Sakamoto, Y.E. Nakagawa, K. Makishima, K. Nakazawa, H. Nishioka, C. François-Martin, Long-term spectral and timing properties of the soft gamma-ray repeater SGR 1833-0832 and detection of extended X-ray emission around the radio pulsar PSR B1830-08. Mon. Not. R. Astron. Soc. **416**, 205–215 (2011). doi:10.1111/j.1365-2966.2011.19022.x

P. Esposito, A. Tiengo, N. Rea, R. Turolla, A. Fenzi, A. Giuliani, G.L. Israel, S. Zane, S. Mereghetti, A. Possenti, M. Burgay, L. Stella, D. Götz, R. Perna, R.P. Mignani, P. Romano, X-ray and radio observations of the magnetar Swift J1834.9-0846 and its dust-scattering halo. Mon. Not. R. Astron. Soc. **429**, 3123–3132 (2013). doi:10.1093/mnras/sts569

J. Faherty, F.M. Walter, J. Anderson, The trigonometric parallax of the neutron star Geminga. Astrophys. Space Sci. **308**, 225–230 (2007). doi:10.1007/s10509-007-9368-0

J. Fang, L. Zhang, Multiband emission from pulsar wind nebulae: a possible injection spectrum. Astron. Astrophys. **515**, 20 (2010). doi:10.1051/0004-6361/200913615

D.F. Figer, F. Najarro, R.P. Kudritzki, The double-lined spectrum of LBV 1806-20. Astrophys. J. Lett. **610**, 109–112 (2004). doi:10.1086/423306

D.A. Frail, S.R. Kulkarni, J.S. Bloom, An outburst of relativistic particles from the soft γ-ray repeater SGR1900+14. Nature **398**, 127–129 (1999). doi:10.1038/18163

D.A. Frail, G. Vasisht, S.R. Kulkarni, The changing structure of the radio nebula around the soft gamma-ray repeater SGR 1806-20. Astrophys. J. Lett. **480**, 129–132 (1997). doi:10.1086/310635

D.D. Frederiks, S.V. Golenetskii, V.D. Palshin, R.L. Aptekar, V.N. Ilyinskii, F.P. Oleinik, E.P. Mazets, T.L. Cline, Giant flare in SGR 1806-20 and its Compton reflection from the Moon. Astron. Lett. **33**, 1–18 (2007). doi:10.1134/S106377370701001X

E. Fuerst, T. Handa, K. Morita, P. Reich, W. Reich, Y. Sofue, Detection of axisymmetric filaments in the filled-center supernova remnant G21.5-0.9. Publ. Astron. Soc. Jpn. **40**, 347–356 (1988)

S. Funk, J.A. Hinton, O.C. de Jager, Energy dependent morphology in the PWN candidate HESS J1825-137, in *ICRC*, vol. 2 (2008), pp. 605–608

B.M. Gaensler, P.O. Slane, The evolution and structure of pulsar wind nebulae. Annu. Rev. Astron. Astrophys. **44**, 17–47 (2006). doi:10.1146/annurev.astro.44.051905.092528

B.M. Gaensler, B.J. Wallace, A multifrequency radio study of supernova remnant G292.0+1.8 and its pulsar wind nebula. Astrophys. J. **594**, 326–339 (2003). doi:10.1086/376861

B.M. Gaensler, K.T.S. Brazier, R.N. Manchester, S. Johnston, A.J. Green, SNR G320.4-01.2 and PSR B1509-58: new radio observations of a complex interacting system. Mon. Not. R. Astron. Soc. **305**, 724–736 (1999). doi:10.1046/j.1365-8711.1999.02500.x

B.M. Gaensler, J. Arons, V.M. Kaspi, M.J. Pivovaroff, N. Kawai, K. Tamura, Chandra imaging of the X-ray nebula powered by pulsar B1509-58. Astrophys. J. **569**, 878–893 (2002). doi:10.1086/339354

B.M. Gaensler, E. van der Swaluw, F. Camilo, V.M. Kaspi, F.K. Baganoff, F. Yusef-Zadeh, R.N. Manchester, The mouse that soared: high-resolution X-ray imaging of the pulsar-powered bow shock G359.23-0.82. Astrophys. J. **616**, 383–402 (2004). doi:10.1086/424906

B.M. Gaensler, C. Kouveliotou, J.D. Gelfand, G.B. Taylor, D. Eichler, R.A.M.J. Wijers, J. Granot, E. Ramirez-Ruiz, Y.E. Lyubarsky, R.W. Hunstead, D. Campbell-Wilson, A.J. van der Horst, M.A. McLaughlin, R.P. Fender, M.A. Garrett, K.J. Newton-McGee, D.M. Palmer, N. Gehrels, P.M. Woods, An expanding radio nebula produced by a giant flare from the magnetar SGR 1806-20. Nature **434**, 1104–1106 (2005). doi:10.1038/nature03498

F.P. Gavriil, M.E. Gonzalez, E.V. Gotthelf, V.M. Kaspi, M.A. Livingstone, P.M. Woods, Magnetar-like emission from the young pulsar in Kes 75. Science **319**, 1802 (2008). doi:10.1126/science.1153465

J.D. Gelfand, P.O. Slane, W. Zhang, A dynamical model for the evolution of a pulsar wind nebula inside a nonradiative supernova remnant. Astrophys. J. **703**, 2051–2067 (2009). doi:10.1088/0004-637X/703/2/2051

J.D. Gelfand, Y.E. Lyubarsky, D. Eichler, B.M. Gaensler, G.B. Taylor, J. Granot, K.J. Newton-McGee, E. Ramirez-Ruiz, C. Kouveliotou, R.A.M.J. Wijers, A rebrightening of the radio nebula associated with the 2004 December 27 giant flare from SGR 1806-20. Astrophys. J. Lett. **634**, 89–92 (2005). doi:10.1086/498643

E.V. Gotthelf, J.A. Tomsick, J.P. Halpern, J.D. Gelfand, F.A. Harrison, S.E. Boggs, F.E. Christensen, W.W. Craig, J.C. Hailey, V.M. Kaspi, D.K. Stern, W.W. Zhang, NuSTAR discovery of a young, energetic pulsar associated with the luminous gamma-ray source HESS J1640-465. Astrophys. J. **788**, 155 (2014). doi:10.1088/0004-637X/788/2/155

J. Granot, E. Ramirez-Ruiz, G.B. Taylor, D. Eichler, Y.E. Lyubarsky, R.A.M.J. Wijers, B.M. Gaensler, J.D. Gelfand, C. Kouveliotou, Diagnosing the outflow from the SGR 1806-20 giant flare with radio observations. Astrophys. J. **638**, 391–396 (2006). doi:10.1086/497680

J. Granot, R. Gill, G. Younes, J. Gelfand, A. Harding, C. Kouveliotou, M.G. Baring, Learning about the magnetar Swift J1834.9-0846 from its wind nebula. Mon. Not. R. Astron. Soc. **464**, 4895–4926 (2017). doi:10.1093/mnras/stw2554

L. Gratton, Source models with electron diffusion. Astrophys. Space Sci. **16**, 81–100 (1972). doi:10.1007/BF00643094

D.A. Green, A revised galactic supernova remnant catalogue. Bull. Astron. Soc. India **37**, 45–61 (2009)

D.A. Green, P.A.G. Scheuer, Upper limits on the infrared flux density of the 'filled-centre' supernova remnant 3C58. Mon. Not. R. Astron. Soc. **258**, 833–840 (1992). doi:10.1093/mnras/258.4.833

M.-H. Grondin, S. Funk, M. Lemoine-Goumard, A. Van Etten, J.A. Hinton, F. Camilo, I. Cognard, C.M. Espinoza, P.C.C. Freire, J.E. Grove, L. Guillemot, S. Johnston, M. Kramer, J. Lande, P. Michelson, A. Possenti, R.W. Romani, J.L. Skilton, G. Theureau, P. Weltevrede, Detection of the pulsar wind nebula HESS J1825-137 with the Fermi Large Area Telescope. Astrophys. J. **738**, 42 (2011). doi:10.1088/0004-637X/738/1/42

M.-H. Grondin, R.W. Romani, M. Lemoine-Goumard, L. Guillemot, A.K. Harding, T. Reposeur, The Vela-X pulsar wind nebula revisited with four years of Fermi Large Area Telescope observations. Astrophys. J. **774**, 110 (2013). doi:10.1088/0004-637X/774/2/110

C.A. Hales, B.M. Gaensler, S. Chatterjee, E. van der Swaluw, F. Camilo, A proper motion for the pulsar wind nebula G359.23-0.82, the "Mouse", associated with the energetic radio pulsar J1747-2958. Astrophys. J. **706**, 1316–1322 (2009). doi:10.1088/0004-637X/706/2/1316

A.K. Harding, I. Contopoulos, D. Kazanas, Magnetar spin-down. Astrophys. J. Lett. **525**, 125–128 (1999). doi:10.1086/312339

F.A. Harrison, W.W. Craig, F.E. Christensen, C.J. Hailey, W.W. Zhang, S.E. Boggs, D. Stern, W.R. Cook, K. Forster, P. Giommi, B.W. Grefenstette, Y. Kim, T. Kitaguchi, J.E. Koglin, K.K. Madsen, P.H. Mao, H. Miyasaka, K. Mori, M. Perri, M.J. Pivovaroff, S. Puccetti, V.R. Rana, N.J. Westergaard, J. Willis, A. Zoglauer, H. An, M. Bachetti, N.M. Barrière, E.C. Bellm, V. Bhalerao, N.F. Brejnholt, F. Fuerst, C.C. Liebe, C.B. Markwardt, M. Nynka, J.K. Vogel, D.J. Walton, D.R. Wik, D.M. Alexander, L.R. Cominsky, A.E. Hornschemeier, A. Hornstrup, V.M. Kaspi, G.M. Madejski, G. Matt, S. Molendi, D.M. Smith, J.A. Tomsick, M. Ajello, D.R. Ballantyne, M. Baloković, D. Barret, F.E. Bauer, R.D. Blandford, W.N. Brandt, L.W. Brenneman, J. Chiang, D. Chakrabarty, J. Chenevez, A. Comastri, F. Dufour, M. Elvis, A.C. Fabian, D. Farrah, C.L. Fryer, E.V. Gotthelf, J.E. Grindlay, D.J. Helfand, R. Krivonos, D.L. Meier, J.M. Miller, L. Natalucci, P. Ogle, E.O. Ofek, A. Ptak, S.P. Reynolds, J.R. Rigby, G. Tagliaferri, S.E. Thorsett, E. Treister, C.M. Urry, The Nuclear Spectroscopic Telescope Array (NuSTAR) high-energy X-ray mission. Astrophys. J. **770**, 103 (2013). doi:10.1088/0004-637X/770/2/103

D.J. Helfand, E.V. Gotthelf, J.P. Halpern, Vela pulsar and its synchrotron nebula. Astrophys. J. **556**, 380–391 (2001). doi:10.1086/321533

H.E.S.S. Collaboration, A. Abramowski, F. Acero, F. Aharonian, A.G. Akhperjanian, G. Anton, A. Balzer, A. Barnacka, U. Barres de Almeida, Y. Becherini, J. Becker, B. Behera, K. Bernlöhr, A. Bochow, C. Boisson, J. Bolmont, P. Bordas, J. Brucker, F. Brun, P. Brun, T. Bulik, I. Büsching, S. Carrigan, S. Casanova, M. Cerruti, P.M. Chadwick, A. Charbonnier, R.C.G. Chaves, A. Cheesebrough, L.-M. Chounet, A.C. Clapson, G. Coignet, G. Cologna, J. Conrad, M. Dalton, M.K. Daniel, I.D. Davids, B. Degrange, C. Deil, H.J. Dickinson, A. Djannati-Ataï, W. Domainko, L. O'C. Drury, F. Dubois, G. Dubus, K. Dutson, J. Dyks, M. Dyrda, K. Egberts, P. Eger, P. Espigat, L. Fallon, C. Farnier, S. Fegan, F. Feinstein, M.V. Fernandes, A. Fiasson, G. Fontaine, A. Förster, M. Füssling, Y.A. Gallant, H. Gast, L. Gérard, D. Gerbig, B. Giebels, J.F. Glicenstein, B. Glück, P. Goret, D. Göring, S. Häffner, J.D. Hague, D. Hampf, M. Hauser, S. Heinz, G. Heinzelmann, G. Henri, G. Hermann, J.A. Hinton, A. Hoffmann, W. Hofmann, P. Hofverberg, M. Holler, D. Horns, A. Jacholkowska, O.C. de Jager, C. Jahn, M. Jamrozy, I. Jung, M.A. Kastendieck, K. Katarzynski, U. Katz, S. Kaufmann, D. Keogh, D. Khangulyan, B. Khélifi, D. Klochkov, W. Kluzniak, T. Kneiske, N. Komin, K. Kosack, R. Kossakowski, H. Laffon, G. Lamanna, D. Lennarz, T. Lohse, A. Lopatin, C.-C. Lu, V. Marandon, A. Marcowith, J. Masbou, D. Maurin, N. Maxted, M. Mayer, T.J.L. McComb, M.C. Medina, J. Méhault, R. Moderski, E. Moulin, C.L. Naumann, M. Naumann-Godo, M. de Naurois, D. Nedbal, D. Nekrassov, N. Nguyen, B. Nicholas, J. Niemiec, S.J. Nolan, S. Ohm, E. de Ona Wilhelmi, B. Opitz, M. Ostrowski, I. Oya, M. Panter, M. Paz Arribas, G. Pedaletti, G. Pelletier, P.-O. Petrucci, S. Pita, G. Pühlhofer, M. Punch, A. Quirrenbach, M. Raue, S.M. Rayner, A. Reimer, O. Reimer, M. Renaud, R. de Los Reyes, F. Rieger, J. Ripken, L. Rob, S. Rosier-Lees, G. Rowell, B. Rudak, C.B. Rulten, J. Ruppel, V. Sahakian, D. Sanchez, A. Santangelo, R. Schlickeiser, F.M. Schöck, A. Schulz, U. Schwanke, S. Schwarzburg, S. Schwemmer, M. Sikora, J.L. Skilton, H. Sol, G. Spengler, L. Stawarz, R. Steenkamp, C. Stegmann, F. Stinzing, K. Stycz, I. Sushch, A. Szostek, J.-P. Tavernet, R. Terrier, M. Tluczykont, K. Valerius, C. van Eldik, G. Vasileiadis, C. Venter, J.P. Vialle, A. Viana, P. Vincent, H.J. Völk, F. Volpe, S. Vorobiov, M. Vorster, S.J. Wagner, M. Ward, R. White, A. Wierzcholska, M. Zacharias, A. Zajczyk, A.A. Zdziarski, A. Zech, H.-S. Zechlin, Discovery of the source HESS J1356-645 associated with the young and energetic PSR J1357-6429. Astron. Astrophys. **533**, 103 (2011). doi:10.1051/0004-6361/201117445

H.E.S.S. Collaboration, A. Abramowski, F. Aharonian, F. Ait Benkhali, A.G. Akhperjanian, E. Angüner, G. Anton, S. Balenderan, A. Balzer, A. Barnacka, et al., HESS J1818-154, a new composite supernova remnant discovered in TeV gamma rays and X-rays. Astron. Astrophys. **562**, 40 (2014). doi:10.1051/0004-6361/201322914

J.A. Hinton, S. Funk, R.D. Parsons, S. Ohm, Escape from Vela X. Astrophys. J. Lett. **743**, 7 (2011). doi:10.1088/2041-8205/743/1/L7

D. Hooper, P. Blasi, P. Dario Serpico, Pulsars as the sources of high energy cosmic ray positrons. J. Cosmol. Astropart. Phys. **1**, 025 (2009). doi:10.1088/1475-7516/2009/01/025

C.Y. Hui, W. Becker, X-ray emission properties of the old pulsar PSR B2224+65. Astron. Astrophys. **467**, 1209–1214 (2007). doi:10.1051/0004-6361:20066562

G.L. Israel, P. Esposito, N. Rea, F. Coti Zelati, A. Tiengo, S. Campana, S. Mereghetti, G.A. Rodriguez Castillo, D. Götz, M. Burgay, A. Possenti, S. Zane, R. Turolla, R. Perna, G. Cannizzaro, J. Pons, The discovery, monitoring and environment of SGR J1935+2154. Mon. Not. R. Astron. Soc. **457**, 3448–3456 (2016). doi:10.1093/mnras/stw008

S.P. Johnson, Q.D. Wang, The pulsar B2224+65 and its jets: a two epoch X-ray analysis. Mon. Not. R. Astron. Soc. **408**, 1216–1224 (2010). doi:10.1111/j.1365-2966.2010.17200.x

P. Kaaret, H.L. Marshall, T.L. Aldcroft, D.E. Graessle, M. Karovska, S.S. Murray, A.H. Rots, N.S. Schulz, F.D. Seward, Chandra observations of the young pulsar PSR B0540-69. Astrophys. J. **546**, 1159–1167 (2001). doi:10.1086/318287

D.L. Kaplan, D.W. Fox, S.R. Kulkarni, E.V. Gotthelf, G. Vasisht, D.A. Frail, Precise Chandra localization of the soft gamma-ray repeater SGR 1806-20. Astrophys. J. **564**, 935–940 (2002). doi:10.1086/324339

N.S. Kardashev, Nonstationarity of spectra of young sources of nonthermal radio emission. Sov. Astron. **6**, 317 (1962)

O. Kargaltsev, G.G. Pavlov, Pulsar wind nebulae in the Chandra era, in *40 Years of Pulsars: Millisecond Pulsars, Magnetars and More*, ed. by C. Bassa, Z. Wang, A. Cumming, V.M. Kaspi, American Institute of Physics Conference Series, vol. 983 (2008), pp. 171–185. doi:10.1063/1.2900138

O. Kargaltsev, B. Rangelov, G.G. Pavlov, Gamma-ray and X-ray Properties of Pulsar Wind Nebulae and Unidentified Galactic TeV Sources. ArXiv e-prints (2013)

O. Kargaltsev, Z. Misanovic, G.G. Pavlov, J.A. Wong, G.P. Garmire, X-ray observations of parsec-scale tails behind two middle-aged pulsars. Astrophys. J. **684**, 542–557 (2008). doi:10.1086/589145

O. Kargaltsev, C. Kouveliotou, G.G. Pavlov, E. Göğüş, L. Lin, S. Wachter, R.L. Griffith, Y. Kaneko, G. Younes, X-ray observations of the new unusual magnetar Swift J1834.9-0846. Astrophys. J. **748**, 26 (2012). doi:10.1088/0004-637X/748/1/26

C.F. Kennel, F.V. Coroniti, Confinement of the Crab pulsar's wind by its supernova remnant. Astrophys. J. **283**, 694–709 (1984a). doi:10.1086/162356

C.F. Kennel, F.V. Coroniti, Magnetohydrodynamic model of Crab Nebula radiation. Astrophys. J. **283**, 710–730 (1984b). doi:10.1086/162357

H.L. Kestenbaum, W. Ku, R. Novick, R.S. Wolff, Measurement of the spatial structure of the X-ray source in the Crab Nebula. II. Observation of the 1974 December 28 lunar occultation. Astrophys. J. Lett. **202**, 21 (1975). doi:10.1086/181971

J.G. Kirk, Y. Lyubarsky, J. Petri, The theory of pulsar winds and nebulae, in *Astrophys. Space Sci. Library*, ed. by W. Becker. Astrophys. Space Sci. Library, vol. 357 (2009), p. 421. doi:10.1007/978-3-540-76965-1_16

S. Klepser, S. Carrigan, E. de Oña Wilhelmi, C. Deil, A. Förster, V. Marandon, M. Mayer, K. Stycz, K. Valerius for the H. E. S. S. collaboration, A Population of Teraelectronvolt Pulsar Wind Nebulae in the H.E.S.S. Galactic Plane Survey. ArXiv e-prints (2013)

N. Klingler, O. Kargaltsev, B. Rangelov, G.G. Pavlov, B. Posselt, C.-Y. Ng, Chandra observations of outflows from PSR J1509-5850. Astrophys. J. **828**, 70 (2016a). doi:10.3847/0004-637X/828/2/70

N. Klingler, B. Rangelov, O. Kargaltsev, G.G. Pavlov, R.W. Romani, B. Posselt, P. Slane, T. Temim, C.-Y. Ng, N. Bucciantini, A. Bykov, D.A. Swartz, R. Buehler, Deep Chandra observations of the pulsar wind nebula created by PSR B0355+54. Astrophys. J. **833**, 253 (2016b). doi:10.3847/1538-4357/833/2/253

S.S. Komissarov, Y.E. Lyubarsky, Synchrotron nebulae created by anisotropic magnetized pulsar winds. Mon. Not. R. Astron. Soc. **349**, 779–792 (2004). doi:10.1111/j.1365-2966.2004.07597.x

B.-C. Koo, C.F. McKee, K.-W. Suh, D.-S. Moon, T. Onaka, M.G. Burton, M. Hiramatsu, M.S. Bessell, B.M. Gaensler, H.-J. Kim, J.-J. Lee, W.-S. Jeong, H.-G. Lee, M. Im, K. Tatematsu, K. Kohno, R. Kawabe, H. Ezawa, G. Wilson, M.S. Yun, D.H. Hughes, IRAS 15099-5856: remarkable mid-infrared source with prominent crystalline silicate emission embedded in the supernova remnant MSH15-52. Astrophys. J. **732**, 6 (2011). doi:10.1088/0004-637X/732/1/6

R. Kothes, T.L. Landecker, W. Reich, S. Safi-Harb, Z. Arzoumanian, DA 495: an aging pulsar wind nebula. Astrophys. J. **687**, 516–531 (2008). doi:10.1086/591653

R. Kothes, X.H. Sun, W. Reich, T.J. Foster, G141.2+5.0, a new pulsar wind nebula discovered in the Cygnus Arm of the Milky Way. Astrophys. J. Lett. **784**, 26 (2014). doi:10.1088/2041-8205/784/2/L26

C. Kouveliotou, A. Tennant, P.M. Woods, M.C. Weisskopf, K. Hurley, R.P. Fender, S.T. Garrington, S.K. Patel, E. Göğüş, Multiwavelength observations of the soft gamma repeater SGR 1900+14 during its 2001 April activation. Astrophys. J. Lett. **558**, 47–50 (2001). doi:10.1086/323496

W. Ku, H.L. Kestenbaum, R. Novick, R.S. Wolff, Energy dependence of the size of the X-ray source in the Crab Nebula. Astrophys. J. Lett. **204**, 77–81 (1976). doi:10.1086/182058

S.M. LaMassa, P.O. Slane, O.C. de Jager, Probing the nature of the Vela X cocoon. Astrophys. J. Lett. **689**, 121 (2008). doi:10.1086/595958

T.L. Landecker, L.A. Higgs, H.J. Wendker, G76.9+1.0, a supernova remnant and unusual properties. Astron. Astrophys. **276**, 522 (1993)

C.C. Lang, Q.D. Wang, F. Lu, K.I. Clubb, The radio properties and magnetic field configuration in the Crab-like pulsar wind nebula G54.1+0.3. Astrophys. J. **709**, 1125–1137 (2010). doi:10.1088/0004-637X/709/2/1125

M. Lemoine-Goumard, V.E. Zavlin, M.-H. Grondin, R. Shannon, D.A. Smith, M. Burgay, F. Camilo, J. Cohen-Tanugi, P.C.C. Freire, J.E. Grove, L. Guillemot, S. Johnston, M. Keith, M. Kramer, R.N. Manchester, P.F. Michelson, D. Parent, A. Possenti, P.S. Ray, M. Renaud, S.E. Thorsett, P. Weltevrede, M.T. Wolff, Discovery of gamma- and X-ray pulsations from the young and energetic PSR J1357-6429 with Fermi and XMM-Newton. Astron. Astrophys. **533**, 102 (2011). doi:10.1051/0004-6361/201117413

F.J. Lu, Q.D. Wang, B. Aschenbach, P. Durouchoux, L.M. Song, batitleChandra Observation of Supernova Remnant G54.1+0.3: A Close Cousin of the Crab Nebula. Astrophys. J. Lett. **568**, 49–52 (2002). doi:10.1086/340137

M. Lyutikov, Mass-loading of pulsar winds. Mon. Not. R. Astron. Soc. **339**, 623–632 (2003). doi:10.1046/j.1365-8711.2003.06141.x

K.K. Madsen, S. Reynolds, F. Harrison, H. An, S. Boggs, F.E. Christensen, W.W. Craig, C.L. Fryer, B.W. Grefenstette, C.J. Hailey, C. Markwardt, M. Nynka, D. Stern, A. Zoglauer, W. Zhang, Broadband X-ray imaging and spectroscopy of the Crab Nebula and pulsar with NuSTAR. Astrophys. J. **801**, 66 (2015). doi:10.1088/0004-637X/801/1/66

R.N. Manchester, L. Staveley-Smith, M.J. Kesteven, The radio structure of supernova remnant 0540-693. Astrophys. J. **411**, 756–760 (1993). doi:10.1086/172877

M. Marelli, A. de Luca, D. Salvetti, N. Sartore, A. Sartori, P. Caraveo, F. Pizzolato, P.M. Saz Parkinson, A. Belfiore, PSR J0357+3205: the tail of the turtle. Astrophys. J. **765**, 36 (2013). doi:10.1088/0004-637X/765/1/36

J. Martín, D.F. Torres, N. Rea, Time-dependent modelling of pulsar wind nebulae: study on the impact of the diffusion-loss approximations. Mon. Not. R. Astron. Soc. **427**, 415–427 (2012). doi:10.1111/j.1365-2966.2012.22014.x
H. Matheson, S. Safi-Harb, The plerionic supernova remnant G21.5-0.9: in and out. Adv. Space Res. **35**, 1099–1105 (2005). doi:10.1016/j.asr.2005.04.050
H. Matheson, S. Safi-Harb, The plerionic supernova remnant G21.5-0.9 powered by PSR J1833-1034: new spectroscopic and imaging results revealed with the Chandra X-ray observatory. Astrophys. J. **724**, 572–587 (2010). doi:10.1088/0004-637X/724/1/572
F. Mattana, D. Götz, R. Terrier, L. Bouchet, G. Ponti, M. Falanga, M. Renaud, I. Caballero, S. Soldi, J.A. Zurita Heras, S. Schanne, Extended hard X-ray emission from the Vela pulsar wind nebula. Astrophys. J. Lett. **743**, 18 (2011). doi:10.1088/2041-8205/743/1/L18
K.E. McGowan, W.T. Vestrand, J.A. Kennea, S. Zane, M. Cropper, F.A. Córdova, Probing the pulsar wind nebula of PSR B0355+54. Astrophys. J. **647**, 1300–1308 (2006). doi:10.1086/505522
S. Mereghetti, J.A. Pons, A. Melatos, Magnetars: properties, origin and evolution. Space Sci. Rev. **191**, 315–338 (2015). doi:10.1007/s11214-015-0146-y
S. Mereghetti, D. Götz, A. von Kienlin, A. Rau, G. Lichti, G. Weidenspointner, P. Jean, The first giant flare from SGR 1806-20: observations using the anticoincidence shield of the spectrometer on INTEGRAL. Astrophys. J. Lett. **624**, 105–108 (2005). doi:10.1086/430669
R.P. Mignani, G.G. Pavlov, O. Kargaltsev, Optical-ultraviolet spectrum and proper motion of the middle-aged pulsar B1055-52. Astrophys. J. **720**, 1635–1643 (2010). doi:10.1088/0004-637X/720/2/1635
K. Mori, D.N. Burrows, J.J. Hester, G.G. Pavlov, S. Shibata, H. Tsunemi, Spatial variation of the X-ray spectrum of the Crab Nebula. Astrophys. J. **609**, 186–193 (2004). doi:10.1086/421011
G. Morlino, M. Lyutikov, M. Vorster, Mass loading of bow shock pulsar wind nebulae. Mon. Not. R. Astron. Soc. **454**, 3886–3901 (2015). doi:10.1093/mnras/stv2189
E. Nakar, T. Piran, R. Sari, Pure and loaded fireballs in soft gamma-ray repeater giant flares. Astrophys. J. **635**, 516–521 (2005). doi:10.1086/497296
C.-Y. Ng, B.M. Gaensler, S. Chatterjee, S. Johnston, Radio polarization observations of G319.9-0.7: a bow-shock nebula with an azimuthal magnetic field powered by pulsar J1509-5850. Astrophys. J. **712**, 596–603 (2010). doi:10.1088/0004-637X/712/1/596
M. Nynka, C.J. Hailey, S.P. Reynolds, H. An, F.K. Baganoff, S.E. Boggs, F.E. Christensen, W.W. Craig, E.V. Gotthelf, B.W. Grefenstette, F.A. Harrison, R. Krivonos, K.K. Madsen, K. Mori, K. Perez, D. Stern, D.R. Wik, W.W. Zhang, A. Zoglauer, NuSTAR study of hard X-ray morphology and spectroscopy of PWN G21.5-0.9. Astrophys. J. **789**, 72 (2014). doi:10.1088/0004-637X/789/1/72
S.A. Olausen, V.M. Kaspi, The McGill magnetar catalog. Astrophys. J. Suppl. Ser. **212**, 6 (2014). doi:10.1088/0067-0049/212/1/6
F. Pacini, M. Salvati, On the evolution of supernova remnants. Evolution of the magnetic field, particles, content, and luminosity. Astrophys. J. **186**, 249–266 (1973). doi:10.1086/152495
S. Park, J.P. Hughes, P.O. Slane, D.N. Burrows, B.M. Gaensler, P. Ghavamian, A half-megasecond Chandra observation of the oxygen-rich supernova remnant G292.0+1.8. Astrophys. J. Lett. **670**, 121–124 (2007). doi:10.1086/524406
L. Pavan, G. Pühlhofer, P. Bordas, M. Audard, M. Balbo, E. Bozzo, D. Eckert, C. Ferrigno, M.D. Filipović, M. Verdugo, R. Walter, Closer view of the IGR J11014-6103 outflows. Astron. Astrophys. **591**, 91 (2016). doi:10.1051/0004-6361/201527703
G.G. Pavlov, S. Bhattacharyya, V.E. Zavlin, New X-ray observations of the Geminga pulsar wind nebula. Astrophys. J. **715**, 66–77 (2010). doi:10.1088/0004-637X/715/1/66
G.G. Pavlov, O. Kargaltsev, W.F. Brisken, Chandra observation of PSR B1823-13 and its pulsar wind nebula. Astrophys. J. **675**, 683–694 (2008). doi:10.1086/525842
G.G. Pavlov, D. Sanwal, V.E. Zavlin, The pulsar wind nebula of the Geminga pulsar. Astrophys. J. **643**, 1146–1150 (2006). doi:10.1086/503250
G.G. Pavlov, M.A. Teter, O. Kargaltsev, D. Sanwal, The variable jet of the Vela pulsar. Astrophys. J. **591**, 1157–1171 (2003). doi:10.1086/375531
Planck Collaboration, M. Arnaud, M. Ashdown, F. Atrio-Barandela, J. Aumont, C. Baccigalupi, A.J. Banday, R.B. Barreiro, E. Battaner, K. Benabed, A. Benoit-Lévy, J.-P. Bernard, M. Bersanelli, P. Bielewicz, J. Bobin, J.R. Bond, J. Borrill, F.R. Bouchet, C.L. Brogan, C. Burigana, J.-F. Cardoso, A. Catalano, A. Chamballu, H.C. Chiang, P.R. Christensen, S. Colombi, L.P.L. Colombo, B.P. Crill, A. Curto, F. Cuttaia, R.D. Davies, R.J. Davis, P. de Bernardis, A. de Rosa, G. de Zotti, J. Delabrouille, F.-X. Désert, C. Dickinson, J.M. Diego, S. Donzelli, O. Doré, X. Dupac, T.A. Enßlin, H.K. Eriksen, F. Finelli, O. Forni, M. Frailis, A.A. Fraisse, E. Franceschi, S. Galeotta, K. Ganga, M. Giard, Y. Giraud-Héraud, J. González-Nuevo, K.M. Górski, A. Gregorio, A. Gruppuso, F.K. Hansen, D.L. Harrison, C. Hernández-Monteagudo, D. Herranz, S.R. Hildebrandt, M. Hobson, W.A. Holmes, K.M. Huffenberger, A.H. Jaffe, T.R. Jaffe, E. Keihänen, R. Keskitalo, T.S. Kisner, R. Kneissl, J. Knoche, M. Kunz,

H. Kurki-Suonio, A. Lähteenmäki, J.-M. Lamarre, A. Lasenby, C.R. Lawrence, R. Leonardi, M. Liguori, P.B. Lilje, M. Linden-Vørnle, M. López-Caniego, P.M. Lubin, D. Maino, M. Maris, D.J. Marshall, P.G. Martin, E. Martínez-González, S. Masi, S. Matarrese, P. Mazzotta, A. Melchiorri, L. Mendes, A. Mennella, M. Migliaccio, M.-A. Miville-Deschênes, A. Moneti, L. Montier, G. Morgante, D. Mortlock, D. Munshi, J.A. Murphy, P. Naselsky, F. Nati, F. Noviello, D. Novikov, I. Novikov, N. Oppermann, C.A. Oxborrow, L. Pagano, F. Pajot, R. Paladini, F. Pasian, M. Peel, O. Perdereau, F. Perrotta, F. Piacentini, M. Piat, D. Pietrobon, S. Plaszczynski, E. Pointecouteau, G. Polenta, L. Popa, G.W. Pratt, J.-L. Puget, J.P. Rachen, W.T. Reach, W. Reich, M. Reinecke, M. Remazeilles, C. Renault, J. Rho, S. Ricciardi, T. Riller, I. Ristorcelli, G. Rocha, C. Rosset, G. Roudier, B. Rusholme, M. Sandri, G. Savini, D. Scott, V. Stolyarov, D. Sutton, A.-S. Suur-Uski, J.-F. Sygnet, J.A. Tauber, L. Terenzi, L. Toffolatti, M. Tomasi, M. Tristram, M. Tucci, G. Umana, L. Valenziano, J. Valiviita, B. Van Tent, P. Vielva, F. Villa, L.A. Wade, D. Yvon, A. Zacchei, A. Zonca, Planck intermediate results. XXXI. Microwave survey of Galactic supernova remnants. Astron. Astrophys. **586**, 134 (2016). doi:10.1051/0004-6361/201425022

T.A. Porter, A.W. Strong, A new estimate of the Galactic interstellar radiation field between 0.1 μm and 1000 μm, in *ICRC*, vol. 4 (2005), p. 77

B. Posselt, G. Spence, G.G. Pavlov, A Chandra search for a pulsar wind nebula around PSR B1055-52. Astrophys. J. **811**, 96 (2015). doi:10.1088/0004-637X/811/2/96

B. Posselt, G.G. Pavlov, P.O. Slane, R. Romani, N. Bucciantini, A.M. Bykov, O. Kargaltsev, M.C. Weisskopf, C.-Y. Ng, Geminga's puzzling pulsar wind nebula. ArXiv e-prints (2016)

S.P. Reynolds, Magnetohydrodynamic Models for the Structure of Pulsar-Wind Nebulae. ArXiv Astrophysics e-prints (2003)

S.P. Reynolds, Synchrotron-loss spectral breaks in pulsar-wind nebulae and extragalactic jets. Astrophys. J. **703**, 662–670 (2009). doi:10.1088/0004-637X/703/1/662

S.P. Reynolds, H.D. Aller, Radio observations of the Crab-like supernova remnant 3C 58. I—Total intensity observations. Astrophys. J. **327**, 845–852 (1988). doi:10.1086/166242

S.P. Reynolds, K.J. Borkowski, A compact X-ray source in the radio pulsar-wind nebula G141.2+5.0. Astrophys. J. Lett. **816**, 27 (2016). doi:10.3847/2041-8205/816/2/L27

S.P. Reynolds, R.A. Chevalier, Evolution of pulsar-driven supernova remnants. Astrophys. J. **278**, 630–648 (1984). doi:10.1086/161831

S.P. Reynolds, F.C. Jones, A Green's function solution for the spherical diffusion equation in the presence of energy losses, and its applications to Crablike supernova remnants, in *International Cosmic Ray Conference*, vol. 2 (1991), p. 400

M.S.E. Roberts, C.R. Tam, V.M. Kaspi, M. Lyutikov, G. Vasisht, M. Pivovaroff, E.V. Gotthelf, N. Kawai, The pulsar wind nebula in G11.2-0.3. Astrophys. J. **588**, 992–1002 (2003). doi:10.1086/374266

R.W. Romani, M.S. Shaw, F. Camilo, G. Cotter, G.R. Sivakoff, The Balmer-dominated bow shock and wind nebula structure of γ-ray pulsar PSR J1741-2054. Astrophys. J. **724**, 908–914 (2010). doi:10.1088/0004-637X/724/2/908

M.M. Romanova, G.A. Chulsky, R.V.E. Lovelace, Winds, B-fields, and magnetotails of pulsars. Astrophys. J. **630**, 1020–1028 (2005). doi:10.1086/431727

C.J. Salter, S.P. Reynolds, D.E. Hogg, J.M. Payne, P.J. Rhodes, 84 gigahertz observations of five Crab-like supernova remnants. Astrophys. J. **338**, 171–177 (1989). doi:10.1086/167191

N.I. Serafimovich, Y.A. Shibanov, P. Lundqvist, J. Sollerman, The young pulsar PSR B0540-69.3 and its synchrotron nebula in the optical and X-rays. Astron. Astrophys. **425**, 1041–1060 (2004). doi:10.1051/0004-6361:20040499

F.D. Seward, F.R. Harnden Jr., A new, fast X-ray pulsar in the supernova remnant MSH 15-52. Astrophys. J. Lett. **256**, 45–47 (1982). doi:10.1086/183793

P. Slane, Y. Chen, N.S. Schulz, F.D. Seward, J.P. Hughes, B.M. Gaensler, Chandra observations of the Crab-like supernova remnant G21.5-0.9. Astrophys. J. Lett. **533**, 29–32 (2000). doi:10.1086/312589

P. Slane, D.J. Helfand, E. van der Swaluw, S.S. Murray, New constraints on the structure and evolution of the pulsar wind nebula 3C 58. Astrophys. J. **616**, 403–413 (2004). doi:10.1086/424814

P. Slane, D.J. Helfand, S.P. Reynolds, B.M. Gaensler, A. Lemiere, Z. Wang, The infrared detection of the pulsar wind nebula in the galactic supernova remnant 3C 58. Astrophys. J. Lett. **676**, 33 (2008). doi:10.1086/587031

S.J. Tanaka, F. Takahara, A model of the spectral evolution of pulsar wind nebulae. Astrophys. J. **715**, 1248–1257 (2010). doi:10.1088/0004-637X/715/2/1248

S.J. Tanaka, F. Takahara, Study of four young TeV pulsar wind nebulae with a spectral evolution model. Astrophys. J. **741**, 40 (2011). doi:10.1088/0004-637X/741/1/40

X. Tang, R.A. Chevalier, Particle transport in young pulsar wind nebulae. Astrophys. J. **752**, 83 (2012). doi:10.1088/0004-637X/752/2/83

T. Temim, R.D. Gehrz, C.E. Woodward, T.L. Roellig, N. Smith, L. Rudnick, E.F. Polomski, K. Davidson, L. Yuen, T. Onaka, Spitzer space telescope infrared imaging and spectroscopy of the Crab Nebula. Astron. J. **132**, 1610–1623 (2006). doi:10.1086/507076

T. Temim, P. Slane, S.P. Reynolds, J.C. Raymond, K.J. Borkowski, Deep Chandra observations of the Crab-like pulsar wind nebula G54.1+0.3 and Spitzer spectroscopy of the associated infrared shell. Astrophys. J. **710**, 309–324 (2010). doi:10.1088/0004-637X/710/1/309

T. Temim, P. Slane, C. Kolb, J. Blondin, J.P. Hughes, N. Bucciantini, Late-time evolution of composite supernova remnants: deep Chandra observations and hydrodynamical modeling of a crushed pulsar wind nebula in SNR G327.1-1.1. Astrophys. J. **808**, 100 (2015). doi:10.1088/0004-637X/808/1/100

C. Thompson, R.C. Duncan, The soft gamma repeaters as very strongly magnetized neutron stars. II. Quiescent neutrino, X-ray, and Alfven wave emission. Astrophys. J. **473**, 322–342 (1996). doi:10.1086/178147

W.W. Tian, Z. Li, D.A. Leahy, Q.D. Wang, VLA and XMM-Newton observations of the SNR W41/TeV gamma-ray source HESS J1834-087. Astrophys. J. Lett. **657**, 25–28 (2007). doi:10.1086/512544

A. Tiengo, G. Vianello, P. Esposito, S. Mereghetti, A. Giuliani, E. Costantini, G.L. Israel, L. Stella, R. Turolla, S. Zane, N. Rea, D. Götz, F. Bernardini, A. Moretti, P. Romano, M. Ehle, N. Gehrels, The dust-scattering X-ray rings of the anomalous X-ray pulsar 1E 1547.0-5408. Astrophys. J. **710**, 227–235 (2010a). doi:10.1088/0004-637X/710/1/227

A. Tiengo, G. Vianello, P. Esposito, S. Mereghetti, The spectacular X-ray echo of a magnetar burst, in *X-ray Astronomy 2009; Present Status, Multi-Wavelength Approach and Future Perspectives*, vol. 1248 (2010b), pp. 81–84. doi:10.1063/1.3475361

H. Tong, On the possible wind nebula of magnetar Swift J1834.9-0846: a magnetism-powered synchrotron nebula. Res. Astron. Astrophys. **16**, 143 (2016). doi:10.1088/1674-4527/16/9/143

D.F. Torres, A rotationally-powered magnetar nebula around Swift J1834.9-0846. ArXiv e-prints (2016)

D.F. Torres, A. Cillis, J. Martín, E. de Oña Wilhelmi, Time-dependent modeling of TeV-detected, young pulsar wind nebulae. J. High Energy Astrophys. **1**, 31–62 (2014). doi:10.1016/j.jheap.2014.02.001

E. van der Swaluw, T.P. Downes, R. Keegan, An evolutionary model for pulsar-driven supernova remnants. A hydrodynamical model. Astron. Astrophys. **420**, 937–944 (2004). doi:10.1051/0004-6361:20035700

E. van der Swaluw, A. Achterberg, Y.A. Gallant, G. Tóth, Pulsar wind nebulae in supernova remnants. Spherically symmetric hydrodynamical simulations. Astron. Astrophys. **380**, 309–317 (2001). doi:10.1051/0004-6361:20011437

A. Van Etten, R.W. Romani, Multi-zone modeling of the pulsar wind nebula HESS J1825-137. Astrophys. J. **742**, 62 (2011). doi:10.1088/0004-637X/742/2/62

D. Viganò, N. Rea, P. Esposito, S. Mereghetti, G.L. Israel, A. Tiengo, R. Turolla, S. Zane, L. Stella, Searching for small-scale diffuse emission around SGR 1806-20. J. High Energy Astrophys. **3**, 41–46 (2014). doi:10.1016/j.jheap.2014.10.001

M. Vigelius, A. Melatos, S. Chatterjee, B.M. Gaensler, P. Ghavamian, Three-dimensional hydrodynamic simulations of asymmetric pulsar wind bow shocks. Mon. Not. R. Astron. Soc. **374**, 793–808 (2007). doi:10.1111/j.1365-2966.2006.11193.x

M.J. Vorster, O. Tibolla, S.E.S. Ferreira, S. Kaufmann, Time-dependent modeling of pulsar wind nebulae. Astrophys. J. **773**, 139 (2013). doi:10.1088/0004-637X/773/2/139

K.W. Weiler, N. Panagia, Are Crab-type supernova remnants (plerions) short-lived? Astron. Astrophys. **70**, 419 (1978)

M.C. Weisskopf, J.J. Hester, A.F. Tennant, R.F. Elsner, N.S. Schulz, H.L. Marshall, M. Karovska, J.S. Nichols, D.A. Swartz, J.J. Kolodziejczak, S.L. O'Dell, Discovery of spatial and spectral structure in the X-ray emission from the Crab Nebula. Astrophys. J. Lett. **536**, 81–84 (2000). doi:10.1086/312733

F.P. Wilkin, Modeling nonaxisymmetric bow shocks: solution method and exact analytic solutions. Astrophys. J. **532**, 400–414 (2000). doi:10.1086/308576

B.J. Williams, K.J. Borkowski, S.P. Reynolds, J.C. Raymond, K.S. Long, J. Morse, W.P. Blair, P. Ghavamian, R. Sankrit, S.P. Hendrick, R.C. Smith, S. Points, P.F. Winkler, Ejecta, dust, and synchrotron radiation in SNR B0540-69.3: a more Crab-like remnant than the Crab. Astrophys. J. **687**, 1054–1069 (2008). doi:10.1086/592139

D.S. Wong, J.M. Cordes, S. Chatterjee, E.G. Zweibel, J.P. Finley, R.W. Romani, M.P. Ulmer, Chandra observations of the Guitar Nebula, in *High Energy Processes and Phenomena in Astrophysics*, ed. by X.D. Li, V. Trimble, Z.R. Wang. IAU Symposium, vol. 214 (2003), p. 135

G. Younes, C. Kouveliotou, O. Kargaltsev, G.G. Pavlov, E. Göğüş, S. Wachter, XMM-Newton view of Swift J1834.9-0846 and its magnetar wind nebula. Astrophys. J. **757**, 39 (2012). doi:10.1088/0004-637X/757/1/39

G. Younes, C. Kouveliotou, O. Kargaltsev, R. Gill, J. Granot, A.L. Watts, J. Gelfand, M.G. Baring, A. Harding, G.G. Pavlov, A.J. van der Horst, D. Huppenkothen, E. Göğüş, L. Lin, O.J. Roberts, The wind nebula around magnetar Swift J1834.9-0846. ArXiv e-prints (2016)

G. Younes, C. Kouveliotou, A. Jaodand, M.G. Baring, A.J. van der Horst, A.K. Harding, J.W.T. Hessels, N. Gehrels, R. Gill, D. Huppenkothen, J. Granot, E. Göğüş, L. Lin, X-ray and radio observations of the magnetar SGR J1935+2154 during its 2014, 2015, and 2016 outbursts. ArXiv e-prints (2017)

F. Yusef-Zadeh, J. Bally, A non-thermal axially symmetric radio wake towards the Galactic centre. Nature **330**, 455–458 (1987). doi:10.1038/330455a0

F. Yusef-Zadeh, B.M. Gaensler, A radio study of the mouse, G359.23-0.82. Adv. Space Res. **35**, 1129–1136 (2005). doi:10.1016/j.asr.2005.03.003

A. Zajczyk, Y.A. Gallant, P. Slane, S.P. Reynolds, R. Bandiera, C. Gouiffès, E. Le Floc'h, F. Comerón, L. Koch Miramond, Infrared imaging and polarimetric observations of the pulsar wind nebula in SNR G21.5-0.9. Astron. Astrophys. **542**, 12 (2012). doi:10.1051/0004-6361/201117194

S.V. Zharikov, D.A. Zyuzin, Y.A. Shibanov, R.E. Mennickent, The PSR J1124-5916 wind nebula in the near-infrared. Astron. Astrophys. **554**, 120 (2013). doi:10.1051/0004-6361/201321283

Space Sci Rev (2017) 207:235–290
DOI 10.1007/s11214-017-0371-7

Pulsar Wind Nebulae with Bow Shocks: Non-thermal Radiation and Cosmic Ray Leptons

A.M. Bykov[1,2,3] · E. Amato[4] · A.E. Petrov[1] · A.M. Krassilchtchikov[1] · K.P. Levenfish[1]

Received: 27 March 2017 / Accepted: 27 April 2017 / Published online: 12 June 2017

Abstract Pulsars with high spin-down power produce relativistic winds radiating a non-negligible fraction of this power over the whole electromagnetic range from radio to gamma-rays in the pulsar wind nebulae (PWNe). The rest of the power is dissipated in the interactions of the PWNe with the ambient interstellar medium (ISM). Some of the PWNe are moving relative to the ambient ISM with supersonic speeds producing bow shocks. In this case, the ultrarelativistic particles accelerated at the termination surface of the pulsar wind may undergo reacceleration in the converging flow system formed by the plasma outflowing from the wind termination shock and the plasma inflowing from the bow shock. The presence of magnetic perturbations in the flow, produced by instabilities induced by the accelerated particles themselves, is essential for the process to work. A generic outcome of this type of reacceleration is the creation of particle distributions with very hard spectra, such as are indeed required to explain the observed spectra of synchrotron radiation with photon indices $\Gamma \lesssim 1.5$. The presence of this hard spectral component is specific to PWNe with bow shocks (BSPWNe). The accelerated particles, mainly electrons and positrons, may end up containing a substantial fraction of the shock ram pressure. In addition, for typical ISM and pulsar parameters, the e^+ released by these systems in the Galaxy are numerous enough to contribute a substantial fraction of the positrons detected as cosmic ray (CR) particles above few tens of GeV and up to several hundred GeV. The escape of ultrarelativistic particles from a BSPWN—and hence, its appearance in the far-UV and X-ray bands—is determined by the relative directions of the interstellar magnetic field, the velocity of the astrosphere and the pulsar rotation axis. In this respect we review the observed appearance and multiwavelength spectra of three different types of BSPWNe: PSR J0437-4715, the Guitar

✉ A.M. Bykov
byk@astro.ioffe.ru

E. Amato
amato@arcetri.astro.it

[1] Ioffe Institute, 194021, St. Petersburg, Russia

[2] St. Petersburg Polytechnic University, St. Petersburg, Russia

[3] International Space Science Institute, Bern, Switzerland

[4] INAF – Osservatorio Astrofisico di Arcetri, Largo E. Fermi, 5, 50125, Firenze, Italy

and Lighthouse nebulae, and Vela-like objects. We argue that high resolution imaging of such objects provides unique information both on pulsar winds and on the ISM. We discuss the interpretation of imaging observations in the context of the model outlined above and estimate the BSPWN contribution to the positron flux observed at the Earth.

Keywords Pulsar wind nebulae · Supernova remnants · Relativistic winds · Shocks · Positrons

1 Introduction

Pulsars are magnetized fast rotating neutron stars, whose rotation energy is carried away by relativistic winds blowing out of their magnetic poles and mainly consisting of electron-positron ($e^{\pm}$) pairs. Multiband emission observed from these winds allows one to study the complex physics of energy transport in space (Arons 2002, 2012; Gaensler and Slane 2006; Kirk et al. 2009; Bucciantini 2014; Kargaltsev et al. 2015). Multi-wavelength observations have revealed a vast variety of radiative efficiencies of pulsar winds (PWs) and pulsar wind nebulae (PWNe) produced by the winds, ranging from some tens percent in the Crab nebula (see e.g. Bühler and Blandford 2014) to much lower values in a number of other PWNe (see, e.g., Li et al. 2008; Kargaltsev and Pavlov 2008). Relativistic particles leaving the observable PWNe interact with the ambient medium on larger scales (upto tens of parsec) and thus provide a tool to reveal the structure of magnetic field and some other parameters of the interstellar medium (ISM), as discussed below in Sect. 6. This is a probable case for the fast enough, supersonically moving pulsars.

An analysis of radio pulsar velocity distribution based on the large scale survey at 0.4 GHz by Arzoumanian et al. (2002) suggested a two-component velocity distribution with characteristic velocities of 90 and 500 $\mathrm{km\,s^{-1}}$ and inferred that $\sim 15\%$ of the pulsars have natal velocities greater than 1000 $\mathrm{km\,s^{-1}}$. The authors also concluded that under some simplified assumptions on the supernova remnant (SNR) evolution, about 10% of the pulsars would leave the host remnant before they are 20 kyr old.

The catalog of 233 pulsars (both young and recycled) compiled by Hobbs et al. (2005) contains derived pulsar speeds measured in one coordinate (1D) and the transverse velocities (2D). No evidence for a bimodal velocity distribution was found. The authors derived the mean 1D pulsar speed of $152 \pm 10\ \mathrm{km\,s^{-1}}$ for the normal pulsars and $54 \pm 6\ \mathrm{km\,s^{-1}}$ for the recycled ones. The corresponding mean 2D velocities in the catalog are $246 \pm 22\ \mathrm{km\,s^{-1}}$ and $87 \pm 13\ \mathrm{km\,s^{-1}}$, with the highest values $\sim 1600\ \mathrm{km\,s^{-1}}$ inferred for the pulsars B2224+64 and B2011+38. The mean 3D natal pulsar velocity was derived to be $400 \pm 40\ \mathrm{km\,s^{-1}}$.

For typical pulsar and ISM parameters, most pulsars are expected to leave their parent SNRs during the Sedov-Taylor phase of remnant expansion. The time $T_{\rm esc}$ at which this occurs can be estimated as $V_{\rm PSR} T_{\rm esc} = R_{\rm ST}(T_{\rm esc}/t_{\rm ST})^{2/5}$, where $V_{\rm PSR}$ is the velocity of pulsar proper motion, $t_{\rm ST}$ and $R_{\rm ST}$ are the time and remnant radius at the beginning of the Sedov-Taylor phase. With an average estimated velocity of about 400 $\mathrm{km\,s^{-1}}$, an estimate of the time at which a pulsar escapes from its SNR can be found as

$$T_{\rm esc} = \left(\frac{R_{\rm ST}}{V_{\rm PSR} T_{\rm ST}^{2/5}}\right)^{\frac{5}{3}} \approx 45\ \mathrm{kyr}\left(\frac{E_{\rm SN}}{10^{51}\ \mathrm{erg}}\right)^{\frac{1}{3}} \left(\frac{n_H}{\mathrm{cm}^{-3}}\right)^{-\frac{1}{3}} \left(\frac{V_{\rm PSR}}{400\ \mathrm{km/s}}\right)^{-\frac{5}{3}}. \tag{1}$$

A pulsar aged a few tens of kyr still has a non-negligible fraction of its original spin-down energy to spend. In general, one can write $\dot{\Omega} \propto \Omega^n$, with n—the pulsar braking index, Ω and $\dot{\Omega}$—the star rotation frequency and its time derivative, respectively. For dipole spin-down one expects $n = 3$, since $\dot{E} = I\Omega\dot{\Omega} \propto B_*^2\Omega^4$, where $I \approx 10^{45}$ g cm^2 is the star momentum of inertia and B_*—its surface magnetic field. While it is true that for the few cases, where the braking index has been actually measured, its value has always been found to be less than 3, with an average of 2.5 (see, e.g., Faucher-Giguère and Kaspi 2006 and references therein), in the following we will assume the standard dipole spin-down law. Hence, one can compute the rotational energy the pulsar is left with at the moment T_{esc} when it leaves the remnant. From the equation:

$$I\Omega\dot{\Omega} = \frac{4}{9}\frac{B_*^2 R_*^6 \Omega^4}{c^3}, \tag{2}$$

where $R_* \approx 10$ km is the neutron star radius, we find:

$$\Omega(t) = \frac{\Omega_0}{(1 + t/\tau_b)^{\frac{1}{2}}}, \quad \text{with } \tau_b = \frac{9Ic^3}{8B_*^2 R_*^6 \Omega_0^2}, \tag{3}$$

and $\Omega_0 = 2\pi/P_0$—the initial star rotation frequency (P_0 is the initial star spin period). Finally, the residual rotation energy of the star at T_{esc} is:

$$E_{res} = \frac{1}{2}I\Omega^2(T_{esc}) = \frac{1}{2}\frac{I\Omega_0^2}{(1 + t_{esc}/\tau_b)}. \tag{4}$$

The average value of E_{res} can be estimated by taking the average values of the pulsar surface magnetic field and initial period. From the work of Faucher-Giguère and Kaspi (2006) we have: $\langle B_* \rangle \approx 4.5 \times 10^{12}$ G and $\langle P_0 \rangle = 300$ ms. Using the estimate of T_{esc} from Eq. (1) one obtains $E_{res} \approx 10^{47}$ erg and $\dot{E}_{res} \approx 5 \times 10^{34}$ erg/s.

The fast average proper motion means that a non-negligible fraction of the pulsars propagating through the ISM may form a cometary-like bow shock structure. Optical and radio observations discovered a few bow shocks associated with fast moving pulsars (see, e.g., Kulkarni and Hester 1988; Cordes et al. 1993; Frail et al. 1996; Ng et al. 2012; Brownsberger and Romani 2014). High resolution X-ray observations of PWNe revealed a rich collection of their morphologies (Kargaltsev and Pavlov 2008) including a number of cometary type objects, some of which harbour bow shocks.

Bow-shock nebulae are known to originate from the interaction of winds from moving stars with the ambient medium and appear in association with very different types of stars: of solar type (see, e.g., Baranov et al. 1971), young massive stars (see, e.g., Weaver et al. 1977), and fast rotating magnetized neutron stars (see, e.g., Bucciantini and Bandiera 2001; Blondin et al. 2001). The dynamics of astrophysical bow shocks has been modelled both analytically, within the thin shell approximation, and by means of numerical hydrodynamics schemes (see, e.g., Bandiera 1993; Wilkin 1996).

The rich set of observed morphological and spectral appearances of PWNe traces the differences in the intrinsic properties of the parent pulsars (spin-down power and inclination), in their proper velocities, and in the evolutionary stages during the interaction of PWNe with the ISM (Gaensler and Slane 2006).

The structure of PWNe can be heavily affected by the supersonic motion of the medium in which the nebula is expanding. This can be either associated with the fast proper motion of the pulsar through the normal ISM or with the reverse shock of the supernova propagating back towards the origin of the explosion. The effect of the external ram pressure of the fast moving ambient medium inside the supernova remnant on the PWN structure was studied in the frame of magnetohydrodynamical models by van der Swaluw (2003), van der Swaluw et al. (2004). The authors applied the model to three SNRs: N157B, G327.1-1.1, and W44, and concluded that the head of the PWN is not bounded by a bow shock in the case of N157B, G327.1-1.1, while in the case of W44 they argued for the supersonic scenario with a bow shock PWN (BSPWN) due to the fast motion of the pulsar. Another case of a bow shock type structure occurs when the PW located inside an SNR is interacting with a mildly supersonic flow of Mach number $\gtrsim 1$ produced in the transition phase when the reverse shock reaches the center of the remnant (Chevalier and Reynolds 2011). A detailed one-zone model of spectral evolution of PWNe inside SNRs was presented by Bucciantini et al. (2011). The authors found that the one-zone model provides a satisfactory description of a number of PWNe assuming that relativistic pairs are injected in these nebulae with a spectral distribution in the form of a broken power law, rather flat at low energy (energy spectral index close to 1) and steepening (spectral index larger than 2) above some break energy. They found that the intrinsic spectral break turns out to occur at similar energies for PWNe with rather different characteristics. A more sophisticated modeling of BSPWN dynamics was undertaken by Bucciantini et al. (2005), who investigated it in the framework of axisymmetric relativistic MHD, highlighting the dependence of the dynamics on the PW magnetization.

The Alfvénic Mach number of a bow shock of speed $v_{\rm sh}$ (at the bow apex it is about the pulsar speed) is

$$\mathcal{M}_{\rm a} = v_{\rm sh}(4\pi\rho_{\rm a})^{1/2}/B_0 \approx 460\, v_8\, n_{\rm a}^{1/2}/B_{-6}, \tag{5}$$

where $n_{\rm a}$ is the ionized ambient gas number density measured in cm^{-3}, B_{-6} is the local magnetic field in the shock upstream measured in μG and v_8 is the shock velocity in 10^8 cm s^{-1}. The sonic Mach number for a shock propagating in the interstellar plasma of the standard (Solar) abundance is

$$\mathcal{M}_{\rm s} \approx 85\, v_8 \cdot \left[T_4 \cdot (1 + f_{\rm ei})\right]^{-1/2}, \tag{6}$$

where T_4 is the plasma ion temperature measured in 10^4 K and $f_{\rm ei} = T_{\rm e}/T_{\rm i}$ is the ratio between the electron and ion temperatures.

Consider a pulsar moving supersonically with a velocity $V_{\rm psr}$ relative to the local rest frame of the ambient medium. The standoff distance $R_{\rm cd}$ of the PW termination shock in this frame can be estimated as

$$R_{\rm cd} \approx \sqrt{\frac{\zeta_{\rm K}\dot{E}}{6\pi\rho_{\rm ism}V_{\rm psr}^2 c}} \approx 2\times 10^{16}\dot{E}_{35}^{1/2} n_{\rm ism}^{-1/2} V_{200}^{-1}\ {\rm cm}, \tag{7}$$

where $\dot{E}$ is the pulsar spin-down power, $\rho_{\rm ism}$ is the ambient ISM density. In fact, $R_{\rm cd}$ is rather the location of the contact discontinuity, than either of the PWN termination shock, or of the bow shock. The parameter $\zeta_{\rm K}$ depends on the magnetic inclination angle of the pulsar; $\zeta_{\rm K} = 1$ if the PW is spherically symmetric (e.g., van der Swaluw et al. 2003). However, observations of the Crab nebula as well as *particle-in-cell* and MHD modeling of the relativistic pulsar wind indicate that the wind power is likely highly anisotropic. Such anisotropy affects the value of $\zeta_{\rm K}$ and, therefore, $R_{\rm cd}$, which scales $\propto \zeta^{1/2}$. This means that if the direction of the pulsar proper motion is close to the equatorial plane of a low-inclination rotator (i.e., the direction of the maximal wind power) the position of the contact discontinuity will be at an angular distance $\geq 6''\zeta_{\rm K}^{1/2} d_{250}^{-1}$ for the most likely pulsar and ISM parameters. The images like the one presented in Fig. 8 may allow us to constrain the wind power anisotropy.

The model of a PWN evolving inside the remnant of the parent supernova developed by van der Swaluw et al. (2004) employed a hydrodynamical approach to simulate the evolution of such a system when the pulsar's velocity is high. The authors considered four different stages of PWN evolution: the supersonic expansion stage, the stage of reverse shock interaction followed by the subsonic expansion stage and, finally, the bow shock stage. Within this model the bow shock stage occurs at roughly half the crossing time, when the pulsar is located at about 0.68 times the radius of the SNR forward shock. The authors suggested that the pulsar in the W44 SNR has a bow shock because of its supersonic proper motion. However, Chevalier and Reynolds (2011) argued that even in the case of the Vela pulsar, which is located closer than 0.68 of the SNR radius to the center of the remnant, and whose measured proper velocity is below 70 $\mathrm{km\,s^{-1}}$ (Dodson et al. 2003b), a bow shock still can be formed. It is likely that the SNR reverse shock has recently passed over the Vela pulsar and a mildly supersonic flow of $\mathcal{M}_s \sim 1.3$ is naturally produced. Chevalier and Reynolds (2011) noted that van der Swaluw et al. (2004) had reached their conclusion having assumed that the SNR flow can be described by the Sedov solution, while there is a transition solution preceding the Sedov stage, which is likely to describe the current evolutionary stage of Vela more accurately. Within such a refined description, Chevalier and Reynolds (2011) found that for Vela-like pulsars the flow velocity downstream of the reverse shock of speed v_{sh} is characterized by a velocity 0.75 v_{sh} which appears to be about 1.3 times the speed of sound.

In what follows, we discuss in some detail how particles are accelerated in PWNe (Sect. 2), with a specific attention to acceleration in the PWNe driving bow shocks. Namely, efficient acceleration of particles in colliding shock flows (CSFs) between the pulsar wind termination surfaces and bow shocks is considered in Sect. 3 and a Monte-Carlo model is discussed in Sect. 4, which is further employed to simulate acceleration, transport, and synchrotron emission of the high energy $e^{\pm}$ pairs in BSPWNe. In Sect. 5 we review far-UV and X-ray observations of emission structures in the vicinities of some fast moving pulsars with BSPWNe and apply the model of Sect. 4 to simulate the synchrotron images and spectra of the sources similar to the nebula around PSSR J0437-4715. Section 6 is devoted to peculiar extended X-ray structures associated with the Geminga PWN, the "Guitar nebula", and the "Lighthouse nebula". The PWNe are moving supersonically and have hard photon indexes of X-ray emission in the vicinity of the bow shocks. The fascinating extended structures observed in X-rays can be associated with multi-TeV regime particles escaping from the bow shock PWN astrospheres to the ISM producing the magnetized ballistic beams of accelerated particles along the interstellar magnetic field lines. In Sect. 7 we discuss the X-ray observations of several BSPWNe possibly interacting with the remnants of their natal supernovae, and discuss the Vela X nebula in Sect. 8 as a generic case, which also includes a broadband modeling of its spectrum. Finally, in Sect. 9, we discuss the role of Bow Shock PWNe as plausible sources of the positron excess recently highlighted by PAMELA and AMS-02 measurements.

2 Particle Acceleration in Pulsar Wind Nebulae

PWNe are powered by highly magnetized relativistic winds which are terminated at some distance from the star. At the termination surface a considerable fraction of the pulsar spin-down power is transferred to accelerated particles producing non-thermal radiation (see Rees and Gunn 1974; Kennel and Coroniti 1984; Chevalier 2000; Arons 2012; Sironi et al. 2015).

The maximal energy of accelerated particles is limited according to a given magnetic luminosity of the source $\mathcal{L}_{\mathcal{M}}$ (see, e.g., Lemoine and Waxman 2009). Such energy can be estimated by the condition on the particle acceleration time scale in the comoving frame of the flow, which should be shorter than the dynamical time of the flow. Within the validity limits of relativistic MHD the acceleration time exceeds the particle gyroperiod (this condition does not apply to the region where magnetic reconnection occurs).

Let us consider acceleration of particles of charge Z in a Poynting flux dominated source of magnetic luminosity $\mathcal{L}_{\mathcal{M}}$. According to the analysis of Norman et al. (1995), Waxman (1995), Lemoine and Waxman (2009), the luminosity required to reach the energy E (in the absence of energy losses) is

$$\mathcal{L}_{\mathcal{M}} \approx 3 \times 10^{33} \cdot \frac{\Gamma_{\rm flow}^2}{\beta_{\rm flow}} \left(\frac{E/Z}{10^{14}\ \rm eV} \right)^2 \ {\rm erg\,s^{-1}}, \quad (8)$$

where $\beta_{\rm flow}$ is the flow speed divided by the speed of light c, and $\Gamma_{\rm flow}$ the flow Lorentz factor. Therefore the maximal energy of a particle which is achievable in the MHD outflow without the account for energy losses is limited by:

$$E_{\rm max} \approx Z \times 10^{14} \cdot \frac{\beta_{\rm flow}^{1/2}}{\Gamma_{\rm flow}} \left(\frac{\dot{E}}{3 \times 10^{33}\ {\rm erg\,s^{-1}}} \right)^{1/2} \ {\rm eV}, \quad (9)$$

This approximate relation indicates that the most favourable sites to achieve the highest energy are indeed in the trans- or sub-relativistic flows of $\Gamma_{\rm flow}\beta_{\rm flow} \sim 1$ located between the ultrarelativistic termination shock and the non-relativistic bow shock which we will discuss in Sect. 3 below.

Sironi et al. (2013) provided a detailed analysis of the maximal energies of the particles accelerated at relativistic shocks. They found that the relativistic perpendicular shocks propagating in $e^{\pm}$ pair plasma (which is the likely case in PWNe) can efficiently accelerate the pairs only if the magnetization in the upstream plasma is $\lesssim 10^{-3}$ (and even lower, $\lesssim 3 \times 10^{-5}$, for the electron-ion plasmas). The maximal Lorentz factor of the pairs accelerated at the PW termination shock was estimated by Sironi et al. (2013) from the constraint that the acceleration time of the highest energy pairs be shorter than the termination shock crossing time. These authors found that

$$\gamma_{\rm max} \approx 1.9 \times 10^7 \cdot \left(\frac{\dot{E}}{10^{38}\ {\rm erg\,s^{-1}}} \right)^{3/4} \left(\frac{\dot{N}}{10^{40}\ {\rm s^{-1}}} \right)^{-1/2}, \quad (10)$$

where $\dot{N}$ is the total particle flux entering the nebula, which is somewhat uncertain and is likely within the range $3 \times 10^{38} \lesssim \dot{N} \lesssim 10^{40}\ {\rm s^{-1}}$. The estimated maximal energy is rather low even for the Crab nebula in spite of its high spin-down power. Therefore Sironi et al. (2013) concluded that Fermi acceleration at the termination shock of PWNe is not a likely mechanism for producing the synchrotron emitting pairs and some other mechanisms should be considered.

A number of pulsars with observed bow shocks, including PSR B2224+65, which powers the Guitar nebula, the old recycled pulsar J0437-4715, PSR J0633+1746 (Geminga) and others, have a spin-down power $\dot{E}_{35} \lesssim 0.3$. Several BSPWNe harbor regions of the extended X-ray emission, where the magnetic field is estimated to be a few tens of μG. In order to produce X-ray synchrotron photons in such a field, pairs with a Lorentz factor $\gamma \gtrsim 10^8$ are required. But according to Eq. (8) such high energies are compatible with the amount of spin-down power mentioned above only if $\Gamma_{\rm flow}^2 \lesssim 3$ while $\beta_{\rm flow} \gtrsim 0.1$, i.e., the outflow is transrelativistic in the particle acceleration zone. The Vela pulsar as well as the pulsar IGR J11014-6103 (which powers the Lighthouse nebula) have much higher spin-down power and the condition for acceleration of pairs above the Lorentz factor $\gamma \gtrsim 10^8$ is relaxed to $\beta_{\rm flow} \gtrsim 10^{-3}$.

In the light of these estimates, we will discuss below the acceleration of particles in fast flows around PW termination shocks and bow shocks.

2.1 Particle Acceleration and Magnetic Field Amplification at Bow Shocks

Diffusive shock acceleration (Axford et al. 1977; Krymskii 1977; Bell 1978; Blandford and Ostriker 1978) has been demonstrated to be a very attractive mechanism of particle acceleration in supersonic flows of a moderate magnetization. There is a growing body of observational evidence in favor of this mechanism acting in SNRs, AGN jets and some other objects (see, e.g., Bell 2013; Helder et al. 2012; Marcowith et al. 2016). Particle acceleration in fast MHD shocks is strongly coupled with magnetic field amplification (MFA) which is due to the growth of seed fluctuations by the cosmic ray (CR) driven instabilities. A number of such instabilities resulting in MFA can be excited during diffusive shock acceleration (see, e.g., Bell 2004; Amato and Blasi 2009; Bykov et al. 2012, 2014; Schure et al. 2012).

The maximal energy $E_{\rm max}$ of a particle of charge Z accelerated by a strong quasi-steady bow shock of velocity $v_{\rm sh}$ and characteristic radius $R_{\rm sh}$ is limited by the shock curvature. It can be roughly estimated from

$$D(E_{\rm max}) \sim v_{\rm sh} R_{\rm sh}, \tag{11}$$

where $D(E)$ is the particle diffusion coefficient in the shock upstream. Particle propagation at a pulsar bow shock depends on the shock obliquity, i.e., on the angle between the proper velocity of the pulsar and the direction of the local mean interstellar magnetic field. If the shock is quasi-parallel, it is determined by the magnetic fluctuations amplified by the CR-driven instabilities, as shown below in Sect. 2.2. On the other hand, a transverse shock geometry may reduce the radial diffusion coefficients and therefore drastically change the conditions of particle confinement.

2.2 Quasi-parallel Bow Shocks

The amplitude $\delta B_{\rm w}$ of the fluctuating magnetic field amplified by CR-driven instabilities in the upstream of a shock can be estimated from the saturation condition of Bell (2004):

$$\frac{\delta B_{\rm w}^2}{8\pi} \approx \frac{v_{\rm sh}}{2c} U_{\rm CR}, \tag{12}$$

where $U_{\rm CR} \approx \zeta_{\rm CR}\, \rho v_{\rm sh}^2/2$ is the energy density of the CRs accelerated at the shock. Then the amplitude of the fluctuating magnetic field is

$$\delta B_{\rm w} \approx \sqrt{2\pi \zeta_{\rm CR} \rho v_{\rm sh}^3/c} \sim 20 \zeta_{\rm CR}^{1/2}\, v_{\rm sh8}^{3/2}\, n^{1/2}\ \mu{\rm G}, \tag{13}$$

where $\zeta_{\rm CR}$ is the fraction of shock ram pressure converted into the energy of the accelerated particles, $v_{\rm sh8}$ is the shock velocity in 10^8 cm s^{-1} and n is the number density of the ISM. This amplitude may reach $\delta B_{\rm w} \gtrsim B_0$, where B_0 is the mean magnetic field in the ISM. In any case, in the vicinity of a bow shock, $\delta B_{\rm w}$ is likely well above the estimated amplitude of the field fluctuations in the background ISM turbulence at the scales below 10^{15} cm, which are important for the scattering of TeV range particles.

The mean free path $\Lambda(E)$ of a relativistic proton in the quiet ISM away from SNRs or other active space objects can be estimated from the galactic CR propagation models as $\Lambda(E) \sim (3\text{–}5) \times 10^{18}\ (E/{\rm GeV})^a$ cm, where the index $a = 0.3\text{–}0.6$ (see, e.g., Strong et al. 2007). Therefore, the background magnetic turbulence in the quiet ISM can not confine the relativistic $e^{\pm}$ pairs accelerated in the PWNe within a region of scale size $\sim R_{\rm cd}$ as given by Eq. (7), for a bow shock speed above 100 km s^{-1}. Depending on the bow shock velocity,

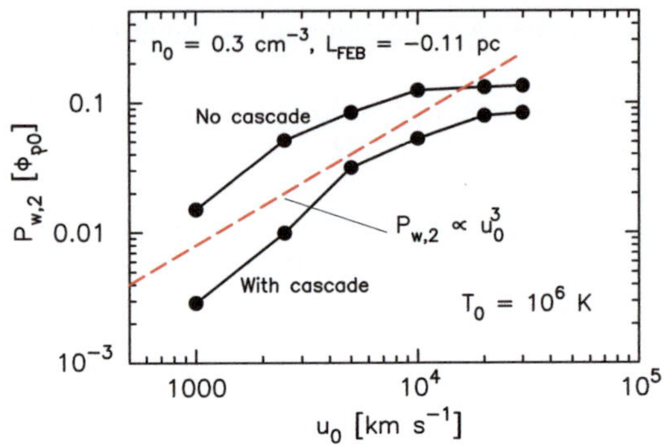

Fig. 1 The magnetic turbulence energy flux in the shock downstream $\zeta_B \equiv P_{w,2}$ as a function of shock velocity simulated with a nonlinear Monte-Carlo model of diffusive shock acceleration (Bykov et al. 2014)

strong MFA by CR-driven instabilities at the bow shock would affect the propagation and confinement of relativistic $e^{\pm}$ pairs up to Lorentz factors $\gtrsim 10^7$.

Nonlinear Monte-Carlo modeling of the diffusive shock acceleration process in the presence of MFA (Bykov et al. 2014) has revealed that the regime consistent with (12) holds for shock velocities below 5000 km s^{-1}. At higher shock velocities the energy density of the fluctuating magnetic field scales as a fraction ζ_B of the shock ram pressure. The MFA efficiency is illustrated in Fig. 1 where the ratio ζ_B of the amplified magnetic field energy density to the shock ram pressure is shown as a function of the shock velocity. In the high shock speed regime one can estimate the amplitude of the enhanced magnetic field as

$$\delta B_w \approx \sqrt{4\pi \zeta_B \rho}\, v_{sh} \sim 50 \left[\frac{\zeta_B}{0.01}\right]^{1/2} v_{sh8}\, n^{1/2}\ \mu\text{G}. \tag{14}$$

The validity of the Bohm diffusion approach requires strong scattering of particles by quasi-resonant fluctuations, i.e., $\delta B_w \sim B_0$. In such a case the diffusion coefficient $D(E) \approx \eta c E/(Ze\,\delta B_w)$, where $\eta \gtrsim 1$ is a numerical factor (see, e.g., Reville et al. 2008).

Models of extended fast shocks demonstrated that diffusive shock acceleration is rather efficient in this framework with $\zeta_{CR} > 0.2$ (see, e.g., Helder et al. 2012). The energy density of the CR-driven magnetic fluctuations in the shock vicinity is dominated by the fluctuations with typical wavenumbers k_* which satisfy the condition $k_* R_g(E_{max}) \lesssim 1$, where R_g is the gyroradius of a particle.

Using Eqs. (11), (7) one can obtain the maximal energy of accelerated ions as

$$\begin{aligned} E_{max} &\approx 3Ze\delta B_w \eta^{-1} R_{sh} \frac{v_{sh}}{c} \approx \frac{Ze}{\eta}\sqrt{\frac{3\zeta_K \zeta_{CR} \dot{E}}{c}} \left[\frac{v_{sh}}{c}\right]^{3/2} \\ &\approx \frac{0.2Z(\zeta_K\zeta_{CR})^{1/2}}{\eta} \dot{E}_{35}^{1/2} v_{sh8}^{3/2}\ \text{TeV} \end{aligned} \tag{15}$$

in the regime which is described by Eq. (13), while in the regime of high shock velocity described by Eq. (14) it can be approximated by

$$\begin{aligned} E_{max} &\approx 3Ze\delta B_w \eta^{-1} R_{sh} \frac{v_{sh}}{c} \approx \frac{Ze}{\eta}\sqrt{\frac{6\zeta_K \zeta_B \dot{E}}{c}}\, \frac{v_{sh}}{c} \\ &\approx \frac{4.5Z(\zeta_K\zeta_B)^{1/2}}{\eta} \dot{E}_{35}^{1/2} v_{sh8}\ \text{TeV}. \end{aligned} \tag{16}$$

Under the approximations described above, the maximal energies of the accelerated nuclei do not depend on the ambient ISM density but depend on the bow shock velocity in the ISM rest frame.

2.3 Propagation of Pulsar Wind-Accelerated $e^{\pm}$ Pairs at a Bow Shock

The protons accelerated at the bow shock produce magnetic fluctuations at wavelengths comparable to their gyroradii in the vicinity of the bow shock. Moreover, magnetic fluctuations of wavelengths up to a scale-length $L_{\rm cor}$ which is much longer than the gyroradii of protons of energy $E_{\rm max}$ can be produced via the inverse turbulent cascading and the non-resonant CR-driven instabilities (see, e.g., Bykov et al. 2013b, and references therein). These CR-driven fluctuations will determine the diffusive transport of $e^{\pm}$ pairs accelerated at the PW termination surface at energies above 10 GeV in the region between the contact discontinuity and the bow shock. The strong long-wavelength fluctuations would provide a Bohm-type diffusion coefficient $D(E) \approx \eta\, D_B(E)$ with $\eta \sim 1$ for $e^{\pm}$ pairs up to a Lorentz factor γ_1 determined by the condition $R_{\rm g0}(mc^2\gamma_1) \approx L_{\rm cor}$, where $R_{\rm g0}(E) = E/eB_0$ is the particle gyroradius in the mean magnetic field B_0.

Accurate estimations of the scale-length $L_{\rm cor}$ are not available yet because of the limited dynamical range of numerical simulations of the CR-driven turbulence. Therefore, we consider $L_{\rm cor}$ as a free parameter which determines the Lorentz factor γ_1 where a transition from the Bohm-like diffusion to the small-scale scattering regime given by Eq. (17) takes place.

For particles with gyroradii larger than the scale-length of the strongly amplified magnetic turbulence $L_{\rm cor}$, the mean free path produced by the short-scale modes can be approximated as

$$\Lambda(p) = \frac{4}{\pi}\frac{R_{\rm st}^2(p)}{L_{\rm cor}} \propto p^2, \tag{17}$$

where $R_{\rm st}(p) = cp/(e\,\delta B_{\rm w})$. In Sect. 4 we will apply the diffusion model to simulate distinctive features of synchrotron images of some BSPWNe. Apart from the effect on propagation of the $e^{\pm}$ pairs, the MHD flow between a fast bow shock and the PW termination shock (a colliding shock flow) may result in a specific way of acceleration of the highest energy $e^{\pm}$ pairs. The hard energy spectra of the pairs produced in the colliding shock flow (CSF) can be responsible for the hard spectra of synchrotron and inverse Compton photons (of photon indexes $\Gamma \sim$ 1.0–1.5) recently observed from the heads of some of BSPWNe in the X-rays and gamma-rays, respectively.

2.4 Quasi-transverse Bow-Shocks

When a shock is highly oblique in respect to the direction of the local mean magnetic field, the transverse diffusion coefficient plays an important role in the diffusive shock acceleration. When the particle mean free path along the magnetic field $\Lambda(p)$ is large compared to the gyroradius $R_{\rm g0}$, the transverse diffusion coefficient $D_\perp$ is much smaller than that along the mean field $D_\parallel$:

$$D_\perp \approx D_\parallel(p)\frac{1}{1+\zeta^2}, \qquad D_\parallel(p) = \frac{v\Lambda(p)}{3}, \qquad \zeta = \Lambda(p)/R_{\rm g0}. \tag{18}$$

When $\zeta \gg 1$ one obtains $D_\perp = [R_{\rm g0}(p)/\Lambda(p)]^2 D_\parallel(p)$. Thus, the orientation of the local magnetic field can significantly affect the transport of particles across the shock front, and, consequently, the efficiency of particle acceleration (see, e.g., Jokipii 1987; Takamoto and Kirk 2015).

The different bow shock obliquities may explain the substantial differences in the luminosities and observed spectra of BSPWNe with otherwise similar properties, in particular, the nebulae associated with PSR J0437-4715 and PSR J1741-2054.

3 Particle Acceleration in Colliding Shock Flows

Diffusive shock acceleration with MFA is rather an efficient way to produce relativistic particles. However, an even more efficient realization of the Fermi acceleration mechanism is acceleration of particles in colliding MHD flows (colliding shock flows, CSFs) carrying magnetic inhomogeneities, which may scatter relativistic particles (Bykov 2005; Bykov et al. 2013). This kind of MHD flows is likely realized in a number of astrophysical objects, in particular, when an expanding shell of a supernova is closely approaching or interacting with a fast powerful wind of a nearby young massive star. Converging MHD flows occur in the vicinity of a contact discontinuity of an astropause, downstream of a fast bow shock produced by a supersonically moving astrosphere driven by a stellar wind (the relativistic PW in the case of a PWN).

Consider a simple analytical test particle kinetic model of particle acceleration by the Fermi mechanism in a CSF (Bykov 2005). Two colliding plane-parallel flows of velocities u_1 and u_2 aligned with the positive and negative directions of axis z and occupying half-spaces $z < 0$ and $z > 0$, respectively, are separated by a discontinuity in the plane $z = 0$. The flows carry fluctuating magnetic fields amplified by anisotropic energetic particle distributions (e.g., the CR current driven instability). The particles propagate in a diffusive medium described by the fast particle diffusion coefficients $D_{\mathrm{i}}(\gamma)$ which depend on the particle Lorentz factor γ.

This simplified model allows us to illustrate the Fermi acceleration in a CSF of a BSPWN and obtain some important estimates. The collision of the flow emerging from the PW termination region with that coming from the bow shock can be viewed within the discussed simple scheme, but in the vicinity of the contact discontinuity the flows have a complex multi-dimensional structure. Note that the high CR pressure in this region may be dynamically important for the flow structure if the MHD turbulence provides a strong coupling of the CRs with the flow. However, the highest energy $e^{\pm}$ pairs accelerated in the PW termination region may have so high Lorentz factors that their mean free paths become larger than the characteristic size of the multi-dimensional bypassing flows area in the vicinity of the contact discontinuity. Let us denote the minimal Lorentz factor corresponding to this regime as γ_0.

Then the accelerated pair distribution function above γ_0 in between the two shocks can be approximated by the solution obtained from the kinetic equations in the discussed simple plane-parallel model:

$$N_i^{\pm}(z,\gamma,t) = \frac{N_0(\gamma_0,\gamma_m)}{\gamma}\cdot H(\gamma-\gamma_0)\cdot H\big(t-\tau_a(\gamma)\big)\cdot \exp\big(-u_{\mathrm{i}}\cdot|z|/D_{\mathrm{i}}(\gamma)\big), \tag{19}$$

where $H(x)$ is the standard Heaviside step function, γ_m is the maximal Lorentz factor of an accelerated particle, and $N_0(\gamma_0,\gamma_m)$ is a normalization constant. The acceleration time of the highest energy $e^{\pm}$ pair is

$$\tau_a = \frac{3}{(u_1+u_2)}\int_{\gamma_0}^{\gamma}\left(\frac{D_1(\gamma)}{u_1}+\frac{D_2(\gamma)}{u_2}\right)\frac{d\gamma}{\gamma}. \tag{20}$$

We take $u_1 \approx c/3$, $u_2 \ll c$ in accordance with the typical flow velocities downstream of the PW termination shock and at the bow shock, respectively, while $D_2(\gamma) = \eta c R_g(\gamma)/3$ ($\eta \geq 1$) is the Bohm diffusion coefficient in the half-space $z > 0$. Here $R_g(\gamma)$ is the particle gyroradius in the magnetic field amplified by the CR-driven instabilities. In the limit $u_2 \ll c$ Eq. (20) simplifies to

$$\tau_a \approx \frac{9}{c} \int_{\gamma_0}^{\gamma} \frac{D_2(\gamma)}{u_2} \frac{d\gamma}{\gamma}. \tag{21}$$

As the mean magnetic field behind the termination shock of a PWN is azimuthal, the radial transport of the pairs at high energies across the mean field is due to the transverse diffusion. The pair propagation at the bow shock depends on the shock obliquity. The transverse shock geometry reduces the radial diffusion coefficients.

3.1 The Maximal Energies of $e^{\pm}$ Pairs Accelerated in CSFs

The maximal Lorentz factors γ_m of the $e^{\pm}$ pairs accelerated in CSFs are mainly limited either (i) by particle escape from the accelerator or (ii) by the particle energy losses.

In the first case, the particle confinement condition in a CSF requires that the acceleration time Eq. (21) does not exceed the time of diffusion through the region within the bow shock. Thus, the maximal Lorentz factor should not exceed γ_{m1} satisfying the equation

$$\frac{9D(\gamma_{m1})}{c u_{\rm sh}} = \frac{R_{\rm cd}^2}{4D(\gamma_{m1})}. \tag{22}$$

Therefore, assuming the Bohm diffusion in the accelerator, the upper limit on γ_{m1} can be obtained from

$$R_g(\gamma_{m1}) \approx \frac{1}{2}\sqrt{\frac{u_{\rm sh}}{c}}\, R_{\rm cd}. \tag{23}$$

However, the estimations based on the analytic one-dimensional solutions given by Eqs. (19)–(20) are rather rough. Hence in Sect. 4 we will discuss test particle Monte-Carlo simulations of the CR acceleration and transport in BSPWNe.

In the second case, the synchrotron-Compton energy loss time of the pairs, τ_s, should be longer than the acceleration time. This provides another constraint on the maximal Lorentz factor $\gamma_m \leq \gamma_{m2}$ given by the equation $\tau_s(\gamma_{m2}) = \tau_a(\gamma_{m2})$. Using a simplified approximation for τ_s which accounts for inverse Compton losses on the cosmic microwave background photons (as appropriate for very high energy pairs) and synchrotron losses in the magnetic field B:

$$\tau_s(\gamma) \approx 7 \times 10^{19} \gamma^{-1} \left[1 + \left(\frac{B}{3\,\mu\mathrm{G}}\right)^2\right]^{-1} \mathrm{s},$$

one obtains

$$\gamma_{m2}^2 \approx 4 \times 10^{17} \eta^{-1} \left(\frac{u_2}{100\ \mathrm{km\,s^{-1}}}\right)\left(\frac{B}{3\,\mu\mathrm{G}}\right)\left[1 + \left(\frac{B}{3\,\mu\mathrm{G}}\right)^2\right]^{-1}. \tag{24}$$

Then the maximal Lorentz factor $\gamma_{\rm m}$ of the pairs accelerated in a CSF can be estimated as $\gamma_{\rm m} = \min(\gamma_{m1}, \gamma_{m2})$.

The spectral upturn at the high energies, namely, the fact that a large fraction of energy is carried by the particles with the longest mean free path, can result in an important non-linear modification of the flow in the region of convergence. Such an effect was studied in the context of CR-modified shocks by Bykov et al. (2013) within a plane-parallel time-dependent model. In the case of interest here, i.e., for the diffusive transport and re-acceleration of the high energy end of the spectrum of the $e^{\pm}$ pairs which were produced at the PW termination shock, the non-planarity is very important. Therefore, in Sect. 4 we discuss a Monte-Carlo model of acceleration and propagation of high energy $e^{\pm}$ pairs in the astrosphere produced by the relativistic PW which moves supersonically through the ambient medium.

Acceleration in a CSF can provide very fast and efficient creation of a nonthermal particle population in between the PW termination region and the bow shock. A very hard energy spectrum of the $e^{\pm}$ pairs is formed, $\propto \gamma^{-b}$ with $b \sim 1$, which may contain a substantial fraction of the kinetic power released by the pulsar in the highest energy particles. The synchrotron X-ray emission from the extended region between the PW termination shock and the bow shock where acceleration of pairs occurs would have a hard photon index, $1 \le \Gamma \le 1.5$. The associated inverse Compton emission may reach the TeV regime and extend to large distances downstream of the bow shock. Such an extended structure of hard emission may be treated as a distinctive feature of BSPWNe.

4 The Monte-Carlo Particle Acceleration and Transport Model

The modeling of the structure of PWNe interacting with supersonic flows either inside or outside of supernova remnants with an adequate account for the magnetic field structure is a complicated problem (see, e.g., Blondin et al. 2001; Bucciantini and Bandiera 2001; van der Swaluw et al. 2004; Bucciantini 2014). The study of particle transport in these systems faces the complexities related to a non-trivial behaviour of magnetized flows in the vicinity of the contact discontinuity. Fortunately, for electrons and positrons emitting synchrotron radiation in the UV and X-ray range this problem is alleviated due to their large mean free paths. Along with the deceleration of the colliding flows and dissipation of the transported magnetic inhomogeneities it significantly simplifies the problem. The transport of high energy $e^{\pm}$ pairs is not very sensitive to the complicated structure of the multidimensional flow in the vicinity of the contact discontinuity. However, the flow advects the low energy pairs along the nebula trail. We discuss here a simplified Monte-Carlo model of the high energy pair transport in BSPWNe.

4.1 The Model Approach

Simulation of the relativistic PW particle acceleration and transport through a PWN interacting with a supersonic flow is performed via a stationary Monte-Carlo approach. The $e^{\pm}$ pairs injected into the system at the PW termination surface are treated as test particles and propagated through the PWN with a fixed set of diffusion parameters, flow velocities, and geometry. An important feature of the approach is the account of the main geometrical properties of the modelled system.

4.2 The Model Geometry

The modelled system is assumed to be axisymmetric and treated as a composition of axisymmetric regions with spatially homogeneous but energy-dependent particle diffusion coefficients. A sketch of the modelled system geometry is presented in Fig. 2. The spherical coordinates r, θ, and ϕ are centered at the pulsar position and the zenith direction is aligned

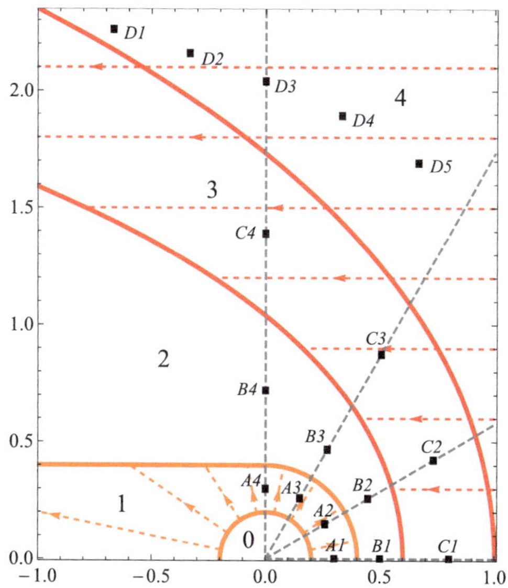

Fig. 2 The geometry of the model: half of the axial section of the system by the plane perpendicular to the line of sight, crossing the pulsar (plane *PoS*). The numbers 0–4 label the regions with different diffusion modes: 0 is the cold PW, 1 (between the orange curves) is the shocked PW, 2 is the zone near the contact discontinuity, 3 (between the red curves) is the postshock flow of the ISM matter, 4 is the unperturbed ISM. The relative sizes of the regions are not to the scale and depend on the particular properties of a BSPWN. The locations (bins) chosen in the following sections to illustrate simulated particle energy distributions are marked by labelled black squares ($A1$–$D5$). The bins are located at the midpoints of the dashed gray segments of the lines connecting the pulsar and the surfaces, which separate regions 0–4 (see Sect. 4.2). The *gray lines* are directed at 0°, 30°, 60°, and 90° with respect to the system symmetry axis

with the pulsar velocity vector. Also the cylindrical coordinates (ρ, α, x) orientated along the symmetry axis of the system (Ox) are used.

The innermost spherical region with radius r_{ts} corresponds to the cold pulsar wind. Through its boundary—the termination shock—the particles are injected into the system. The region labeled "1" corresponds to the shocked pulsar wind (the PWN). Its outer boundary has the following shape:

$$r(\theta) = \begin{cases} a_{in}, & \text{if } \theta \leq \pi/2 \\ a_{in}/\sin\theta, & \text{if } \theta > \pi/2. \end{cases} \tag{25}$$

The scaling factor a_{in} is a model parameter. The shocked PW flow velocity is directed radially and its value is $u_w = \frac{c}{3} r_{ts}^2 / r^2$ at distance r from the pulsar. The region labeled "3" corresponds to the postshock flow of the ambient ISM. Its boundaries are described by the equation for the bow shock shape derived by Wilkin (1996) in the thin shell approximation:

$$r(\theta, \sigma) = \sigma R_{cd} \csc\theta \sqrt{3(1 - \theta \cot\theta)}. \tag{26}$$

Here $R_{\rm cd}$ is the standoff distance defined by Eq. (7). The positions of inner and outer boundaries of region "3" are determined by the model parameter σ. The apexes of these surfaces are denoted as $a_{\rm out}$ and $a_{\rm sys}$, respectively. The velocity of the flow through the bow shock in the pulsar rest frame is fixed and directed oppositely to the pulsar velocity in both regions

"3" and "4". In region "2", located near the contact discontinuity, advection is neglected. Region "4" corresponds to the unperturbed ambient ISM. The system is bounded by a cylinder whose axis coincides with the system symmetry axis and whose height is $h_{\rm sys} = x_{\rm FEB} + a_{\rm sys}$, where $x = -x_{\rm FEB}$ is the coordinate of the tail-oriented base of the cylinder. The cylinder radius $\rho_{\rm sys}$ matches the radius of section of region "3" outer boundary by the $x = -x_{\rm FEB}$ plane. At the cylinder, free escape boundary conditions are imposed.

4.3 Particle Propagation

In each of the regions "1"–"4" a spatially uniform mean value of magnetic induction B is specified. The parameters of particle diffusion are chosen as follows. In the PWN we adopt the mean free path λ_1 proportional to the particle gyroradius R_g:

$$\lambda_1(B,\eta,\gamma) = \eta R_g(B,\gamma) = \eta \frac{m_e c^2 \gamma}{eB}. \tag{27}$$

For the ambient medium the mean free path is taken in the form:

$$\lambda_{\rm am}(\gamma) = 3 \times 10^{18} \left(\frac{E}{1\ {\rm GeV}} \right)^{1/3} \ {\rm cm}. \tag{28}$$

In regions "2" and "3" the mean free path is chosen in the form:

$$\lambda_{\rm bow}(B,\eta,\gamma) = \begin{cases} \eta R_g(B,\gamma), & \text{if } \gamma \le \gamma_1 \\ \eta R_g(B,\gamma)\gamma/\gamma_1, & \text{if } \gamma_1 < \gamma \le \gamma_2 \\ \lambda_{\rm am}, & \text{if } \gamma > \gamma_2. \end{cases} \tag{29}$$

In Eqs. (27)–(29) e and m_e are the charge and mass of an electron or a positron, and $E = m_e c^2 \gamma$ is its energy. The values of $\eta \ge 1$ and γ_1 are free parameters of the scattering model, while γ_2 is derived from the continuity of the mean free path value.

At the beginning of a simulation the initial spectrum of particles, with a given shape, is generated. Then, in the course of Monte-Carlo modeling of particle acceleration and transport, the generated particles are injected into the system one by one. In the developed numerical code the injection is implemented as follows. Two random numbers are generated for every given particle to define its initial position on the termination shock. Two more are generated to specify the initial velocity direction. The particle then propagates for a distance equal to its mean free path in region "1" and after that is scattered for the first time.

The scatterings of the particles are isotropic in the local plasma rest frame. At each scattering the value of particle momentum in the local plasma rest frame is calculated. Then two random numbers which define two scattering angles are generated and the particle momentum is rotated according to the paper of Ostrowski (1991). After that the new particle momentum in the pulsar rest frame is calculated. The value of $\delta\Omega_{\rm max}$, introduced by Ostrowski (1991) is taken equal to π (due to that, η actually should not be less than 2).

After injection, the particle is propagated through the system. The particle moves along a straight line during the time interval $dt_{\rm mfp} = \Lambda/v$ (where Λ is its mean free path at the current location, and $v \approx c$ is its velocity), and then is scattered. This cycle repeats until the particle finally crosses the free escape boundary, and, therefore, leaves the system. Then the next particle is injected and its propagation starts. When the particle crosses the border between two regions with different scattering parameters, the crossing point is determined and the particle is moved from this point in the same direction for a distance equal to $\lambda_{\rm new}(1 - \Delta x/\lambda_{\rm old})$, where $\lambda_{\rm old}$ is the mean free path in the region it has left, $\lambda_{\rm new}$ is the mean free path in the region it has entered and Δx is the path of the particle in the region it has left after the last scattering. If the particle enters the cold PW region, it is reflected from its boundary in the case when its gyroradius in the magnetic field downstream of the PW is smaller than r_{ts} and crosses it without scattering otherwise.

The phase space of the system is divided into bins—small volumes of phase space, which are used to 'detect' the particle spatial and momentum distribution. The coordinate space of the system is permeated by a cylindrical 'grid' of bins. The cylinder containing the system is splitted into N_x equal cylinders along the Ox axis. Each of these smaller cylinders is divided into N_ρ coaxial cylindrical layers, whose outer radii grow linearly as $\rho_n = n\rho_{\rm sys}/N_\rho$, where $n = 1, 2, \ldots, N_\rho$. Each cylindrical layer is divided into N_α equal segments by semi-planes $\alpha = 2\pi k/N_\alpha$, $k = 1, \ldots, N_\alpha$. To simulate the particle energy distribution, recording of the absolute values of particle momenta at each detector is required. In order to do this, we introduce a momentum range which is somewhat wider than the range of particle momenta in the initial (injected) particle distribution. This 'detector' range is logarithmically binned; hence N_m momentum bins are introduced. In case of considerable anisotropy of particle energy distribution function detection of particle angular distribution is also of some interest. Thus, the ranges of $\mu = \cos\theta$ and ϕ values are splitted into N_μ and N_ϕ bins, respectively. The particle is detected by the bin with a given set of numbers $(n_x, n_\rho, n_\alpha, n_m, n_\mu, n_\phi)$ if its energy corresponds to the n_m-th momentum bin, direction of velocity to n_μ-th and n_ϕ-th angular bins, and it crosses the left boundary of the spatial bin (n_x, n_ρ, n_α) corresponding to $x = -x_{\rm FEB} + (n_x - 1)h_{\rm sys}/N_x$. To obtain the particle distribution function from the 'detection' data, we employ the formulae analogous to those presented in Sect. 3.1.4 of Vladimirov (2009). In order to compensate for the growth of the bin size with the increasing distance from the axis, the values measured by the detectors are divided by the ratio of the current bin volume to the volume of the bin at $n = 1$.

To take into account the effect of energy losses due to synchrotron radiation, after each straight line piece of trajectory, the particle momentum value is re-evaluated as $p_2 = p_1/(1 + p_1 S dt)$. Here p_1 is the momentum in the local plasma rest frame at the beginning of the current piece of trajectory, p_2 is the new momentum value in the same frame, dt is the propagation time, and S is the synchrotron loss coefficient:

$$\frac{dp}{dt} = -Sp^2 = -\frac{4}{9}\frac{e^4 B^2}{m^4 c^6} p^2.$$

Since for all the explored sets of model parameters the particles do not suffer strong energy losses over a diffusion length, the particle energy is effectively decreased due to these losses only after the particle has crossed all the 'detectors' along a mean free path. If a particle crosses the boundary between regions with different values of B, the decrease of its energy is calculated consequently for pieces of trajectory in each region.

4.4 The Model Spectra

The Monte-Carlo procedure described in Sect. 4.3 allows us to simulate particle spectra at certain locations inside the BSPWN system as well as spectra of synchrotron emission radiated by the accelerated particles. The intensity of synchrotron emission at various photon energies can be calculated for the lines of sight crossing these locations, thus allowing us to simulate emission maps in the sky plane. The locations chosen to present the particle spectrum modeling results are specified as follows. We choose a certain axial section plane of the axisymmetric BSPWN system. A right-handed Cartesian coordinate system (xyz) centered at the point P, the pulsar position, is introduced. The x axis matches the symmetry axis and the plane Pxz matches the fixed axial section. We choose four groups of points in the Pxz plane. Namely, we take four rays starting at P, lying in $z \geq 0$ and oriented at angles 0, 30, 60 and 90 degrees with respect to the symmetry axis. Along each ray we select three locations, each in the center of the interval delimited by the intersections of the ray with the surfaces separating regions with different diffusion properties. Thus, we select $3 \times 4 = 12$ points in regions "1"–"3". We include five additional points in region "4". Five rays parallel to the Pz axis and lying in the Pxz plane ($z \geq 0$) are chosen at x coordinates $x_j = -x_{\rm FEB} + j(a_{sys} + x_{\rm FEB})/6$, $j = 1..5$. The locations are taken at the centers of the intervals between the points of intersection of these rays with the outer boundary of region "3" and with the outer boundary of the whole system.

The orientation of the considered system with respect to the direction to the observer is defined as follows. We introduce an axis Y, directed from the observer towards the pulsar (point P_0) and the plane of the sky (PoS), orthogonal to this axis and containing P_0. The orientation of the source is determined by three angles Φ, Θ, and Ψ. Φ is the angle between the Py axis and the line l_Y of intersection of the planes Pzy and PoS (chosen in the positive direction: counterclockwise when looking at the Pzy plane against the Px axis direction). Θ is the angle between Y and Px axes. Ψ is the angle between l_Y and the North direction in the PoS. Thus, one may introduce a coordinate system (XYZ) centered at P_0, X and Z axes in the PoS, where the Z axis is aligned with the direction to the North. Then the angles Φ, Θ, and Ψ would be the three Euler angles defining the orientation of the XYZ system with respect to the xyz system. In the following and in Fig. 2 we use $\Theta = 90°$.

5 Observations and Modeling of X-ray and Far-UV Emission from BSPWNe

Multi-wavelength observations in the radio, optical and X-ray bands have revealed the presence of bow-shock type structures in a number of PWNe (see, e.g., Pellizzoni et al. 2004; Gaensler and Slane 2006).

An example of an apparent bow shock PWN (BSPWN) is the Mouse nebula powered by the radio pulsar J1747-2958, which is moving with transverse velocity $(306 \pm 43) \times (d/5\ \mathrm{kpc})\ \mathrm{km\,s^{-1}}$ (Gaensler et al. 2004; Hales et al. 2009). The spin-down power of J1747-2958 is about $2.5 \times 10^{36}\ \mathrm{erg\,s^{-1}}$, while the X-ray luminosity of the nebula in the 0.5–8 keV range is about $5 \times 10^{34}\ \mathrm{erg\,s^{-1}}$, and the observed PWN structure shows a narrow $45''$-long X-ray tail (the length is ~ 1.1 pc at the 5 kpc distance). The best-fit power law photon index of the X-ray spectrum $\Gamma = 2.0 \pm 0.3$ was derived from *Chandra* observations by Gaensler et al. (2004), who also noticed some steepening of the spectrum from the halo to the tail of the PWN.

High resolution observations of PWNe in H_α (see, e.g., Kulkarni and Hester 1988; Cordes et al. 1993) have provided images of cometary-like bow shock structures ahead of fast moving pulsars. A recent survey of such bow shocks (Brownsberger and Romani 2014) listed 9 pulsars with resolved H_α emission at the apex and a number of pulsars with upper limits set on their H_α flux.

5.1 Synchrotron Spectra and Images of the Far-UV Bow Shock of PSR J0437-4715

One of the most interesting and best studied objects in the compilation of Brownsberger and Romani (2014) is the bow shock of the nearest millisecond pulsar J0437-4715. The pulsar is located at a reliably estimated distance of about 160 pc in a binary system with measured parallax and transverse velocity $\sim 104\ \mathrm{km\,s^{-1}}$ derived from the motion of the pulsar companion and the bow shock apex measured over 17 yrs. The estimated spin-down power of J0437-4715 is $\dot{E} \sim 6 \times 10^{33}\ \mathrm{erg\,s^{-1}}$.

Extended X-ray emission from the $\sim 5''$ vicinity of J0437-4715 was detected with *Chandra* by Rangelov et al. (2016). This emission indicates the presence of a faint ($L \sim 3 \times 10^{28}\ \mathrm{erg\,s^{-1}}$) PWN produced by J0437-4715. The X-ray photon index of the PWN is $\Gamma = 1.8 \pm 0.4$.

Far-ultraviolet (FUV) imaging of the vicinity of J0437-4715 performed by Rangelov et al. (2016) with the *Hubble Space Telescope* has revealed a bow-like structure which stands $10''$ ahead of the pulsar and is coincident with the apex part of the H_α bow shock earlier observed by Brownsberger and Romani (2014). The observed 1250–2000 Å luminosity of the FUV bow is about $5 \times 10^{28}\ \mathrm{erg\,s^{-1}}$—an order of magnitude higher than the H_α luminosity of the same structure. The FUV observations also revealed a likely flux enhancement in the extended (about $3''$) region at the limb of the bow, which is not seen in the wide-band optical/near-UV images.

Both H_α and FUV emission at the bow of J0437-4715 can be associated with line and continuum radiation of shocked interstellar plasma. In particular, even a simple one-dimensional model of gas flowing through a fast enough interstellar shock (see, e.g., Bykov et al. 2013a) can be used to explain the fluxes of broadband FUV and H_α emission observed by Rangelov et al. (2016). Within such a model the 1250–2000 Å band emission is generated in the hot downstream of the shock and dominated by several individual lines such as C IV 1549 Å, O IV 1403 Å, Si IV 1397 Å, C II 1335 Å, and He II 1640 Å rather than by continuum UV radiation.

Alternatively, synchrotron radiation of accelerated electrons and positrons in the magnetic field amplified at the bow shock could also provide the observed optical and FUV fluxes. As illustrated in Fig. 4, even the observed morphology of the FUV and X-ray emission around J0437-4715 can be understood within such an interpretation.

Modeling of broadband synchrotron emission from PWNe with a single power law particle injection spectrum fails to reproduce the observed spectral energy distributions (e.g., Reynolds and Chevalier 1984; Atoyan and Aharonian 1996; Amato et al. 2000). Hence, a spectral break is inferred from the data of radio and X-ray observations of young PWNe in supernova remnants (see, e.g., Gelfand et al. 2009; Bucciantini et al. 2011). Namely, Bucciantini et al. (2011) found that injection of a particle spectrum in the form of a broken power law with a hard low-energy spectral index ≤ 1.7 (for $3 \times 10^4 \lesssim \gamma \lesssim 10^6$) and the high energy injection index ≥ 2.14 can yield a satisfactory description of the available observational data for a number of PWNe. The maximal Lorentz factors of $e^\pm$ pairs estimated from the voltage of the total magnetospheric potential of a pulsar are $\sim 10^8$ for the spin-down power $\dot{E} \sim 6 \times 10^{33}\ \mathrm{erg\,s^{-1}}$.

To model the observed optical, FUV and X-ray emission of the J0437-4715, a power law spectrum of $e^\pm$ pairs is injected at the PW termination surface with the index 2.2–2.3 (for the Lorentz factors $\gamma_0 \leq \gamma \leq \gamma_\mathrm{m}$). This corresponds to the common models of particle acceleration at relativistic shocks (see, e.g., Achterberg et al. 2001; Ellison and Double 2004; Keshet and Waxman 2005; Pelletier et al. 2009; Bykov et al. 2012). While it is not clear that the standard Fermi acceleration might be effective at the PW termination shock, unless

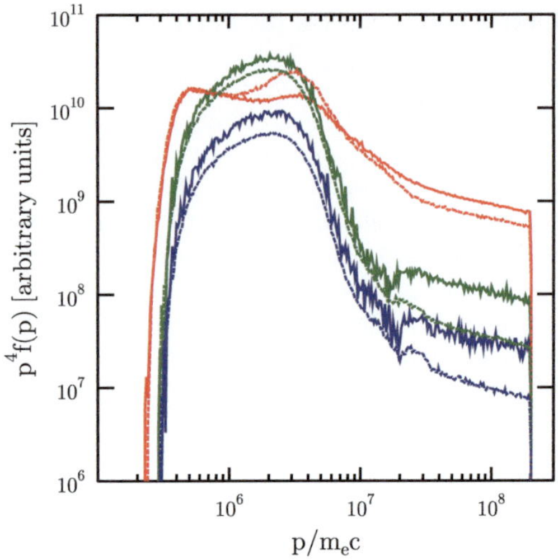

Fig. 3 Non-smoothed Monte-Carlo simulated local spectra of $e^{\pm}$ pairs in different zones of a BSPWN (see Sect. 4) for parameters similar to those of the source associated with PSR J0437-4715. The *red curves* correspond to the PW termination shock downstream: the *dashed one* refers to the $A2$ bin, the *solid one* refers to the $A4$ bin viewed along the ray directed to the observer. The *blue curves* correspond to the bow shock region (bin $C1$—the *solid curve*, bin $C3$—the *dashed curve*). The *green curves* correspond to the region between the two shocks (bin $B1$—the *solid curve*, bin $B3$—the *dashed curve*). The locations of all the bins are shown in Fig. 2. Worse statistics for bins $B1$ and $C1$ is due to the smaller squares of 'detectors' near the system symmetry axis (see Sect. 4.3)

highly efficient dissipation of the PW magnetic field occurs before (or at) the termination shock surface, it is true, nevertheless, that this spectral slope is what is inferred from X-ray observations for the majority of PWNe. The shape of the particle spectrum below γ_0 does not need to be specified in the present modeling. The maximal Lorentz factor of the $e^{\pm}$ pairs accelerated at the PW termination surface is likely scaled with the pulsar spin-down power as $\gamma_{\rm m} \propto \dot{E}^{0.5}$ (as the size of the termination shock) and thus may reach (3–5) $\times\ 10^7$ for J0437-4715, i.e., it is just below the value achievable in the magnetospheric potential gap. Pair (re-)acceleration in the colliding shock flow (CSF) of the PW termination shock and the bow shock would result in substantial spectral hardening up to the maximal energy given by Eq. (23). For PSR J0437-4715, with $u_{\rm sh} \sim 100$ km s^{-1} and $R_{\rm cd} \sim 2 \times 10^{16}$ cm, from the simplified estimation of Eq. (23) one can obtain $\gamma_{m1} \approx 3 \times 10^6$. Indeed, the modeled spectra of $e^{\pm}$ pairs re-accelerated in the CSF demonstrate a break at $\gamma \sim$ (2–3) $\times\ 10^6$ in all the three considered spatial zones (Fig. 3).

In Fig. 3 the accelerated $e^{\pm}$ pair spectra in different zones of a BSPWN for parameters similar to those of the source associated with PSR J0437-4715 are shown. They have been simulated within a Monte-Carlo model, where a power law spectrum $\propto \gamma^{-s}$ with $s = 2.3$ upto $\gamma_{\rm max} = 2 \times 10^8$ (which is less than the value given by the magnetospheric potential $\sqrt{3\dot{E}/2c}$) is injected in the immediate downstream of the PW termination surface. As described in Sect. 4, the injected pairs propagate through the colliding flow where the efficient re-acceleration reshapes the initial spectrum. The spectra are presented for a few bins, whose locations are shown in Fig. 2. Namely, the red curves correspond to the PW termination shock downstream: the dashed one refers to the $A2$ bin, while the solid one refers to the bin located at the same distance from the pulsar as the $A4$ bin, but lying on the ray directed from the pulsar to observer. The blue curves correspond to the bow shock region (bin $C1$—the

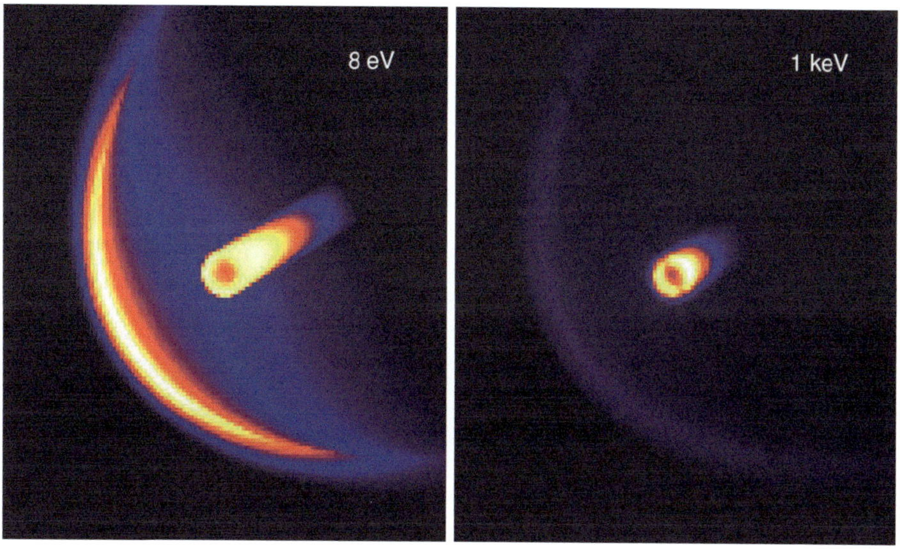

Fig. 4 Simulated synchrotron images of an object similar to J0437-4715, for $E = 8$ eV (≈ 1550 Å, *left*) and 1 keV (*right*). The absolute intensity is given in normalized units to show the contrast change through the image

solid curve, bin $C3$—the dashed curve). The green curves correspond to the region between the two shocks (bin $B1$—the solid curve, bin $B3$—the dashed curve). Worse statistics for bins $B1$ and $C1$ is related to the smaller squares of 'detectors' near the system symmetry axis (see Sect. 4.3).

To confront the modeling with the observational data, one needs to construct images and spectra of synchrotron emission produced by the modelled distributions of accelerated $e^{\pm}$. The intensity of the synchrotron radiation averaged over the magnetic field orientations was obtained using the approximations given by Zirakashvili and Aharonian (2007). The modifications of the synchrotron emission spectra related to the Doppler beaming were taken into account.

Simulated synchrotron images of an object similar to J0437-4715 at $E = 8$ eV and 1 keV are shown in Fig. 4. On the left panel of Fig. 4 a bright bow-like structure can be seen, which corresponds to the inner part of region "3" in Fig. 2. A very faint counterpart of the bow is hardly seen in this zone on the right panel, where the 1 keV emission is shown. Some details of the simulated source morphology can be affected by varying the parameters of modeling, such as the pulsar axes inclination angles and others, over the allowed parameter space. However, the presence of the bright bow-like structure in the FUV regime and its faint appearance in the X-rays remains.

The spectra of synchrotron emission produced by the re-accelerated pairs in the modelled BSPWN are shown in Fig. 5. The simulated synchrotron spectra are shown for a source similar to J0437-4715. The orientation of the source is given by angles $\Phi = 90^\circ$, $\Theta = 90^\circ$ and $\Psi = 149.5^\circ$ (see Sect. 4.4), i.e. the pulsar proper velocity vector lies in the plane of sky perpendicular to the line of sight crossing the pulsar (plane *PoS*) and the position angle $\theta_{pos} = 270^\circ - \Psi = 120.5^\circ$. The curves show the spectral energy distribution (νF_ν) calculated by integration of the spectral emissivity along the lines of sight crossing the bins which were chosen to show the local particle spectra. The colors and styles of the curves are the same as in Fig. 3.

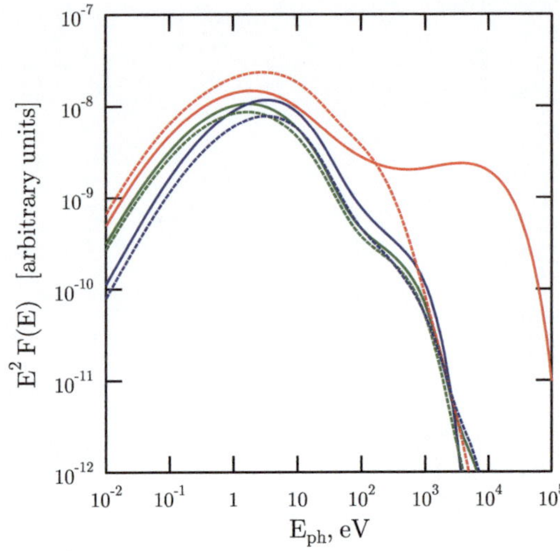

Fig. 5 Simulated synchrotron spectra of a source similar to J0437-4715. The orientation of the source is given by the angles $\Phi = 90°$, $\Theta = 90°$, and $\Psi = 149.5°$ (see Sect. 4.4), i.e., the pulsar proper velocity vector lies in the plane of sky perpendicular to the line of sight crossing the pulsar (plane *PoS*) and the position angle $\theta_{pos} = 270° - \Psi = 120.5°$. The *curves* show the spectral energy distribution (νF_ν) calculated by integration of the spectral emissivity along the lines of sight, which cross the bins, indicated in the caption to Fig. 3. The colors and styles of the curves are the same as in Fig. 3

The ratio of the modelled FUV luminosity of the bow-like structure to the modelled X-ray luminosity of the prominent feature in the vicinity of the TS is about 2. The modelled luminosity of the J0437-4715 bow in the H_α band is lower than the measured value, so most of the observed H_α emission in the WIYN filter W012 (Brownsberger and Romani 2014) cannot be attributed to the synchrotron radiation. In general, the simulated synchrotron spectra are consistent with the data of multiband observations of J0437-4715 mentioned above.

The extended (about 3″) FUV flux enhancement (dubbed "blob") which is apparent at the limb of the bow, but not seen in the wide-band optical/near-UV images of Rangelov et al. (2016), can be explained in the frame of the synchrotron model. The optical-UV continuum spectrum is steep at the bow as it is clearly seen in Fig. 5. This means that even relatively modest fluctuations of the magnetic field may produce a sharp structure at higher photon energies, which is not prominent at lower energies. The intermittent clumpy images are a characteristic feature of the systems with steep synchrotron spectra, as it was shown by Bykov et al. (2008).

Another case of a bow shock driven by a supersonically moving pulsar is the nebula of the gamma-ray pulsar J1741-2054, whose estimated velocity is $\sim 150\ \mathrm{km\,s^{-1}}$ (assuming the uncertain distance of about 400 pc). The H_α bow shock standing at 1.″5 from the pulsar position was reported by Romani et al. (2010). The pulsar spin-down power is about $9.5 \times 10^{33}\ \mathrm{erg\,s^{-1}}$. Deep *Chandra* observations of PSR J1741-2054 performed by Auchettl et al. (2015) have revealed a PWN with an extended ($\sim 2'$) trail, whose power law photon index is $\Gamma = 1.67 \pm 0.06$ and the total 0.5–10 keV luminosity is about $2.4 \times 10^{30}\ \mathrm{erg\,s^{-1}}$. The trail shows no for synchrotron cooling. Note that the total X-ray luminosity of PSR J1741-2054 is two orders of magnitude higher than that of PSR J0437-4715, though their estimated spin-down powers and proper velocities are comparable. The measured X-ray photon index of J1741-2054 is consistent with the model discussed above, which predicts harder synchrotron spectra in the optical-UV regime followed by asymptotical softening towards the X-ray band. The shape of the simulated energy distribution of the high-energy X-ray emitting particles reproduces the shape of the spectra of $e^\pm$ pairs accelerated at a relativistic termination shock. For the power law spectra $\propto \gamma^{-s}$, where $s \sim 2.2$ it gives the photon indices close to 1.6. As the spin-down powers and velocities of the two PSRs are similar, the break energy between the optical-UV component with the hard spectrum produced in the

CSF between the termination and bow shocks and that produced at the termination shock of PSR J1741-2054 is expected to be at a few eV. Therefore, the FUV synchrotron luminosity of the bow shock of PSR J1741-2054 is expected to be about a few times 10^{28} erg s^{-1}.

6 Extended X-Ray Structures Associated with Bow Shock Pulsars

High resolution observations with the *Chandra* X-ray observatory (Tananbaum et al. 2014) have revealed a great diversity of X-ray appearances of PWNe (Gaensler and Slane 2006; Kargaltsev and Pavlov 2008). For the subset of PWNe with bow shocks (BSPWNe), born due to fast supersonic motion of pulsars, the reasons for diversity are even more numerous than for the rest of the class, as in addition to the pulsar and PW properties (such as pulsar obliquity and wind magnetization), which determine the wind anisotropy (e.g., Bühler and Giomi 2016), the angle between the star proper velocity and its rotation axis as well as the direction of the local interstellar magnetic field are also important, and greatly extend the parameter space. Some particular cases and configurations will be discussed below.

Peculiar extended X-ray structures associated with PWNe—the "Guitar nebula" and the "Lighthouse nebula"—were observed around two apparently fast supersonically moving pulsars PSR 2224+65 (see, e.g., Romani et al. 1997; Hui and Becker 2007; Johnson and Wang 2010; Hui et al. 2012) and IGR J11014-6103 (see, e.g., Pavan et al. 2011, 2014, 2016). Also, *Chandra* observations revealed a highly extended X-ray feature likely associated with the nearby γ-ray pulsar PSR J0357+3205 (see, e.g., De Luca et al. 2013), whose space velocity is estimated to be about 390 km s^{-1} (assuming the ~ 500 pc distance) and is well aligned with the X-ray feature.

The observed structures of the Guitar and Lighthouse nebulae, as well as the appearance of extended X-ray structures around the Geminga pulsar may be understood within the two jets scenario combined with re-acceleration of the hard-spectrum $e^{\pm}$ component in colliding shock flows (CSFs).

6.1 The Geminga PWN

A bright nearby gamma-ray pulsar, PSR J0633+1746 ("Geminga"), is located at the estimated distance of 250^{+230}_{-80} pc (Verbiest et al. 2012). Its age is about 340 kyr and its spin-down power is about 3.3×10^{34} erg s^{-1}. The estimated transverse velocity of Geminga is about 211 km s^{-1}, which is supersonic for the most widespread ISM phases, and thus it should make a BSPWN. Pulsed X-ray emission up to 20 keV was recently detected from PSR J0633+1746 with the *Nuclear Spectroscopic Telescope Array* (NuSTAR) by Mori et al. (2014). These authors found that two models can fit the observed spectrum. One of the models consists of a power law component with photon index about 1.7 and two blackbody components of temperatures $kT \sim 44$ eV and 195 eV. An equally good model consists of a blackbody component of $kT \sim 42$ eV and a broken power law with photon indexes $\Gamma \sim 2.0$ and 1.4 with a break at 3.4 keV. Much harder photon indexes are needed to model the X-ray spectrum of the Geminga PWN.

With a deep *X-ray Multi-Mirror Mission-Newton (XMM-Newton)* observation of the PWN, Caraveo et al. (2003) revealed an extended structure with two $\sim 2'$ elongated, nearly parallel, X-ray tails trailing behind the fast moving pulsar PSR J0633+1746. However, no H_α emission from the nebula was found in a 5-hour exposure of the ESO VLT/FORS1 spectrograph. In addition to the two lateral ~ 0.2 pc long tails, an axial tail of ~ 0.05 pc length resolved into a few individual segments was found in deep *Chandra* observations by Posselt et al. (2017). The X-ray spectra of the lateral tails, which can be modelled with power law photon indexes $\Gamma \approx 1$, are much harder than that of the axial tail, where $\Gamma \approx 1.6$. The lateral tails are connected to the pulsar, Posselt et al. (2017) found some indications of apparent motions of their footpoints.

The Milagro collaboration (Abdo et al. 2009) reported a detection of a several degrees size TeV emission region in the vicinity of the Geminga PWN. Recently the *High Altitude Water Cherenkov Observatory (HAWC)* confirmed the TeV source to be about 2 degrees large and found that its TeV spectral index $\Gamma \sim 2$ (Abeysekara et al. 2017). Contribution of TeV range positrons from the Geminga pulsar is considered as a potential source of the excess over the standard predictions of secondary production in the interstellar matter (see, e.g., Hooper et al. 2017), which will be discussed in detail in Sect. 9.

The observed lack of H_α emission from the PWN may indicate that PSR J0633+1746 is moving through a rarefied ambient medium: an analysis of the geometry of the PWN tails allowed Posselt et al. (2017) to conclude that the ambient number density around the nebula is below 0.01 cm^{-3}. For the bow shock moving at about 210 $km\,s^{-1}$ such an extremely low density would indicate that the surrounding ISM is hot and thus the Mach number of the shock is moderate. The shock velocity is somewhat higher than that of PSR J0437-4715 and, more importantly, the estimated spin-down power of PSR J0633+1746 is about an order of magnitude higher than that of PSR J0437-4715. Hence, the hard spectrum of $e^\pm$ pairs re-accelerated in the colliding shock flow between the bow shock and the PSR wind termination shock of Geminga can be extended to sufficiently high energy to produce synchrotron emission in the X-ray band (according to the maximal energies of the $e^\pm$ pairs estimated in Sect. 3.1). The X-ray emitting accelerated pairs may flow away through the two magnetic jets and form the apparent lateral tails bent behind due to the pulsar proper motion. Within such a scenario the axial tail may be filled with the pairs accelerated at the PW termination shock and therefore their emission spectrum would have a steeper spectrum, possibly with a photon index close to the observed $\Gamma \approx 1.6$.

An important constraint for the colliding flow model is the spatial distribution of spectral hardness of the observed X-ray emission. The spectra of the "ring" region (Posselt et al. 2017) and of the axial tail are softer than those of the lateral tails. The observed photon indexes of the lateral tails ($\Gamma \sim 1$) are substantially harder than those from the other regions of the PWN ($\Gamma > 1.5$) as well as that of the pulsar. Assuming a synchrotron origin of the hard emission, one may locate the sources of radiating electrons in the region where the relativistic outflow behind the termination shock is colliding with the bow shock downstream flow. In this region the multi-TeV electrons emitting X-rays can have mean free paths long enough to be accelerated via the Fermi mechanism between the converging flows. This would result in the spectrum of particles $\propto \gamma^{-s}$ with $s \gtrsim 1$.

If the pulsar velocity is close to the normal to the "wind plane", the model of particle acceleration in the colliding flows of the anisotropic wind and of the bow shock downstream could explain the hard X-ray spectral indexes in the lateral tails of the Geminga PWN: here the accelerated electrons are advected through the region of the contact discontinuity between the shocked PW and the interstellar plasma. Numerous simulations of planetary and heliospheric bow shocks predict the enhanced plasma density and magnetic field magnitude in the vicinity of the extended contact discontinuity. The accelerated X-ray radiating electrons would produce the "two tails" configuration as a projection of the rotation figure with a thin radiating shell. The ratio of X-ray surface brightness in the middle of the nebula, at a distance z from the pulsar, to that in the tails would be then $\propto [\delta(z)/R_{\rm cd}(z)]^{1/2}$, where $\delta(z)$ is the radiating shell thickness.

In the frame of the discussed model, the axial tail of the Geminga PWN may be attributed to the synchrotron X-rays from the magnetic tail where the magnetic field reconnection regions and plasma sheets are similar to planetary magnetotails (e.g., Eastwood et al. 2016) An important difference, however, is that, contrary to the planetary and heliospheric magnetotails, reconnecting magnetic flows in PWNe are expected to be composed of moderately magnetized relativistic plasma. Recent two-dimensional particle-in-cell (PIC) simulations of relativistic reconnection in such objects (Sironi et al. 2016) have revealed a phenomenon of formation of quasi-spherical plasmoids, which may show up as blobs apparent in the X-ray images of the axial tail of Geminga. To be consistent with the observed spectra of the blobs, the simulated spectra of the plasmoids derived by Sironi et al. (2016) must have moderate magnetization parameter $3 \leq \sigma \leq 10$.

6.2 Extended X-Ray Structures in the Guitar Nebula

The "Guitar nebula" is associated with the bow shock from PSR B2224+65 (see, e.g., Cordes et al. 1993; Romani et al. 1997; Chatterjee and Cordes 2004). The pulsar is likely one of the fastest pulsars observed with a radio measured proper motion. Harrison et al. (1993) estimated its transverse velocity to be about 986 $\mathrm{km\,s^{-1}}$ at the distance of 1.1 kpc, derived from the dispersion measure. The multi-wavelength study of the nebula by Chatterjee and Cordes (2004) resulted in the distance estimation between 1 and 2 kpc suggesting even faster velocity of PSR B2224+65. The pulsar has a modest spin-down power $\sim 10^{33}$ $\mathrm{erg\,s^{-1}}$. Its observed X-ray spectrum is well described by a photon index $\Gamma = 1.58^{+0.43}_{-0.33}$ (see Hui and Becker 2007).

The X-ray observations of PSR B2224+65 revealed a jet-like feature extended to $\sim 2.4'$ (i.e. of about a parsec length) away to the north-west from the pulsar and offset by about 118° from the pulsar direction of motion (see, e.g., Hui and Becker 2007; Johnson and Wang 2010; Hui et al. 2012, and the references therein). The observed spectrum of the extended X-ray filament is much harder than of that of the pulsar. Johnson and Wang (2010) derived a photon index of $\Gamma = 1.00^{+0.53}_{-0.47}$ in the extended X-ray filament, while a softer spectrum, with $\Gamma = 1.69^{+0.39}_{-0.34}$, was observed from the head segment of the PWN. Hui et al. (2012) discussed the possible progenitor star of PSR B2224+65, which is running away from the Cygnus OB9 association. The authors also suggested that the inverse Compton scattering of $e^\pm$ pairs of $\gamma \sim 10^4$ could produce the extended X-ray feature. The total magnetospheric potential of PSR B2224+65 can provide a maximum particle energy $\sqrt{3\dot{E}/2c} \lesssim 7.5 \times 10^{13}$ eV, which is likely below the energy of the X-ray emitting pairs in the observed extended filament, assuming synchrotron emission in the typical ISM magnetic field ~ 3 μG. We will discuss further the acceleration of the $e^\pm$ pairs in the PWN associated with PSR B2224+65.

A scenario to explain the complex phenomenology of the extended X-ray filament was proposed by Bandiera (2008). The author suggested that the highest energy electrons of Lorentz factor $\sim 10^8$ are accelerated at the termination shock and then escape from the bow shock into the ISM. They emit synchrotron X-rays in a magnetic field of strength ~ 45 μG directed along the filament. Such magnetic field is higher than typical ISM values and must have been amplified in some way (e.g., by a streaming instability). The idea behind this scenario is to attribute the observed X-ray emission to collective radiation of accelerated pairs diffusing into the ISM, rather than interpreting the filament as a particle beam. It seems indeed difficult to keep a parsec size approximately linear structure of the beam without a strong bending, despite the high ram pressure due to the fast motion of the pulsar. For example, the X-ray emitting tails of the Geminga pulsar are highly bent, while the estimated proper velocity of PSR B2224+65 is much larger than that of Geminga.

For viability of this scenario a hard spectrum of accelerated pairs escaping the PWN with Lorentz factors $\sim 10^8$ is needed. Indeed, the X-ray luminosity of the filament is rather high, corresponding to about a percent of the pulsar spin-down power (if the distance to PSR B2224+65 is about 1 kpc, or even higher for a 2 kpc distance). The photon index of the X-ray filament, estimated from *Chandra* observations, is also hard: $\Gamma = 1.00^{+0.53}_{-0.47}$.

Diffusive shock acceleration at the highly relativistic termination shock would provide a pair spectrum of index 2.2–2.3 and corresponding photon index of synchrotron radiation softer than 1.6, which is consistent with that obtained by Johnson and Wang (2010) for the head segment. With the moderate spin-down power derived for PSR B2224+65, the maximal Lorentz factor of $e^{\pm}$ estimated from Eq. (8) is $\sim 10^8$. This maximum energy can only be achieved in a transrelativistic outflow with $\Gamma^2_{\rm flow}/\beta_{\rm flow} \sim 1$. This excludes as the acceleration sites both the highly relativistic wind before the termination shock and the region around the non-relativistic bow shock. On the other hand, acceleration of pairs up to $\gamma \sim 10^8$ may occur in the transrelativistic colliding flow behind the termination shock and the bow shock.

A model of the long linear X-ray filament in PSR B2224+65 also has to explain why it is seen just in the north-west direction from the pulsar. In this context it is important that a similar X-ray filament was reported for the pulsar IGR J11014-6103 by Pavan et al. (2011, 2014, 2016). Recently, Yoon and Heinz (2017) presented hydrodynamic simulations illustrating that the assumption of a density profile of the ISM with inhomogeneities in the form of a series of density discontinuities ahead of the fast moving PSR B2224+65 can reproduce the form of the bow shock and the multiple bubbles seen in H_α observations of the Guitar nebula. Such a model may also help to understand the peculiar appearance of the H_α bow shock of the recycled millisecond pulsar J2124-3358.

6.3 Extended X-Ray Structures in the Lighthouse Nebula

IGR J11014-6103 is a source discovered with the *IBIS/ISGRI* camera aboard the gamma-ray telescope *INTEGRAL* (Bird et al. 2010). With follow up *Chandra* observations, Pavan et al. (2011) revealed, in addition to the point-like source, an extended X-ray structure and a helical type tail of $\sim 4'$ extension. The tail is nearly transverse to the system proper motion (see Fig. 6) and somewhat resembles a similar structure in PSR B2224+65. The authors associated IGR J11014-6103 with the PWN (the "Lighthouse nebula") which is produced by the high-velocity ($\sim 1000\ \mathrm{km\,s^{-1}}$) pulsar of estimated spin-down power $\dot{E} \sim 10^{37}\ \mathrm{erg\,s^{-1}}$. The pulsar might originate from the supernova remnant MSH 11-61A at an estimated distance of 7 ± 1 kpc. At this distance the physical extension of the tail would be about 10 pc. The spectral map of the extended filament shows some patchy patterns with spectrum softer than that of PSR B2224+65. If the tail extension is indeed well above 1 pc, the synchrotron burn-off effects may play a role. Deep *Chandra* observations indicated large deviations from a simple helical model at small and large distances, and the possible presence of an apparent brightness dip at about $50''$ distance from the pulsar (Pavan et al. 2016). Pavan et al. (2016) concluded that both a ballistic jet scenario and the scenario of Bandiera (2008), which considers synchrotron radiation by high energy pairs propagating along pre-existing interstellar magnetic field lines, can explain only some of the observed features.

Apart from the Guitar and Lighthouse nebulae, several other elongated structures associated with pulsars are known. A very extended ($\gtrsim 9'$) X-ray filament in the trail of the radio-quiet middle-aged gamma-ray pulsar J0357+3205 was revealed by De Luca et al. (2013). The space velocity of the pulsar was estimated as $\sim 390\ \mathrm{km\,s^{-1}}$ if it is at a distance of about 500 pc, though no H_α emission was detected. Marelli et al. (2016) found two tail-type nebulae associated with the old pulsar PSR J2055+2539. The pulsar has a modest

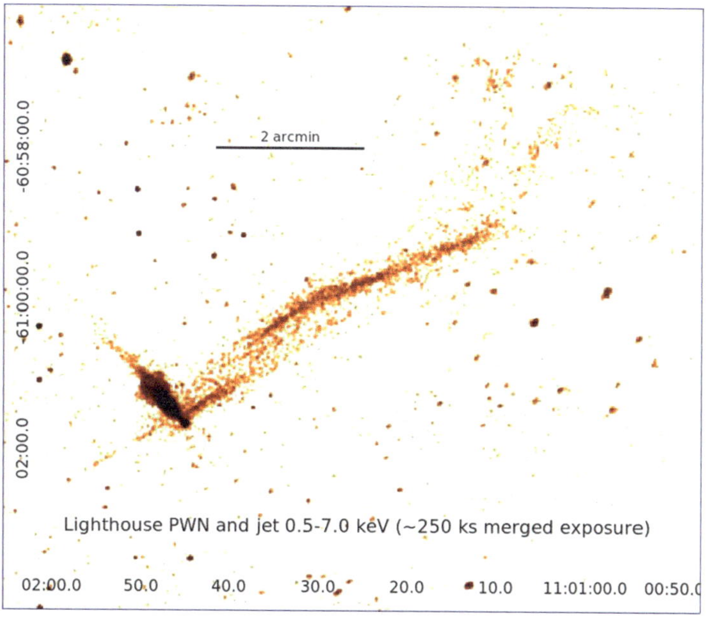

Fig. 6 A 0.5–7 keV *Chandra* image of IGR J11014-6103 (the Lighthouse nebula) likely powered by PSR J1101-6101. The image shows the BSPWN as well as large-scale jet-like features, all launched by IGR J11014-6103 (see Pavan et al. 2016)

spin-down power, $\dot{E} \sim 5 \times 10^{33}$ erg s^{-1}, a characteristic age of about 1.2 Myr, and the estimated distance is ~ 600 pc. The bright tail has an angular size of $12' \times 20''$ corresponding to $\sim (2.1 \times 0.05)$ pc, while the faint one is of $250'' \times 30''$ size, i.e. (0.7×0.09) pc. The X-ray luminosity of the bright tail in the 0.3–10 keV range is $\sim 2 \times 10^{-3} \dot{E}$ (assuming a 600 pc distance). No information on the proper motion of PSR J2055+2539 is available yet, but Marelli et al. (2016) suggested that it is likely to produce a bow-shock.

6.4 Magnetic Jets in BSPWNe

The structure of an astrosphere produced by a relativistic pulsar wind confined by a counter propagating supersonic flow has much in common with that of the heliosphere. Recent progress in the heliosphere observations with *Voyager* 1 and 2, *Cassini*, and *Interstellar Boundary Explorer* satellites, have initiated a new model of the structure of the heliosphere developed by Opher et al. (2015) and Drake et al. (2015). The model suggests that a two-jet structure might provide a better description of the system than the standard comet-like shape. The oppositely directed jets are eventually deflected into the tail region by the motion of the Sun through the ISM. In the models of Drake et al. (2015) and Golikov et al. (2017) the heliosphere is axisymmetric and the structure of the subsonic flows downstream of the termination shock, in the heliosheath and heliopause, is governed by the solar magnetic field. Tension of the solar magnetic field produces a drop in the total pressure between the termination shock and the heliopause. The pressure drop accelerates the plasma flows downstream of the termination shock into the north and south directions to form the collimated jets.

The fact that the magnetic pressure of the azimuthal field may dominate the structure of the subsonic flow in the inner heliosheath was revealed in the early models discussed by Axford (1972). Later, Begelman and Li (1992) proposed that the pinching effect of the toroidal magnetic field in the subsonic flow downstream of the termination shock can be responsible for the observed elongation of the Crab nebula. Similar axisymmetric structures extended in the polar direction were advocated by Chevalier and Luo (1994) for the magnetic shaping of planetary nebulae. Finally, a number of studies (see, e.g., Lyubarsky 2002; Komissarov and Lyubarsky 2004; Del Zanna et al. 2004; Porth et al. 2014; Olmi et al. 2016) demonstrated that magnetic collimation of the plasma downstream of a PW termination shock can be a plausible mechanism behind formation of jets in PWNe.

The double jet structure proposed for the heliosphere has not been studied in the regime relevant for supersonically moving PWNe. However, there are reasons to speculate that the tension of the azimuthal magnetic field between the PWN termination shock and the bow shock may drive formation of a double jet structure in this system as well. There is no reason to believe that the orientation of the interstellar magnetic field with respect to the pulsar rotation axis is the same for all BSPWNe. In the case when the plane defined by the directions of the interstellar field and the proper velocity of the pulsar is nearly perpendicular to the pulsar rotation axis, one may expect that the jets would be confined to the region between the apex of the bow shock and the downstream of the termination shock. This constraint may be fulfilled for a narrow subset of bow shock PWNe. For instance, in the Vela PWN (which is likely a BSPWN), shown in the left panel of Fig. 8 the two apparent X-ray jets are connected to the pulsar position. This may be not too surprising given that the bow shock formed in the Vela PWN due to the mildly supersonic flow behind the supernova reverse shock (Chevalier and Reynolds 2011) is likely quasi-parallel. The available PWN jet models are axially symmetric relative to the pulsar rotation axis, so the scenario described above needs to be confirmed with more complex simulations.

The magnetic jets would channel away some fraction of the high energy particles accelerated by the colliding shock flow in the equatorial wind plane. Magnetically dominated force-free structures are observed in the solar atmosphere and in terrestrial laboratories in the form of magnetic flux ropes (see, e.g., Russell et al. 1990; Birn and Priest 2007; Daughton et al. 2011; Vinogradov et al. 2016). Magnetic flux ropes typically have helical field structure with the maximum of the (untwisted) field at the rope axis. The magnetic field lines in the jets may reconnect with the interstellar magnetic field lines providing a way for the magnetized ultrarelativistic particles of energies $\gtrsim 10$ TeV to escape into the ISM. If the particles are (re)accelerated in CSF, they would have hard spectra of indexes below 2 and therefore contain most of the energy in the highest energy end of their distribution. This is an important point in regard of explaining the observed high efficiencies of the X-ray emission from the extended structures in the Guitar nebula (see, e.g., Johnson and Wang 2010) and in the Lighthouse nebula (see, e.g, Pavan et al. 2016). The X-ray efficiency may reach a percent of the pulsar spin-down power, a fact that requires (i) *hard spectra of accelerated pairs* and (ii) *an amplified magnetic field* in the extended X-ray emitting filaments. We shall discuss below, in Sect. 6.5, a possible mechanism of magnetic field amplification by a CR beam propagating along the local magnetic field.

Reconnection of magnetic field lines in the jet with the oppositely directed interstellar field would happen for only one of the two jets associated with the bow shock PWN. This may explain the apparent asymmetry of the X-ray emitting filaments in both the Guitar and Lighthouse nebulae. The X-ray appearance would depend on the orientation of the pulsar rotation axis (providing the jet directions) as well as the direction of the pulsar proper motion. In both the Guitar and the Lighthouse nebulae the pulsar proper motion is directed nearly transverse to the interstellar magnetic field.

6.5 Magnetic Field Amplification in Extended Ballistic Beams

Relativistic particles leaving the accelerator may produce streaming instabilities which would amplify fluctuations of the interstellar magnetic fields to high magnitudes (e.g., Wentzel 1974; Cesarsky 1980; Blandford and Eichler 1987; Berezinskii et al. 1990; Lucek and Bell 2000).

A fast, non-resonant, almost purely-growing instability driven by CR current was discovered by Bell (2004, 2005). The growth rate for the Bell's mode propagating along the mean field is

$$\Gamma_{\rm B} = v_{\rm a}\sqrt{k_{\rm c}|k| - k^2} \tag{30}$$

where

$$k_{\rm c} = \frac{4\pi J_{\rm CR}}{cB}, \qquad k_{\rm c} R_g > 1. \tag{31}$$

Here $J_{\rm CR}$ is the electric current carried by the ultrarelativistic particles in the beam escaping the accelerator. Note that in the case of CRs leaving a BSPWN the CR beam is composed of ultrarelativistic $e^{\pm}$ pairs and protons. The highest energy CRs can be accelerated in colliding shock flows (see Sect. 3) which energize the highest energy end of the CR spectrum. While the pairs are likely injected into CSF from the PW, the relativistic protons can be injected due to the diffusive shock acceleration at the bow shock. The exact CR composition in the PWNe is somewhat uncertain: both injectures are subject of the intensive studies. However, the CR current $J_{\rm CR}$ exists in the beam leaving the accelerator if the net charge is nonzero in the CR beam.

The saturation level $B_{\rm b}$ of the short scale magnetic field in the beam amplified by Bell's instability can be estimated from the condition $k_{\rm c} R_g \sim 1$ (Bell 2004). Then using Eq. (31) one can get

$$B_{\rm b} \approx \frac{\sqrt{\eta_{\rm CR} \dot{E}/c}}{R_{\rm b}}, \tag{32}$$

where $\eta_{\rm CR}$ is the fraction of pulsar spin-down power converted into the accelerated high energy particles forming the beam and $R_{\rm b}$ is the radius of the beam which is $R_g(\gamma_{\rm max}) \lesssim R_{\rm b} < R_{\rm cd}$.

The scale length $l_{\rm b}$ of the beam $\sim c/V_{\rm psr} \times R_g$, unless the length is limited by the particle energy losses which is the case for a small $V_{\rm psr}$. The leading edge of the beam is located at a distance $\delta_{\rm bmf} \sim c{\Gamma_{\rm B}}^{-1}$ ahead of the region filled with the amplified magnetic field. The maximal growth rate is $\Gamma_{\rm B}^{\rm max} \sim v_{\rm a} k_{\rm c}/2$, and therefore

$$\delta_{\rm bmf} \approx \frac{2c}{v_{\rm a} k_{\rm c}} = \frac{c^2\sqrt{\rho}}{\sqrt{\pi} J_{\rm CR}}. \tag{33}$$

The condition $\delta_{\rm bmf} < l_{\rm b}$ must be fulfilled to fill the beam with the magnetic field amplified by the CR driven short scale Bell's instability. Bell's instability amplifies highly only the magnetic field fluctuations of short scales $kR_g > 1$, which are inefficient for scattering the highest energy pairs. The known long-wavelength instabilities (see, e.g., Bykov et al. 2013b) are too slow to be able to fill the beam with long wavelength turbulence. Therefore the high energy pair mean free paths are determined by the background interstellar magnetic fluctuations. The synchrotron-Compton losses of $\gtrsim 10$ TeV pairs in the amplified magnetic field can limit the X-ray filament length not to exceed greatly ~ 10 pc.

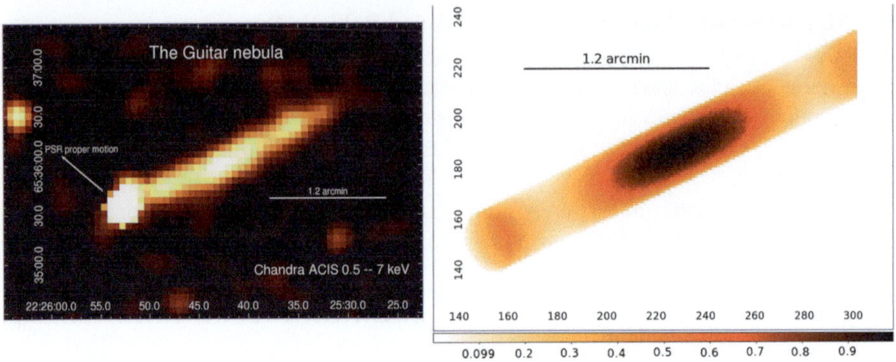

Fig. 7 *Left panel*: a 0.5–7 keV *Chandra* image of the Guitar nebula constructed from the merged archive observations of B2224+65 (PI: D.Q. Wang). *Right panel*: a simulated synchrotron image of the nebula. The simulated X-ray synchrotron structure is produced by a beam of particles accelerating in the colliding shock flows of the BSPWN, then escaping from the acceleration site and propagating along the ISM magnetic field lines. The ISM magnetic field is supposed to be amplified by the CR current-driven instability (see Sect. 6.5 for details)

The mean free path $\Lambda(p)$ of a 10 TeV energy particle along the interstellar magnetic field is a few parsecs (e.g., Strong et al. 2007), while its gyroradius R_g is about 10^{16} cm. Therefore the particles are highly magnetized and their diffusion coefficient transverse to the mean magnetic field estimated from Eq. (18) is $D_\perp \approx vR_g/3 \times (R_g/\Lambda)$. The transverse diffusion coefficient is small enough to keep the transverse size of the beam $l_{b\perp}$ to be about a few R_g.

The main X-ray filament in the Lighthouse nebula extends for about 11 pc at the estimated distance of about 7 kpc (see, e.g., Pavan et al. 2016), while the scale size of the X-ray filament in the Guitar nebula is of about a parsec length (see, e.g., Johnson and Wang 2010). The X-ray filament length l_b in this simplified model scales as $\sqrt{\dot{E}}/V_{psr}$. In the case of the Lighthouse nebula the synchrotron-Compton losses of the X-ray emitting pairs are likely to limit the filament scale size, while in the Guitar nebula, the filament size is roughly consistent with the discussed scaling.

The essential feature of the model described above is formation of a hard spectrum of particles thanks to acceleration in the CSF between the PW termination shock and the bow shock: these particles may power the extended X-ray filaments. The mechanism efficiently converts a sizeable fraction of the wind power into CR luminosity dominated by the highest energy pairs. The accelerated pairs can escape the system through the jets which can be formed in the magnetized region between the shocks. These jets, bent by the ram pressure due to the supersonic motion of the pulsar, can be an explanation of the hard tails observed in the Geminga PWN. If a fast pulsar moves transverse to the interstellar magnetic field, then the reconnection of the ISM magnetic field with the field in a pulsar jet can launch a beam of ultrarelativistic pairs along the interstellar field, which may explain the extended X-ray features observed around PSR B2224+65 and IGR J11014-6103.

In Fig. 7 (right panel) a simulated synchrotron image of the magnetized ballistic beam is presented. A simplified kinetic modeling is performed for a set of parameters likely relevant for the elongated structure of the Guitar PWN. In the discussed scenario only the particles at the highest energies achieved in the colliding shock flow can escape the accelerator (see Fig. 2 in Bykov et al. 2013) to be injected into the beam. The angular distribution of these particles is anisotropic. While the particle pitch-angle distribution is nearly isotropic inside the acceleration region, only particles with pitch-angles $\theta \leq \theta_c < \pi/2$ are injected into the beam. The magnetic field amplification by the CR-current driven instability provided $B_{aism} = 30$ μG along the beam. An apparent bright region in the central part of the observed structure is related to an inhomogeneity of the field in the magnetic rope. In the presented simulation a circularly polarized field variation with the wavelength 1.3 pc and amplitude $\delta B_{aism} = 15$ μG propagating along the beam was imposed onto the background field. The beam was supposed to be directed towards the observer at an angle of 45° relative to the line of sight. The beam had a cylindrical shape and its radius was 3×10^{17} cm. The distance to the source was $d_{\rm beam} = 1.7$ kpc. The particle pitch angle evolution along the beam conserves the adiabatic invariant $p_{\perp}^2/B$, since the mean free path of the X-ray emitting pairs is larger than the beam length.

7 BSPWNe Possibly Associated with SNRs

Some of the fast moving young pulsars with PWNe reviewed by Kargaltsev and Pavlov (2008) are located within young supernova remnants (SNRs). Radio observations of the young active pulsar PSR B1853+01 by Frail et al. (1996) revealed that the pulsar, of the spin-down power about 4.3×10^{35} erg s^{-1}, is located at the apex of an extended feature which has a cometary morphology. The pulsar has a derived transverse velocity of about 375 km s^{-1}. Petre et al. (2002) observed in X-rays the pulsar PSR B1853+01 and its associated PWN, embedded into the supernova remnant W44. The authors found the photon index of the extended X-ray nebula to be about 2.2, much steeper than that of PSR B1853+01 which is ~ 1.4.

A similar cometary shape PWN around the putative pulsar CXOU J061705.3+222127 was observed in the young supernova remnant IC 443 (see, e.g., Bocchino and Bykov 2001; Swartz et al. 2015). Deep X-ray observations reported by Swartz et al. (2015) revealed a ring-like structure surrounding the pulsar and a jet-like feature. The authors found no evidence for a bow shock or contact discontinuity neither spectrally nor morphologically, since the nearly circular ring is not distorted by motion through the ambient medium. They placed an upper limit on the proper motion of the pulsar of about 310 km s^{-1} at a distance of 1.5 kpc. Earlier Bocchino and Bykov (2001) revealed a significant spectral softening along the extended cometary shape X-ray structure, with the power law photon index of about 1.6 in the head and about 2.2 in the tail of the nebula. This is consistent with the expectations based on the simple one zone model of PWNe (see e.g. Chevalier 2000). However, the softening is not typical for many of the bow shock PWNe which we discuss here.

The PWN energized by PSR B1951+32 is located in the field of the supernova remnant CTB 80 (G69.0+2.7). This is another interesting case of a BSPWN. PSR B1951+32 is an energetic pulsar at a 2 kpc distance with a spin-down power $\dot{E} = 3.7 \times 10^{36}$ erg s^{-1}, an estimated age $\sim 10^5$ yrs, and a moderate proper velocity of 240 ± 40 km s^{-1} (Moon et al. 2004; Li et al. 2005). Radio and optical H_α observations revealed a bow shock and Moon et al. (2004) estimated that the contact discontinuity between the shocked PW and the shocked ambient ISM is at a distance about 0.05 pc from the pulsar.

Spatially resolved X-ray spectra of the main components of the PWN were studied with *Chandra* by Li et al. (2005). They found that the photon index of the emission from the bow shock region is $1.6^{+0.1}_{-0.2}$ and it is harder than that of the entire nebula, 1.73 ± 0.03. Moreover, Mattana et al. (2009) reported a significant detection of hard X-ray emission from CTB 80/PSR B1951+32 up to 70 keV with the *IBIS/ISGRI* camera aboard the *INTEGRAL* mission.

Li et al. (2005) pointed out that the observed hard emission spectrum may be due either to the acceleration of PW particles at the termination shock or to a contribution from a second $e^{\pm}$ component. They proposed that the new component of emitting $e^{\pm}$ may originate at the bow shock, which was suggested by Hester and Kulkarni (1988) to explain the optical filaments in CTB 80. PSR B1951+32 has spin-down power and proper velocity not very different from those of the Vela pulsar and we will argue in Sect. 8 that the hard component may naturally arise due to the acceleration of relativistic $e^{\pm}$ pairs in a CSF between the termination and bow shocks.

An extended (arcminute size) faint X-ray nebula to the south of PSR J0633+0632 was detected by Ray et al. (2011) with *Chandra*. The pulsar, which is a gamma-ray active *Fermi* source (see Abdo et al. 2013), was resolved in the X-ray observation, while its optical counterpart is not yet found (Mignani et al. 2016). The X-ray spectrum of the large scale extended PWN has a very hard photon index of 0.74–1.29 for a fixed $N_H = 0.15 \times 10^{22}\ \mathrm{cm\,s^{-1}}$ ($0.9^{+0.5}_{-0.4}$ for a varied column density). Such a hard spectrum is very difficult to explain as synchrotron emission of the relativistic $e^{\pm}$ accelerated only at the PW termination shock. The radio-quiet pulsar has an estimated spin-down power $\dot{E} \sim 10^{35}\ \mathrm{erg\,s^{-1}}$ and the pulsar position suggests the Rosette nebula as its possible birth place (see for a discussion Danilenko et al. 2015). In this case the distance to the pulsar would be ~ 1.5 kpc and then its proper velocity is $\sim 1000\ \mathrm{km\,s^{-1}}$.

Now we will discuss a representative case of a PWN interacting with its SNR. The Vela PWN is a bright nearby object, well studied in multi-wavelength observations. The Vela pulsar is located within its maternal SNR and thus the PWN is interacting with the SNR.

8 The Vela PWN and Vela-like Objects

The bright radio and gamma-ray pulsar PSR B0833-45 (the Vela pulsar) has a period about 89.3 ms, characteristic spin-down age of 11000 yr, and a spin-down power of $\sim 7 \times 10^{36}\ \mathrm{erg\,s^{-1}}$. It is located at a distance of 287^{+19}_{-17} pc, derived from the parallax measured through the very long baseline interferometry observation by Dodson et al. (2003b). The authors derived the proper transverse space velocity of pulsar PSR B0833-45 to be $61 \pm 2\ \mathrm{km\,s^{-1}}$. This velocity is likely subsonic being the pulsar located inside the Vela SNR, where Bocchino et al. (1999) derived from *ROSAT* observations the presence of a multi-temperature inhomogeneous plasma with a temperature range from about 0.1 to ~ 1 keV and a typical number density 0.1–0.5 $\mathrm{cm^{-3}}$.

The nearby Vela supernova remnant appears in multi-wavelength images as a nearly circular outer shell of about 8° in diameter. The bright Vela X region, of about 2° × 3° scale size, is located near the center of the remnant to the south from the pulsar PSR B0833-45. This region, which has a high degree of radio polarization and a radio spectral index (0.08 ± 0.10) much flatter than the main SNR shell, is likely powered by the pulsar (see e.g. Weiler and Panagia 1980; Frail et al. 1997). VLA radio observations of the Vela X region by Frail et al. (1997) revealed a network of fine, extended overlapping linear filaments over the region. The authors also estimated an equipartition magnetic field ~ 20 μG in the region.

Radio observations of the vicinity of the Vela pulsar by Dodson et al. (2003a) with *Australia Telescope Compact Array* revealed a highly polarized structure of a few arcminutes extent with two emission lobes to the south and north of the Vela pulsar. While the radio structure surrounds the pulsar, no excess at the position of the X-ray bright nebula was detected. The observed polarization vectors show a symmetry with respect to the spin axis of the Vela pulsar. Dodson et al. (2003a) suggested that the complex emission from the degree scale size Vela X region is the result of the integrated evolutionary history of the Vela pulsar, which makes it different from the few arcminutes vicinity of the pulsar, where emission is dominated by the currently injected particle population (see also the high frequency radio observations of Vela X by Hales et al. 2004). The observed polarized radio structure was modeled by Chevalier and Reynolds (2011) as an initially thick toroidal structure produced by the PW which is interacting with a mildly supersonic inward flow from the supernova remnant. The flow can be produced in a transient phase of the SNR evolution when the reverse shock reaches the center of the remnant.

The high resolution *Chandra* X-ray observations have revealed a fascinating synchrotron nebula around the Vela pulsar with a bright structured torus and two jets (Helfand et al. 2001; Pavlov et al. 2001a,b, 2003). The total 1–8 keV luminosity of the bright nebula within 42″ from the pulsar is about 6×10^{32} erg s^{-1} (see e.g. Pavlov et al. 2003) (assuming a distance of 300 pc) which is a very small fraction ($\sim 10^{-4}$) of the pulsar spin-down power. The photon indexes of the power law fits vary across the nebula being about 1.3 ± 0.1 for the outer jet of X-ray luminosity about 3×10^{30} erg s^{-1}, close to 1.1–1.2 for the inner jet and between 1.2 and 1.5 within the bright torii (e.g. Pavlov et al. 2003). A two arcminute scale size northern jet demonstrated helical type dynamical behavior (Durant et al. 2013) and a number of fine scale structures in the nebula were identified by Levenfish et al. (2013) in deep *Chandra* images of the Vela PWN. The few arcminutes size nebula with polarized radio lobes and bright synchrotron X-ray structures is most likely a result of the recent activity of the Vela pulsar over a few decades time scale after the reverse shock passed through the nebula (see e.g. Chevalier and Reynolds 2011). A few degrees size extended Vela X region in this scenario is produced by the integrated past activity of the Vela pulsar with account of the relativistic particle escape, their transport and magnetic field structure (see e.g. Aharonian et al. 2006; de Jager et al. 2008; Hinton et al. 2011; Grondin et al. 2013).

Hard X-ray and gamma-ray observations are very useful to study the radiation of the highest energy particles escaped from the Vela PWN. *Suzaku/XIS* observations just outside the few degrees size radio emitting Vela X region were reported by Katsuda et al. (2011) who found a large scale hard X-ray emission of constant power law photon index ≈ 2.4 extending throughout the field of view. While the photon index remains constant, the observed flux is decreasing with increasing distance from the Vela pulsar. Therefore, the authors suggested that the hard X-ray emission is associated with the Vela pulsar wind nebula and that distinct populations of relativistic $e^{\pm}$ pairs are needed to explain both the X-ray PWN and Vela X region.

A wide field hard X-ray image of the Vela X region was obtained by Mattana et al. (2011) with the *IBIS/ISGRI* camera aboard the *INTEGRAL* satellite. The image revealed a significant extended emission of $\sim$ a degree scale size both in the northern and southern sides of the region around the Vela pulsar. The spectrum of the Vela nebula in the 18–400 keV energy range was measured by *IBIS/ISGRI* and *SPI* on board of the *INTEGRAL* satellite. Given the point spread function of the coded mask instrument *IBIS/ISGRI* it was possible to obtain the spectrum within 6′ from the Vela pulsar. The fit to the combined *INTEGRAL* and *Suzaku XIS* data yielded a broken power law with the break energy 27 ± 3 keV and the photon indexes $1.642^{+0.005}_{-0.006}$ and 2.07 ± 0.05 below and above the break respectively. The plasma temperature in the hot Vela supernova remnant environment was derived as $kT = 0.214^{+0.003}_{-0.005}$ keV. The authors pointed out that the observed change of the slope at ~ 25 keV is compatible with the standard value of 0.5 due to the cooling break in models with a continuous particle injection and radiative losses (Chevalier 2000). The slope before the break is consistent with the value of photon index 1.6 which is expected in the model of diffusive particle acceleration at relativistic shock. However, this slope is steeper than the photon index 1.3–1.4 observed with *Chandra* within the bright $\sim$ arcminute size torus-jet nebula around the Vela pulsar. This very hard photon spectrum is indeed reminiscent of those observed at keV photon energies from a number of bow shock PWNe (Kargaltsev and Pavlov 2008).

The *AGILE* satellite studied the region and the particle population there in the energy range from 100 MeV to 3 GeV (Pellizzoni et al. 2010). The source AGL J0834-4539 was detected at the significance level about 5σ and the spectrum sampled in three energy bands (0.1–0.5 GeV, 0.5–1 GeV, 1–3 GeV) yields a power law fit with a photon index 1.67 ± 0.25.

Gamma-ray emission from the Vela X region was detected by *the Fermi Large Area Telescope*, and the spectrum in the 200 MeV to 20 GeV range was measured in the first year of observations (Abdo et al. 2010). With a detailed morphological and spectral analysis of the Vela X region based on four years of observations in the 0.3–100 GeV range, Grondin et al. (2013) detected the emission from the northern wing (from the pulsar) of the Vela X region.

The total gamma-ray luminosity of the Vela X region detected by *Fermi LAT* above 0.2 GeV is 2.4×10^{33} erg s^{-1} which is 0.03% of the pulsar spin-down power (Grondin et al. 2013). These authors studied separately the spectra of the northern wing (to the Vela pulsar position) of the *Fermi LAT* image which can be fitted with a single power law of photon index $2.25 \pm 0.07 \pm 0.20$. Fitting the southern wing required instead two power laws, with a break at an energy of $2.1 \pm 0.5 \pm 0.6$ GeV: the photon index before the break was found to be $1.81 \pm 0.10 \pm 0.24$ while after the break it would steepen to $2.90 \pm 0.25 \pm 0.07$. Grondin et al. (2013) pointed out that while the energy break at 2.1 GeV is significant, the spectral difference between the northern and southern wings is only marginal. *Fermi LAT* data analysis by Abdo et al. (2010) favored a scenario with two distinct electron populations to understand the multiwavelength spectrum of Vela X. Fast diffusion of particles accelerated in the vicinity of the pulsar was discussed by Hinton et al. (2011) and Grondin et al. (2013) to explain the origin of the steep *Fermi LAT* spectrum. Moreover, Hinton et al. (2011) suggested that if the accelerated particles being released early in the evolution of the nebula then travel with a large diffusion coefficient to avoid severe adiabatic losses, this could help to understand the rising positron fraction in the local CR spectrum, an issue which we shall discuss in some detail below.

Very high energy gamma-ray emission from 0.75 to 70 TeV was detected with the groundbased Cherenkov telescope *H.E.S.S.* from the Vela X region (Aharonian et al. 2006; Abramowski et al. 2012). The source HESS J0835-455 is extended to the south from the pulsar PSR B0833-45 and the morphology of the source was characterized as an ellipse with a Gaussian width along the major axis of $0.52° \pm 0.02°$, while $0.12° \pm 0.02°$ is the width along the minor axis. The source shape is different from that obtained by Grondin et al. (2013) with the four years of *Fermi LAT* data, where the northern wing of the emission was detected. However the source is positionally coincident with the source AGL J0834-4539 imaged by the *AGILE* telescope in the energy range from 0.1 to 3 GeV. The spectral model of the entire Vela X region (about 1.2° circle around the central position) in range from 0.75 to 70 TeV is the power law with a photon index $1.32 \pm 0.06_{\mathrm{stat}} \pm 0.11_{\mathrm{sys}}$ and exponential cut off at $14.0 \pm 1.6_{\mathrm{stat}} \pm 2.6_{\mathrm{sys}}$ TeV, while the index is even somewhat harder for the external ring from 0.8 to 1.2 degrees (Abramowski et al. 2012).

Recently Marubini et al. (2015) discussed optical observations of the Vela X region using *Digital Sky Survey* and SAAO 1.0 m telescope observations in an attempt to constrain the multi-component models of the high energy lepton injection suggested by multiwavelength studies of Vela X.

A PWN that is rather similar to Vela (Chevalier and Reynolds 2011) is that powered by the pulsar J2229+6114 of spin-down power $\sim 2.2 \times 10^{37}$ erg s^{-1} and associated with SNR

G106.6+2.9 (also known as the "Boomerang"). The comet-shaped structure of G106.6+2.9 appears as a highly polarized incomplete radio shell with a flat spectrum and an X-ray source of photon index ~ 1.5 (Halpern et al. 2001a,b). The authors suggested that the X-ray source may be a wind-blown bubble or bow-shock nebula produced by a young pulsar. A harder X-ray photon index $\Gamma = 1.3 \pm 0.1$ and an X-ray luminosity of 8.2×10^{32} erg s^{-1} were estimated for the PWN (assumed at a distance of 3 kpc as estimated from X-ray absorption) by Kargaltsev and Pavlov (2008).

A study by Kothes et al. (2001) of the kinematics of the neutral hydrogen and molecular material associated with the SNR has suggested a distance of ~ 0.8 kpc to the SNR and a size of about 14×6 pc. The J2229+6114 PWN would have a diameter of 0.8 pc. This estimate of the distance is supported by the highly polarized emission. On the other hand the distance derived from X-ray absorption (about 3 kpc) and the pulsar dispersion measure (~ 7 kpc) are much larger, making the distance still very uncertain (Chevalier and Reynolds 2011).

8.1 A Model of X-Ray Emission from the Vela BSPWN

An analysis of the *ROSAT HRI* X-ray image by Markwardt and Ögelman (1998) suggested that the compact ~ 2 arcmin "kidney bean" shaped X-ray nebula is due to the bow shock produced by the fast moving Vela pulsar, with an inferred velocity of 260 km s^{-1}. However, as we referred above, the derived proper transverse space velocity of pulsar PSR B0833-45 is only 61 ± 2 km s^{-1} (Dodson et al. 2003b), while the estimated plasma temperature inside the Vela SNR is $kT \gtrsim 0.2$ keV (though inhomogeneous). Therefore the proper velocity of PSR B0833-45 is too slow to launch a bow shock in such a hot medium. On the other hand the flow behind the SNR reverse shock (see e.g. Truelove and McKee 1999) may produce the bow shock. Chevalier and Reynolds (2011) noted that there is a transition period before the Sedov solution is reached and this would allow to produce the mildly supersonic flow for the centrally located pulsar PSR B0833-45. They suggested that the reverse shock has recently passed over the Vela pulsar and the plasma flow is mildly supersonic with sonic Mach number $\mathcal{M}_s \sim 1.3$ at the pulsar position.

As we discussed in Sect. 3, the colliding shock flow is expected to occur between the termination surface of the highly relativistic PW and the pulsar bow shock. The magnetized PW is likely highly anisotropic (see e.g. Komissarov and Lyubarsky 2003; Del Zanna et al. 2004; Kirk et al. 2009; Porth et al. 2014). Relativistic magnetohydrodynamic simulations, supplied with some recipe for the transport and spectral evolution of the accelerated particles, have been extensively used to assess the relation between the wind properties and nebular emission. This has been done through synthetic maps of PWNe, representing their synchrotron emissivity, spectral index and polarization (see e.g. Del Zanna et al. 2006; Volpi et al. 2008; Olmi et al. 2016; Porth et al. 2016).

Recently Bühler and Giomi (2016) performed axisymmetric relativistic magnetohydrodynamic simulations to test the dependence of the PWN morphology appearance on the PW magnetization and obliquity of the pulsar. The simulations suggested that in order to reproduce qualitatively the double rings appearance of the Vela PWN, the pulsar obliquity angle should be about $\pi/4$ and the wind magnetization should be rather high, with $\sigma_0 \sim 3$ before annihilation of the magnetic field in the striped region, and $\sigma \sim 0.12$ after that (see e.g. Sironi and Spitkovsky 2011). The authors simulated the PWN synchrotron emission following the evolution of the PW and assuming that the photon emissivity depends on the Doppler factor of the local flow and magnetic field transverse to the line of sight. The origin and spectral evolution of the accelerated particles were not directly modeled in these

RMHD simulations: particles were just assumed to be there, due to acceleration either at the termination shock or further downstream in the nebula. The relatively large residual magnetization of the flow in the simulation by Bühler and Giomi (2016) even after the field annihilation region makes diffusive shock acceleration of the pairs at the ultrarelativistic termination shock rather problematic (Sironi et al. 2013), and the maximal energies estimated according to Eq. (10) are not enough to provide the extended hard X-ray emission observed by Katsuda et al. (2011), Mattana et al. (2011).

Since we aim to discuss here mostly the effect of the bow shock on the Vela PWN high energy emission spectrum, let us assume that a power law distribution of accelerated pairs $\propto Q_0\gamma^{-s}$, with $s = 2.2$ up to $\gamma_{\max} = 10^9$ (which is below the value estimated from the magnetospheric potential $\sim \sqrt{3\dot{E}/2c}$), is injected at the pulsar wind termination surface. Then we model the particle propagation and re-acceleration in the bow shock PWN through the Monte-Carlo model which described in Sect. 4.

Several studies have addressed the propagation of accelerated pairs across a PWN and the consequent variations of spectral index. Tang and Chevalier (2012) concluded that a purely diffusive model can well reproduce the spectral index maps as well as the nebular size for PWNe without an apparent bow shock signature, like the Crab nebula, G21.5-0.9 and 3C 58. This is because in these PWNe one-dimensional advection of the pairs dominates only close to the pulsar, while on larger scales the streamlines become so twisted that the motion can be described as turbulent diffusion. Porth et al. (2016) recently modeled the transport of high-energy accelerated pairs in PWNe using a three-dimensional magnetohydrodynamic simulation of the nebular dynamics and a test-particle propagation model for the accelerated pairs. The authors found that strong fluctuations of the flow velocity resulted in energy independent diffusion in PWNe with Pèclet number of the order of unit. The bow shock flows were not modeled in these papers.

As extensively discussed above, in the case of bow shock PWNe, the MHD flows are important further away from the pulsar and the colliding flows may result in pair re-acceleration causing a spectral hardening. This is clearly seen in Fig. 10 where the local spectra in the different parts of the bow shock PWN are presented. The simulation run is for a magnetic field strength of 200 μG immediately downstream of the PW termination shock (Sect. 1 in Fig. 2), 120 μG at the bow shock (Sect. 3 in Fig. 2), 20 μG in the region between the two shocks (Sect. 2 in Fig. 2). The flow velocity is taken to be at 500 $\mathrm{km\,s^{-1}}$.

In Fig. 10 we illustrate the simulated local spectra of $e^{\pm}$ pairs in different zones of a BSPWN (see Sect. 4) for parameters similar to those estimated for the Vela PWN. Spectra are shown for a few bins of the Monte-Carlo model, whose locations are given in Fig. 2. The red curves correspond to the PW termination shock downstream: the solid one refers to the $A2$ bin, the dashed one—to the $A4$ bin. The blue curves correspond to the bow shock region (bin $C2$—the solid curve, bin $C4$—the dashed curve), while the green ones—to the region between the two shocks (bin $B2$—the solid curve, bin $B4$—the dashed curve). Because of the efficient $e^{\pm}$ pairs re-acceleration in the colliding flows the pairs spectra in the latter two regions are much harder than what was injected at the wind termination surface.

The radiation spectra in the multi-zone model of a Vela-like PWN are shown in Fig. 11. The orientation of the source is given by the angles $\Phi = 90°$, $\Theta = 90°$, and $\Psi = 270°$ (see Sect. 4.4), i.e., the pulsar velocity in the local ambient matter rest frame lies in the plane of sky perpendicular to the line of sight crossing the pulsar (plane PoS) and is directed to the north. The curves show the spectral energy distribution (νF_ν) calculated by integration of the spectral emissivity along the lines of sight crossing the same locations. The colors and styles of the curves are the same as in Fig. 10.

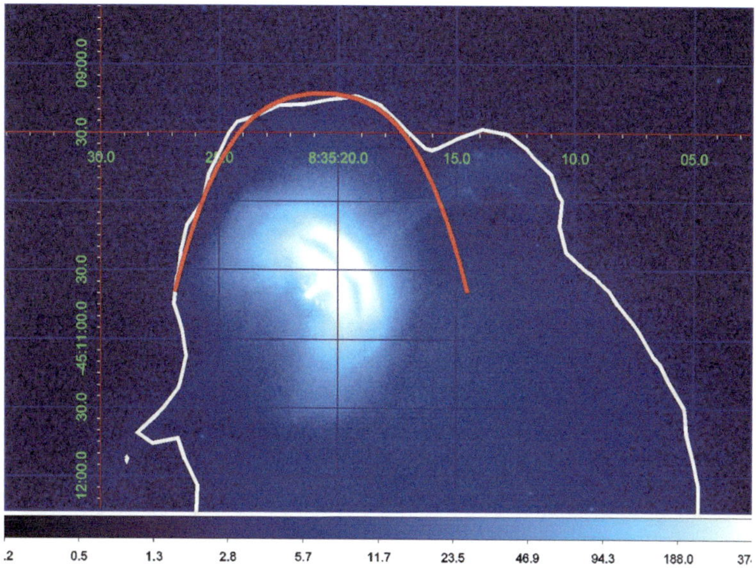

Fig. 8 The X-ray image of the Vela PWN in the 0.5–7 keV energy band. The image is composed of about 250 ks of merged *Chandra ACIS* observations at different epochs. The *white contour* shows the boundary of the 0.5–7 keV diffuse emission. The *red curve* is an overimposed shape of the bow shock of moderate Mach number 1.3, as suggested by Chevalier and Reynolds (2011). The image illustrates the apparent structure of the asymmetric bow shock nebula surrounding the double ring structure of the PWN around PSR B0833-45 (Helfand et al. 2001; Pavlov et al. 2001a) and the jet-like structure discussed in detail by Durant et al. (2013)

It should be noted here that the radiation spectra presented in Fig. 11 are simulated for the outer regions of the PWN where the flow velocity typically does not exceed 0.3 c. The effect of particle acceleration in the colliding plasma flows between the downstream of the relativistic termination shock and the bow shock results in a very hard spectrum of the diffuse X-ray emission in the vicinity of the bow shock. The bright X-ray arcs, rings, and knots dominating the emission from the inner part of the nebula in Fig. 8 are likely highly enhanced by the Doppler boosting in the relativistic outflow in the vicinity of the termination shock (see, e.g., Komissarov and Lyubarsky 2004; Del Zanna et al. 2006; Bühler and Giomi 2016). Hence, to derive the flux of the $e^{\pm}$ pairs injected into the CSF acceleration from the observed X-ray image shown in Fig. 8 one has to use a quantitative model of the Doppler boosted synchrotron X-ray radiation from the transrelativistic outflow at the wind termination surface. Having this in mind, we show in Fig. 10 the spectra of $e^{\pm}$ pairs simulated for a given arbitrarily normalized spectrum of the pairs, which were injected into the downstream of the termination shock. The absolute flux of the injected pairs (the red curve in Fig. 10) determines the fluxes of the pairs accelerated at the CSF (shown as blue and green curves in Fig. 10). The spectra were simulated within the test particle approximation, which assumes that the pressure of the accelerated particles is below the magnetic pressure at the bow shock.

The spectral shape above a few keV is sensitive to the position of the apex and the magnetic field strength at the bow shock. The spectra presented in Fig. 11 illustrating the modeling of a Vela-like PWN correspond to a rather large value of the bow shock apex.

The acceleration time to reach in CSF the maximal $e^{\pm}$ energies shown in Fig. 10 (blue and green curves) according to Eq. (21) is about 20–30 yrs for the bow shock upstream flow velocity $\sim 500\ \mathrm{km\,s^{-1}}$ in a Vela-like PWN.

In Fig. 12 the power law photon spectral indexes obtained by fitting the simulated synchrotron spectra in a wide energy range are presented. In the *Chandra ACIS* energy range, at both the bow shock and in the region between the termination and bow shocks, we find photon indexes $\Gamma < 1.5$ consistent with what observed in the arcminute-size X-ray nebula (Helfand et al. 2001; Pavlov et al. 2001a). These indexes are substantially harder than what one would expect for accelerated $e^{\pm}$ pairs injected at the wind termination surface. The magenta line in Fig. 12 corresponds to a position outside the PWN, before the bow shock (the northern region), but inside the Vela SNR where large scale X-ray emission of power law photon index ≈ 2.4 was found in the 2–10 keV *Suzaku/XIS* observations reported by Katsuda et al. (2011).

The red curve in Fig. 11 shows the spectrum of the synchrotron radiation on the lines of sight which are crossing the PW termination shock downstream. The spectrum was simulated assuming a regular magnetic field in the emitting volume. It is likely, however, that the magnetic fields in PWNe are fluctuating. This would result in a particularly strong variability in the steep gamma-ray tail of the spectrum, where modest fluctuations of the magnetic field may lead to strong flares of spectral flux (Bykov et al. 2012). *eASTROGAM*, a space mission dedicated to the gamma-ray observations with sensitivity $\sim 10^{-12}$ erg cm^{-2} s^{-1} in the energy range 0.2–100 MeV, extending up to GeV energies (see, e.g., Tatischeff et al. 2016), can be used to detect or constrain the flux variations in MeV energy regime from the Vela PWN.

It is worth to note that the Vela SNR, being a nearby young SNR hosting a PWN is expected to be a major contributor to the observed leptonic cosmic ray component (Hinton et al. 2011). The spectrum of $e^{\pm}$ injected by the Vela SNR consists of both particles accelerated at the SNR forward shock and pairs accelerated in the Vela PWN which escape from the nebula and its bow shock. The spectral index of the $e^{\pm}$ escaped the BSPWN is likely harder than that produced at the SNR forward shock and extends up to ~ 50 TeV. Therefore the spectrum of $e^{\pm}$ injected from the Vela SNR would have a spectral break at some energy below 50 TeV with a softer spectrum above the break energy. A break at much lower energy might actually be required not to overpredict the flux of CR leptons, as constrained from AMS-02 measurements (Aguilar et al. 2014) and the latest analysis of Fermi/LAT (Di Mauro et al. 2017). We shall discuss more extensively the role of PWNe as the CR positron sources in the Galaxy in Sect. 9.

The ultrarelativistic $e^{\pm}$ pair spectra shown in Fig. 10 were produced in the colliding flow behind the bow shock. If the Bohm diffusion model is a good approximation for the transport of pairs in region "3" of the Monte-Carlo model (see Sect. 4.2), then, from Eq. (21) the time needed to accelerate the pairs to $\gamma \gtrsim 3 \times 10^8$ in a magnetic field of $\gtrsim 10$ μG is below 30 yrs. Therefore, within this model the high energy pair spectrum of the Vela PWN is indeed multi-component because of the colliding flow produced by the bow shock. This study supports the interpretation of the Vela PWN as the result of the interaction of the PW with a mildly supersonic MHD inflow (Chevalier and Reynolds 2011). The flow bypasses the pulsar with speed $\gtrsim 500$ km s^{-1}, as might happen in a transient phase while the reverse shock approaches the center of the progenitor supernova remnant.

The multi-component model outlined above may help to understand the observed X-ray morphology and spectra on the few arcminutes size nebula. We discussed in the Sect. 6 the problem of the escape of $e^{\pm}$ pairs as well as protons accelerated at a bow shock PWN. The structure of the large scale magnetic field around the PWN can determine the escape of some of the accelerated particles. In the case of the Guitar nebula and the Lighthouse nebula, as we have argued earlier, the interstellar magnetic field is likely nearly transverse to the pulsar proper motion.

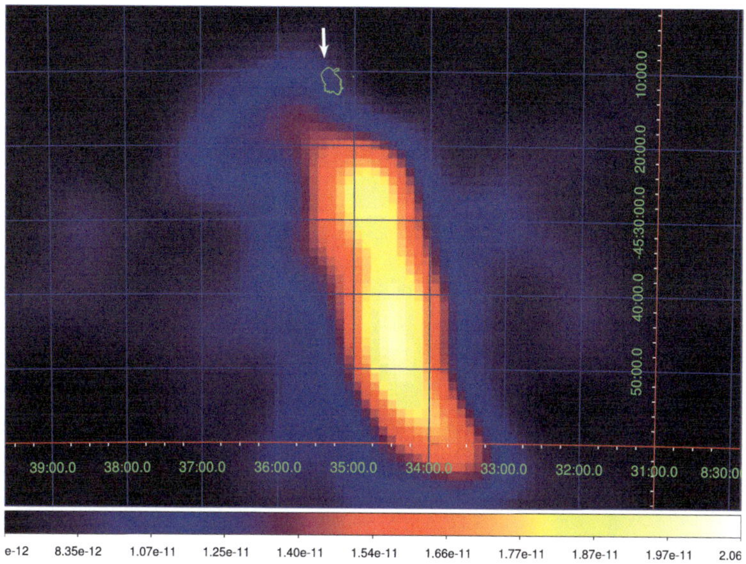

Fig. 9 The very high energy gamma-ray (0.75 to 70 TeV) image of Vela X region as observed by the H.E.S.S. telescope (HESS J0835-455, Aharonian et al. 2006; Abramowski et al. 2012). The *small green contours* at the north correspond to the X-ray *Chandra ACIS* image in 0.5–7 keV energy band of the arcminute size Vela PWN shown in Fig. 8. The *white arrow* shows the expected direction of the velocity of the large scale flow behind the reverse shock of the Vela SNR

Fig. 10 The simulated local spectra of $e^{\pm}$ pairs in different zones of a BSPWN (see Fig. 2) for parameters similar to those estimated for the Vela PWN. The *red curves* correspond to the PW termination shock downstream: the *solid one* refers to the $A2$ bin, the *dashed one* refers to the $A4$ bin. The *blue curves* correspond to the bow shock region (bin $C2$—the *solid curve*, bin $C4$—the *dashed curve*). The *green curves* correspond to the region between the two shocks (bin $B2$—the *solid curve*, bin $B4$—the *dashed curve*). The locations of all the bins are shown in Fig. 2

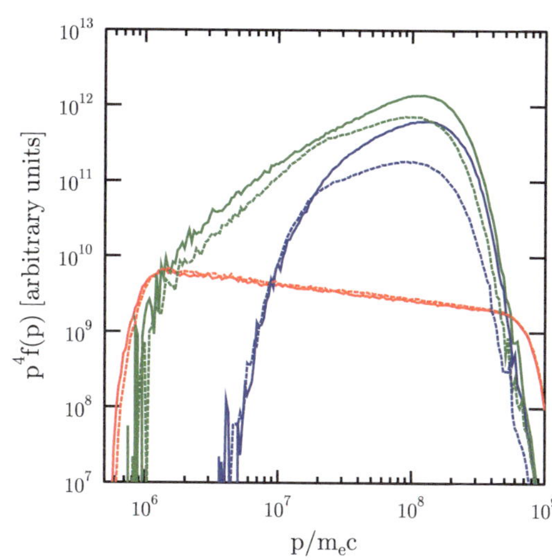

The local magnetic field in the degree scale size region around the Vela PWN is determined by the structure of the MHD flow inside the Vela SNR. If the magnetic field is nearly radial and aligned with the flow velocity behind the SNR reverse shock, then one may understand what observed in the very high energy gamma-rays by H.E.S.S. and shown in Fig. 9. Indeed, pairs with very hard spectra, with particle spectrum index ~ 1.1–1.5, accelerated up to $\gamma \gtrsim 10^8$ at the colliding flow in the region between the wind termination surface and

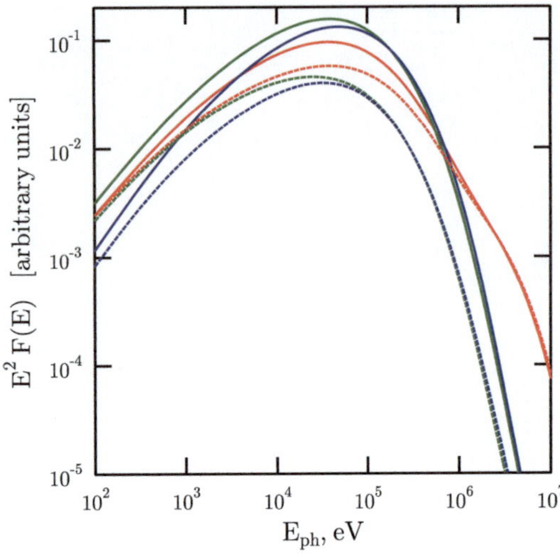

Fig. 11 The simulated synchrotron spectra of a source similar to the Vela PWN, derived from the spectra of $e^{\pm}$ pairs illustrated in Fig. 10. The orientation of the source is given by the angles $\Phi = 90°$, $\Theta = 90°$, and $\Psi = 270°$ (see Sect. 4.4), i.e., the pulsar velocity in the local ambient matter rest frame lies in the plane of sky perpendicular to the line of sight crossing the pulsar (plane *PoS*) and is directed to the north. The *curves* show the spectral energy distribution (νF_ν) calculated by integration of the spectral emissivity along the lines of sight, which cross the bins, indicated in the caption to Fig. 10. The colors and styles of the curves are the same as in Fig. 10

Fig. 12 The power law photon spectral indexes obtained by fitting of the simulated synchrotron spectra of a source similar to the Vela PWN in the wide energy range. The *magenta line* corresponds to a line of sight crossing the SNR outside the BSPWN. This line of sight crosses the bin *D*3 (see Fig. 2). Other curves correspond to the lines of sight defined in the caption to Fig. 11. The colors and styles of those curves are the same as in Fig. 10

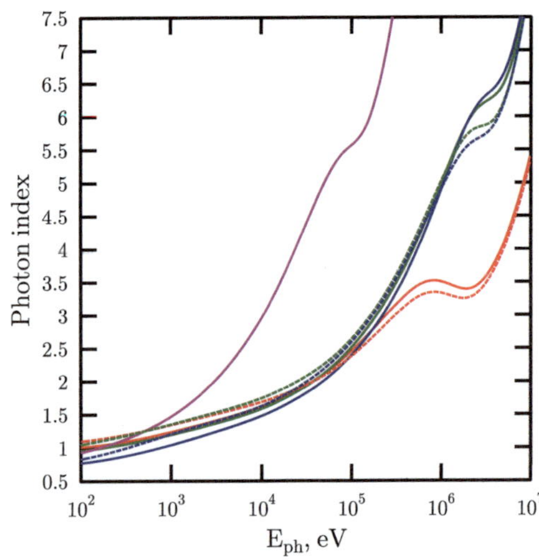

the bow shock (see the blue curve in Fig. 10) can escape the source. The escaped particles would propagate along the mean field to a few parsec scale distances without appreciable losses (assuming the large scale magnetic field to be below $\sim 50\ \mu$G). They would produce the TeV range emission by inverse Compton scattering of the background photons. Note that earlier X-ray studies by Bamba et al. (2010) of PWNe of different ages, up to ~ 100 kyr, but all characterised by low surface brightness, suggested that the accelerated electrons up to ~ 80 TeV can escape from the PWNe without relevant energy losses. Many authors (see, e.g., de Jager et al. 2008; Aharonian et al. 2006; Abdo et al. 2010; Pellizzoni et al. 2010; Grondin et al. 2013; Abramowski et al. 2012), who studied different aspects of the extended Vela X region discussed above, have concluded that multiple populations of accelerated particles are needed to explain the object. In the case of bow shock PWNe, acceleration of pairs in the colliding flow structure can naturally produce an extra particle population, in addition to acceleration at the termination shock.

Some Vela-like objects show hard indexes of the X-ray emission. The pulsar wind nebula G106.6+2.9, with a thick toroidal structure, may have much in common with the Vela PWN, and the scenario discussed above can explain the hard photon index $\Gamma = 1.3 \pm 0.1$ observed in X-rays (Kargaltsev and Pavlov 2008). Also the PWN of the young pulsar PSR B1823-13 showing properties similar to the Vela pulsar has a rather hard photon index, $\Gamma = 1.3 \pm 0.4$ (see e.g. Kargaltsev and Pavlov 2008), while the progenitor SNR has not been detected yet (Gaensler et al. 2003). The radio pulsar B1046-58 is associated with an asymmetric PWN, which may imply a supersonic proper motion (Gonzalez et al. 2006). The X-ray emission from this source also shows a hard photon index $\Gamma = 1 \pm 0.2$ (Kargaltsev and Pavlov 2008). The Vela-like object, PWN G284.0-1.8, associated with pulsar J1016-5857, is characterized by an X-ray spectrum which can be fitted by a power law with index $\Gamma = 1.32 \pm 0.25$ (Camilo et al. 2004). The PWN around the fast-spinning energetic pulsar PSR J0855-4644 likely produced an elongated axisymmetric X-ray nebula with a longitudinal extent of $10''$, with a very hard spectrum of photon index $\Gamma \sim 1.1$ (Maitra et al. 2016). Moreover, H.E.S.S. Collaboration et al. (2016) showed that at least a part of the shell emission observed by H.E.S.S. from the supernova remnant RX J0852.0-4622 (also dubbed Vela Junior) is likely due to the presence of a PWN around PSR J0855-4644 located near the south-eastern rim of the SNR. The origin and age of PSR J0855-4644 are not yet established (see e.g. Acero et al. 2013a): it can be interacting with the ejecta of RX J0852.0-4622 SNR, likely located at a distance of about 1 kpc and may be associated with the Vel OB1 association (0.8 kpc) (see e.g. Allen et al. 2015). Among the PWNe sources which demonstrated a prominent GeV-TeV emission listed in Kargaltsev et al. (2013), Acero et al. (2013b) only very few have photon indexes of the TeV emission below 2.0. All of these very interesting objects need further multi-wavelength studies to see if there is indeed a subclass of Vela-like PWNe.

9 Positrons from BSPWNe

As already mentioned in previous sections, BSPWNe are likely to play an important role in explaining one of the most intriguing anomalies recently discovered in the field of cosmic ray physics, the so-called *positron excess*. In the following we briefly review what this excess is, why it is both mysterious and potentially very important, and try to assess what role BSPWNe might play in explaining it.

9.1 The *Positron Excess*

Galactic Cosmic Rays (CRs) are known to be mainly made of protons (90% by number), with a minor fraction of heavier nuclei (about 10% of Helium in the GeV range and minor percentages of more massive species), a tiny fraction, about 1%, of electrons, and an even smaller fraction of antimatter, positrons and antiprotons. In recent years a few experiments have been devoted to precise measurements of the composition and spectrum of CRs,

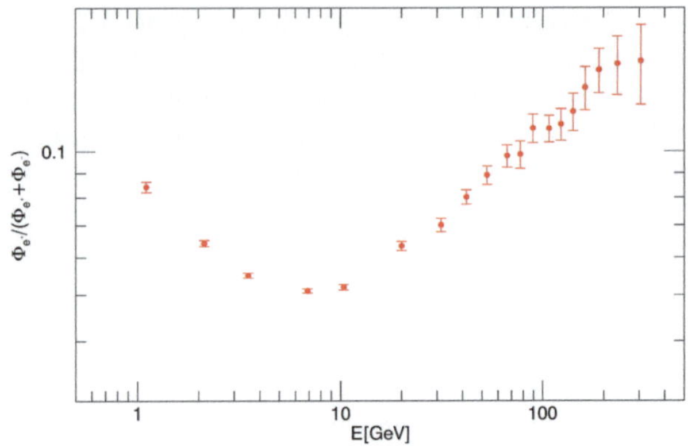

Fig. 13 The positron fraction measured by AMS-02 (Aguilar et al. 2013) as a function of energy. The figure has been created using the data published by Aguilar et al. (2013) as supplemental material

finding intriguing results on both their hadronic and leptonic components. The most relevant results for the discussion in this section concern the latter and have been first obtained with sufficiently high significance by PAMELA (Adriani et al. 2009) and more recently confirmed with even higher statistics by AMS02 (Aguilar et al. 2013). What these two experiments found is that the ratio between the positron flux (Φ_{e^+}) and the total electron (Φ_{e^-}) plus positron flux in CRs, $\chi = \Phi_{e^+}/(\Phi_{e^+} + \Phi_{e^-})$, increases with energy for energies above ~ 6 GeV. The results published by AMS-02 (Aguilar et al. 2013) are reported in Fig. 13.

A growing positron fraction is in plain contrast with the standard scenario of CR origin and propagation and indeed, its unequivocal detection by PAMELA (following earlier, but statistically less significant claims by CAPRICE (Boezio et al. 2001), HEAT (Beatty et al. 2004) and AMS-01 (AMS-01 Collaboration et al. 2007)), raised considerable interest in the scientific community, one of the main reasons being the fact that the excess positrons might be revealing dark matter related processes.

Before discussing how this excess could be related to the class of sources being discussed in this article, let us briefly review why it appears so puzzling and interesting.

In the standard model for the origin and propagation of Galactic CRs, these particles are accelerated in the blast waves of Supernova explosions and propagate diffusively throughout the Galaxy interacting with the interstellar gas. The SN blast wave picks up the particles to be accelerated from the Interstellar Medium (ISM), and hence only ordinary matter enters the acceleration process. Electrons are accelerated together with protons and heavier nuclei, but positrons (and antiprotons) present in CRs have long been considered to be of purely secondary origin, namely deriving from the nuclear collisions of CR protons with the ISM during their journey through the Galaxy. The most relevant chain of reactions for positron production involves p-p scattering leading to the generation of charge pions. These subsequently decay into muons that finally produce positrons and neutrinos:

$$p + p \rightarrow p + n + \pi^+ + \cdots, \qquad \pi^+ \rightarrow \mu^+ + \nu_\mu, \qquad \mu^+ \rightarrow e^+ + \nu_e + \bar{\nu}_\mu. \tag{34}$$

A similar reaction involving negatively charged pions lead to electron production. In the standard scenario, all these secondary leptons are much less numerous than primary electrons. In addition, their ratio to primaries is expected to monotonically decrease with increasing energy. The reason for this expectation is more easily understood in the case of the ratio between secondaries and primaries that do not suffer energy losses, while the actual demonstration in the case of leptons that do suffer appreciable energy losses in the relevant energy range is more cumbersome.

Let us therefore consider, just for illustrative purposes, the ratio between secondary and primary nuclei that do not suffer relevant energy losses during propagation. In relation to CR confinement, let us describe the Galaxy as a cylinder of radius $R_d \approx 15$ kpc (the star and gas disk radius) and height $H \approx 3$ kpc (as the most recent propagation models suggest for the extent of the magnetized halo of the Galaxy, see e.g. Trotta et al. 2011). Let us further assume that CRs diffuse through the volume of the Galaxy with a spatially uniform diffusion coefficient $D(E) \propto E^{\delta}$, where E is the CR energy and $\delta > 0$. Finally, particles are considered free to leave the Galaxy through the vertical bounds at $\pm H$. This is the so-called *Leaky Box Scenario* for CR propagation.

In the absence of energy losses, the equilibrium spectrum of primary CRs, $N_{\text{prim}}(E)$, will then simply result from balance between injection and escape from the Galaxy

$$N_{\text{prim}}(E) = Q_{\text{prim}}(E)\tau_{\text{esc}} \propto Q_{\text{prim}}(E)\frac{H^2}{D(E)} \propto E^{-(\gamma_{\text{inj}}+\delta)}, \tag{35}$$

where we have taken the number of particles injected by the sources per unit time and energy interval to scale with energy as $Q_{\text{prim}}(E) \propto E^{-\gamma_{\text{inj}}}$.

Following the same reasoning we can also derive the equilibrium spectrum of secondaries. Now the injection relies on the spallation reactions of primary nuclei and is then described by

$$Q_{\text{sec}}(E) \propto N_{\text{prim}}(E)\sigma_{\text{sp}} c n_H \propto E^{-(\gamma_{\text{inj}}+\delta)}, \tag{36}$$

where σ_{sp} is the spallation cross-section (assumed to be independent of energy, which is a good approximation in the GeV-TeV energy range), c is the speed of light and n_H is the density of target material.

As a result, the equilibrium spectrum of secondaries will be:

$$N_{\text{sec}}(E) = Q_{\text{sec}}(E)\tau_{\text{esc}} \propto E^{-(\gamma_{\text{inj}}+2\delta)} \tag{37}$$

and the secondary-to-primary flux ratio will scale as:

$$\frac{\Phi_{\text{sec}}(E)}{\Phi_{\text{prim}}(E)} \propto E^{-\delta}. \tag{38}$$

While absence of energy losses is a good approximation for protons and nuclei of all energies between a few GeV and the knee, the same is not true for leptons, which lose energy through synchrotron and Inverse Compton radiation while propagating through the Galaxy. If we take the standard value of magnetic and radiation energy density in the Galaxy, $U_B \approx U_{\text{rad}} \approx 1$ eV cm^{-3}, the radiative loss time, $\tau_{\text{loss}}(E) \approx 2.5 \times 10^8$ yr $(E/\text{GeV})^{-1}$, is shorter than the propagation time $\tau_{\text{esc}}(E) \approx 2 \times 10^8$ yr $(E/\text{GeV})^{\delta}$ already at $E >$ few GeV, for $0.2 < \delta < 0.8$, as inferred from recent studies of CR propagation (Trotta et al. 2011). In this case, in the energy range of our interest, the equilibrium spectrum of electrons, which we approximate as all primaries, is described by:

$$N_{\text{e}^-}(E) \propto E^{-\gamma_{\text{inj,el}}} E^{-(1+\delta)/2}, \tag{39}$$

where the first term on the *rhs* comes from injection and the second from propagation. The form of the latter is determined by the combination of diffusive propagation and radiative energy losses that limit the number of contributing sources with respect to the case of nuclei, imposing constraints in terms of distance in space and time.

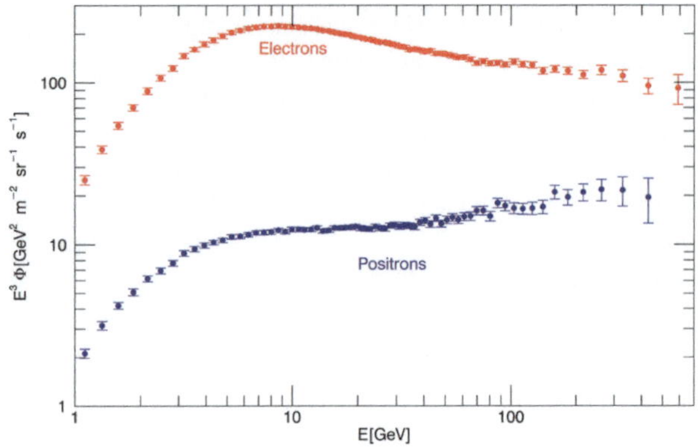

Fig. 14 The spectra of electrons and positrons measured by AMS-02 (Aguilar et al. 2014) as a function of energy. The figure has been built from the data published by Aguilar et al. (2014) as supplemental material

Analogously, for positrons, that are all secondaries, we find:

$$N_{e^+}(E) \propto E^{-(\gamma_{\rm inj,p}+\delta)} E^{-(1+\delta)/2}, \tag{40}$$

where what appears in the first term of the *rhs* is the equilibrium proton spectrum ($\propto E^{-\gamma_p}$). The result for the positron fraction is then:

$$\chi = \frac{\Phi_{e^+}}{\Phi_{e^+} + \Phi_{e^-}} \approx \frac{\Phi_{e^+}}{\Phi_{e^-}} \propto E^{-(\gamma_{\rm inj,p}-\gamma_{\rm inj,el}+\delta)}, \tag{41}$$

and in the standard scenario, with $\gamma_{\rm inj,p} = \gamma_{\rm inj,el}$, $\chi \propto E^{-\delta}$ is expected to monotonically decrease with energy. This is why the observed increase was found to be puzzling and stimulated a plethora of attempts at explaining it.

Before discussing the possible explanations and how BSPWNe enter the scene, an important remark is in order. The prediction of a decreasing positron fraction is a strong one only in a scenario that assumes a spatially uniform distribution of sources. This is indeed the assumption underlying Eqs. (39) and (40), but we know that this assumption does not apply to SNRs, the putative sources of primary electrons: these are discrete sources in space and time and their distribution in the Galaxy is not uniform at all. The effects of the discreteness of the sources and their rapidly decreasing density with Galactocentric distance were analysed for CR nuclei by Blasi and Amato (2012a), and in that case they were found to be poor: the resulting spectrum of all different species still respected on average the predictions of the simplified *leaky box model* (LBM hereafter). With electrons, however, due to radiative losses, some difference might appear, since especially at the highest energies the number of sources that might contribute to the flux at Earth is likely to be very small, and strongly dependent on the recent SN explosion history in the solar neighborhood. A possibility would then be that it is actually the electron flux that steepens with energy with respect to the LBM prediction, rather than a flattening occurring in the positron flux. This possibility has been ruled out by more recent analysis of the PAMELA (Adriani et al. 2011) and AMS-02 data (Aguilar et al. 2014), which clearly show a flattening of the positron flux, a real *positron excess*. The spectra of electrons and positrons obtained by AMS-02 (Aguilar et al. 2014) are shown in Fig. 14.

9.2 Qualitative Evaluation of the PWN Contribution

An intriguing suggestion immediately made after this discovery was that these extra positrons could come from dark matter annihilation in the Galaxy (see e.g. Serpico 2012; Di

Mauro et al. 2016 for a review). While a dark matter related origin of the rise of χ is difficult to rule out completely, astrophysical explanations seem favored at the current time and the most natural place where to look for a source of extra positrons is the magnetosphere of a pulsar, where a large number of electron-positron pairs are created by every electron that leaves the star surface.

These $e^- - e^+$ pairs end up in the PW and are then accelerated at its termination shock. The efficiency of conversion of the pulsar spin-down power into particle acceleration is usually thought to be very high, and measured to reach about 30% in the case of the Crab nebula, the prototype PWN. In addition, we know from observations of many such objects that the particle spectrum is usually a broken power law: from radio observations we infer a particle spectral index γ_R in the range $1 < \gamma_R < 2$, while the X-rays come from particles with spectral index $\gamma_X > 2$. The boundary between the two regimes is inferred to occur in the several hundreds GeV range. It is apparent then that these are extremely promising candidates to explain the excess at least from the spectral point of view: the excess is observed in the energy region between few and 300 GeV (no data are available at higher energies) and in this interval PWNe could well enter the scene with their flat positron spectrum.

Of course as long as the particles are part of the PWN they are confined within the remnant of the parent supernova explosion. Their Larmor radii in the nebular magnetic field (of typical strength around a few × 100 μG) are very small and diffusive escape is not effective. The situation is different when the pulsar breaks out of the remnant and finds itself in the ISM. The time at which this happens and the energy that is left in the system at that time were derived in Sect. 1, where we found, for typical values of the pulsar and ISM parameters, $T_{\rm esc} \approx$ few $\times\ 10^4$ yr and $E_{\rm res} \approx 10^{47}$ erg.

After time $T_{\rm esc}$ the supersonic motion of the pulsar through the ISM transforms its PWN in a Bow Shock PWN. As we extensively discussed in earlier sections, the plasma in front of the pulsar is confined by the bow shock, but particles flowing through the tail are free to leave the system and become part of the cosmic ray pool.

These particles are the interesting ones in view of explaining the *positron excess*. Their spectrum, at the energies of interest, is shown to be flat, similar to what described above for their younger counterparts. In particular, radio emission again shows a flat particle spectrum in the up to hundreds of GeV. In addition, numerical simulations support the idea that these low energy particles can easily escape from the system along the tail with very little energy losses (Bucciantini et al. 2005).

These particles are easily shown to be numerous enough to explain the excess. This can be done through the following back-of-the-envelope calculation.

Let us go back to the rough description of the electron spectrum in Eq. (39), and let us assume that these particles are accelerated in SNRs with the same spectrum as protons: given the protons' spectrum measured at Earth (Eq. 35) $\gamma_{\rm inj,p} + \delta \approx 2.7$, one can promptly derive the spectrum that must be injected by the sources as soon as the energy dependence of the diffusion coefficient is known. For the latter we assume a slope $\delta = 1/3$, in agreement with a Kolmogorov's model for the turbulence in the Galaxy (which is preferred to steeper energy dependencies by considerations concerning the CR anisotropy Blasi and Amato 2012b). We then obtain that electrons must be injected with $\gamma_{\rm inj,el} \approx 2.4$. The fraction of SN explosion energy that is usually assumed to be converted into CR acceleration is $\xi_{\rm CR} = 0.1$ and most of it is carried by protons. The ratio between accelerated electrons and protons that is needed to interpret the CR measurements at Earth is usually $f_{\rm e,p} \approx 0.01$. We can use this information to find the number of electrons that SNRs inject in the Galaxy per unit time and unit energy interval:

$$Q_{e^-}(E) \approx (\gamma_{e^-} - 2)\mathcal{R}\frac{\xi_{\rm CR} f_{e,p} E_{\rm SN}}{E_0^2}\left(\frac{E}{E_0}\right)^{-\gamma_{el}}, \tag{42}$$

with $\mathcal{R}$ the SN rate in the Galaxy and $E_0 = 1$ GeV corresponding to the particles that carry most of the energy in the distribution. If we now follow a similar line of reasoning for the positrons, assuming that they are injected by BSPWNe, which convert some fraction η of $E_{\rm res}$ in a distribution of electrons and positrons characterized by a flat spectrum $E^{-\gamma_{pos}}$ with $\gamma_{pos} < 2$, extending up to $E_{\rm max} \approx 500$ GeV, we can write:

$$Q_{e^+}(E) \approx (2 - \gamma_{pos}) f_{\rm II}\mathcal{R}\frac{\eta E_{\rm res}}{E_{\rm max}^2}\left(\frac{E}{E_{\rm max}}\right)^{-\gamma_{pos}}, \tag{43}$$

where $E_{\rm res}$ is given in Eq. (4) and $f_{\rm II}$ takes into account the fact that only $\approx 80\%$ of the SN explosions give birth to a pulsar. Since propagation affects electrons and positrons in the same way, the ratio χ will simply scale as the injection, and for energies large enough that the secondary positrons can be neglected, will read

$$\chi \approx \frac{Q_{e^+}}{Q_{e^-}} = \frac{2-\gamma_{pos}}{\gamma_{el}-2}\frac{f_{\rm II}\eta E_{\rm res}}{\xi_{\rm CR} f_{e,p} E_{\rm SN}}\frac{E^{(\gamma_{el}-\gamma_{pos})}}{E_0^{(\gamma_{el}-2)} - E_{\rm max}^{(\gamma_{pos}-2)}} \approx 0.035\left(\frac{E}{30\ {\rm GeV}}\right)^{1/2} \tag{44}$$

where the last approximate equality holds for the typical values of the parameters mentioned above. At 30 GeV this value of χ is within a factor of 2 of the AMS-02 measurement reported in Fig. 13, which strongly suggest that the contribution of PWNe must definitely be taken into account when wondering about the positron excess. In addition the energy dependence in Eq. (44) also approximately agrees with the data.

9.3 Positrons from BSPWNe

A quantitative calculation of the flux of electrons and positrons expected from BSPWNe was carried out by Blasi and Amato (2011), who showed that BSPWNe are likely to contribute a large fraction of the needed extra positrons, if not all of them. That work was adopting some simplifying assumptions about the spatial distribution of the SN explosions in the Galaxy (the presence of arms was not considered) and was associating to every explosion a universal pulsar with typical values of the parameters, rather than accounting for the spread of parameter across the pulsar population. In the following we show the results of a more refined version of the same exercise, including the above mentioned missing ingredients and the most recent information available from AMS-02 on the spectrum of primary protons (Aguilar et al. 2015) and on the diffusion coefficient (Aguilar et al. 2016). We first provide the analytical expression for the contribution at Earth of a single source and then sum up the contribution of all the potentially relevant sources in the Galaxy, generated at random from an appropriate distribution. Some other studies along the same line were performed by Di Mauro et al. (2014), who considered the contribution of putative nebulae associated with pulsars from the ATNF catalogue, rather than a generic pulsar population.

The equation describing the transport of particles in the presence of spatial diffusion, energy losses and catastrophic losses reads:

$$\frac{\partial f(E,\mathbf{r},t)}{\partial t} - D(E)\nabla^2 f(E,\mathbf{r},t) + \frac{\partial}{\partial E}\big(B(E)f(E,\mathbf{r},t)\big) + \frac{f(E,\mathbf{r},t)}{T} = Q(E,\mathbf{r},t) \tag{45}$$

where f is the number of particles of energy E per unit volume and energy interval at position $\mathbf{r}$ and time t, $D(E)$ is the particle diffusion coefficient, $B(E) = dE/dt$ represents the particles' energy losses, T is the timescale for catastrophic losses (particle destruction) and $Q(E, \mathbf{r}, t)$ is a source term (number of particles with energy E injected in the system per unit time and energy interval at position $\mathbf{r}$ and time t). The solution of Eq. (45) can be found through the Green function method (Syrovatskii 1959), and allows one to compute the particle distribution function at any location once $D(E)$, $B(E)$, T and $Q(E, \mathbf{r}, t)$ are known. In the absence of catastrophic losses, the Green function reads:

$$G(\mathbf{r}, t, E; \mathbf{r}_0, t_0, E_0) = \frac{\exp[-\frac{(\mathbf{r}-\mathbf{r}_0)^2}{4\lambda^2(E,E_0)}]}{(4\pi\lambda^2(E, E_0))^{3/2}} \frac{\delta(t - t_0 - \tau(E, E_0))}{|B(E)|}, \tag{46}$$

where δ is the Dirac delta function, τ is the timescale of energy losses, and λ is the particle diffusion length. In formulae we have:

$$\tau(E, E_0) = \int_{E_0}^{E} \frac{dy}{B(y)} \quad \text{and} \quad \lambda^2(E, E_0) = \int_{E_0}^{E} \frac{D(y)}{B(y)} dy. \tag{47}$$

The full solution of Eq. (45) can be found through convolution of the Green function in Eq. (46) with the appropriate source term. Sources such as SNRs and BSPWNe are localised in space and time. The injection by one such source, located at position $\mathbf{r}_s$ at time t_s, can be described as

$$Q(E, \mathbf{r}, t) = N_{\text{inj}}(E)\delta(\mathbf{r} - \mathbf{r}_s)\delta(t - t_s). \tag{48}$$

Convolution of the Green function in Eq. (46) with the source term in Eq. (48) provides an expression for the particle density per unit volume and unit energy interval contributed at $\mathbf{r} = (x, y, z)$ by a single source of space-time coordinates $(x_s, y_s, z_s; t_s)$:

$$f(E, \mathbf{r}, t) = N_s(E_0) \frac{|B(E_0)|}{|B(E)|} \frac{\exp[\frac{-(x-x_s)^2+(y-y_s)^2}{4\lambda^2(E,E_0)}]}{(4\pi\lambda^2(E, E_0))^{\frac{3}{2}}} \times \sum_{n=-\infty}^{\infty} (-1)^n \exp\left[\frac{-(z - z'_n)^2}{4\lambda^2(E, E_0)}\right] \tag{49}$$

where

$$E_0 = \frac{E}{1 - B(E)(t - t_s)/E}. \tag{50}$$

In Eq. (49) $z'_n = (-1)^n z_s + 2nH$ and the sum on the *rhs* is introduced to take into account the boundary condition of particle escape from $\pm H$ (see e.g. Blasi and Amato (2012a) for a discussion).

Equation (49) can be used with any distribution of sources. In particular, when a spatially uniform distribution of sources is considered, it is easy to see that its integration over the source volume leads to a particle spectrum as described by Eq. (39), which we used for our qualitative discussion.

However for a quantitative evaluation of the electron and positron spectrum contributed by SNRs and PWNe the procedure to follow is different. SN explosions must be generated at random places in the Galaxy and at random times. In order to be sure to correctly reproduce the spectrum of low energy CRs, one needs to include all the explosions occurring in a

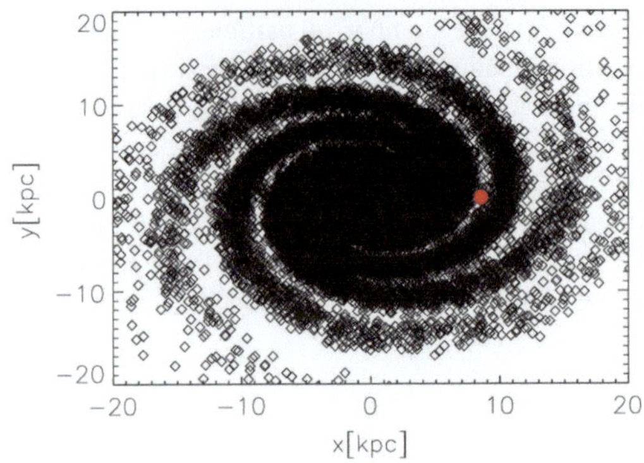

Fig. 15 The projected positions in the Galactic plane of about 30000 SN explosions in one of the considered realisations of the Galaxy. The *red dot* represents the position of the Earth

time span corresponding to the escape-time of these particles. To do this, we assume (as we always will, hereafter) a Kolmogorov diffusion coefficient, in agreement with the estimate by the AMS-02 team based on the B/C ratio (Aguilar et al. 2016). The normalisation is taken according to the most recent GALPROP analysis (Trotta et al. 2011), resulting into:

$$D(E) = 1.33 \times 10^{28}\ \mathrm{cm^2\,s^{-1}}\,\frac{H_0}{\mathrm{kpc}}\left(\frac{E}{3\ \mathrm{GeV}}\right)^{\frac{1}{3}}, \tag{51}$$

where $H_0 = 3$ kpc is the halo size. The escape time of 10 GeV particles is then $\tau_{\mathrm{esc}}(10\ \mathrm{GeV}) = H_0^2/D(10\ \mathrm{GeV}) = 45$ Myr. With an explosion rate as described by Faucher-Giguère and Kaspi (2006), namely a Gaussian distribution around an average rate of 2.8 explosions/century and a standard deviation of 0.1 events/century, the number of explosions that have to be taken into account is $\approx 1.2 \times 10^6$. The explosions are assumed to occur at random places in the Galaxy, spatially distributed as:

$$n_{\mathrm{SN}}(\rho_s, z_s) \propto \left(\frac{\rho_s}{R_\odot}\right)^{\alpha} \exp\left[-\frac{\beta(\rho_s - R_\odot)}{R_\odot}\right] \exp\left[-\frac{z_s}{h_{\mathrm{SN}}}\right] \tag{52}$$

where a cylindrical coordinate system has been adopted, with ρ_s the radial coordinate of the explosion and z_s its height above or below the Galactic plane. The parameters appearing in Eq. (52) have the following values: the sun radial distance from the Galactic centre is $R_\odot = 8.5$ kpc, $h_{\mathrm{SN}} = 50$ pc (form Faucher-Giguère and Kaspi (2006)), $\alpha = 1.09$ and $\beta = 3.87$ from Green (2015). The presence of Galactic arms has been included on top of this profile, according to the description given by Faucher-Giguère and Kaspi (2006). The resulting distribution of SN explosions can be seen in Fig. 15, where the coordinates of about 30000 events are reported in a Galactocentric frame.

All these events are considered as actual contributors to the primary electron flux at Earth, while only 80% of them, randomly selected, are assumed to be associated to type II explosions giving origin to a pulsar and hence included in the count of positron contributors. The flux of electrons and positrons is obtained as the sum over all relevant sources of the single source contribution described in Eq. (49), where $N_s(E) = Q_s(E)/\mathcal{R}$ with $Q_s(E)$ given by Eqs. (42) and (43) for electrons and positrons respectively. The electrons' injection spectral index is assumed to be the same as that of protons, and the spectrum of the latter is taken equal to the fit to the newest data provided by the AMS-02 collaboration (Aguilar et al. 2015). The SN explosion energy that enters the normalisation is assumed to be $E_{\mathrm{SN}} = 10^{51}$ erg for all events. As far as the positrons are concerned, observations of PWNe suggest

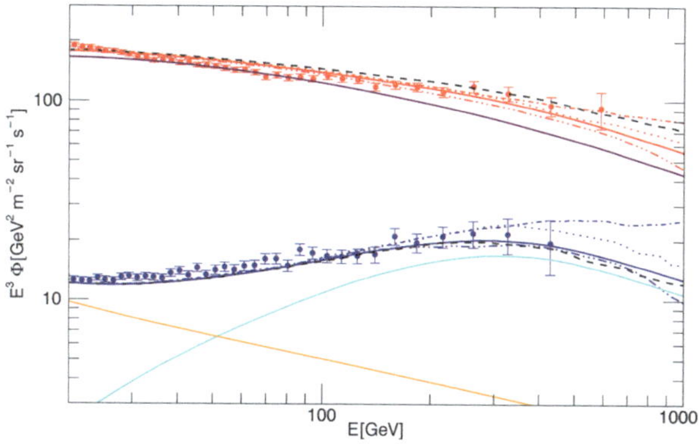

Fig. 16 The electrons (*red*) and positrons (*blue*) spectrum at Earth as computed within the described model is compared with AMS-02 measurements. The legend for the different curves is as follows. The *solid thick curves* refer to ensemble averages over the 30 different realisations of the Galaxy, while the other curves correspond to individual realisations. Averages are shown for the total electron (*red*) and positron (*blue*) spectra, and also for the different contributions: electrons and positrons of secondary origin (*orange*); electrons and positrons from Bow Shock PWNe (*light blue*); primary electrons from SNRs (*purple*). The *shed black curve* refers to a particular realisation that seems to fit especially well the positron fraction data

that γ_{pos} (and accompanying γ_{el}) is different from source to source but without an obvious relation to any other observable. The results that are shown in the following refer to the choice $\gamma_{pos} = 1.2$ up to a maximal energy $E_{\max} = 500$ GeV. A steeper power law with index 2.3 is assumed at higher energies.

$E_{\rm res}$ will also vary from source to source depending both on the pulsar properties and on the environment in which the explosion occurs. In the absence of a good and easy recipe for how to vary the latter, previous work can be improved in realism by taking into account variations of the former. The surface magnetic field is assumed to be the same for all pulsars and corresponding to the average value proposed by Faucher-Giguère and Kaspi (2006), $B_* = 4.5 \times 10^{12}$, but the pulsar's spin period at birth and the proper motion vary among different objects according to the distributions provided by the same authors.

In addition to primary positrons from BSPWNe and primary electrons from the same sources plus SNRs, we also account for secondary electrons, generated from CR interactions with the ISM. The density of the latter is taken to correspond to 1 particle/cm^3 and the incident proton spectrum is taken again from AMS-02 (Aguilar et al. 2015).

In Fig. 16 the spectrum of electrons and positrons resulting from this calculation is shown together with the AMS-02 data (Aguilar et al. 2014). Some of the individual curves, corresponding to different realisations, are also shown in order to give an idea of how different the electrons' and positrons' spectra can be at high energy depending on the source configuration. A conversion efficiency of $\eta = 22\%$ of the pulsar spin-down energy into acceleration of pairs has been adopted.

The difference between the different realisations at high energies depends on the strong decrease with increasing electron energy of the number of contributing sources. Figure 17 serves the purpose of highlighting this effect. The two panels of the figure show the sources that in this calculation actually make a contribution to the flux of positrons (and electrons) of different energies at Earth. It is apparent that while at 10 GeV the number of relevant sources (green diamonds) is large (more than 2000 in the particular realisation of the Galaxy to which the plots refer), it strongly decreases with increasing energy: at 100 GeV about 300 sources are relevant (black triangles in the plots), and only a handful at 1 TeV (blue

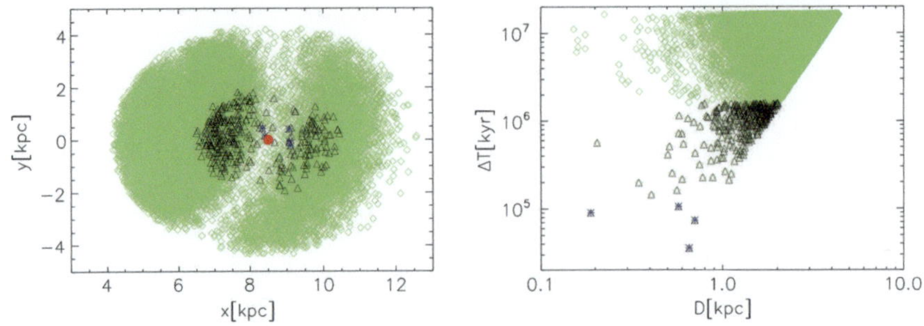

Fig. 17 In the *left panel* we show the projection onto the Galactic plane of the positions of all sources that contribute positrons to Earth at three different energies: 10 GeV (*green diamonds*), 100 GeV (*black triangles*), 1 TeV (*blue asterisks*). The Earth position is shown as a *red circle*. In the *right panel*, the distance in time and space of the contributing sources is shown with the same notation as on the left

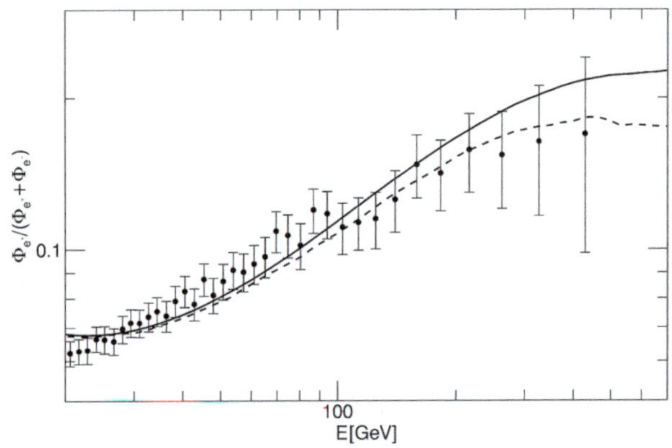

Fig. 18 The positron fraction as a function of energy compared with AMS02 data. The *thick solid curve* corresponds to the average spectra shown as the solid (blue and red) curves of Fig. 16. The *dashed curve* refers to the same configuration shown by the corresponding curves of Fig. 16

asterisks). The panel on the left shows the projected position in the Galactic plane. The fact that the Earth is in an inter-arm region, poorly populated with sources is evident from the plot and plays an important role: this is likely the main reason for the need of a larger value of η with respect to the calculation by Blasi and Amato (2011).

From the left panel of Fig. 17 one sees instead what a small range of distances in space and time can contribute to positrons at the highest energies: already at 100 GeV all the leptons that reach Earth come from sources closer than 2 kpc and younger than 2 Myr.

In Fig. 18, finally, the rise of the positron fraction with energy that this model provides is compared with AMS-02 data. The different curves are drawn following the same notation as in Fig. 16. A generally good agreement is seen between the model and the data.

Further refinement of this kind of calculation are possible and desirable, possibly with inclusion of some model for the density distribution of the ISM. However the main point this section aims at making is that bow shock PWNe are expected to play an important role as contributors of cosmic ray positrons in the Galaxy. That their contribution to the positron excess be non-negligible seems unavoidable and this must be properly assessed and taken into account before using positrons as a diagnostics of other phenomena. In this sense, achieving a better understanding of the physics of these sources is a task with potential far reaching implications for cosmic ray and dark matter physics.

Acknowledgements The authors thank a referee for a constructive comment. A.M. Bykov and E. Amato thank the staff of ISSI for their generous hospitality and assistance. A.M. Bykov and A.E. Petrov were supported by the RSF grant 16-12-10225. A.B. thanks A.V. Artemyev, S.M. Osipov, and G.G. Pavlov for discussions. Some of the modeling was performed at the "Tornado" subsystem of the St. Petersburg Polytechnic University supercomputing center.

References

A.A. Abdo, B.T. Allen, T. Aune, et al., Milagro observations of multi-TeV emission from galactic sources in the Fermi bright source list. Astrophys. J. **700**, 127–131 (2009). doi:10.1088/0004-637X/700/2/L127

A.A. Abdo, M. Ackermann, M. Ajello, et al., Fermi Large Area Telescope observations of the Vela-X pulsar wind nebula. Astrophys. J. **713**, 146–153 (2010). doi:10.1088/0004-637X/713/1/146

A.A. Abdo, M. Ajello, A. Allafort, et al., The second Fermi Large Area Telescope catalog of gamma-ray pulsars. Astrophys. J. Suppl. Ser. **208**, 17 (2013). doi:10.1088/0067-0049/208/2/17

A.U. Abeysekara, A. Albert, R. Alfaro, et al., The 2HWC HAWC observatory gamma ray catalog. ArXiv e-prints (2017)

A. Abramowski, F. Acero, F. Aharonian, et al., Probing the extent of the non-thermal emission from the Vela X region at TeV energies with H.E.S.S. Astron. Astrophys. **548**, 38 (2012). doi:10.1051/0004-6361/201219919

F. Acero, Y. Gallant, J. Ballet, et al., A new nearby pulsar wind nebula overlapping the RX J0852.0-4622 supernova remnant. Astron. Astrophys. **551**, 7 (2013a). doi:10.1051/0004-6361/201220799

F. Acero, M. Ackermann, M. Ajello, et al., Constraints on the galactic population of TeV pulsar wind nebulae using Fermi Large Area Telescope observations. Astrophys. J. **773**, 77 (2013b). doi:10.1088/0004-637X/773/1/77

A. Achterberg, Y.A. Gallant, J.G. Kirk, et al., Particle acceleration by ultrarelativistic shocks: theory and simulations. Mon. Not. R. Astron. Soc. **328**, 393–408 (2001). doi:10.1046/j.1365-8711.2001.04851.x

O. Adriani, G.C. Barbarino, G.A. Bazilevskaya, et al., An anomalous positron abundance in cosmic rays with energies 1.5–100 GeV. Nature **458**, 607–609 (2009). doi:10.1038/nature07942

O. Adriani, G.C. Barbarino, G.A. Bazilevskaya, et al., Cosmic-ray electron flux measured by the PAMELA experiment between 1 and 625 GeV. Phys. Rev. Lett. **106**(20), 201101 (2011). doi:10.1103/PhysRevLett.106.201101

M. Aguilar, G. Alberti, B. Alpat, et al., First result from the alpha magnetic spectrometer on the International Space Station: precision measurement of the positron fraction in primary cosmic rays of 0.5–350 GeV. Phys. Rev. Lett. **110**(14), 141102 (2013). doi:10.1103/PhysRevLett.110.141102

M. Aguilar, D. Aisa, A. Alvino, et al., Electron and positron fluxes in primary cosmic rays measured with the Alpha Magnetic Spectrometer on the International Space Station. Phys. Rev. Lett. **113**(12), 121102 (2014). doi:10.1103/PhysRevLett.113.121102

M. Aguilar, D. Aisa, B. Alpat, et al., Precision measurement of the proton flux in primary cosmic rays from rigidity 1 GV to 1.8 TV with the alpha magnetic spectrometer on the International Space Station. Phys. Rev. Lett. **114**(17), 171103 (2015). doi:10.1103/PhysRevLett.114.171103

M. Aguilar, L. Ali Cavasonza, G. Ambrosi, et al., Precision measurement of the boron to carbon flux ratio in cosmic rays from 1.9 GV to 2.6 TV with the alpha magnetic spectrometer on the International Space Station. Phys. Rev. Lett. **117**(23), 231102 (2016). doi:10.1103/PhysRevLett.117.231102

F. Aharonian, A.G. Akhperjanian, A.R. Bazer-Bachi, et al., First detection of a VHE gamma-ray spectral maximum from a cosmic source: HESS discovery of the Vela X nebula. Astron. Astrophys. **448**, 43–47 (2006). doi:10.1051/0004-6361:200600014

G.E. Allen, K. Chow, T. DeLaney, et al., On the expansion rate, age, and distance of the supernova remnant G266.2-1.2 (Vela Jr.). Astrophys. J. **798**, 82 (2015). doi:10.1088/0004-637X/798/2/82

E. Amato, P. Blasi, A kinetic approach to cosmic-ray-induced streaming instability at supernova shocks. Mon. Not. R. Astron. Soc. **392**, 1591–1600 (2009). doi:10.1111/j.1365-2966.2008.14200.x

E. Amato, M. Salvati, R. Bandiera, et al., Inhomogeneous models for plerions: the surface brightness profile of the Crab Nebula. Astron. Astrophys. **359**, 1107–1110 (2000)

AMS-01 Collaboration, M. Aguilar, J. Alcaraz, et al., Cosmic-ray positron fraction measurement from 1 to 30 GeV with AMS-01. Phys. Lett. B **646**, 145–154 (2007). doi:10.1016/j.physletb.2007.01.024

J. Arons, Theory of pulsar winds, in *Neutron Stars in Supernova Remnants*, ed. by P.O. Slane, B.M. Gaensler Astronomical Society of the Pacific Conference Series, vol. 271, 2002, p. 71

J. Arons, Pulsar wind nebulae as cosmic pevatrons: a current sheet's tale. Space Sci. Rev. **173**, 341–367 (2012). doi:10.1007/s11214-012-9885-1

Z. Arzoumanian, D.F. Chernoff, J.M. Cordes, The velocity distribution of isolated radio pulsars. Astrophys. J. **568**, 289–301 (2002). doi:10.1086/338805

A.M. Atoyan, F.A. Aharonian, On the mechanisms of gamma radiation in the Crab Nebula. Mon. Not. R. Astron. Soc. **278**, 525–541 (1996). doi:10.1093/mnras/278.2.525

K. Auchettl, P. Slane, R.W. Romani, et al., X-ray analysis of the proper motion and pulsar wind nebula for PSR J1741-2054. Astrophys. J. **802**, 68 (2015). doi:10.1088/0004-637X/802/1/68

W.I. Axford, The interaction of the solar wind with the interstellar medium. NASA Spec. Publ. **308**, 609 (1972)

W.I. Axford, E. Leer, G. Skadron, The acceleration of cosmic rays by shock waves, in *Proc. 15th ICRC (Plovdiv)*, vol. 11, 1977, p. 132

A. Bamba, T. Anada, T. Dotani, et al., X-ray evolution of pulsar wind nebulae. Astrophys. J. **719**, 116–120 (2010). doi:10.1088/2041-8205/719/2/L116

R. Bandiera, Modelling non-axisymmetric bow shocks. Astron. Astrophys. **276**, 648 (1993)

R. Bandiera, On the X-ray feature associated with the Guitar nebula. Astron. Astrophys. **490**, 3–6 (2008). doi:10.1051/0004-6361:200810666

V.B. Baranov, K.V. Krasnobaev, A.G. Kulikovskii, A model of the interaction of the solar wind with the interstellar medium. Sov. Phys. Dokl. **15**, 791 (1971)

J.J. Beatty, A. Bhattacharyya, C. Bower, et al., New measurement of the cosmic-ray positron fraction from 5 to 15 GeV. Phys. Rev. Lett. **93**(24), 241102 (2004). doi:10.1103/PhysRevLett.93.241102

M.C. Begelman, Z.-Y. Li, An axisymmetric magnetohydrodynamic model for the Crab pulsar wind bubble. Astrophys. J. **397**, 187–195 (1992). doi:10.1086/171778

A.R. Bell, The acceleration of cosmic rays in shock fronts. I. Mon. Not. R. Astron. Soc. **182**, 147–156 (1978)

A.R. Bell, Turbulent amplification of magnetic field and diffusive shock acceleration of cosmic rays. Mon. Not. R. Astron. Soc. **353**, 550–558 (2004)

A.R. Bell, The interaction of cosmic rays and magnetized plasma. Mon. Not. R. Astron. Soc. **358**, 181–187 (2005)

A.R. Bell, Cosmic ray acceleration. Astropart. Phys. **43**, 56–70 (2013). doi:10.1016/j.astropartphys.2012.05.022

V.S. Berezinskii, S.V. Bulanov, V.A. Dogiel, et al., *Astrophysics of Cosmic Rays* 1990

A.J. Bird, A. Bazzano, L. Bassani, et al., The fourth IBIS/ISGRI soft gamma-ray survey catalog. Astrophys. J. Suppl. Ser. **186**, 1–9 (2010). doi:10.1088/0067-0049/186/1/1

J. Birn, E.R. Priest, *Reconnection of Magnetic Fields* 2007

R. Blandford, D. Eichler, Particle acceleration at astrophysical shocks: a theory of cosmic ray origin. Phys. Rep. **154**, 1–75 (1987). doi:10.1016/0370-1573(87)90134-7

R.D. Blandford, J.P. Ostriker, Particle acceleration by astrophysical shocks. Astrophys. J. **221**, 29–32 (1978)

P. Blasi, E. Amato, *Positrons from Pulsar Winds*. Astrophysics and Space Science Proceedings, vol. 21, 2011, p. 624

P. Blasi, E. Amato, Diffusive propagation of cosmic rays from supernova remnants in the Galaxy. I: spectrum and chemical composition. J. Cosmol. Astropart. Phys. **1**, 010 (2012a). doi:10.1088/1475-7516/2012/01/010

P. Blasi, E. Amato, Diffusive propagation of cosmic rays from supernova remnants in the Galaxy. II: anisotropy. J. Cosmol. Astropart. Phys. **1**, 011 (2012b). doi:10.1088/1475-7516/2012/01/011

J.M. Blondin, R.A. Chevalier, D.M. Frierson, Pulsar wind nebulae in evolved supernova remnants. Astrophys. J. **563**, 806–815 (2001). doi:10.1086/324042

F. Bocchino, A.M. Bykov, The plerion nebula in IC 443: the XMM-Newton view. Astron. Astrophys. **376**, 248–253 (2001). doi:10.1051/0004-6361:20010882

F. Bocchino, A. Maggio, S. Sciortino, ROSAT PSPC observation of the NE region of the VELA supernova remnant. III. The two-component nature of the X-ray emission and its implications on the ISM. Astron. Astrophys. **342**, 839–853 (1999)

M. Boezio, G. Barbiellini, V. Bonvicini, et al., Measurements of cosmic-ray electrons and positrons by the Wizard/CAPRICE collaboration. Adv. Space Res. **27**, 669–674 (2001). doi:10.1016/S0273-1177(01)00108-9

S. Brownsberger, R.W. Romani, A survey for Hα pulsar bow shocks. Astrophys. J. **784**, 154 (2014). doi:10.1088/0004-637X/784/2/154

N. Bucciantini, Review of the theory of pulsar-wind nebulae. Astron. Nachr. **335**, 234–239 (2014). doi:10.1002/asna.201312024

N. Bucciantini, R. Bandiera, Pulsar bow-shock nebulae. I. Physical regimes and detectability conditions. Astron. Astrophys. **375**, 1032–1039 (2001). doi:10.1051/0004-6361:20010900

N. Bucciantini, E. Amato, L. Del Zanna, Relativistic MHD simulations of pulsar bow-shock nebulae. Astron. Astrophys. **434**, 189–199 (2005). doi:10.1051/0004-6361:20042205

N. Bucciantini, J. Arons, E. Amato, Modelling spectral evolution of pulsar wind nebulae inside supernova remnants. Mon. Not. R. Astron. Soc. **410**, 381–398 (2011). doi:10.1111/j.1365-2966.2010.17449.x

R. Bühler, R. Blandford, The surprising Crab pulsar and its nebula: a review. Rep. Prog. Phys. **77**(6), 066901 (2014). doi:10.1088/0034-4885/77/6/066901

R. Bühler, M. Giomi, The imprint of pulsar parameters on the morphology of pulsar wind nebulae. Mon. Not. R. Astron. Soc. **462**, 2762–2776 (2016). doi:10.1093/mnras/stw1773

A.M. Bykov, Energetic processes and nonthermal emission of starforming complexes, in *Massive Stars and High-Energy Emission in OB Associations*, ed. by G. Rauw, Y. Nazé, R. Blomme, et al., 2005, pp. 95–98

A.M. Bykov, D.C. Ellison, M. Renaud, Magnetic fields in cosmic particle acceleration sources. Space Sci. Rev. **166**, 71–95 (2012). doi:10.1007/s11214-011-9761-4

A.M. Bykov, P.E. Gladilin, S.M. Osipov, Non-linear model of particle acceleration at colliding shock flows. Mon. Not. R. Astron. Soc. **429**, 2755–2762 (2013). doi:10.1093/mnras/sts553

A.M. Bykov, Y.A. Uvarov, D.C. Ellison, Dots, clumps, and filaments: the intermittent images of synchrotron emission in random magnetic fields of young supernova remnants. Astrophys. J. **689**, 133 (2008). doi:10.1086/595868

A.M. Bykov, G.G. Pavlov, A.V. Artemyev, et al., Twinkling pulsar wind nebulae in the synchrotron cut-off regime and the γ-ray flares in the Crab Nebula. Mon. Not. R. Astron. Soc. **421**, 67–71 (2012). doi:10.1111/j.1745-3933.2011.01208.x

A.M. Bykov, M.A. Malkov, J.C. Raymond, et al., Collisionless shocks in partly ionized plasma with cosmic rays: microphysics of non-thermal components. Space Sci. Rev. **178**, 599–632 (2013a). doi:10.1007/s11214-013-9984-7

A.M. Bykov, A. Brandenburg, M.A. Malkov, et al., Microphysics of cosmic ray driven plasma instabilities. Space Sci. Rev. (2013b). doi:10.1007/s11214-013-9988-3

A.M. Bykov, D.C. Ellison, S.M. Osipov, et al., Magnetic field amplification in nonlinear diffusive shock acceleration including resonant and non-resonant cosmic-ray driven instabilities. Astrophys. J. **789**, 137 (2014). doi:10.1088/0004-637X/789/2/137

A. Bykov, N. Gehrels, H. Krawczynski, et al., Particle acceleration in relativistic outflows. Space Sci. Rev. **173**, 309–339 (2012). doi:10.1007/s11214-012-9896-y

F. Camilo, B.M. Gaensler, E.V. Gotthelf, et al., Chandra detection of a synchrotron nebula around the Vela-like pulsar J1016-5857. Astrophys. J. **616**, 1118–1123 (2004). doi:10.1086/424924

P.A. Caraveo, G.F. Bignami, A. De Luca, et al., Geminga's tails: a pulsar bow shock probing the interstellar medium. Science **301**, 1345–1348 (2003). doi:10.1126/science.1086973

C.J. Cesarsky, Cosmic-ray confinement in the galaxy. Annu. Rev. Astron. Astrophys. **18**, 289–319 (1980). doi:10.1146/annurev.aa.18.090180.001445

S. Chatterjee, J.M. Cordes, Smashing the Guitar: an evolving neutron star bow shock. Astrophys. J. **600**, 51–54 (2004). doi:10.1086/381498

R.A. Chevalier, A model for the X-ray luminosity of pulsar nebulae. Astrophys. J. **539**, 45–48 (2000). doi:10.1086/312835

R.A. Chevalier, D. Luo, Magnetic shaping of planetary nebulae and other stellar wind bubbles. Astrophys. J. **421**, 225–235 (1994). doi:10.1086/173640

R.A. Chevalier, S.P. Reynolds, Pulsar wind nebulae with thick toroidal structure. Astrophys. J. **740**, 26 (2011). doi:10.1088/2041-8205/740/1/L26

J.M. Cordes, R.W. Romani, S.C. Lundgren, The Guitar nebula—a bow shock from a slow-spin, high-velocity neutron star. Nature **362**, 133–135 (1993). doi:10.1038/362133a0

A. Danilenko, P. Shternin, A. Karpova, et al., The γ-ray pulsar J0633+0632 in X-rays. Publ. Astron. Soc. Aust. **32**, 038 (2015). doi:10.1017/pasa.2015.40

W. Daughton, V. Roytershteyn, H. Karimabadi, et al., Role of electron physics in the development of turbulent magnetic reconnection in collisionless plasmas. Nat. Phys. **7**, 539–542 (2011). doi:10.1038/nphys1965

O.C. de Jager, P.O. Slane, S. LaMassa, Probing the radio to X-ray connection of the Vela X pulsar wind nebula with Fermi LAT and H.E.S.S. Astrophys. J. **689**, 125 (2008). doi:10.1086/595959

A. De Luca, R.P. Mignani, M. Marelli, et al., PSR J0357+3205: a fast-moving pulsar with a very unusual X-ray trail. Astrophys. J. **765**, 19 (2013). doi:10.1088/2041-8205/765/1/L19

L. Del Zanna, E. Amato, N. Bucciantini, Axially symmetric relativistic MHD simulations of pulsar wind nebulae in supernova remnants. On the origin of torus and jet-like features. Astron. Astrophys. **421**, 1063–1073 (2004). doi:10.1051/0004-6361:20035936

L. Del Zanna, D. Volpi, E. Amato, et al., Simulated synchrotron emission from pulsar wind nebulae. Astron. Astrophys. **453**, 621–633 (2006). doi:10.1051/0004-6361:20064858

M. Di Mauro, F. Donato, N. Fornengo, et al., Interpretation of AMS-02 electrons and positrons data. J. Cosmol. Astropart. Phys. **4**, 006 (2014). doi:10.1088/1475-7516/2014/04/006

M. Di Mauro, F. Donato, N. Fornengo, et al., Dark matter vs. astrophysics in the interpretation of AMS-02 electron and positron data. J. Cosmol. Astropart. Phys. **5**, 031 (2016). doi:10.1088/1475-7516/2016/05/031

M. Di Mauro, S. Manconi, A. Vittino, et al., Theoretical interpretation of Pass 8 *Fermi*-LAT $e^+ + e^-$ data. ArXiv e-prints (2017)

R. Dodson, D. Lewis, D. McConnell, et al., The radio nebula surrounding the Vela pulsar. Mon. Not. R. Astron. Soc. **343**, 116–124 (2003a). doi:10.1046/j.1365-8711.2003.06653.x

R. Dodson, D. Legge, J.E. Reynolds, et al., The Vela pulsar's proper motion and parallax derived from VLBI observations. Astrophys. J. **596**, 1137–1141 (2003b). doi:10.1086/378089

J.F. Drake, M. Swisdak, M. Opher, A model of the heliosphere with jets. Astrophys. J. **808**, 44 (2015). doi:10.1088/2041-8205/808/2/L44

M. Durant, O. Kargaltsev, G.G. Pavlov, et al., The helical jet of the Vela pulsar. Astrophys. J. **763**, 72 (2013). doi:10.1088/0004-637X/763/2/72

J. Eastwood, M. Goldman, D. Newman, et al., Ion and electron kinetic physics associated with magnetotail dipolarization fronts, in *EGU General Assembly Conference Abstracts*. EGU General Assembly Conference Abstracts, vol. 18, 2016, p. 3535

D.C. Ellison, G.P. Double, Diffusive shock acceleration in unmodified relativistic, oblique shocks. Astropart. Phys. **22**, 323–338 (2004). doi:10.1016/j.astropartphys.2004.08.005

C.-A. Faucher-Giguère, V.M. Kaspi, Birth and evolution of isolated radio pulsars. Astrophys. J. **643**, 332–355 (2006). doi:10.1086/501516

D.A. Frail, M.F. Bietenholz, C.B. Markwardt, A radio/X-ray comparison of the Vela X region. Astrophys. J. **475**, 224–230 (1997)

D.A. Frail, E.B. Giacani, W.M. Goss, et al., The pulsar wind nebula around PSR B1853+01 in the supernova remnant W44. Astrophys. J. **464**, 165 (1996). doi:10.1086/310103

B.M. Gaensler, P.O. Slane, The evolution and structure of pulsar wind nebulae. Annu. Rev. Astron. Astrophys. **44**, 17–47 (2006). doi:10.1146/annurev.astro.44.051905.092528

B.M. Gaensler, N.S. Schulz, V.M. Kaspi, et al., XMM-Newton observations of PSR B1823-13: an asymmetric synchrotron nebula around a Vela-like pulsar. Astrophys. J. **588**, 441–451 (2003). doi:10.1086/368356

B.M. Gaensler, E. van der Swaluw, F. Camilo, et al., The mouse that soared: high-resolution X-ray imaging of the pulsar-powered bow shock G359.23-0.82. Astrophys. J. **616**, 383–402 (2004). doi:10.1086/424906

J.D. Gelfand, P.O. Slane, W. Zhang, A dynamical model for the evolution of a pulsar wind nebula inside a nonradiative supernova remnant. Astrophys. J. **703**, 2051–2067 (2009). doi:10.1088/0004-637X/703/2/2051

E.A. Golikov, V.V. Izmodenov, D.B. Alexashov, et al., Two-jet astrosphere model: effect of azimuthal magnetic field. Mon. Not. R. Astron. Soc. **464**, 1065–1076 (2017). doi:10.1093/mnras/stw2402

M.E. Gonzalez, V.M. Kaspi, M.J. Pivovaroff, et al., Chandra and XMM-Newton observations of the Vela-like pulsar B1046-58. Astrophys. J. **652**, 569–575 (2006). doi:10.1086/507125

D.A. Green, Constraints on the distribution of supernova remnants with galactocentric radius. Mon. Not. R. Astron. Soc. **454**, 1517–1524 (2015). doi:10.1093/mnras/stv1885

M.-H. Grondin, R.W. Romani, M. Lemoine-Goumard, et al., The Vela-X pulsar wind nebula revisited with four years of Fermi Large Area Telescope observations. Astrophys. J. **774**, 110 (2013). doi:10.1088/0004-637X/774/2/110

H.E.S.S. Collaboration, H. Abdalla, A. Abramowski, et al., Deeper H.E.S.S. observations of Vela Junior (RX J0852.0-4622): morphology studies and resolved spectroscopy. ArXiv e-prints (2016)

A.S. Hales, S. Casassus, H. Alvarez, et al., Vela X at 31 GHz. Astrophys. J. **613**, 977–985 (2004). doi:10.1086/422598

C.A. Hales, B.M. Gaensler, S. Chatterjee, et al., A proper motion for the pulsar wind nebula G359.23-0.82, the "Mouse," associated with the energetic radio pulsar J1747-2958. Astrophys. J. **706**, 1316–1322 (2009). doi:10.1088/0004-637X/706/2/1316

J.P. Halpern, E.V. Gotthelf, K.M. Leighly, et al., A possible X-ray and radio counterpart of the high-energy gamma-ray source 3EG J2227+6122. Astrophys. J. **547**, 323–333 (2001a). doi:10.1086/318361

J.P. Halpern, F. Camilo, E.V. Gotthelf, et al., PSR J2229+6114: discovery of an energetic young pulsar in the error box of the EGRET source 3EG J2227+6122. Astrophys. J. **552**, 125–128 (2001b). doi:10.1086/320347

P.A. Harrison, A.G. Lyne, B. Anderson, New determinations of the proper motions of 44 pulsars. Mon. Not. R. Astron. Soc. **261**, 113–124 (1993). doi:10.1093/mnras/261.1.113

E.A. Helder, J. Vink, A.M. Bykov, et al., Observational signatures of particle acceleration in supernova remnants. Space Sci. Rev. **173**, 369–431 (2012). doi:10.1007/s11214-012-9919-8

D.J. Helfand, E.V. Gotthelf, J.P. Halpern, Vela pulsar and its synchrotron nebula. Astrophys. J. **556**, 380–391 (2001). doi:10.1086/321533

J.J. Hester, S.R. Kulkarni, The origin and energetics of CTB 80. Astrophys. J. **331**, 121–125 (1988). doi:10.1086/185249

J.A. Hinton, S. Funk, R.D. Parsons, et al., Escape from Vela X. Astrophys. J. **743**, 7 (2011). doi:10.1088/2041-8205/743/1/L7

G. Hobbs, D.R. Lorimer, A.G. Lyne, et al., A statistical study of 233 pulsar proper motions. Mon. Not. R. Astron. Soc. **360**, 974–992 (2005). doi:10.1111/j.1365-2966.2005.09087.x

D. Hooper, I. Cholis, T. Linden, et al., HAWC observations strongly favor pulsar interpretations of the cosmic-ray positron excess. ArXiv e-prints (2017)

C.Y. Hui, W. Becker, X-ray emission properties of the old pulsar PSR B2224+65. Astron. Astrophys. **467**, 1209–1214 (2007). doi:10.1051/0004-6361:20066562

C.Y. Hui, R.H.H. Huang, L. Trepl, et al., XMM-Newton observation of PSR B2224+65 and its jet. Astrophys. J. **747**, 74 (2012). doi:10.1088/0004-637X/747/1/74

S.P. Johnson, Q.D. Wang, The pulsar B2224+65 and its jets: a two epoch X-ray analysis. Mon. Not. R. Astron. Soc. **408**, 1216–1224 (2010). doi:10.1111/j.1365-2966.2010.17200.x

J.R. Jokipii, Rate of energy gain and maximum energy in diffusive shock acceleration. Astrophys. J. **313**, 842–846 (1987). doi:10.1086/165022

O. Kargaltsev, G.G. Pavlov, Pulsar wind nebulae in the Chandra era, in *40 Years of Pulsars: Millisecond Pulsars, Magnetars and More*, ed. by C. Bassa, Z. Wang, A. Cumming, et al.American Institute of Physics Conference Series, vol. 983, 2008, pp. 171–185. doi:10.1063/1.2900138

O. Kargaltsev, B. Rangelov, G.G. Pavlov, Gamma-ray and X-ray properties of pulsar wind nebulae and unidentified galactic TeV sources. ArXiv e-prints (2013)

O. Kargaltsev, B. Cerutti, Y. Lyubarsky, et al., Pulsar-wind nebulae. Recent progress in observations and theory. Space Sci. Rev. **191**, 391–439 (2015). doi:10.1007/s11214-015-0171-x

S. Katsuda, K. Mori, R. Petre, et al., Suzaku detection of diffuse hard X-ray emission outside Vela X. Publ. Astron. Soc. Jpn. **63**, 827–836 (2011). doi:10.1093/pasj/63.sp3.S827

C.F. Kennel, F.V. Coroniti, Confinement of the Crab pulsar's wind by its supernova remnant. Astrophys. J. **283**, 694–709 (1984). doi:10.1086/162356

U. Keshet, E. Waxman, Energy spectrum of particles accelerated in relativistic collisionless shocks. Phys. Rev. Lett. **94**(11), 111102 (2005). doi:10.1103/PhysRevLett.94.111102

J.G. Kirk, Y. Lyubarsky, J. Petri, The theory of pulsar winds and nebulae, in *Astrophysics and Space Science Library*, ed. by W. Becker Astrophysics and Space Science Library, vol. 357, 2009, p. 421. doi:10.1007/978-3-540-76965-1_16

S.S. Komissarov, Y.E. Lyubarsky, The origin of peculiar jet-torus structure in the Crab nebula. Mon. Not. R. Astron. Soc. **344**, 93–96 (2003). doi:10.1046/j.1365-8711.2003.07097.x

S.S. Komissarov, Y.E. Lyubarsky, Synchrotron nebulae created by anisotropic magnetized pulsar winds. Mon. Not. R. Astron. Soc. **349**, 779–792 (2004). doi:10.1111/j.1365-2966.2004.07597.x

R. Kothes, B. Uyaniker, S. Pineault, The supernova remnant G106.3+2.7 and its pulsar-wind nebula: relics of triggered star formation in a complex environment. Astrophys. J. **560**, 236–243 (2001). doi:10.1086/322511

G.F. Krymskii, A regular mechanism for the acceleration of charged particles on the front of a shock wave. Akad. Nauk SSSR Dokl. **234**, 1306–1308 (1977)

S.R. Kulkarni, J.J. Hester, Discovery of a nebula around PSR1957+20. Nature **335**, 801–803 (1988). doi:10.1038/335801a0

M. Lemoine, E. Waxman, Anisotropy vs chemical composition at ultra-high energies. J. Cosmol. Astropart. Phys. **11**, 009 (2009). doi:10.1088/1475-7516/2009/11/009

K.P. Levenfish, A.M. Bykov, M. Durant, et al., Finest persistent structures in the Vela PWN. Mem. Soc. Astron. Ital. **84**, 588 (2013)

X.H. Li, F.J. Lu, T.P. Li, X-ray spectroscopy of PSR B1951+32 and its pulsar wind nebula. Astrophys. J. **628**, 931–937 (2005). doi:10.1086/430941

X.-H. Li, F.-J. Lu, Z. Li, Nonthermal X-ray properties of rotation-powered pulsars and their wind nebulae. Astrophys. J. **682**, 1166–1176 (2008). doi:10.1086/589495

S.G. Lucek, A.R. Bell, Non-linear amplification of a magnetic field driven by cosmic ray streaming. Mon. Not. R. Astron. Soc. **314**, 65–74 (2000)

Y.E. Lyubarsky, On the structure of the inner Crab nebula. Mon. Not. R. Astron. Soc. **329**, 34–36 (2002). doi:10.1046/j.1365-8711.2002.05151.x

C. Maitra, F. Acero, C. Venter, Constraining the geometry of PSR J0855-4644: a nearby pulsar wind nebula with double torus/jet morphology. ArXiv e-prints (2016)

A. Marcowith, A. Bret, A. Bykov, et al., The microphysics of collisionless shock waves. Rep. Prog. Phys. **79**(4), 046901 (2016). doi:10.1088/0034-4885/79/4/046901

M. Marelli, D. Pizzocaro, A. De Luca, et al., The tale of the two tails of the oldish PSR J2055+2539. Astrophys. J. **819**, 40 (2016). doi:10.3847/0004-637X/819/1/40

C.B. Markwardt, H.B. Ögelman, The X-ray bow shock nebula of the Vela pulsar. Mem. Soc. Astron. Ital. **69**, 927 (1998)
T.E. Marubini, R.R. Sefako, C. Venter, et al., A search for optical counterparts of the complex Vela X system. ArXiv e-prints (2015)
F. Mattana, D. Götz, R. Terrier, et al., The emerging population of pulsar wind nebulae in hard X-rays, in *American Institute of Physics Conference Series*, ed. by J. Rodriguez, P. Ferrando American Institute of Physics Conference Series, vol. 1126, 2009, pp. 259–262. doi:10.1063/1.3149427
F. Mattana, D. Götz, R. Terrier, et al., Extended hard X-ray emission from the Vela pulsar wind nebula. Astrophys. J. **743**, 18 (2011). doi:10.1088/2041-8205/743/1/L18
R.P. Mignani, N. Rea, V. Testa, et al., Observations of three young γ-ray pulsars with the Gran Telescopio Canarias. Mon. Not. R. Astron. Soc. **461**, 4317–4328 (2016). doi:10.1093/mnras/stw1629
D.-S. Moon, J.-J. Lee, S.S. Eikenberry, et al., PSR B1951+32: a bow shock-confined X-ray nebula, a synchrotron knot, and an optical counterpart candidate. Astrophys. J. **610**, 33–36 (2004). doi:10.1086/423238
K. Mori, E.V. Gotthelf, F. Dufour, et al., A broadband X-ray study of the Geminga pulsar with NuSTAR and XMM-Newton. Astrophys. J. **793**, 88 (2014). doi:10.1088/0004-637X/793/2/88
C.-Y. Ng, N. Bucciantini, B.M. Gaensler, et al., An extreme pulsar tail protruding from the frying pan supernova remnant. Astrophys. J. **746**, 105 (2012). doi:10.1088/0004-637X/746/1/105
C.A. Norman, D.B. Melrose, A. Achterberg, The origin of cosmic rays above 10 18.5 eV. Astrophys. J. **454**, 60 (1995). doi:10.1086/176465
B. Olmi, L. Del Zanna, E. Amato, et al., Multi-D magnetohydrodynamic modelling of pulsar wind nebulae: recent progress and open questions. J. Plasma Phys. **82**(6), 635820601 (2016). doi:10.1017/S0022377816000957
M. Opher, J.F. Drake, B. Zieger, et al., Magnetized jets driven by the Sun: the structure of the heliosphere revisited. Astrophys. J. **800**, 28 (2015). doi:10.1088/2041-8205/800/2/L28
M. Ostrowski, Monte Carlo simulations of energetic particle transport in weakly inhomogeneous magnetic fields. I—particle acceleration in relativistic shock waves with oblique magnetic fields. Mon. Not. R. Astron. Soc. **249**, 551–559 (1991)
L. Pavan, E. Bozzo, G. Pühlhofer, et al., IGR J11014-6103: a newly discovered pulsar wind nebula? Astron. Astrophys. **533**, 74 (2011). doi:10.1051/0004-6361/201117379
L. Pavan, P. Bordas, G. Pühlhofer, et al., The long helical jet of the Lighthouse nebula, IGR J11014-6103. Astron. Astrophys. **562**, 122 (2014). doi:10.1051/0004-6361/201322588
L. Pavan, G. Pühlhofer, P. Bordas, et al., Closer view of the IGR J11014-6103 outflows. Astron. Astrophys. **591**, 91 (2016). doi:10.1051/0004-6361/201527703
G.G. Pavlov, V.E. Zavlin, D. Sanwal, et al., The X-ray spectrum of the Vela pulsar resolved with the Chandra X-ray observatory. Astrophys. J. **552**, 129–133 (2001a). doi:10.1086/320342
G.G. Pavlov, O.Y. Kargaltsev, D. Sanwal, et al., Variability of the Vela pulsar wind nebula observed with Chandra. Astrophys. J. **554**, 189–192 (2001b). doi:10.1086/321721
G.G. Pavlov, M.A. Teter, O. Kargaltsev, et al., The variable jet of the Vela pulsar. Astrophys. J. **591**, 1157–1171 (2003). doi:10.1086/375531
G. Pelletier, M. Lemoine, A. Marcowith, On Fermi acceleration and magnetohydrodynamic instabilities at ultra-relativistic magnetized shock waves. Mon. Not. R. Astron. Soc. **393**, 587–597 (2009)
A. Pellizzoni, F. Mattana, S. Mereghetti, et al., Pulsar bow-shocks. Mem. Soc. Astron. Ital. **5**, 195 (2004)
A. Pellizzoni, A. Trois, M. Tavani, et al., Detection of gamma-ray emission from the Vela pulsar wind nebula with AGILE. Science **327**, 663 (2010). doi:10.1126/science.1183844
R. Petre, K.D. Kuntz, R.L. Shelton, The X-ray structure and spectrum of the pulsar wind nebula surrounding PSR B1853+01 in W44. Astrophys. J. **579**, 404–410 (2002). doi:10.1086/342672
O. Porth, S.S. Komissarov, R. Keppens, Three-dimensional magnetohydrodynamic simulations of the Crab nebula. Mon. Not. R. Astron. Soc. **438**, 278–306 (2014). doi:10.1093/mnras/stt2176
O. Porth, M.J. Vorster, M. Lyutikov, et al., Diffusion in pulsar wind nebulae: an investigation using magnetohydrodynamic and particle transport models. Mon. Not. R. Astron. Soc. **460**, 4135–4149 (2016). doi:10.1093/mnras/stw1152
B. Posselt, G.G. Pavlov, P.O. Slane, et al., Geminga's puzzling pulsar wind nebula. Astrophys. J. **835**, 66 (2017). doi:10.3847/1538-4357/835/1/66
B. Rangelov, G.G. Pavlov, O. Kargaltsev, et al., First detection of a pulsar bow shock nebula in far-UV: PSR J0437-4715. Astrophys. J. **831**, 129 (2016). doi:10.3847/0004-637X/831/2/129
P.S. Ray, M. Kerr, D. Parent, et al., Precise γ-ray timing and radio observations of 17 Fermi γ-ray pulsars. Astrophys. J. Suppl. Ser. **194**, 17 (2011). doi:10.1088/0067-0049/194/2/17
M.J. Rees, J.E. Gunn, The origin of the magnetic field and relativistic particles in the Crab nebula. Mon. Not. R. Astron. Soc. **167**, 1–12 (1974). doi:10.1093/mnras/167.1.1

B. Reville, S. O'Sullivan, P. Duffy, et al., The transport of cosmic rays in self-excited magnetic turbulence. Mon. Not. R. Astron. Soc. **386**, 509–515 (2008). doi:10.1111/j.1365-2966.2008.13059.x
S.P. Reynolds, R.A. Chevalier, Evolution of pulsar-driven supernova remnants. Astrophys. J. **278**, 630–648 (1984). doi:10.1086/161831
R.W. Romani, J.M. Cordes, I.-A. Yadigaroglu, X-ray emission from the Guitar nebula. Astrophys. J. **484**, 137–140 (1997). doi:10.1086/310796
R.W. Romani, M.S. Shaw, F. Camilo, et al., The Balmer-dominated bow shock and wind nebula structure of γ-ray pulsar PSR J1741-2054. Astrophys. J. **724**, 908–914 (2010). doi:10.1088/0004-637X/724/2/908
C.T. Russell, E.R. Priest, L.C. Lee, *Physics of Magnetic Flux Ropes*. Washington DC American Geophysical Union Geophysical Monograph Series, vol. 58, 1990. doi:10.1029/GM058
K.M. Schure, A.R. Bell, L. O'C Drury, et al., Diffusive shock acceleration and magnetic field amplification. Space Sci. Rev. **173**, 491–519 (2012)
P.D. Serpico, Astrophysical models for the origin of the positron "excess". Astropart. Phys. **39**, 2–11 (2012). doi:10.1016/j.astropartphys.2011.08.007
L. Sironi, A. Spitkovsky, Acceleration of particles at the termination shock of a relativistic striped wind. Astrophys. J. **741**, 39 (2011). doi:10.1088/0004-637X/741/1/39
L. Sironi, D. Giannios, M. Petropoulou, Plasmoids in relativistic reconnection, from birth to adulthood: first they grow, then they go. Mon. Not. R. Astron. Soc. **462**, 48–74 (2016). doi:10.1093/mnras/stw1620
L. Sironi, U. Keshet, M. Lemoine, Relativistic shocks: particle acceleration and magnetization. Space Sci. Rev. **191**, 519–544 (2015). doi:10.1007/s11214-015-0181-8
L. Sironi, A. Spitkovsky, J. Arons, The maximum energy of accelerated particles in relativistic collisionless shocks. Astrophys. J. **771**, 54 (2013). doi:10.1088/0004-637X/771/1/54
A.W. Strong, I.V. Moskalenko, V.S. Ptuskin, Cosmic-ray propagation and interactions in the galaxy. Annu. Rev. Nucl. Part. Sci. **57**, 285–327 (2007)
D.A. Swartz, G.G. Pavlov, T. Clarke, et al., High spatial resolution X-ray spectroscopy of the IC 443 pulsar wind nebula and environs. Astrophys. J. **808**, 84 (2015). doi:10.1088/0004-637X/808/1/84
S.I. Syrovatskii, The distribution of relativistic electrons in the galaxy and the spectrum of synchrotron radio emission. Sov. Astron. **3**, 22 (1959)
M. Takamoto, J.G. Kirk, Rapid cosmic-ray acceleration at perpendicular shocks in supernova remnants. Astrophys. J. **809**, 29 (2015). doi:10.1088/0004-637X/809/1/29
H. Tananbaum, M.C. Weisskopf, W. Tucker, et al., Highlights and discoveries from the Chandra X-ray observatory. Rep. Prog. Phys. **77**(6), 066902 (2014). doi:10.1088/0034-4885/77/6/066902
X. Tang, R.A. Chevalier, Particle transport in young pulsar wind nebulae. Astrophys. J. **752**, 83 (2012). doi:10.1088/0004-637X/752/2/83
V. Tatischeff, M. Tavani, P. von Ballmoos, et al., The e-ASTROGAM gamma-ray space mission, in *Society of Photo-Optical Instrumentation Engineers (SPIE) Conference Series*. Proc. SPIE, vol. 9905, 2016, p. 99052. doi:10.1117/12.2231601
R. Trotta, G. Jóhannesson, I.V. Moskalenko, et al., Constraints on cosmic-ray propagation models from a global Bayesian analysis. Astrophys. J. **729**, 106 (2011). doi:10.1088/0004-637X/729/2/106
J.K. Truelove, C.F. McKee, Evolution of nonradiative supernova remnants. Astrophys. J. Suppl. Ser. **120**, 299–326 (1999). doi:10.1086/313176
E. van der Swaluw, Interaction of a magnetized pulsar wind with its surroundings. MHD simulations of pulsar wind nebulae. Astron. Astrophys. **404**, 939–947 (2003). doi:10.1051/0004-6361:20030452
E. van der Swaluw, T.P. Downes, R. Keegan, An evolutionary model for pulsar-driven supernova remnants. A hydrodynamical model. Astron. Astrophys. **420**, 937–944 (2004). doi:10.1051/0004-6361:20035700
E. van der Swaluw, A. Achterberg, Y.A. Gallant, et al., Interaction of high-velocity pulsars with supernova remnant shells. Astron. Astrophys. **397**, 913–920 (2003). doi:10.1051/0004-6361:20021488
J.P.W. Verbiest, J.M. Weisberg, A.A. Chael, et al., On pulsar distance measurements and their uncertainties. Astrophys. J. **755**, 39 (2012). doi:10.1088/0004-637X/755/1/39
A.A. Vinogradov, I.Y. Vasko, A.V. Artemyev, et al., Kinetic models of magnetic flux ropes observed in the Earth magnetosphere. Phys. Plasmas **23**(7), 072901 (2016). doi:10.1063/1.4958319
A. Vladimirov, Modeling magnetic field amplification in nonlinear diffusive shock acceleration, PhD thesis, North Carolina State University, 2009
D. Volpi, L. Del Zanna, E. Amato, et al., Non-thermal emission from relativistic MHD simulations of pulsar wind nebulae: from synchrotron to inverse Compton. Astron. Astrophys. **485**, 337–349 (2008). doi:10.1051/0004-6361:200809424
E. Waxman, Cosmological gamma-ray bursts and the highest energy cosmic rays. Phys. Rev. Lett. **75**, 386–389 (1995). doi:10.1103/PhysRevLett.75.386
R. Weaver, R. McCray, J. Castor, et al., Interstellar bubbles. II—structure and evolution. Astrophys. J. **218**, 377–395 (1977). doi:10.1086/155692
K.W. Weiler, N. Panagia, VELA X and the evolution of plerions. Astron. Astrophys. **90**, 269–282 (1980)

D.G. Wentzel, Cosmic-ray propagation in the galaxy—collective effects. Annu. Rev. Astron. Astrophys. **12**, 71–96 (1974). doi:10.1146/annurev.aa.12.090174.000443

F.P. Wilkin, Exact analytic solutions for stellar wind bow shocks. Astrophys. J. **459**, 31 (1996). doi:10.1086/309939

D. Yoon, S. Heinz, Bow-shock pulsar-wind nebulae passing through density discontinuities. Mon. Not. R. Astron. Soc. **464**, 3297–3308 (2017). doi:10.1093/mnras/stw2590

V.N. Zirakashvili, F. Aharonian, Analytical solutions for energy spectra of electrons accelerated by non-relativistic shock-waves in shell type supernova remnants. Astron. Astrophys. **465**, 695–702 (2007). doi:10.1051/0004-6361:20066494

Space Sci Rev (2017) 207:291–317
DOI 10.1007/s11214-017-0376-2

Magnetoluminescence

R. Blandford[1] · **Y. Yuan**[2] · **M. Hoshino**[3] · **L. Sironi**[4]

Received: 22 April 2017 / Accepted: 5 May 2017 / Published online: 12 June 2017

Abstract Pulsar Wind Nebulae, Blazars, Gamma Ray Bursts and Magnetars all contain regions where the electromagnetic energy density greatly exceeds the plasma energy density. These sources exhibit dramatic flaring activity where the electromagnetic energy distributed over large volumes, appears to be converted efficiently into high energy particles and γ-rays. We call this general process magnetoluminescence. Global requirements on the underlying, extreme particle acceleration processes are described and the likely importance of relativistic beaming in enhancing the observed radiation from a flare is emphasized. Recent research on fluid descriptions of unstable electromagnetic configurations are summarized and progress on the associated kinetic simulations that are needed to account for the acceleration and radiation is discussed. Future observational, simulation and experimental opportunities are briefly summarized.

Keywords Plasmas · Acceleration of particles · Radiation mechanisms: non-thermal · Magnetic fields

✉ Y. Yuan
yajiey@astro.princeton.edu

R. Blandford
rdb3@stanford.edu

M. Hoshino
hoshino@eps.s.u-tokyo.ac.jp

L. Sironi
lsironi@astro.columbia.edu

1 KIPAC, Stanford University, Stanford, CA 94305, USA

2 Dept. Astrophysical Sciences, Princeton University, Princeton, NJ 08540, USA

3 University of Tokyo, 7-3-1 Hongo, Bunkyo, Tokyo 113-0033, Japan

4 Department of Astronomy, Columbia University, New York, NY, 10027, USA

1 Introduction and Context

The discovery of cosmic rays over a century ago (Hess 1912), of double radio sources and radio supernova remnants in the 1940s (Reber and Greenstein 1947), of X-ray Binaries (Giacconi et al. 1962), quasars (Schmidt 1963), pulsars (Hewish et al. 1968) and Gamma Ray Bursts (Klebesadel et al. 1973) in the 1960s, magnetars in the 1970s and Fast Radio Bursts (Lorimer et al. 2007) a decade ago have revealed the nonthermal universe. This is a very different world from the thermal universe studied by stellar astronomers. Roughly ten percent of the total gravitational energy that is released in the universe—principally by black holes and neutron stars—(and nearly one percent of the nuclear energy) is carried off or radiated by relativistic particles and these have to be accelerated. The only way to accelerate charges is through the application of electric field. However, it is hard to sustain large enough electrostatic field for this purpose and moving or rapidly changing magnetic field is generally invoked to generate the necessary potential differences.

While the prodigious power of high energy processes has long been appreciated (e.g., Burbidge 1956), it is their rapid variability that has drawn more attention in recent years. This takes us into the regime of *extreme astrophysics*. The associated particle acceleration has to be extremely potent and the steady, stochastic energy conversion of traditional Fermi acceleration is not an option. New mechanisms that can operate in a great variety of sites are needed. A common characteristic of many of these cosmic sources is the extraction of rotational and gravitational energy from a *prime mover* by large scale electromagnetic field. In the case of a pulsar or a black hole it is rotational energy. In the case of a magnetar it is ultimately gravitational energy. In the case of an accretion disk, it is both. We are not confident that this is always the case, but it is beginning to look that way.

Inevitably, this electromagnetic field is accompanied by plasma, at the very least sufficient to supply the space charge and current, and this plasma is ultimately expected to share the electromagnetic energy. Two quite different pathways for this energy conversion are available. The first is *dissipation*. Electromagnetic energy is converted ("prodigally") to particle energy at a rate $\mathbf{E} \cdot \mathbf{j}$ per unit volume, and entropy is created. The second is *plasma acceleration*. The electromagnetic force density $\rho\mathbf{E} + \mathbf{j} \times \mathbf{B}$ increases the bulk kinetic energy of all of the plasma ("industriously"). This may be very efficient but does not directly produce dramatic particle acceleration and rapid variability. This may happen later, for example when the bulk flow passes through a strong shock front. (Rapidly variation from the solar wind stimulated Gold (1955) to make the radical proposal that collisionless plasma could form essentially discontinuous shock fronts.) Of course, electromagnetic field must be present at the shock (in order to mediate the shock in the absence of collisions), but this is probably independent of the electromagnetic field associated with the prime mover.

Observations of rapidly variable sources, especially by γ-ray telescopes, strongly suggest the presence of regions of relatively high electromagnetic energy density which is efficiently dissipated and where electrons and positrons are efficiently accelerated up to their radiation reaction limits on a light crossing timescale. The available electromagnetic energy is seized, "plutocratically", by a small minority of particles and not, "democratically", by the bulk of the plasma. Additional features, such as bulk relativistic motion, may be present and needed to account for the observations. We call this general phenomenon *magnetoluminescence* (Blandford et al. 2014, 2015) by analogy with the flashes of light seen from collapsing bubbles created by ultrasound—sonoluminescence (Brenner et al. 2002). (This phenomenon can also be produced naturally by pistol shrimp (Lohse et al. 2001).) Similar mechanisms may have to be invoked to account for the acceleration of cosmic rays with PeV (10^{15} eV), EeV (10^{18} eV) and even ZeV (10^{21} eV) energy.

Our goal in this chapter is to complement some of the other chapters in this volume by providing a general description of extreme particle acceleration that brings out the common features present in the many sources where magnetoluminescence is seen. In the rest of the chapter, we first give a quick summary of the challenges brought up by various extreme particle accelerators in Sect. 2, then discuss some basic principles of particle acceleration in Sect. 3 and global considerations of electromagnetic dissipation in Sect. 4. We show relativistic MHD and kinetic modeling of the possible processes underlying magnetoluminescence in Sects. 5 and 6, respectively. We conclude with a discussion of future observational, simulation and experimental prospects in Sect. 7.

2 Extreme Particle Accelerators

2.1 Pulsars and Their Nebulae

A striking and relatively recent example of extreme particle acceleration is provided by the γ-ray flares in the Crab Nebula, best observed by Fermi (e.g., Buehler and Blandford 2014). The nebula itself, which has to be our best laboratory for high energy astrophysics, has been observed over the entire electromagnetic spectrum from $\lesssim 100$ MHz to $\gtrsim 1.5$ TeV (see, e.g., Hester 2008, for a review). It has been powered for the past 963 y. by the central pulsar with its ~ 30 Hz rotation frequency. The pulsar loses energy at a rate $\sim 5 \times 10^{31}$ W, roughly four times the current bolometric power ($\sim 1.3 \times 10^{31}$ W) of the nebula, which peaks around ~ 10 eV. The power assuredly takes the form of an electromagnetic Poynting flux leaving the surface of the neutron star, becoming a relativistic electromagnetic wind with properties that are best defined by the ratio of the electromagnetic energy flux to the particle energy flux (usually assumed to comprise pairs, though ion-dominated winds are possible), σ, the outflow Lorentz factor and a description of the variation of the outflow with latitude. The Crab Nebula demonstrates that electromagnetic conversion of prime-mover power into relativistic electrons and positrons can be very efficient. However, it has not told us where this conversion takes place and there are viable models that locate this conversion all the way from the pulsar light cylinder to the outer nebula (e.g., Kirk et al. 2009; Arons 2012).

The Crab pulsar wind must decelerate from the speed of light to the expansion speed of the supernova remnant $\sim 1500\ \mathrm{km\,s^{-1}}$. This is commonly supposed to begin at a strong, relativistic, collisionless shock front located where the outflow momentum flux matches the ambient nebula pressure at a radius $\sim 3 \times 10^{15}$ m ~ 0.1 pc (Rees and Gunn 1974; Kennel and Coroniti 1984). If, as expected, the strength of the wind decreases significantly with latitude, the shock surface will be quite oblate (e.g., Komissarov and Lyubarsky 2003). This is compatible with the observation of an equatorial torus at X-ray energies (Weisskopf et al. 2000).

However, there is no observation that suggests the presence of a shock outside the equatorial zone. The "inner knot", a compact feature located $0.65''$ southeast of the Crab pulsar in infrared and optical images (Hester et al. 1995; Rudy et al. 2015), has been modeled as a point on the oblique part of the termination shock in the polar region (Komissarov and Lyubarsky 2004; Komissarov and Lyutikov 2011), but there is some evidence that it may be associated with the low latitude section of the pulsar wind where the effective magnetization becomes low because of the alternating magnetic flux in a "striped" wind (Lyutikov et al. 2016a; Yuan and Blandford 2015). At higher latitudes, it is consistent with $\sigma \gg 1$ at this radius (as we must assume for the mechanisms we are reviewing here) so that the shock is weak—there may be several of them—and most of the deceleration happens gradually. Given this model, the magnetic field strength is ~ 100 nT here, decreasing to ~ 30 nT in the body of the nebula.

The observed γ-ray flares (e.g., Buehler and Blandford 2014) happen with a cadence ~ 1 y and have only been seen at energies ~ 300 MeV and there are no associated pulsar timing glitches. Variation, on timescales as short as a few h has been reported. The peak isotropic luminosity is roughly 10^{29} W and the energy radiated is $\sim 10^{34}$ J. The flares and secular observations (Wilson-Hodge et al. 2011) demonstrate that the energy conversion is intermittent not steady and that the mechanism can be locally cataclysmic.

2.2 Blazars

Most galaxies, at least as massive as our Galaxy, possess nuclei surrounding a massive ($\sim 10^6 - 10^{10}$ $M_\odot$) black hole which become *active* either when they are supplied with gas that is efficiently accreted or the holes are spun up and the rotational energy is tapped (e.g., Meier 2012). A common expression of this activity is the formation of a pair of antiparallel *relativistic jets*, with Lorentz factors $\Gamma \sim 10$, presumably directed along the hole's rotation axis, at least initially. The presence of these jets was originally inferred from the observations of large double radio sources straddling the optical image of the host galaxy, which they supply with energy, momentum and mass. However, jets are now observed throughout the electromagnetic spectrum, most notably at γ-ray energies (see, e.g., Madejski and Sikora 2016, for a review). As the outflows are relativistic close enough to the hole, jet emission is strongly beamed and a small minority of sources are observed to be unusually bright as they are directed towards us. These are called *blazars* and they can dominate a flux-limited sample. Observations made using Very Long Baseline Interferometry (VLBI) at cm and mm wavelengths have confirmed that these jets are already collimated on scales $\lesssim 100$–1000 black hole gravitational radii (Marscher et al. 2008) and we are now on the threshold of resolving the curved spacetime around the black hole (Broderick et al. 2015; Fish et al. 2016). Relativistic jets are now acknowledged as being powered by the inner accretion disk and the black hole spin energy but the relative importance of these two sources is debated.

Blazars are frequently divided into two types, *BL Lac objects*, or BLL, and *Flat Spectrum Radio Quasars* or FSRQ. BLL are low total power objects and mostly local; FSRQ are high power quasars which can outshine the host galaxy and are mostly at cosmological redshifts. Both classes exhibit remarkable GeV/TeV flares with source variation times as short as ~ 2–3 minutes (e.g., Albert et al. 2007; Aharonian et al. 2007; Aleksić et al. 2011; Ackermann et al. 2016). This suggests that the emission originates very close to the jet source. However, there is a lower bound on the radius where the particle acceleration and emission can occur due to absorption involving pair production on soft photons. This radius can be calculated assuming QED (in which we should have absolute confidence!) and this defines the *gammasphere* (Blandford and Levinson 1995). As with the Crab Nebula, there are viable models with emission radii ranging from the gammasphere to the sites of recollimation located $\gtrsim 100$ pc from the hole. There is now good evidence that the γ-rays are produced before and within the radio emission (Max-Moerbeck et al. 2014).

2.3 Gamma Ray Bursts

Most *Gamma Ray Bursts*, GRB, are cosmologically distant and release the energy equivalent of a significant fraction of a stellar mass (see, e.g., Kouveliotou et al. 2012, for a review). There are also two types: the *short bursts* lasting $\lesssim 2$ s are conjectured to be neutron star binaries, merging under gravitational radiation reaction and potentially observable by LIGO/VIRGO; the *long bursts* are convincingly associated with a subset of core-collapse supernovae tentatively identified with progenitors that have lost their helium and hydrogen envelopes although other factors could easily be relevant. GRBs are also relativistic jets (with $\Gamma \sim 100$–1000). They are collimated by the collapsing star in the long bursts; the jets from short bursts may be collimated by a very dense accretion disk.

The prompt γ-ray emission from GRBs is often found to be variable on timescales as short as ~ 10 ms. The *afterglow*, observed throughout the electromagnetic spectrum, is formed later when the jets are decelerated by interaction with the circumstellar medium.

One common feature of models of long and short bursts and AGN (and, also, Galactic superluminal sources, Mirabel and Rodríguez 1999) is the realization that the accreting gas near the hole cannot cool requiring the formation of a thick orbiting torus at small radius which defines two funnels that may define the initial jet shape (Yuan and Narayan 2014). (Jets are also formed by pulsars like the Crab pulsar, and here the hoop stress of the toroidal magnetic field downstream of the termination shock may be responsible for the confinement see, e.g., Komissarov and Lyubarsky 2003.) There are actually three accretion regimes. At low mass supply rates the electrons may be much cooler than the ions and unable to cool the gas on an inflow timescale. Conversely, at a high mass supply rate, the gas can radiate but the Thomson opacity is so high that the photons are convected inward faster than they can escape outward. It is only at an intermediate mass supply rate that a thin accretion disk will form. On quite general grounds, the low and high supply regimes require powerful winds to carry off the energy that is released during accretion (Blandford and Begelman 1999).

2.4 Magnetars

A minority of neutron stars, called *magnetars* have surface field strengths in the 10–100 GT range over two orders of magnitude more than typical radio or X-ray pulsars (see Kaspi and Beloborodov 2017, for a review). They are found to be rotating with periods $\sim$ 1–10 s range but could have been borne with millisecond periods and, if so, they would be strong candidates for the prime-movers of GRBs. Magnetars show dramatic flares (Evans et al. 1980; Hurley et al. 1999, 2005; Mereghetti et al. 2005; Boggs et al. 2007) which are likely to be caused by magnetic instability associated with either the crust or the magnetosphere (Thompson and Duncan 1995, 1996). (The energy release is far too much to be derived from rotation.) The emission sites are likely to be electromagnetically-dominated. The underlying electromagnetic instability may involve rapid, relativistic reconnection of flux tubes, or may be simple untangling with small changes of magnetic helicity. Shocks and non-relativistic reconnection may account for the acceleration needed for more slowly varying emission from these sources.

3 General Principles

3.1 Steady vs. Impulsive Acceleration

We have emphasized the rapid variability in these sources because it provides an important clue as to the underlying physical conditions (Begelman et al. 2008). However, much of the power is associated with slowly varying source components and extreme particle acceleration is not required. Nonetheless, it is still the case that the radiative efficiency must be high which implies that the associated cooling timescale should be short compared with the expansion timescale in an outflow and the timescales for other non-radiative losses.

The most commonly invoked means for converting large scale electromagnetic energy into high energy particles and radiation is through magnetic reconnection (Biskamp 2000; Priest and Forbes 2000). There has been much effort devoted to understanding nonrelativistic reconnection especially in the context of space and solar physics and also for Tokamaks. The basic idea is that magnetic field is convected by the flow, more or less incompressibly with little dissipation almost everywhere. However, small regions are created where the field reverses abruptly and the current density increases until sufficient dissipation is produced to allow the magnetic field lines to "exchange partners", or "reconnect", with a loss of magnetic energy. It is generally the case that surprisingly energetic particles are produced but with low efficiency. Reconnection need not be steady; various instabilities, like the *tearing mode*, can occur (e.g., Furth et al. 1963). Also the microscopic description requires the tensor character of the electrical conductivity to be included (e.g., Wang et al. 2000). In Sect. 3.2, we will discuss some basic particle acceleration mechanisms in reconnection.

Relativistic reconnection is now being simulated and is clearly much more promising as an agency of extreme particle acceleration (e.g. Zenitani and Hoshino 2001; Guo et al. 2014; Sironi and Spitkovsky 2014; Werner et al. 2016). However, non-relativistic reconnection may be a slow process, and as such would be better suited to steady acceleration to modest energy, rather than the extreme, "impulsive" acceleration needed for the most dramatic flares.

3.2 Basic Particle Acceleration in Reconnection

There has been much simulation study of the particle acceleration in magnetic reconnection, and several different acceleration processes operating in various regions in the multi-scale reconnection system have been proposed. Since the dominant electric field responsible for particle acceleration is the inductive electric field during time evolution of reconnection, the magnetic diffusion region, which coincides with the so-called X-type region, would be the principal particle acceleration site. As shown in Fig. 1(A), the charged particles, whose motions are described by *Speiser* (meandering) motion (Speiser 1965), can almost directly resonate with the inductive electric field and can be quickly accelerated in the diffusion region where the magnetic field is weak. This acceleration may be almost free from synchrotron radiation loss. However, the volume size of the X-type region is small for the standard *Petschek* reconnection model associated with a pair of slow mode shocks (Petschek 1964), and the total energy release rate would not be necessarily large. For the case of *Sweet-Parker* reconnection with an elongated diffusion region in plasma sheet (Parker 1963), there may be enough volume for direct acceleration in the meandering motion by the electric field, but the magnitude of the electric field is weak compared to Petschek reconnection, because the dimensionless reconnection rate in the Sweet-Parker model is estimated as $1/\sqrt{R_m}$, which should be contrasted with the reconnection rate in Petschek model $\sim 1/\ln(R_m)$, where $R_m = LV_A/\eta$ is the magnetic Reynolds number based on the inflow Alfvén velocity V_A, which is also frequently referred to as Lundquist number, L is the length of the reconnection region and η is the magnetic diffusivity.

In addition to the *Speiser* motion in the diffusion region, particles can be accelerated in the pileup magnetic field region produced by compression of the reconnection outflow. The collision of the reconnection outflow emanating from the diffusion region with the pre-existing plasma at rest will produce the pileup region associated with a large jump of magnetic field. Figure 1(B) shows that the ∇B and/or curvature B drift motion for ions/electrons,

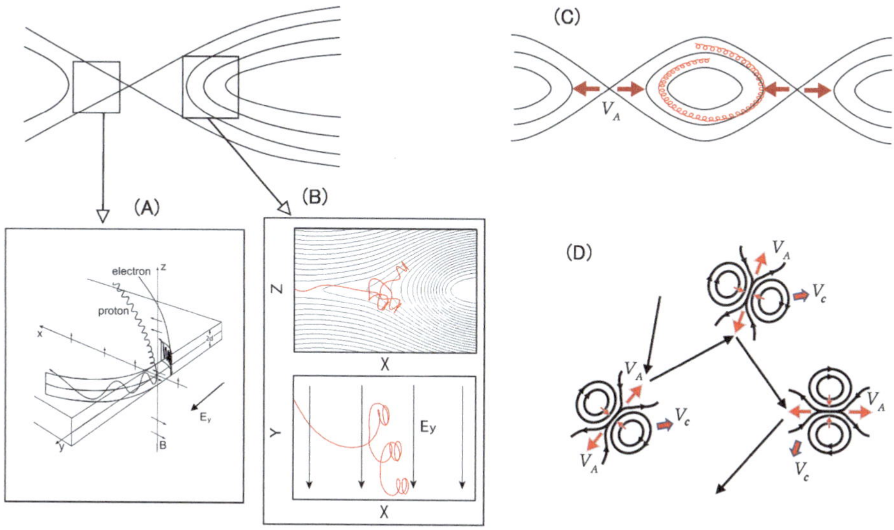

Fig. 1 Typical particle orbits accelerated during magnetic reconnection. (**A**) *Speiser* motion/meandering motion, adapted from (Speiser 1965). Particles can get accelerated during the meandering/bouncing motion along the electric field, under the anti-parallel magnetic field B_x, (**B**) the gradient/curvature B drift motion in the magnetic field pileup region, (**C**) the trapped particles in the shrinking magnetic islands, (**D**) the stochastic Fermi acceleration in multiple magnetic islands

which is parallel/anti-parallel to the inductive electric field, contributes to the energization (Hoshino et al. 2001). If the particles are magnetized and their gyro-radii are small compared with the scale height of the pileup magnetic field, the first adiabatic invariant $p_\perp^2/B$ is conserved, and a large energy gain can be expected due to the large jump of the magnetic field intensity.

Not only the X-type and the B pileup regions in the single reconnection site, but magnetic islands surrounded by the loop-like magnetic field lines, which situation can happen for multiple reconnection/tearing mode instability in a long current sheet, can invoke particle energization, shown in Fig. 1(C). Because of the shrinking magnetic field lines bounded by two X-type regions at each end, the particles trapped by the magnetic island can gain energy by reflection from both ends of the contracting magnetic field line (Drake et al. 2006). The energy gain in this Fermi acceleration process can be estimated from the conservation of the second adiabatic invariant $p_\parallel L$, where L is the loop length of the magnetic island. However, as being energized by the contracting magnetic field line, the particles will escape from the island if their gyro radii become comparable with the size of the island and the particle acceleration will cease.

However, it is possible that high energy particles can be further accelerated in a much larger plasma environment with many magnetic islands. By analogy to the original Fermi acceleration model, where particles gain energy stochastically during head-on and head-tail collisions of particles with many magnetic clouds, let us replace the magnetic clouds with magnetic islands, as shown in Fig. 1(D). In the original Fermi acceleration, the increase in particle energy is known to be of second order in V_c/c, where V_c is the speed of random motion of magnetic clouds. However, if there are many magnetic islands with the Alfvénic plasma outflow, it is suggested that the particles can preferentially interact with the reconnection outflow region, because the energetic particles have a tendency to be situated outside the magnetic island during acceleration. As a result, the increase in energy becomes first order of V_A/c, where V_A is the Alfvén speed (Hoshino 2012).

Regardless of the acceleration mechanisms in details, another key issue for a rapid magnetic energy dissipation and particle acceleration is how the large electric field is generated during reconnection. The average electric field in the reconnection system is basically determined by $V \times B/c$ induced by the inflow and outflow motions, and roughly speaking the outflow speed can reach the Alfvén speed for both non-relativistic and relativistic reconnections. Since the Alfvén speed is given by $V_A = c\sqrt{\sigma/(\sigma+1)}$, the fast reconnection flow and the large electric field can be realized in a strong magnetized plasma with a large σ value. The spontaneous vs. driven reconnection is another key issue to control the reconnection rate. It is known that small but finite Poynting flux injection from the inflow boundary into the system can drive a fast reconnection for both linear and nonlinear stages. While the linear growth rate for the spontaneous tearing mode is proportional to $R_m^{-3/5}$ (Furth et al. 1963), the growth rate for the driven case is $R_m^{-1/3}$ (Horton and Tajima 1988). The enhancement of the reconnection rate can be also found in the double tearing mode, which is characterized by the presence of two (or more) tearing current sheets whose surfaces are attached to each other. The reason of enhancement of reconnection rate is simply because the surface can share the inflow and outflow boundary condition for two tearing islands, which is the same situation as arises in driven reconnection.

3.3 Pairs vs. Hadrons

There has been much discussion of the nature of the positive charge carriers especially in jets. Are they positrons or protons? Given the large EMFs generated, the former are readily created through γ–γ or even γ–B processes although this may be balanced by pair annihilation at the base of the jet. However hadrons are likely to be incorporated into the flow through entrainment from the surrounding medium, which might comprise gas accreting at high latitude, gas clouds in the interstellar medium, a thick, confining torus or an outflowing wind. Probably all four possibilities are relevant. It may be that this is the explanation for the famous Fanaroff and Riley (1974) dichotomy of double radio sources, with the low power, Type 1 sources being identified with jets that are efficiently decelerated by their surrounding and Type 2 sources containing jets that are not slowed down. If the proton density is high, shocks are likely to be very important particle accelerators but conditions will then not be propitious for extreme acceleration/magnetoluminescence.

3.4 Emission Mechanisms

Although we surely trust the physics of the underlying radiative processes, we are not completely confident that we understand which of them operates and where. The "Bactrian" (two-humped) spectrum characteristic of blazars has been commonly interpreted as a two process—synchrotron and Compton—emission from a single, homogeneous source. However, jets are observed to radiate all along their lengths and it seems more likely that both humps comprise emission from different radii. Interestingly, there is a maximum synchrotron photon energy that can be radiated in a simple electromagnetic source. If we balance the rate of energy gain $\sim eEc$ by the radiative loss and impose the condition $E \lesssim B$, then the maximum photon energy is found to be $\sim m_e c^2/\alpha$, where α is the fine structure constant. This is ~ 70 MeV.

It is common to distinguish two types of Compton radiation with externally produced photons dominating in the FSRQs and internal synchrotron photons dominating in the BLLs. Again, it seems more reasonable to suppose that both options are exercised along the jet.

The self-absorbed synchrotron radio emission seen from jets, known as the *core*, sometimes seems to vary so rapidly that the emission mechanism may have to be coherent, perhaps a cyclotron maser (Begelman et al. 2005). An important constraint is that the radio waves have to avoid absorption due to the induced Compton effect or stimulated Raman scattering which sets an upper limit on the electron column density along the line of sight (Levinson and Blandford 1995).

There are more options if hadrons are present. There have been several suggestions that the γ-rays are produced as a consequence of pion production following proton-proton collisions. The neutral pions decay efficiently into gamma rays. However, more of the energy goes into charged pions where the decays involve muon and electron neutrinos and may produce particles that are insufficiently energetic to radiate efficiently. A variation which deserves more attention is for a proton be accelerated to very high energy so that it can photo-produce electron-positron pairs which quickly radiate by the synchrotron process into the "Compton" hump at energies above the conventional synchrotron limit. This might be important in a high radiation density environment in a GRB or a quasar.

3.5 Relativistic Beaming

Doppler beaming is an essential feature of electromagnetic flows as the characteristic signal speed is necessarily relativistic. The easiest way to handle this is to introduce a *Doppler factor* $\mathcal{D}=[\Gamma(1-\mathbf{V}\cdot\mathbf{n}/c)]^{-1}$, where $\Gamma=(1-V^2/c^2)^{-1/2}$, for Lorentz transformation from a (primed) frame in which we can evaluate the emissivity $j'_{\nu'\Omega'}(\nu',\mathbf{n}')$ and the absorption coefficient $\kappa'(\nu',\mathbf{n}')$. $\mathbf{V}$ is the velocity of the primed frame in the frame of a distant, though not cosmologically distant, observer along a direction $\mathbf{n}$. Primed frame quantities are transformed into observer frame quantities according to $dt=\mathcal{D}^{-1}dt'$, $\nu=\mathcal{D}\nu'$, $\mathbf{n}'=\mathcal{D}\mathbf{n}-\Gamma(\mathcal{D}+1)\mathbf{n}/(\Gamma+1)$, $j_{\nu\Omega}(\nu,\mathbf{n})=\mathcal{D}^2 j_{\nu'\Omega'}(\nu',\mathbf{n}')$, $\kappa_{\nu\Omega}(\nu,\mathbf{n})=\mathcal{D}^{-1}\kappa_{\nu'\Omega'}(\nu',\mathbf{n}')$ (e.g., Blandford and Königl 1979). The intensity, which transforms according to $I_{\nu\Omega}(\nu,\mathbf{n})=\mathcal{D}^3 I_{\nu'\Omega'}(\nu',\mathbf{n}')$, is best considered in the observer frame and can be corrected for cosmological expansion if appropriate. For a source that is beamed towards us, it is a good start to approximate these formulae by setting $\mathcal{D}\sim\Gamma$.

With the increasing sophistication of simulations of electromagnetic sources, it is worth carrying out more careful radiative transfer calculations. These are best prosecuted in the unprimed, observer frame, transforming the emissivity and absorption coefficient into this frame.

4 Global Issues

4.1 Voltages and Currents

Before discussing details, it is helpful to consider the global energy balance in a relativistic, electromagnetic source. In the most simple-minded model of a spinning conductor with angular frequency Ω and threaded by magnetic flux Φ, the potential difference generated is $V_{\rm eff}\sim\Omega\Phi/2\pi$. There will be a current flow $I_{\rm eff}$ outside the body and the effective impedance will be roughly that of free space $Z_{\rm eff}\sim 100\ \Omega$ so that $I_{\rm eff}\sim V_{\rm eff}/Z_{\rm eff}$. The power that can be effectively dissipated as particle acceleration is then $L_{\rm eff}\sim V_{\rm eff}I_{\rm eff}\sim\Omega^2\Phi^2/4\pi^2 Z_{\rm eff}$. In the case of a magnetar where the energy release is gravitational, Ω should be replaced by $\Omega_{\rm eff}$ the reciprocal of the characteristic timescale associated with the release of seismic energy.

Let us give some examples. For the Crab pulsar, $V_{\rm eff} \sim 50$ PV $\sim 5 \times 10^{16}$ V and $L_{\rm eff} \sim 3 \times 10^{31}$ W, comparable with the bolometric luminosity of the Crab Nebula. For a quasar like 3C279, $V_{\rm eff} \sim 300$ EV $\sim 3 \times 10^{20}$ V and $L_{\rm eff} \sim 10^{39}$ W and for a powerful GRB, $V_{\rm eff} \sim 100$ ZV $\sim 10^{23}$ V and $L_{\rm eff} \sim 10^{44}$ W. Note that there is sufficient potential difference to accelerate Ultra High Energy Cosmic Rays (UHECR) in powerful AGN though whether or not these particles can escape without suffering catastrophic photopion production loss is debatable. Newly born magnetars, with 10^{11} T surface magnetic field and an initial spin rate close to the centrifugal breakup limit, could generate an EMF $V_{\rm eff} \sim 10$ ZV $\sim 10^{22}$ V; they, too, have been considered as viable accelerators for UHECR (Arons 2003).

A very simple effective circuit description raises the question of where do these currents close. If this is closeby, the flow becomes gas dynamical and remote acceleration is likely to be associated with shocks and shear flows. Alternatively, the current may persist to the observed extremity of the source, for example the hot spots associated with a powerful double radio source. In this case, the processes discussed here are likely to be relevant to most of the emission.

4.2 Flow of Flux

The voltage $V_{\rm eff}$, measured in V is also the flow of magnetic flux measured in Wb $\rm s^{-1}$. If as is the case with a pulsar wind nebula, the flux is confined within a volume that is expanding much slower than the speed of light then the flux must be dissipated at almost the rate at which is supplied. Particle acceleration in one form or another is inevitable. (If this did not happen and the walls were perfect conductors, then the electromagnetic field would react back on the source.) The only questions are where does it occur and what are the resulting particle energy distributions. For an axisymmetric rotating source, the two most promising possibilities are the symmetry axis where the polar current will be concentrated and along the path of the return current, specifically along the equatorial plane in a pulsar wind nebula and along the jet walls in the case of AGN jets. There is observational evidence for both of these (Hester et al. 1995; Weisskopf et al. 2000; Reid et al. 1989).

5 Relativistic MHD Description

5.1 Knots and Tangles

Although there may be some validity to describing the electromagnetic field under the force-free approximation close to the source, the inertia of the plasma is likely to become significant quite quickly and relativistic MHD will be the preferred fluid description although we have little understanding of how the particle density builds up and is regulated. Nonetheless it is reasonable to expect that MHD instabilities will break any axisymmetry and lead to the formation of current sheets. As is observed in the solar corona, "hairy" magnetic ropes are likely to form, twisted by field-parallel (zero-stress) current. The extent to which this is a fair description of these magnetically dominated sources is unknown but if we adopt it, it is easy to believe that the writhing *magnetic ropes* can become both tangled and knotted.

The distinction is important. A knot, which contributes to the magnetic helicity $H = \int dV \mathbf{A} \cdot \mathbf{B}$ (Moffatt 1978; Bellan 2000), is topologically distinct from a tangle or "unknot" (Fig. 2). If the magnetic field that is created by the prime mover is required to evolve according to the precepts of perfect MHD, it will remain unknotted. Even if H were non-zero initially, it would decrease in inverse proportion to the scale size of the flow. However, if

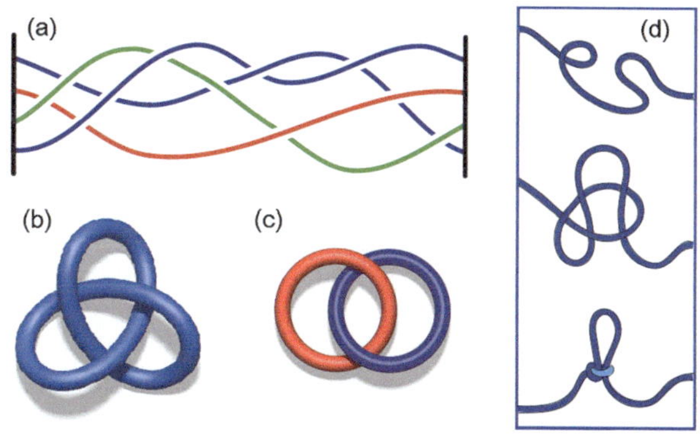

Fig. 2 An illustration of topologies the magnetic "ropes" may take. (**a**) Braid. (**b**) Knot. (**c**) Link. (**d**) Hitch, which is topologically equivalent to a straight line

magnetic ropes form, it is an inevitable consequence of instability in these outflows that there will be places where they will be pulled or pressed against one another. Reconnection of pairs of ropes at unrelated locations seems to be inevitable with an overall release of energy through relaxation of magnetic tension and transverse expansion. The knots that are then created topologically will be initially quite loose but are likely to tighten under the action of magnetic tension with the passage of time. Additional instances of forced reconnection will resolve these knots, leading to further energy release and particle acceleration. (Knots are very important in the history of physics as their *aetherial* expression was once the most popular model of atoms (Tait 1907). They have also played a prominent role in modern field theory.)

However, it also seems to be quite likely that tangles of individual ropes will develop and tighten into *slip knots* or *hitches* that may be resolved without large-scale reconnection simply through an increase in magnetic tension. As with conventional ropes the ease with which this can happen depends upon the surface friction. In the case of magnetic ropes this will depend upon small-scale electromagnetic stresses associated with the "hair"—wandering field lines near the boundary of the flux ropes. If the friction is large, then large tangles with large latent energy can be created.

From a particle acceleration standpoint, tangles can untangle at effectively the speed of light in a relativistic source and can be responsible for the flares; knots require reconnection, which may be slower to resolve, and may be responsible for the higher power, steady emission.

5.2 Sheets, Shocks and Caustics

Many of the relativistic sources that we have been discussing, such as blazar jets, are observed to be expanding relativistically. Others, such as pulsar wind nebulae, have relativistic sound speeds and can therefore sustain internal relativistic motions. Doppler boosting with Lorentz factor Γ, as summarized in Sect. 3.4, is therefore likely to be present and to contribute to the observed flux. If we idealize the source as a sphere moving with a single velocity, then we may see variation as the source expands in its own reference frame. Alternatively, the velocity may turn through an angle $\gtrsim \Gamma^{-1}$ and this will produce a roughly symmetric pulse. Parallel acceleration should produce an asymmetric pulse. Changes in the optical depth at either radio or γ-ray energy can also produce rapid variation.

However, there are additional possibilities if the emission comes from a surface, such as a current sheet or a shock front that is moving ultrarelativistically. An element of the surface may beam the synchrotron or Compton radiation in a narrow cone about its direction of motion. (Among several factors which could complicate a detailed model, the speed of the emitting surface might differ from the speed of the emitting plasma, as happens in a shock front. In addition the electrons and positrons might be carrying a current and moving with different speeds.) Now, for a generic, corrugated surface, the mapping from the moving surface onto a distant sphere might be many to one which can lead to the formation of caustics (cf., Lyutikov et al. 2012). The simplest generic caustic is a *fold* which produces an asymmetric flux variation $\propto |t - t_0|^{-1/2}$ before or after the time t_0 when a pair of images either disappears or appears. Effects such as these could enhance the observed variability from sources where extreme particle acceleration might be taking place.

5.3 MHD Simulations

In recent years much progress has been made in relativistic MHD simulations to address the plasma dynamics in pulsar wind nebulae, the formation, propagation and acceleration of jets from black holes/neutron stars, the propagation of shocks/blast waves in GRBs, and so on. We are gradually gaining a reasonable picture of how the magnetized outflow from the engine shapes its environment, as described in many other chapters of this book. Meanwhile, a lot more work is needed to understand the actual dissipation process. Here, as a good starting point, we can focus on a local region in the plasma outflow that has a tangled magnetic field topology, to see whether a rapid conversion of electromagnetic energy to particle kinetic energy can occur.

As one example, East et al. (2015) considered a family of force-free equilibria in a 3D periodic Cartesian box. These configurations have the topology of multiple flux ropes packed in a complex manner. They are solutions to the force-free equation $\nabla \times \boldsymbol{B} = \lambda \boldsymbol{B}$, where λ is a constant throughout the space, so each of them has a single characteristic wavelength (Chandrasekhar and Kendall 1957; Moffatt 1986; Rosenbluth and Bussac 1979; Bellan 2000):

$$\boldsymbol{B} = \sum_{|\boldsymbol{k}|=|\lambda|} \hat{n}_{\boldsymbol{k}} \times \nabla \chi_{\boldsymbol{k}} + \frac{1}{\lambda} \nabla \times (\hat{n}_{\boldsymbol{k}} \times \nabla \chi_{\boldsymbol{k}}) \tag{1}$$

Here $\chi_{\boldsymbol{k}}$ is the solution to the scalar Helmholtz equation $\nabla^2 \chi + \lambda^2 \chi = 0$, $|\boldsymbol{k}| = \lambda$, and $\hat{n}_{\boldsymbol{k}}$ is a constant unit vector. At a fixed total helicity H, if a configuration has a shorter wavelength, namely, it contains a larger number of smaller flux ropes, then its total magnetic energy is higher.

Constant-λ force-free configurations have historical importance, because Taylor's conjecture suggests that a closed, magnetically dominated plasma tends to relax to lower energies while conserving the total helicity, and the relaxed state is just one such constant-λ configuration (Taylor 1974, 1986).

MHD/force-free simulations by East et al. (2015) found that, typically, the shorter wavelength equilibria are unstable to ideal MHD modes; the instability grows on Alfvén crossing time scales, and in the nonlinear regime the system goes through a turbulent relaxation process, dissipating a finite amount of energy within just one dynamic time scale. Eventually the plasma settles into the longest possible wavelength.

Figure 3 shows a few examples of the force-free equilibria and the corresponding maximally growing unstable mode. It suggests that the instability can be viewed as shearing and merging of the flux ropes. The velocity field appears to have non-smooth features, reminiscent of spontaneous current sheets that occur at the flux tube boundaries. In Fig. 4, it is shown that different volume averaged magnetization values $\sigma \equiv \langle B^2/(4\pi w)\rangle$ (where w is the plasma enthalpy) give the same intermediate and final energy levels, consistent with conservation of magnetic helicity. The amount of magnetic energy dissipated during the

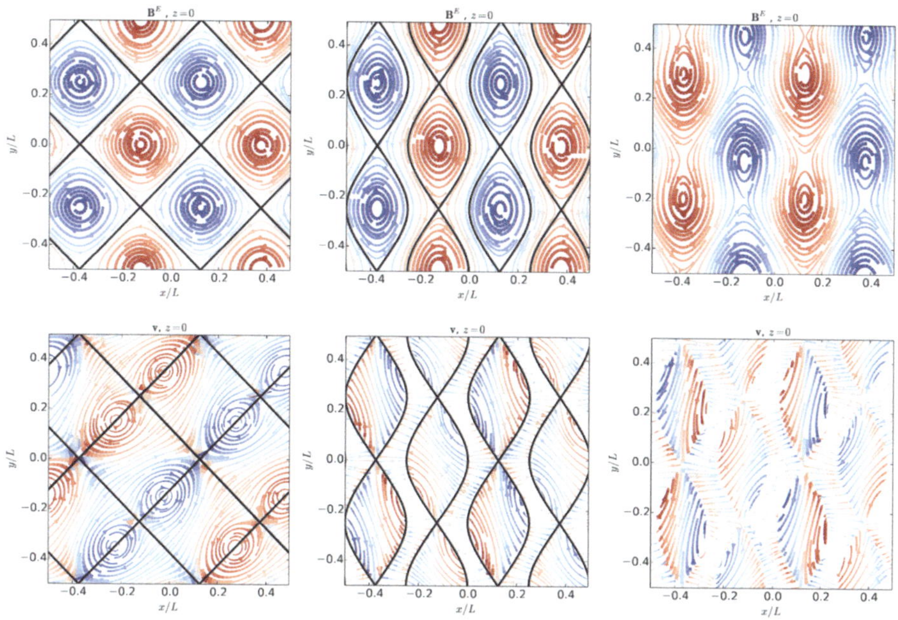

Fig. 3 Streamlines of the magnetic field in the initial force-free equilibria for three different examples (*top row*), and the corresponding velocity field $v = E \times B/B^2$ of the maximally growing unstable mode found from the force-free simulations (*bottom row*) (from East et al. 2015). The first two equilibria are 2D while the third one is 3D. All plots show streamlines on the $z = 0$ plane. The color indicates the perpendicular vector component with *red* and *blue* representing, respectively, out of the page and into the page. The thickness of the streamline is proportional to the vector magnitude. The *black lines* indicate the location of the separatrices in the equilibrium solutions

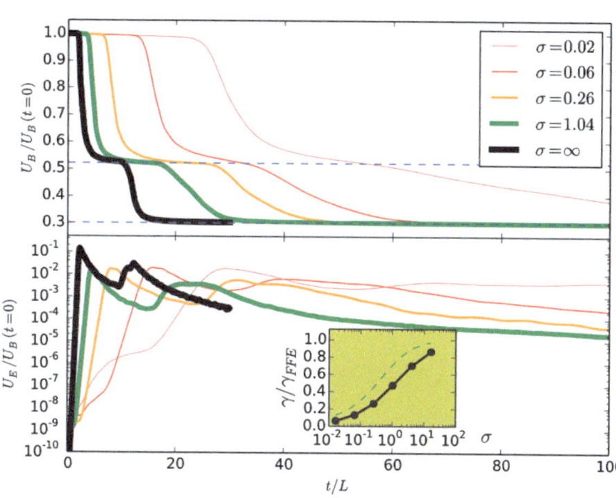

Fig. 4 A comparison of the decay of an $\lambda^2 = 11$ equilibrium in simulations with different values of volume averaged magnetization $\sigma \equiv \langle B^2/(4\pi w)\rangle$, where w is the plasma enthalpy (from East et al. 2015). Shown is the magnetic energy (*top*) and kinetic/electric field energy (*bottom*). The *horizontal dashed lines* in the *top panel* indicate the magnetic energy of $\lambda^2 = 3$ and $\lambda^2 = 1$ states with the same helicity. The *bottom inset* shows the linear growth rate γ measured for runs having different magnetization parameters, along with the Alfvén speed (*dashed line*) for comparison

evolution thus corresponds to the *free* magnetic energy, defined as the energy difference between the initial configuration and the ground state (longest wavelength state), compared at the same helicity.

These systematic numerical investigations suggest that, in general, the generic short wavelength, constant-λ force-free states in 3D periodic boxes are unstable; the instability is characterized by an exponentially growing electric field in the linear phase, and eventually leads to current sheet formation where significant amount of the free magnetic energy is dissipated within just a single light crossing time.

Lyutikov et al. (2016b) also studied independently a subset of the force-free equilibria—2D "ABC" field (after Arnol'd (1965), Beltrami (1902), Childress (1970), e.g. Moffatt (1986))—to explore the viability of impulsive magnetic energy release. They reach the same conclusion that these configurations are unstable to ideal modes and collapse over dynamic time scales. They show that when the structures are initially compressed or sheared (driven system), the development of the instability can be accelerated.

Although the above plasma configurations are quite artificial, they teach us a lot about the generic behavior of highly magnetized, relativistic plasmas. Similarly complex, current-carrying plasma states with plenty of free energy may form in the pulsar wind (e.g., Zrake 2016), or downstream of the oblique termination shock in PWN where the magnetization is still high (Lyutikov et al. 2016b), and also in jets due to kink instability. The insights gained here are instructive to envisage the possibility of catastrophic conversion of large scale electromagnetic energy into particle kinetic energy.

6 Kinetic Description

6.1 Distribution Function

In order to understand what are the underlying dissipation mechanisms, how the released electromagnetic energy is partitioned among the particles, and what emission the accelerated particles produce, we need to follow the evolution of particle distribution in phase space. For relativistic plasma, we use the following covariant definition of distribution function over the 6-dimensional phase space: $F(\boldsymbol{x}, \boldsymbol{u}, t) = dN/(d^3x d^3u)$, where $\boldsymbol{x}$ is the position and $\boldsymbol{u}$ is the spatial part of the 4-velocity of the particles. In many of the aforementioned astrophysical environments, the plasma is collisionless, meaning that the Coulomb collisional time scale is much longer than the relevant dynamic time scales. The evolution of the distribution function for each species F_s then follows the collisionless Boltzmann equation, or Vlasov equation:

$$\frac{\partial F_s}{\partial t} + \nabla_{\boldsymbol{x}} \cdot (\boldsymbol{v} F_s) + \nabla_{\boldsymbol{u}} \cdot \left(\frac{d\boldsymbol{u}}{dt} F_s \right) = 0, \tag{2}$$

where $\boldsymbol{v} = \boldsymbol{u}/\gamma$ is the 3-velocity, $\gamma = \sqrt{1 + u^2/c^2}$ is particle Lorentz factor, and the acceleration of individual particles is determined by

$$m_s \frac{d\boldsymbol{u}}{dt} = q_s \left(\boldsymbol{E} + \frac{\boldsymbol{v}}{c} \times \boldsymbol{B} \right). \tag{3}$$

Other forces, like radiation reaction force, can be included as well. The electromagnetic field determines the motion of particles, at the same time its evolution is determined by the charge and current density provided by the particles

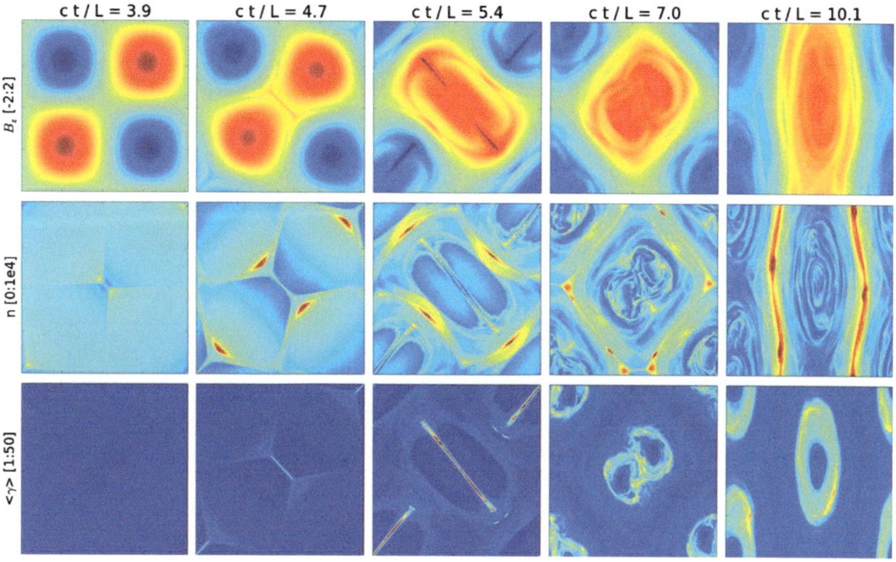

Fig. 5 Snapshots from a 2D PIC simulation of one of the lowest order unstable force-free equilibria (from Nalewajko et al. 2016). *Top panels*: out of plane component of the magnetic field (there is also in-plane magnetic field); *middle panels*: number density n of electrons and positrons; *bottom panels*: average Lorentz factor $\langle\gamma\rangle$ of electrons and positrons

$$\rho = \sum_s q_s \int F_s d^3u, \tag{4}$$

$$\mathbf{J} = \sum_s q_s \int F_s \frac{\mathbf{u}}{\gamma} d^3u, \tag{5}$$

This is a complex system to solve. The current state-of-the-art technique is Particle-In-Cell (PIC) simulations, which exploit a fixed spatial grid to evolve the electromagnetic field, while the particle distribution function is sampled by a large number of particles. PIC method has been successfully applied to study collisionless shocks (e.g., Spitkovsky 2008; Sironi and Spitkovsky 2009, 2011a,b), plane current sheet reconnection (e.g. Kagan et al. 2015), magneto-rotational instability (MRI, e.g. Hoshino 2015; Riquelme et al. 2012), global pulsar magnetospheres (e.g., Chen and Beloborodov 2014; Belyaev 2015; Cerutti et al. 2015; Philippov et al. 2015), and so on, with very fruitful outcomes.

In the following, we summarize some key results from PIC simulations of the rapid energy release processes mentioned in Sect. 5.3. These simple yet instructive models serve as a good testbed for understanding the details of electromagnetic dissipation, particle acceleration and high energy radiation.

6.2 Relaxation of Force-Free Equilibria

A few groups, including Nalewajko et al. (2016), Lyutikov et al. (2016b), Yuan et al. (2016), have carried out 2D PIC simulations of the unstable force-free equilibria. In PIC simulations, the overall evolution of the system is quite similar to MHD/force-free simulations (Fig. 5): initially the ideal instability grows on Alfvén wave crossing time scales, producing current layers at the interface of merging flux ropes—this is consistent with the X-point collapse scenario for current sheet formation. The current layers are short-lived though, and the

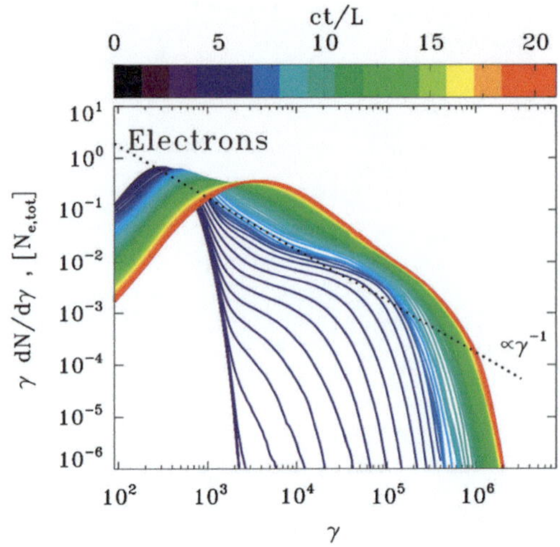

Fig. 6 Particle spectrum from a 2D PIC simulation of ABC instability with initial temperature $kT/mc^2 = 10^2$, in-plane magnetization $\sigma_{\text{in}} = 42$ (defined with respect to the enthalpy w), and $L = 126 r_{\text{L,hot}}$, where L is the linear size of one flux tube, $r_{\text{L,hot}}$ is the Larmor radius of high energy electrons heated/accelerated by reconnection (from Lyutikov et al. 2016b). The spectrum during the initial phase of rapid acceleration (*blue* to *cyan lines*) can be harder than $\gamma dN/d\gamma \propto \gamma^{-1}$ (compare with the *dotted line*), as indeed required by the observations of the Crab flares

system enters a turbulent relaxation phase, eventually settling into the longest wavelength configuration, with almost all the available magnetic free energy converted to particle kinetic energy. The total helicity is also approximately conserved in PIC simulations.

However, PIC simulations show richer structures during the evolution, especially the kinetic effect on current sheet formation—plasmoid generation during the thinning and stretching of the current sheets, and magnetic reconnection at these sheets. It is observed that the thickness of the current layer at its maximal stretch scales as the Larmor radius $r_{\text{L,hot}}$ of the high energy electrons accelerated/heated by reconnection. The reconnection rate is measured to be in the range $v_{\text{rec}}/c \sim 0.2$–$0.5$, which increases with the magnetization and saturates at around 0.5 at high magnetization limit (Lyutikov et al. 2016b). There appear to be two different phases of particle acceleration in the whole process. The first phase corresponds to the formation of the first, biggest current layers where reconnection of in-plane field takes place. $\boldsymbol{E} \cdot \boldsymbol{B} \neq 0$ in the current sheet, with the reconnection electric field $\boldsymbol{E}$ directed primarily out of plane, which causes run-away particle acceleration in the current sheet. This acceleration is fast, producing a relatively hard, high energy component in the particle distribution (Fig. 6). The second phase follows when the first current layers dissolve and the system evolves chaotically; the oscillating field structures scatter the particles around; there is evidence of second order Fermi acceleration smoothing out the high energy component to a softer power law, also bulk heating that increases the temperature of the thermal component.

It turns out that the particle acceleration efficiency depends on the mean magnetization of the configuration, in a similar fashion as plane current sheet reconnection scenarios (Guo et al. 2014; Sironi and Spitkovsky 2014; Werner et al. 2016). The particle spectrum gets harder as the mean magnetization increases; both the non-thermal particle fraction and the maximum particle energy increase with the magnetization (Lyutikov et al. 2016b; Nalewajko et al. 2016).

When we look at the synchrotron radiation signals from this evolving plasma, we find that, since the highest energy particles are first accelerated in the current layers by the parallel electric field, they do not radiate much when they are inside the sheet, because the curvature of their trajectory is small (despite the presence of guide field in the current layer). Most of the radiation is produced when particles are ejected from the current layers—their trajectories start to bend significantly in the ambient magnetic field which changes direction

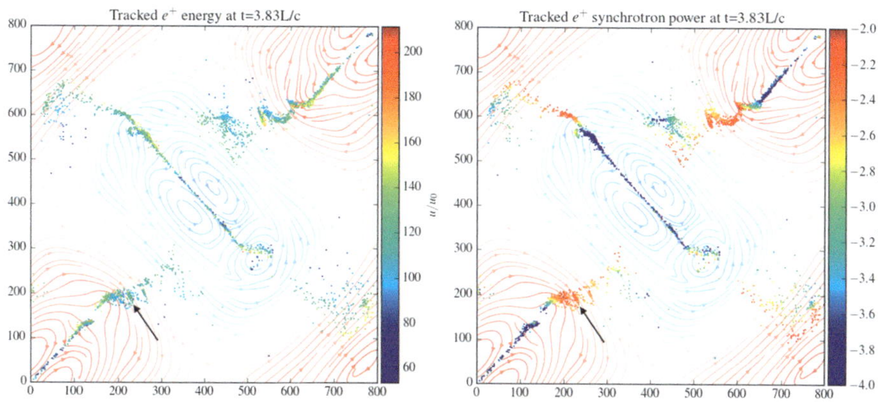

Fig. 7 A snapshot of the location of tracked high energy positrons, plotted over the instantaneous field structure (Yuan et al. 2016). In the *left panel*, the particles are color-coded by their energy while in the *right panel* they are color-coded by the synchrotron power. The *arrows* locate the ejection of plasmoids

at the end of the current layer (Fig. 7). Such a separation of acceleration site and radiative loss site could in principle facilitate acceleration beyond the synchrotron radiation reaction limit (Uzdensky et al. 2011). Fast variability of observed photon flux can be produced when compact plasmoids that contain high energy particles are ejected from the ends of the current layers and get destroyed. These give beamed radiation. An observer sees high intensity radiation when the beam happens to be aligned with the line of sight. This is a picture similar to the kinetic beaming seen in plane current sheet reconnection simulations (Cerutti et al. 2013, 2014). As a result, the high energy radiation is much more variable than the low energy radiation, and these emission peaks are accompanied by an increase in the polarization degree and rapid change of polarization angle in the high energy band (Yuan et al. 2016). The variability timescale is determined by the spatial extent of the emitting structure, e.g. the plasmoids, thus can be much shorter than the light crossing time of the region that collapses.

6.3 X-point Collapse

As we have discussed above, the initial rapid phase of particle acceleration in 2D PIC simulations of unstable force-free equilibria occurs at the current sheets created in between flux ropes. There, the evolution of the field line geometry resembles the X-point collapse studied by Syrovatsky in the non-relativistic regime (Syrovatskii 1966; Imshennik and Syrovatskiĭ 1967) and by Lyutikov et al. (2016b) in the relativistic regime. The vector potential for the X-point geometry is $A_z = -1/2(x^2/a(t)^2 - y^2/b(t)^2)$, where $b(t) = \lambda/a(t)$. The unstressed case of $\lambda = 1$ is stable. Below, we present results for the stressed case $\lambda = 1/\sqrt{2}$, for magnetizations $\sigma \gg 1$ and magnetic fields initialized only in the simulation plane (i.e., in the absence of the so-called "guide field"). As shown in Fig. 8, the collapse proceeds self-similarly: the macroscopic distribution of E^2/B^2 (and hence that of the drift velocity) at later times is a scaled copy of that at previous times, with the overall length scale increasing linearly with time (at the speed of light). This implies that the reconnection rate over the whole configurations remains fixed in time (and we find that for highly magnetized plasmas, the reconnection rate $v_{\rm rec}$ approaches the speed of light on macroscopic scales).

Interestingly, the electric field increases linearly with time. This is ultimately a manifestation of the self-similar macroscopic evolution of the system. Indeed, since in the initial

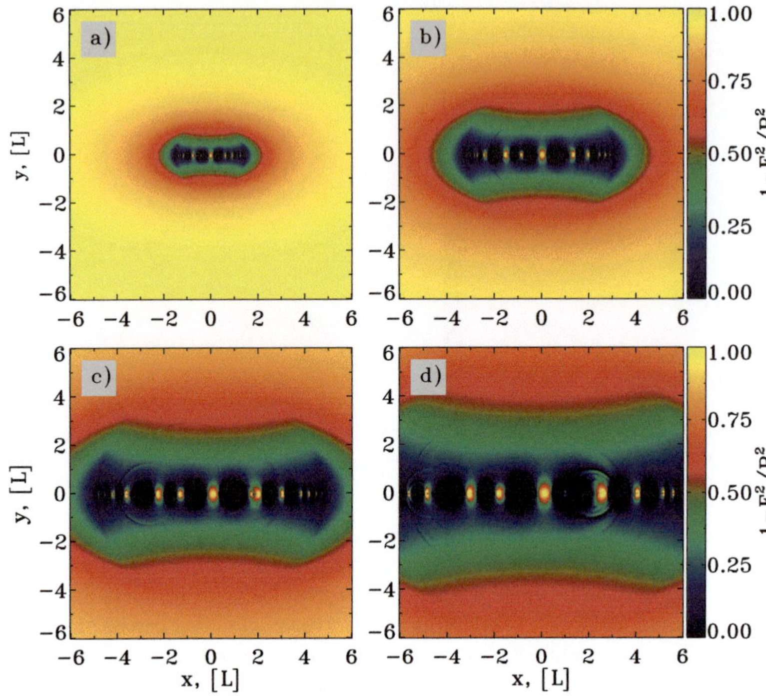

Fig. 8 Late time evolution of the X-point collapse in PIC simulations with zero guide field (Lyutikov et al. 2016b), for $\sigma = 4 \times 10^4$ (the magnetization is measured in the initial setup at unit distance from the center). The plots show the quantity $1 - E^2/B^2$ (strictly speaking, we plot $\max[0, 1 - E^2/B^2]$) at $ct/L = 1.5, 3, 4.5, 6$, from panel (**a**) to (**d**)

configuration the magnetic field strength grows linearly with distance from the origin (i.e., the center of the X-point) and the current sheet size grows linearly with time, the mean magnetic and electric fields in the volume surrounding the current sheet must also grow linearly, with their scaled distributions unchanged. The temporal evolution of the electric field has a direct impact on the maximum particle energy. Quite generally, its time evolution will be

$$\gamma_{\rm max} \propto Et \propto v_{\rm rec} Bt \tag{6}$$

Since both E and B in the reconnection region are scaling linearly with time, one expects $\gamma_{\rm max} \propto t^2$. This is confirmed by Fig. 9, where we follow the trajectories of a number of particles in a simulation with $\sigma = 4 \times 10^2$. The particles are selected such that their Lorentz factor exceeds a given threshold $\gamma_0 = 30$ within the time interval $1.4 \le ct_0/L \le 1.7$, as indicated by the vertical dashed lines in the top panel. The temporal evolution of the Lorentz factor of such particles, presented in the top panel for the 30 positrons reaching the highest energies, follows the track $\gamma \propto t^2 - t_0^2$ that is expected from $d\gamma/dt \propto E(t) \propto t$. Here, t_0 is the injection time, when the particle Lorentz factor γ first exceeds the threshold γ_0. The individual histories of single positrons might differ substantially, but overall the top panel of Fig. 9 suggests that the acceleration process is dominated by direct acceleration by the reconnection electric field. We find that the particles presented in the top panel of Fig. 9 are too energetic to be confined within the small-scale plasmoids in the current sheet (see the small scale structures in Fig. 8), so any acceleration mechanism that relies on plasmoid mergers is found to be unimportant, in this setup.

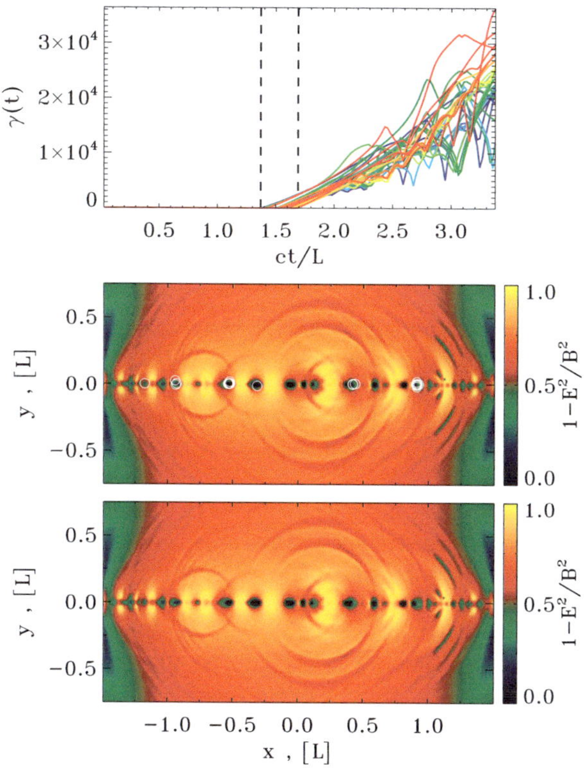

Fig. 9 Physics of particle injection into the acceleration process, from a PIC simulation of stressed X-point collapse with vanishing guide field and $\sigma = 4 \times 10^2$ (Lyutikov et al. 2016b). *Top panel*: we select all the particles that exceed the threshold $\gamma_0 = 30$ within a given time interval (chosen to be $1.4 \le ct_0/L \le 1.7$, as indicated by the *vertical dashed lines*), and we plot the temporal evolution of the Lorentz factor of the 30 particles that at the final time reach the highest energies. The particle Lorentz factor increases as $\gamma \propto t^2 - t_0^2$, where t_0 marks the onset of acceleration (i.e., the time when γ first exceeds γ_0). *Middle panel*: for the same particles as in the *top panel*, we plot their locations at the onset of acceleration with *open white circles*, superimposed over the 2D plot of $1 - E^2/B^2$ (more precisely, of $\max[0, 1 - E^2/B^2]$). Comparison of the *middle panel* with the *bottom panel* shows that particle injection is localized in the vicinity of the X-points in the current sheet (i.e., the blue regions where $E > B$)

Particle injection into the acceleration process happens in the charge-starved regions where $E > B$, i.e., in the small-scale X-points that separate each pair of secondary plasmoids in the current sheet. Indeed, for the same particles as in the top panel, the middle panel in Fig. 9 presents their locations at the onset of acceleration with open white circles, superimposed over the 2D plot of $1 - E^2/B^2$ (more precisely, of $\max[0, 1 - E^2/B^2]$). Comparison of the middle panel with the bottom panel shows that particle injection is localized in the vicinity of the small-scale X-points in the current sheet (i.e., the blue regions where $E > B$). Despite occupying a relatively small fraction of the overall volume, such regions are of paramount importance for particle acceleration.

The explosive stage of X-point collapse produces non-thermal tails (in analogy to the relaxation of unstable force-free structures) whose hardness depends on the average magnetization. For sufficiently high magnetizations and vanishing guide field, the non-thermal particle spectrum consists of two components: a low-energy population with soft spectrum, that dominates the number census; and a high-energy population with hard spectrum, that possesses all the properties needed to explain the Crab flares (Lyutikov et al. 2016b). The particle distribution is highly anisotropic, with high-energy particles beamed primarily along the direction of the accelerating electric field.

6.4 Merging Lundquist Flux Ropes

Above, we have described the evolution of unstable force-free configurations (with the ABC geometry being a special case). There are two key features of the preceding model that are specific to the initial set-up: (i) each flux tube carries non-zero poloidal current; (ii) the initial configuration is an unstable equilibrium. However, as we now show, the evolution is generic, regardless of these conditions. Lyutikov et al. (2016b) investigated a merger of two flux tubes with zero total current, with MHD and PIC simulations. Thus, two flux tubes are not attracted to each other—at least initially. Here, we present the results for Lundquist's force-free cylinders surrounded by uniform magnetic field,

$$\mathbf{B}_L(r \le r_\mathrm{j}) = J_1(r\alpha)\mathbf{e}_\phi + J_0(r\alpha)\mathbf{e}_z, \tag{7}$$

Here, J_0, J_1 are Bessel functions of zeroth and first order and the constant $\alpha \simeq 3.8317$ is the first root of J_0. We chose to terminate this solution at the first zero of J_1, which we denote as r_j and hence continue with $B_z = B_z(r_j)$ and $B_\phi = 0$ for $r > r_j$. Thus the total current of the flux tube is zero. As the result, the azimuthal field vanishes at the boundary of the rope, whereas the poloidal one changes sign inside the rope. The evolution is very slow, given the fact that at the contact the reconnecting field vanishes (i.e., the initial configuration is dynamically stable). To speed things up, the ropes are pushed towards each other.

In Fig. 10, we present the 2D pattern of the out-of-plane field B_z (left column) and of the in-plane magnetic energy fraction $\epsilon_{B,\mathrm{in}} = (B_x^2 + B_y^2)/8\pi nmc^2$ (right column; with superimposed magnetic field lines), from a PIC simulation with $kT/mc^2 = 10^{-4}$, $\sigma_\mathrm{in} = 43$ (only defined with the in-plane fields) and $r_\mathrm{j} = 61 r_{\mathrm{L,hot}}$. As the two magnetic ropes slowly approach, driven by the initial velocity push, reconnection is triggered in the plane $x = 0$, as indicated by the formation and subsequent ejection of small-scale plasmoids. Until $ct/r_\mathrm{j} \sim 4.5$, the cores of the two islands have not significantly moved (black line in the middle panel of Fig. 11, indicating the x_c location of the center of the rightmost island), the reconnection speed is quite small (red line in the middle panel of Fig. 11) and no significant energy exchange has occurred from the fields to the particles (compare the in-plane magnetic energy, shown by the dashed blue line in the top panel of Fig. 11, with the particle kinetic energy, indicated with the red line).

As a result of reconnection, an increasing number of field lines, that initially closed around one of the ropes, are now engulfing both magnetic islands. Their tension force causes the two ropes to approach and merge on a quick (dynamical) timescale, starting at $ct/r_\mathrm{j} \sim 4.5$ and ending at $ct/r_\mathrm{j} \sim 7.5$ (see that the distance of the rightmost island from the center rapidly decreases, as indicated by the black line in the middle panel of Fig. 11). The tension force drives the particles in the flux ropes toward the center, with a fast reconnection speed peaking at $v_\mathrm{rec}/c \sim 0.3$ (red line in the middle panel of Fig. 11). The reconnection layer at $x = 0$ stretches up to a length of $\sim 2r_\mathrm{j}$, and secondary plasmoids are formed. In the central current sheet, it is primarily the in-plane field that gets dissipated (compare the dashed and solid blue lines in the top panel of Fig. 11), driving an increase in the electric energy (green) and in the particle kinetic energy (red). In this phase of evolution, the fraction of initial energy released to the particles is small ($\epsilon_\mathrm{kin}/\epsilon_\mathrm{tot}(0) \sim 0.1$), but the particles advected into the central X-point experience a dramatic episode of acceleration. As shown in

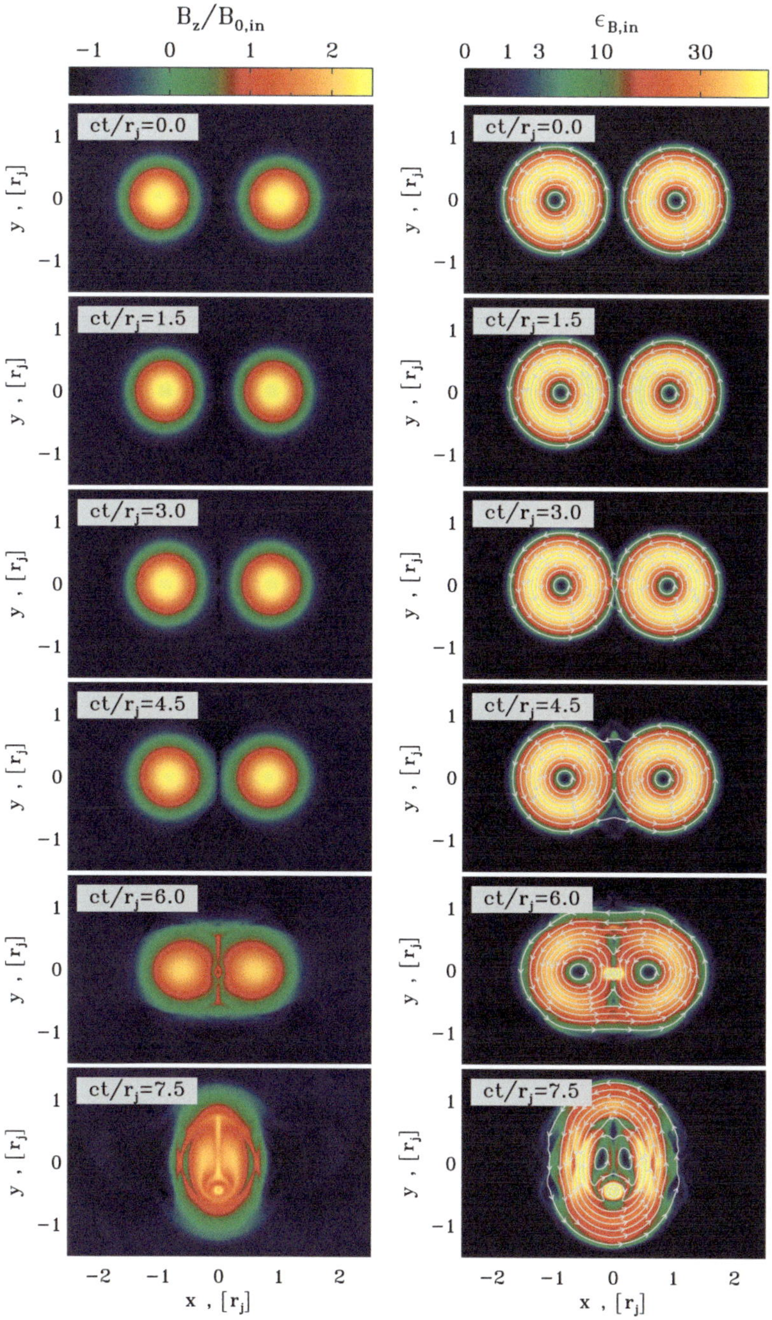

Fig. 10 Temporal evolution of 2D Lundquist ropes (time is measured in c/r_j and indicated in the *grey box* of *each panel*, increasing from top to bottom), from Lyutikov et al. (2016b). The plot presents the 2D pattern of the out-of-plane field B_z (*left column*) and of the in-plane magnetic energy fraction $\epsilon_{B,\mathrm{in}} = (B_x^2 + B_y^2)/8\pi nmc^2$ (*right column*; with superimposed magnetic field lines), from a PIC simulation with $kT/mc^2 = 10^{-4}$, $\sigma_{\mathrm{in}} = 43$ and $r_j = 61 r_{\mathrm{L,hot}}$

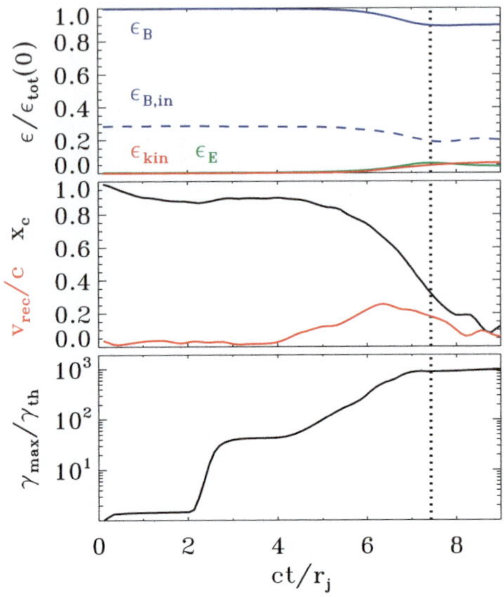

Fig. 11 Temporal evolution of various quantities, from a 2D PIC simulation of Lundquist ropes with $kT/mc^2 = 10^{-4}$, $\sigma_{in} = 43$ and $r_j = 61 r_{L,hot}$ (the same as in Fig. 10), from Lyutikov et al. (2016b). *Top panel*: fraction of energy in magnetic fields (*solid blue*), in-plane magnetic fields (*dashed blue*), electric fields (*green*) and particles (*red*; excluding the rest mass energy), in units of the total initial energy. Middle panel: reconnection rate v_{rec}/c (*red*), and location x_c of the core of the rightmost flux rope (*black*), in units of r_j. *Bottom panel*: evolution of the maximum Lorentz factor γ_{max}

the bottom panel of Fig. 11, the cutoff Lorentz factor γ_{max} of the particle spectrum presents a dramatic evolution, increasing up to $\gamma_{max}/\gamma_{th} \sim 10^3$ within a couple of dynamical times. This phase of extremely fast particle acceleration on a dynamical timescale is analogous to the relaxation of unstable force-free structures discussed above.

7 Future Directions

7.1 Observation

Much has been learned in recent years from observations throughout the electromagnetic spectrum (radio through TeV γ-rays) and beyond (gravitational radiation, neutrinos and cosmic rays). Even more discoveries are anticipated from observations with new facilities that should come on line in the next ~ 5 years. Perhaps the most exciting opportunity is to make a (temporal) match of a short GRB with a LIGO event. Equally interesting would be a temporal association of an“cosmogenic” VHE neutrino with some other cosmic event. We may also learn about the details of jet production in AGN from Event Horizon Telescope observations of M87, as discussed above. Fast Radio Bursts are starting to reveal their secrets and they, too, may be shown to have some physical affinity with the other examples of release of energy, in particular magnetars. Comparisons of radio pulsar observations with corresponding γ-ray observations and studies of radio scintillation are providing a much better understanding of neutron star magnetospheres which may help us calibrate our models of other magnetically dominant sources. Similar remarks apply to solar flares where there can be a rapid release of magnetic energy in a very short space of time.

There will also be new optical/near infrared survey telescopes devoted to transient astronomy including the Zwicky Transient Factory, LSST and Euclid, greatly increasing our ability to monitor familiar types of transient, as well as discover new classes of which we are now ignorant.

7.2 Simulation

With the advance in algorithms, simulation techniques and computational power, much more can be understood about plasma physics under extreme conditions using simulations. On one hand, fluid level simulations are now able to handle complex, realistic astrophysical situations, like the merging of neutron stars/black holes, tidal disruption events, colliding winds in binary systems, accretion disk and jet formation, etc. These provide more physical interpretation of the observed phenomena, also point to possible sites and mechanisms of dissipation and particle acceleration.

On the other hand, kinetic simulations are getting close to modeling multi-scale systems. For example, the role of reconnection and particle acceleration in accretion disk has been recently discussed in PIC simulations (e.g. Hoshino 2015; Riquelme et al. 2012). PIC simulations are also used to study global pulsar magnetospheres (e.g., Chen and Beloborodov 2014; Belyaev 2015; Cerutti et al. 2015; Philippov et al. 2015), relativistic turbulence (Zhdankin et al. 2017), etc. Improved hybrid methods are also bridging between the kinetic and macroscopic length scales (Kunz et al. 2016). A major challenge is to increase the dynamic range of particle energy that can be covered in these simulations to approach that found in the sources discussed above. Furthermore, new physics is being added to traditional PIC simulations, including pair production (Chen and Beloborodov 2014), general relativistic effect (Philippov et al. 2015), radiative feedback (Uzdensky 2016), and hadronic interactions.

7.3 Experiment

Another new development is the prospect of performing novel experiments at the many powerful light sources around the world that are already operational or are expected to become such over the next five years. There are various schemes being explored. For example, γ-rays can be created by Bethe–Heitler bremsstrahlung when a relativistic electron beam interacts with a foil target. Electron-positron pairs will also be produced. However if the γ-rays are above threshold they can also pair-produce on an oncoming, coherent X-ray beam from a powerful laser by the Breit–Wheeler process, just like in cosmic sources. Achieving this pair production will be a major experimental milestone but will have no significance for QED as the cross-section is not in doubt! However, it will open the door to explorations of nonlinear quantum electrodynamics where calculations are hard and deserve validation. In addition and of even more interest for astrophysics are the many body processes where it is the collective properties of the pair plasma that are important. There are additional processes that can be investigated in the same spirit involving large magnetostatic fields.

These explorations, although partly inspired by the challenges of extreme astrophysical acceleration, will be prosecuted in the spirit of comparing experiments that can be performed and diagnosed with numerical simulations. It is an exciting prospect that new effects will be discovered and that some of these may have a role in understanding the processes we have discussed in this chapter.

Acknowledgements We thank Andrei Bykov for his patience and for organizing a very interesting meeting, the other participants for helping us develop our ideas, and many other colleagues, including William East, Serguei Komissarov, Maxim Lyutikov, Krzysztof Nalewajko, Oliver Porth and Jonathan Zrake for their collaboration on research described in this chapter.

References

M. Ackermann, R. Anantua, K. Asano et al., Minute-timescale > 100 MeV γ-ray variability during the giant outburst of quasar 3C 279 observed by Fermi-LAT in 2015 June. Astrophys. J. Lett. **824**, L20 (2016). arXiv:1605.05324. doi:10.3847/2041-8205/824/2/L20

F. Aharonian, A.G. Akhperjanian, A.R. Bazer-Bachi et al., An exceptional very high energy gamma-ray flare of PKS 2155-304. Astrophys. J. Lett. **664**, L71–L74 (2007). arXiv:0706.0797. doi:10.1086/520635

J. Albert, E. Aliu, H. Anderhub et al., Variable very high energy γ-ray emission from Markarian 501. Astrophys. J. **669**, 862–883 (2007). arXiv:astro-ph/0702008. doi:10.1086/521382

J. Aleksić, L.A. Antonelli, P. Antoranz et al., MAGIC discovery of very high energy emission from the FSRQ PKS 1222+21. Astrophys. J. Lett. **730**, L8 (2011). arXiv:1101.4645. doi:10.1088/2041-8205/730/1/L8

V. Arnol'd, Sur la topologie des bcoulements stationnaires des fluides parfaits. C. R. Acad. Sci. Paris **261**, 17–20 (1965)

J. Arons, Magnetars in the metagalaxy: an origin for ultra-high-energy cosmic rays in the nearby universe. Astrophys. J. **589**, 871–892 (2003). arXiv:astro-ph/0208444. doi:10.1086/374776

J. Arons, Pulsar wind nebulae as cosmic pevatrons: a current sheet's tale. Space Sci. Rev. **173**, 341–367 (2012). arXiv:1208.5787. doi:10.1007/s11214-012-9885-1

M.C. Begelman, R.E. Ergun, M.J. Rees, Cyclotron maser emission from blazar jets? Astrophys. J. **625**, 51–59 (2005). arXiv:astro-ph/0502151. doi:10.1086/429550

M.C. Begelman, A.C. Fabian, M.J. Rees, Implications of very rapid TeV variability in blazars. Mon. Not. R. Astron. Soc. **384**, L19–L23 (2008). arXiv:0709.0540. doi:10.1111/j.1745-3933.2007.00413.x

P.M. Bellan, *Spheromaks: A Practical Application of Magnetohydrodynamic Dynamos and Plasma Self-Organization* (Imperial College Press, London, 2000)

E. Beltrami, *Opere Matematiche* (U. Hoepli, Milan, 1902)

M.A. Belyaev, Dissipation, energy transfer, and spin-down luminosity in 2.5D PIC simulations of the pulsar magnetosphere. Mon. Not. R. Astron. Soc. **449**, 2759–2767 (2015). arXiv:1412.2819. doi:10.1093/mnras/stv468

D. Biskamp, *Magnetic Reconnection in Plasmas* (Cambridge University Press, Cambridge, 2000)

R.D. Blandford, M.C. Begelman, On the fate of gas accreting at a low rate on to a black hole. Mon. Not. R. Astron. Soc. **303**, L1–L5 (1999). arXiv:astro-ph/9809083. doi:10.1046/j.1365-8711.1999.02358.x

R.D. Blandford, A. Königl, Relativistic jets as compact radio sources. Astrophys. J. **232**, 34–48 (1979). doi:10.1086/157262

R.D. Blandford, A. Levinson, Pair cascades in extragalactic jets. 1: Gamma rays. Astrophys. J. **441**, 79–95 (1995). doi:10.1086/175338

R. Blandford, P. Simeon, Y. Yuan, Cosmic ray origins: an introduction. Nucl. Phys. B, Proc. Suppl. **256**, 9–22 (2014). doi:10.1016/j.nuclphysbps.2014.10.002. http://adsabs.harvard.edu/abs/2014NuPhS.256....9B

R. Blandford, W. East, K. Nalewajko, Y. Yuan, J. Zrake, Active Galactic Nuclei: The TeV Challenge (2015). arXiv:1511.07515 [astro-ph]

S.E. Boggs, A. Zoglauer, E. Bellm et al., The giant flare of 2004 December 27 from SGR 1806-20. Astrophys. J. **661**, 458–467 (2007). arXiv:astro-ph/0611318. doi:10.1086/516732

M.P. Brenner, S. Hilgenfeldt, D. Lohse, Single-bubble sonoluminescence. Rev. Mod. Phys. **74**, 425–484 (2002). doi:10.1103/RevModPhys.74.425. https://link.aps.org/doi/10.1103/RevModPhys.74.425

A.E. Broderick, R. Narayan, J. Kormendy et al., The event horizon of M87. Astrophys. J. **805**, 179 (2015). arXiv:1503.03873. doi:10.1088/0004-637X/805/2/179

R. Buehler, R. Blandford, The surprising Crab pulsar and its nebula: a review. Rep. Prog. Phys. **77**(6), 066901 (2014). arXiv:1309.7046. doi:10.1088/0034-4885/77/6/066901

G.R. Burbidge, On synchrotron radiation from Messier 87. Astrophys. J. **124**, 416 (1956). doi:10.1086/146237

B. Cerutti, G.R. Werner, D.A. Uzdensky, M.C. Begelman, Simulations of particle acceleration beyond the classical synchrotron burnoff limit in magnetic reconnection: an explanation of the Crab flares. Astrophys. J. **770**, 147 (2013). arXiv:1302.6247. doi:10.1088/0004-637X/770/2/147

B. Cerutti, G.R. Werner, D.A. Uzdensky, M.C. Begelman, Three-dimensional relativistic pair plasma reconnection with radiative feedback in the Crab nebula. Astrophys. J. **782**, 104 (2014). arXiv:1311.2605. doi:10.1088/0004-637X/782/2/104

B. Cerutti, A. Philippov, K. Parfrey, A. Spitkovsky, Particle acceleration in axisymmetric pulsar current sheets. Mon. Not. R. Astron. Soc. **448**, 606–619 (2015). arXiv:1410.3757. doi:10.1093/mnras/stv042

S. Chandrasekhar, P.C. Kendall, On force-free magnetic fields. Astrophys. J. **126**, 457 (1957). doi:10.1086/146413

A.Y. Chen, A.M. Beloborodov, Electrodynamics of axisymmetric pulsar magnetosphere with electron-positron discharge: a numerical experiment. Astrophys. J. Lett. **795**, L22 (2014). arXiv:1406.7834. doi:10.1088/2041-8205/795/1/L22

S. Childress, New solutions of the kinematic dynamo problem. J. Math. Phys. **11**, 3063–3076 (1970). doi:10.1063/1.1665095

J.F. Drake, M. Swisdak, H. Che, M.A. Shay, Electron acceleration from contracting magnetic islands during reconnection. Nature **443**, 553–556 (2006). doi:10.1038/nature05116

W.E. East, J. Zrake, Y. Yuan, R.D. Blandford, Spontaneous decay of periodic magnetostatic equilibria. Phys. Rev. Lett. **115**(9), 095002 (2015). arXiv:1503.04793. doi:10.1103/PhysRevLett.115.095002

W.D. Evans, R.W. Klebesadel, J.G. Laros et al., Location of the gamma-ray transient event of 1979 March 5. Astrophys. J. Lett. **237**, L7–L9 (1980). doi:10.1086/183222

B.L. Fanaroff, J.M. Riley, The morphology of extragalactic radio sources of high and low luminosity. Mon. Not. R. Astron. Soc. **167**, 31P–36P (1974). doi:10.1093/mnras/167.1.31P

V.L. Fish, M.D. Johnson, S.S. Doeleman et al., Persistent asymmetric structure of Sagittarius A* on event horizon scales. Astrophys. J. **820**, 90 (2016). arXiv:1602.05527. doi:10.3847/0004-637X/820/2/90

H.P. Furth, J. Killeen, M.N. Rosenbluth, Finite-resistivity instabilities of a sheet pinch. Phys. Fluids **6**, 459–484 (1963). doi:10.1063/1.1706761

R. Giacconi, H. Gursky, F.R. Paolini, B.B. Rossi, Evidence for x rays from sources outside the solar system. Phys. Rev. Lett. **9**, 439–443 (1962). doi:10.1103/PhysRevLett.9.439

T. Gold, Concluding remarks on turbulence in the interstellar gas, in *Gas Dynamics of Cosmic Clouds*. IAU Symposium, vol. 2 (1955), p. 238

F. Guo, H. Li, W. Daughton, Y.H. Liu, Formation of hard power laws in the energetic particle spectra resulting from relativistic magnetic reconnection. Phys. Rev. Lett. **113**(15), 155005 (2014). arXiv:1405.4040. doi:10.1103/PhysRevLett.113.155005

V.F. Hess, Observations of the penetrating radiation on seven balloon flights. Phys. Z. **13**, 1084–1091 (1912)

J.J. Hester, The Crab nebula: an astrophysical chimera. Annu. Rev. Astron. Astrophys. **46**, 127–155 (2008). doi:10.1146/annurev.astro.45.051806.110608

J.J. Hester, P.A. Scowen, R. Sankrit et al., WFPC2 studies of the Crab nebula. I. HST and ROSAT imaging of the synchrotron nebula. Astrophys. J. **448**, 240 (1995). doi:10.1086/175956

A. Hewish, S.J. Bell, J.D.H. Pilkington, P.F. Scott, R.A. Collins, Observation of a rapidly pulsating radio source. Nature **217**, 709–713 (1968). doi:10.1038/217709a0

W. Horton, T. Tajima, Linear theory of driven reconnection. J. Geophys. Res. **93**, 2741–2748 (1988). doi:10.1029/JA093iA04p02741

M. Hoshino, Stochastic particle acceleration in multiple magnetic islands during reconnection. Phys. Rev. Lett. **108**(13), 135003 (2012). arXiv:1201.0837. doi:10.1103/PhysRevLett.108.135003

M. Hoshino, Angular momentum transport and particle acceleration during magnetorotational instability in a kinetic accretion disk. Phys. Rev. Lett. **114**(6), 061101 (2015). arXiv:1502.02452. doi:10.1103/PhysRevLett.114.061101

M. Hoshino, T. Mukai, T. Terasawa, I. Shinohara, Suprathermal electron acceleration in magnetic reconnection. J. Geophys. Res. **106**, 25979–25998 (2001). doi:10.1029/2001JA900052

K. Hurley, T. Cline, E. Mazets et al., A giant periodic flare from the soft γ-ray repeater SGR1900+14. Nature **397**, 41–43 (1999). arXiv:astro-ph/9811443. doi:10.1038/16199

K. Hurley, S.E. Boggs, D.M. Smith et al., An exceptionally bright flare from SGR 1806-20 and the origins of short-duration γ-ray bursts. Nature **434**, 1098–1103 (2005). arXiv:astro-ph/0502329. doi:10.1038/nature03519

V.S. Imshennik, S.I. Syrovatskiĭ, Two-dimensional flow of an ideally conducting gas in the vicinity of the zero line of a magnetic field. Sov. Phys. JETP **25**, 656 (1967)

D. Kagan, L. Sironi, B. Cerutti, D. Giannios, Relativistic magnetic reconnection in pair plasmas and its astrophysical applications. Space Sci. Rev. **191**, 545–573 (2015). arXiv:1412.2451. doi:10.1007/s11214-014-0132-9

V.M. Kaspi, A. Beloborodov, Magnetars. Annu. Rev. Astron. Astrophys. (2017, in press). arXiv:1703.00068

C.F. Kennel, F.V. Coroniti, Magnetohydrodynamic model of Crab nebula radiation. Astrophys. J. **283**, 710–730 (1984). doi:10.1086/162357

J.G. Kirk, Y. Lyubarsky, J. Petri, The theory of pulsar winds and nebulae, in *Neutron Stars and Pulsars*, ed. by W. Becker. Astrophysics and Space Science Library, vol. 357 (2009), p. 421. arXiv:astro-ph/0703116

R.W. Klebesadel, I.B. Strong, R.A. Olson, Observations of gamma-ray bursts of cosmic origin. Astrophys. J. Lett. **182**, L85 (1973). doi:10.1086/181225

S.S. Komissarov, Y.E. Lyubarsky, The origin of peculiar jet-torus structure in the Crab nebula. Mon. Not. R. Astron. Soc. **344**, L93–L96 (2003). arXiv:astro-ph/0306162. doi:10.1046/j.1365-8711.2003.07097.x

S.S. Komissarov, Y.E. Lyubarsky, Synchrotron nebulae created by anisotropic magnetized pulsar winds. Mon. Not. R. Astron. Soc. **349**, 779–792 (2004). doi:10.1111/j.1365-2966.2004.07597.x

S.S. Komissarov, M. Lyutikov, On the origin of variable gamma-ray emission from the Crab nebula. Mon. Not. R. Astron. Soc. **414**, 2017–2028 (2011). arXiv:1011.1800. doi:10.1111/j.1365-2966.2011.18516.x

C. Kouveliotou, R.A.M.J. Wijers, S. Woosley, *Gamma-Ray Bursts* (Cambridge University Press, Cambridge, 2012). http://adsabs.harvard.edu/abs/2012grb..book.....K

M.W. Kunz, J.M. Stone, E. Quataert, Magnetorotational turbulence and dynamo in a collisionless plasma. Phys. Rev. Lett. **117**(23), 235101 (2016). arXiv:1608.07911. doi:10.1103/PhysRevLett.117.235101

A. Levinson, R. Blandford, Raman scattering in high-radio-brightness astrophysical systems: application to active galactic nuclei. Mon. Not. R. Astron. Soc. **274**, 717–729 (1995). doi:10.1093/mnras/274.3.717

D. Lohse, B. Schmitz, M. Versluis, Snapping shrimp make flashing bubbles. Nature **413**(6855), 477–478 (2001). http://www.nature.com/nature/journal/v413/n6855/abs/413477a0.html. doi:10.1038/35097152

D.R. Lorimer, M. Bailes, M.A. McLaughlin, D.J. Narkevic, F. Crawford, A bright millisecond radio burst of extragalactic origin. Science **318**, 777 (2007). arXiv:0709.4301. doi:10.1126/science.1147532

M. Lyutikov, D. Balsara, C. Matthews, Crab GeV flares from the corrugated termination shock. Mon. Not. R. Astron. Soc. **422**, 3118–3129 (2012). arXiv:1109.1204. doi:10.1111/j.1365-2966.2012.20831.x

M. Lyutikov, S.S. Komissarov, O. Porth, The inner knot of the Crab nebula. Mon. Not. R. Astron. Soc. **456**, 286–299 (2016a). arXiv:1506.07282. doi:10.1093/mnras/stv2570

M. Lyutikov, L. Sironi, S. Komissarov, O. Porth, Particle acceleration in explosive relativistic reconnection events and Crab Nebula gamma-ray flares (2016b). ArXiv e-prints arXiv:1603.05731

G. Madejski, M. Sikora, Gamma-ray observations of active galactic nuclei. Annu. Rev. Astron. Astrophys. **54**, 725–760 (2016). doi:10.1146/annurev-astro-081913-040044

A.P. Marscher, S.G. Jorstad F.D. D'Arcangelo et al., The inner jet of an active galactic nucleus as revealed by a radio-to-γ-ray outburst. Nature **452**, 966–969 (2008). doi:10.1038/nature06895

W. Max-Moerbeck, T. Hovatta, J.L. Richards et al., Time correlation between the radio and gamma-ray activity in blazars and the production site of the gamma-ray emission. Mon. Not. R. Astron. Soc. **445**, 428–436 (2014). arXiv:1408.6264. doi:10.1093/mnras/stu1749

D.L. Meier, *Black Hole Astrophysics: the Engine Paradigm* (Springer, Berlin, 2012). http://adsabs.harvard.edu/abs/2012bhae.book.....M

S. Mereghetti, D. Götz, I.F. Mirabel, K. Hurley, INTEGRAL discovery of persistent hard X-ray emission from the soft gamma-ray repeater SGR 1806-20. Astron. Astrophys. **433**, L9–L12 (2005). arXiv:astro-ph/0411695. doi:10.1051/0004-6361:200500088

I.F. Mirabel, L.F. Rodríguez, Sources of relativistic jets in the galaxy. Annu. Rev. Astron. Astrophys. **37**, 409–443 (1999). arXiv:astro-ph/9902062. doi:10.1146/annurev.astro.37.1.409

H.K. Moffatt, *Magnetic Field Generation in Electrically Conducting Fluids* (Cambridge University Press, Cambridge, 1978). http://adsabs.harvard.edu/abs/1978mfge.book.....M

H.K. Moffatt, Magnetostatic equilibria and analogous Euler flows of arbitrarily complex topology. II—Stability considerations. J. Fluid Mech. **166**, 359–378 (1986). doi:10.1017/S0022112086000198

K. Nalewajko, J. Zrake, Y. Yuan, W.E. East, R.D. Blandford, Kinetic simulations of the lowest-order unstable mode of relativistic magnetostatic equilibria. Astrophys. J. **826**, 115 (2016). arXiv:1603.04850. doi:10.3847/0004-637X/826/2/115

E.N. Parker, The solar-flare phenomenon and the theory of reconnection and annihiliation of magnetic fields. Astrophys. J. Suppl. **8**, 177 (1963). doi:10.1086/190087

H.E. Petschek, Magnetic Field Annihilation. NASA Special Publication 50:425 (1964)

A.A. Philippov, B. Cerutti, A. Tchekhovskoy, A. Spitkovsky, Ab initio pulsar magnetosphere: the role of general relativity. Astrophys. J. Lett. **815**, L19 (2015). arXiv:1510.01734. doi:10.1088/2041-8205/815/2/L19

E. Priest, T. Forbes, *Magnetic Reconnection: MHD Theory and Applications* (Cambridge University Press, Cambridge, 2000). http://ebooks.cambridge.org/ebook.jsf?bid=CBO9780511525087

G. Reber, J.L. Greenstein, Radio-frequency investigations of astronomical interest. Observatory **67**, 15–26 (1947)

M.J. Rees, J.E. Gunn, The origin of the magnetic field and relativistic particles in the Crab nebula. Mon. Not. R. Astron. Soc. **167**, 1–12 (1974)

M.J. Reid, J.A. Biretta, W. Junor, T.W.B. Muxlow, R.E. Spencer, Subluminal motion and limb brightening in the nuclear jet of M87. Astrophys. J. **336**, 112–120 (1989). doi:10.1086/166998

M.A. Riquelme, E. Quataert, P. Sharma, A. Spitkovsky, Local two-dimensional particle-in-cell simulations of the collisionless magnetorotational instability. Astrophys. J. **755**, 50 (2012). arXiv:1201.6407. doi:10.1088/0004-637X/755/1/50

M.N. Rosenbluth, M.N. Bussac, MHD stability of spheromak. Nucl. Fusion **19**, 489–498 (1979)

A. Rudy, D. Horns, A. DeLuca et al., Characterization of the inner knot of the Crab: the site of the gamma-ray flares? Astrophys. J. **811**, 24 (2015). arXiv:1504.04613. doi:10.1088/0004-637X/811/1/24

M. Schmidt, 3C 273: a star-like object with large red-shift. Nature **197**, 1040 (1963). doi:10.1038/1971040a0

L. Sironi, A. Spitkovsky, Particle acceleration in relativistic magnetized collisionless pair shocks: dependence of shock acceleration on magnetic obliquity. Astrophys. J. **698**, 1523–1549 (2009). arXiv:0901.2578. doi:10.1088/0004-637X/698/2/1523

L. Sironi, A. Spitkovsky, Acceleration of particles at the termination shock of a relativistic striped wind. Astrophys. J. **741**, 39 (2011a). arXiv:1107.0977. doi:10.1088/0004-637X/741/1/39

L. Sironi, A. Spitkovsky, Particle acceleration in relativistic magnetized collisionless electron-ion shocks. Astrophys. J. **726**, 75 (2011b). arXiv:1009.0024. doi:10.1088/0004-637X/726/2/75

L. Sironi, A. Spitkovsky, Relativistic reconnection: an efficient source of non-thermal particles. Astrophys. J. Lett. **783**, L21 (2014). arXiv:1401.5471. doi:10.1088/2041-8205/783/1/L21

T.W. Speiser, Particle trajectories in model current sheets. 1. Analytical solutions. J. Geophys. Res. **70**, 4219–4226 (1965). doi:10.1029/JZ070i017p04219

A. Spitkovsky, Particle acceleration in relativistic collisionless shocks: Fermi process at last? Astrophys. J. Lett. **682**, L5 (2008). arXiv:0802.3216. doi:10.1086/590248

S.I. Syrovatskii, Dynamic dissipation of a magnetic field and particle acceleration. Sov. Astron. **10**, 270 (1966)

P.G. Tait, *Properties of Matter* (Black, London, 1907)

J.B. Taylor, Relaxation of toroidal plasma and generation of reverse magnetic fields. Phys. Rev. Lett. **33**(19), 1139–1141 (1974). doi:10.1103/PhysRevLett.33.1139

J.B. Taylor, Relaxation and magnetic reconnection in plasmas. Rev. Mod. Phys. **58**(3), 741–763 (1986). doi:10.1103/RevModPhys.58.741

C. Thompson, R.C. Duncan, The soft gamma repeaters as very strongly magnetized neutron stars—I. Radiative mechanism for outbursts. Mon. Not. R. Astron. Soc. **275**, 255–300 (1995). doi:10.1093/mnras/275.2.255

C. Thompson, R.C. Duncan, The soft gamma repeaters as very strongly magnetized neutron stars. II. Quiescent neutrino, X-ray, and Alfven wave emission. Astrophys. J. **473**, 322 (1996). doi:10.1086/178147

D.A. Uzdensky, Radiative magnetic reconnection in astrophysics, in *Magnetic Reconnection: Concepts and Applications*. Astrophysics and Space Science Library, vol. 427 (2016), p. 473. arXiv:1510.05397. doi:10.1007/978-3-319-26432-5_12

D.A. Uzdensky, B. Cerutti, M.C. Begelman, Reconnection-powered linear accelerator and gamma-ray flares in the Crab nebula. Astrophys. J. Lett. **737**, L40 (2011). arXiv:1105.0942. doi:10.1088/2041-8205/737/2/L40

X. Wang, A. Bhattacharjee, Z.W. Ma, Collisionless reconnection: effects of Hall current and electron pressure gradient. J. Geophys. Res. **105**, 27633–27648 (2000). doi:10.1029/1999JA000357

M.C. Weisskopf, J.J. Hester, A.F. Tennant et al., Discovery of spatial and spectral structure in the X-ray emission from the Crab nebula. Astrophys. J. Lett. **536**, L81–L84 (2000). arXiv:astro-ph/0003216. doi:10.1086/312733

G.R. Werner, D.A. Uzdensky, B. Cerutti, K. Nalewajko, M.C. Begelman, The extent of power-law energy spectra in collisionless relativistic magnetic reconnection in pair plasmas. Astrophys. J. Lett. **816**, L8 (2016). arXiv:1409.8262. doi:10.3847/2041-8205/816/1/L8

C.A. Wilson-Hodge, M.L. Cherry, G.L. Case et al., When a standard candle flickers. Astrophys. J. Lett. **727**, L40 (2011). arXiv:1010.2679. doi:10.1088/2041-8205/727/2/L40

Y. Yuan, R.D. Blandford, On the implications of recent observations of the inner knot in the Crab nebula. Mon. Not. R. Astron. Soc. **454**, 2754–2769 (2015). arXiv:1506.08121. doi:10.1093/mnras/stv2093

F. Yuan, R. Narayan, Hot accretion flows around black holes. Annu. Rev. Astron. Astrophys. **52**, 529–588 (2014). arXiv:1401.0586. doi:10.1146/annurev-astro-082812-141003

Y. Yuan, K. Nalewajko, J. Zrake, W.E. East, R.D. Blandford, Kinetic study of radiation-reaction-limited particle acceleration during the relaxation of unstable force-free equilibria. Astrophys. J. **828**, 92 (2016). arXiv:1604.03179. doi:10.3847/0004-637X/828/2/92

S. Zenitani, M. Hoshino, The generation of nonthermal particles in the relativistic magnetic reconnection of pair plasmas. Astrophys. J. Lett. **562**, L63–L66 (2001). arXiv:1402.7139. doi:10.1086/337972

V. Zhdankin, G.R. Werner, D.A. Uzdensky, M.C. Begelman, Kinetic turbulence in relativistic plasma: from thermal bath to nonthermal continuum. Phys. Rev. Lett. **118**(5), 055103 (2017). arXiv:1609.04851. doi:10.1103/PhysRevLett.118.055103

J. Zrake, Crab flares due to turbulent dissipation of the pulsar striped wind. Astrophys. J. **823**, 39 (2016). arXiv:1512.05426. doi:10.3847/0004-637X/823/1/39

Space Sci Rev (2017) 207:319–360
DOI 10.1007/s11214-017-0364-6

Towards Understanding the Physics of Collisionless Relativistic Shocks

Relativistic Collisionless Shocks

Guy Pelletier[1] · **Andrei Bykov**[2,3,4] · **Don Ellison**[5] · **Martin Lemoine**[6]

Received: 28 November 2016 / Accepted: 4 April 2017 / Published online: 19 April 2017

Abstract Relativistic astrophysical collisionless shocks represent outstanding dissipation agents of the huge power of relativistic outflows produced by accreting black holes, core collapsed supernovae and other objects into multi-messenger radiation (cosmic rays, neutrinos, electromagnetic radiation). This article provides a theoretical discussion of the fundamental physical ingredients of these extreme phenomena. In the context of weakly magnetized shocks, in particular, it is shown how the filamentation type instabilities, which develop in the precursor of pair dominated or electron-ion shocks, provide the seeds for the scattering of high energy particles as well as the agent which preheats and slows down the incoming precursor plasma. This analytical discussion is completed with a mesoscopic, non-linear model of particle acceleration in relativistic shocks based on Monte Carlo techniques. This Monte Carlo model uses a semi-phenomenological description of particle scattering which allows it to calculate the back-reaction of accelerated particles on the shock structure on length

✉ M. Lemoine
lemoine@iap.fr

G. Pelletier
guy.pelletier@obs.ujf-grenoble.fr

A. Bykov
ambykov@yahoo.com

D. Ellison
don_ellison@ncsu.edu

1 UJF-Grenoble 1, Institut de Planétologie et d'Astrophysique de Grenoble (IPAG), UMR 5274, CNRS-INSU, 38041 Grenoble, France

2 Ioffe Institute for Physics and Technology, 194021 St-Petersburg, Russia

3 International Space Science Institute, 3012 Bern, Switzerland

4 St-Petersburg State Polytechnical University, 195251 St-Petersburg, Russia

5 Physics Department, North Carolina State University, Box 8202, Raleigh, NC 27694, USA

6 Institut d'Astrophysique de Paris, UMR 7095, CNRS-UPMC, 98 bis boulevard Arago, 75014 Paris, France

and momentum scales which are currently beyond the range of microscopic particle-in-cell (PIC) simulations.

Keywords Shock waves · Particle acceleration

1 Introduction

Relativistic collisionless shocks are extreme phenomena of plasma physics which form at the interface of the relativistic outflows of powerful astrophysical sources (e.g. gamma-ray bursts, pulsar wind nebulae and active galactic nuclei) and their environment. These shock fronts are not only outstanding dissipation agents, which convert into relativistic heat and disorder a well ordered relativistic flow, they also appear to produce quasi-power law spectra of accelerated particles up to very high energies through a relativistic variant of the Fermi process. These particles either escape the acceleration site and thus become cosmic rays, possibly very high energy cosmic rays, and/or they interact with ambient backgrounds to produce high energy photons and possibly neutrinos over a wide range of energies. In this regard, the physics of relativistic shocks is a key element in the arena of astroparticle physics and high energy astrophysics.

How such shocks form, how particle acceleration takes place, how magnetic turbulence is excited and to what level, remain fundamental open questions. Reviews on the physics, the numerical simulation and the phenomenology of relativistic collisionless shocks exist in the literature, which describe the progress achieved so far, see in particular Bykov et al. (2012b), Sironi et al. (2015), Marcowith et al. (2016). The present review takes a different stance and tries to describe analytically the physics of these phenomena to some level of detail. Due to the limited space, not all details could be presented nor could all regimes be investigated. The emphasis has been deliberately placed on unmagnetized relativistic shocks, which are representative, for example, of external shocks in gamma-ray bursts; other types of shocks are nonetheless briefly addressed to provide a more general picture.

The paper is laid out as follows; in Sect. 2, we provide essential notions on relativistic collisionless shock waves, discussing in particular the shock jump conditions and the particle kinematics; Sect. 3 describes how a relativistic collisionless shock forms, depending on the ambient magnetization and shock velocity; Sect. 4 discusses the physics of the precursor, in which the superthermal particle population mixes with the background plasma, and where the latter is pre-heated and slowed down; Sect. 5 presents some phenomenological consequences and observational tests of the theory. Finally, we complete this analytical model with a description of Monte Carlo numerical simulation techniques in Sect. 6, which provides a complementary point of view on the physics of relativistic shocks. While Monte Carlo simulations must use as input some microphysics describing particle kinematics, which can be derived from the analytical theory presented earlier, they also provide a mesoscopic non-linear picture which is useful to extrapolate microphysical results to astrophysical scales of interest. Our main results are summarized in Sect. 8. Some details of a more technical nature are presented in Appendices A.1, A.2 and A.3.

2 The Basics of Relativistic Shocks

Just as any other shock front, a relativistic shock in a plasma corresponds to a transition between two plasma states at a "front" which moves at velocity close to the velocity of light relative to some unshocked medium (the background plasma). Such a shock transition requires that the shock front moves at a velocity $v_{\rm sh}$ which is faster than the velocity at which causal disturbances can be transmitted in the background unshocked plasma. In the hydrodynamical limit, this velocity is $c_{\rm s}$, the sound velocity; if the background plasma is magnetized, however, various modes can propagate in the plasma, at different velocities, hence several types of shock waves can be envisaged; see Kirk and Duffy (1999) for a detailed discussion. This review will be interested in fast shock waves, which move faster than the fast magnetosonic velocity $(c_{\rm s}^2+v_{\rm A}^2)^{1/2}$, $v_{\rm A}$ corresponding to the Alfvén velocity in the background plasma. For simplicity, it will also be assumed that the shock front is planar and that it propagates into the background plasma along its normal.

Upon crossing the shock front, the ordered kinetic energy is converted into thermal energy mostly. A description of the physics of shock fronts is conveniently made in the so-called front frame that separates two homogeneous plasmas, namely the *upstream plasma*, the unshocked background plasma that inflows into the shock front, and the *downstream plasma*, which flows away from the shock at a subsonic velocity as a consequence of intense heating in the shock transition. In a collisional plasma, this transition is usually quite sharp, its size being measured in units of the mean free path of particles. In collisionless, non-relativistic plasmas, with an oblique magnetic field (i.e. meaning that the field lines have a finite angle with respect to the shock normal), the size of the transition is measured in term of the gyration radius of ions. In parallel shocks (i.e. with a magnetic field parallel to the shock normal), the transition is generally much more extended, because strong disturbances are generated and the absence of collisional dissipation allows them to travel over larger distances. In this case, the upstream and downstream plasmas close to the shock front need not be uniform. Nevertheless, on some scale, provided the downstream flow has been isotropized, the so-called Rankine-Hugoniot jump relations, which express the balance of the fluxes of matter, momentum and energy, are fulfilled.

In the next section, we recall these jump conditions and discuss the kinematics of particles around this shock front, which are peculiar to relativistic shock waves.

2.1 Shock Jump Conditions

In the case of a relativistic shock, it is convenient to derive these jump relations in the shock front frame by using the energy-momentum tensor $T_{\mu\nu}$ (e.g. Blandford and McKee 1976; Double et al. 2004). A word of caution regarding notations: except otherwise noted, and except for proper quantities such as density n, energy density e and pressure p or enthalpy h, quantities will be expressed in the shock front rest frame; if expressed in the upstream or downstream rest frames, these quantities will be indexed with $_{\rm u}$ or $_{\rm d}$ respectively. Figure 1 illustrates the various frames used in this text and the corresponding flow velocities.

Assuming a plane front and a normal coordinate x, the relations express that the momentum flux T^x_x and the energy flux T^t_x keep the same values upstream and downstream. The momentum flux conservation reads:

$$(e_{\rm d}+p_{\rm d})u_{\rm d}^2+p_{\rm d}=n_{\rm u}mc^2u_{\rm u}^2, \tag{1}$$

and the energy flux conservation reads:

$$(e_{\rm d}+p_{\rm d})\Gamma_{\rm d}u_{\rm d}=n_{\rm u}mc^2\Gamma_{\rm u}u_{\rm u}, \tag{2}$$

$u_{\rm u}$ (resp. $u_{\rm d}$) denote the x-component of the upstream (resp. downstream) plasma 4-velocity in the shock frame, with corresponding Lorentz factor $\Gamma_{\rm u}$ (resp. $\Gamma_{\rm d}$). To make contact with

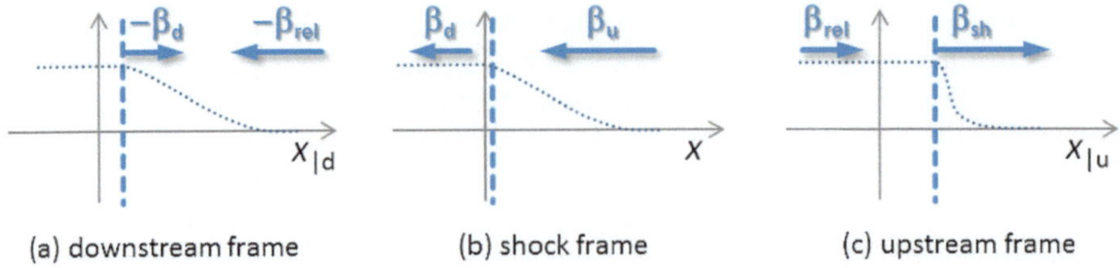

Fig. 1 Sketch of the shock transition in various frames, along with the flow velocities. The *dotted line* represents a sketch of the density distribution of superthermal particles; the extension of this distribution in the upstream delimits the shock precursor. In the shock rest frame (**b**), the downstream moves at velocity β_d relative to the shock front, while the upstream is incoming at velocity β_u. In the upstream rest frame (**c**), the downstream plasma moves at velocity β_{rel}, while the shock front moves at $\beta_{sh} = -\beta_u$. Note that the extension of the precursor is reduced by the Lorentz contraction in this rest frame, relative to the shock rest frame. Finally, in the downstream rest frame (**a**), the shock front moves away at velocity $-\beta_d$, while the usptream is moving at $-\beta_{rel}$

our notations above, the shock velocity relative to the unshocked background plasma is of course $\beta_{sh} = -\beta_u$, with associated Lorentz factor $\gamma_{sh} = (1 - \beta_{sh}^2)^{-1/2}$.

The above equations assume a cold upstream plasma. We introduce an equation of state $p_d = w_d e_d$ for the downstream plasma; since the downstream plasma is relativistically hot, $w_d = 1/3$. Note that this does not require that the distribution function reflect thermal equilibrium, since this ratio holds for any isotropic distribution for which most of the particles are ultra-relativistic; in particular the distribution function can be composed of a thermal part with a high energy isotropized tail. One can generalize the above equation of state to that of a plasma containing a magnetic component set in the transverse directions to the shock normal (i.e. a magnetic field lying in the shock front plane), carrying a fraction σ_d of the energy density of the downstream plasma. Then,

$$w_d = \frac{1}{3}\frac{1+3\sigma_d}{1+\sigma_d}. \tag{3}$$

Now, the downstream velocity β_d is obtained by solving the following algebraic equation derived from the two conservation equations:

$$\beta_d^2 - (1+w_d)\beta_u\beta_d + w_d = 0. \tag{4}$$

For $\beta_u^2 \simeq 1$, corresponding to the ultra-relativistic limit, the two roots are 1 and w_d. The downstream pressure of a strong relativistic shock is then

$$p_d = (1-w_d)\gamma_{sh}^2 n_u mc^2. \tag{5}$$

From the mass flux conservation law, one derives easily that the *apparent* density in the shock front frame is amplified by a factor w_d^{-1}, as measured in the front rest-frame; however the Lorentz transform from the background plasma rest-frame to the shock front frame amplifies the *proper* density by a factor γ_{sh}, while the Lorentz transform from the shock frame to the downstream plasma rest-frame reduces the proper density by a factor $\gamma_d \simeq 3/(2\sqrt{2})$ in the hydrodynamic limit $\sigma_d \ll 1$. Thus the proper density eventually suffers an amplification by a compression ratio corresponding to the jump of the 4-velocity vector $r \equiv u_u/u_d = 2\sqrt{2}\gamma_{sh}$ (the latter equality applying to the hydrodynamical limit).

Combining this result with Eq. (5), one derives a mean energy per particle downstream of the shock of the order of $e_d/n_d \simeq \gamma_{rel}$, with $\gamma_{rel} = \gamma_u\gamma_d(1-\beta_u\beta_d) \simeq \gamma_{sh}/\sqrt{2}$ the relative Lorentz factor between the up- and downstream frames. Superthermal particles share this mean energy per particle of order γ_{sh}.

Let us consider now a mean field having some obliquity $\theta_{B|\mathrm{u}}$ (measured in the background frame). In the ideal magnetohydrodynamics (MHD) approximation, the normal (x-) component of the electric field that compensates the electromotive force $-\boldsymbol{\beta}_{\mathrm{u}} \times \boldsymbol{B}$ vanishes and thus the normal (x-) component B_x of the magnetic field remains unchanged in the shock transition. In contrast, the transverse component of the electric field is continuous at the shock front, therefore $\beta_{\mathrm{u}}\boldsymbol{B}_{\mathrm{u},\perp} = \beta_{\mathrm{d}}\boldsymbol{B}_{\mathrm{d},\perp}$, the $\perp$ symbol denoting the component transverse to the shock normal; $B_\perp$ is thus amplified by a factor w_{d}^{-1} when crossing the shock front, in the shock front frame. However, taking into account the Lorenz factors involved when transforming from the flow co-moving frames to the front frame and vice-versa, the global amplification of the transverse magnetic field component is by the compression ratio $r \equiv u_{\mathrm{u}}/u_{\mathrm{d}} = 2\sqrt{2}\gamma_{\mathrm{sh}}$ as for the density.

2.2 Particle Kinematics

Because of this tremendous amplification of the transverse component of the magnetic field in ultra-relativistic shocks, relativistic shocks generically have a "superluminal" configuration, which means that one cannot find a Lorentz transform with a sub-luminal velocity along the shock surface that would cancel the electric field and have the flow moving along the magnetic field (the so-called de Hoffmann Teller frame, see de Hoffmann and Teller 1950 as well as Begelman and Kirk 1990). Consider indeed a frame moving at the velocity $\boldsymbol{\beta}_{\mathrm{dHT}}$ along the direction of the projection $\boldsymbol{B}_{\mathrm{u},\perp}$ on the shock surface; then in order to cancel the electric field, one needs to boost to a frame moving with a velocity $\boldsymbol{\beta}_{\mathrm{dHT}}$ such that $\boldsymbol{E} + \boldsymbol{\beta}_{\mathrm{dHT}} \times \boldsymbol{B}_{\mathrm{u},\parallel} = 0$; inserting $\boldsymbol{E} = -\boldsymbol{\beta}_{\mathrm{u}} \times \boldsymbol{B}_{\mathrm{u},\perp}$, one obtains

$$\beta_{\mathrm{dHT}} = -\beta_{\mathrm{u}}\frac{B_{\mathrm{u},\perp}}{B_{\mathrm{u},\parallel}} = \beta_{\mathrm{sh}}\gamma_{\mathrm{sh}}\tan\theta_{B,|\mathrm{u}}. \tag{6}$$

Since β_{dHT} must be smaller than one, the condition for having a Lorenz transform which wipes out the electric field in the front is very restrictive, because it requires

$$\gamma_{\mathrm{sh}}\sin\theta_{B,|\mathrm{u}} < 1. \tag{7}$$

When $\gamma_{\mathrm{sh}}\sin\theta_{B,|\mathrm{u}} > 1$, the intersection point of a field line on the shock plane moves at a velocity larger than the velocity of light, hence the word "super-luminal" for qualifying such a shock. Quite generically, ultra-relativistic shocks are super-luminal and the matter flow always crosses the field lines and experiences an electric field.

As for superthermal particles, if they cross the shock front from downstream to upstream, it means that the x-component of their 3-velocity β_x is positive in the shock front frame, or equivalently larger than β_{sh} in the background plasma rest frame. Thus, in this frame, the pitch angle α of the superthermal particle motion with respect to the shock normal is such that $\sin\alpha < 1/\gamma_{\mathrm{sh}}$; the particle distribution of superthermal particles is extremely anisotropic in the rest frame of the background plasma, although it can be considered as nearly isotropic in the shock rest frame.

The trajectory of superthermal particles upstream of the shock is a crucial point in the theory of relativistic shocks, since its extension determines the size of the precursor, which corresponds to that region in which the superthermal and background populations intermix. To a good approximation, one can consider that a superthermal particle, while traveling upstream with momentum $p_{|\mathrm{u}}$, is caught up by the shock front once its momentum is deflected by an angle $\gtrsim 1/\gamma_{\mathrm{sh}}$, because the x-component of the particle 3-velocity then falls below β_{sh} and the shock is trailing right behind the particle. In the presence of an oblique background magnetic field, this takes place on a timescale $t_{u|u} \simeq \omega_{L,0}^{-1}/\gamma_{sh}$, with $\omega_{L,0} \equiv eB_{|u}/p_{|u}$ the gyrofrequency in the background field. Hence the extension of the precursor (Achterberg et al. 2001; Milosavljević and Nakar 2006; Pelletier et al. 2009):

$$\ell_{prec|u} = (1 - \beta_{sh}) t_{u|u} \simeq \frac{\omega_{L,0}^{-1}}{2\gamma_{sh}^3}. \tag{8}$$

As $B_{|u} \to 0$, the size of the precursor becomes controlled by the scattering of the superthermal particles in the self-generated turbulence; this case will be discussed further on.

3 What Is a Collisionless Relativistic Shock?

The concept of collisionless shock relies on the development of collective effects, including the Landau effect, that control the dissipation process, i.e. the transformation of the flux of kinetic energy through the generation of several kinds of entropy. The main entropy generation is the formation of a thermal distribution of ultra-relativistic temperature, with (proper) temperature downstream $T_d \sim \gamma_{sh} mc^2$ (in units in which $k_B = 1$) or merely the isotropization of the distribution function, as mentioned previously in the case of ultra-relativistic shocks.[1] This isotropic distribution function can contain a superthermal component in the form of a quasi-power-law tail; this is what is expected for the generation of very high energy particles in astrophysics to account for gamma-ray radiation, for the production of very high energy cosmic rays, especially ultra-high energy cosmic rays, or even high energy neutrinos. The generation of a superthermal tail does not require more energy, of course; for a given global energy budget, a fraction of the incoming energy is transferred into the pressure of a cosmic ray component corresponding to a significant fraction of the thermal pressure:

$$P_{cr} = \xi_{cr} \gamma_{sh}^2 n_u mc^2. \tag{9}$$

Note that the above equation neglects factors of order unity (see in particular Eq. (5)). Typically a conversion factor ξ_{cr} of order 10 percent is found in numerical simulations and is expected to account for the observations.

Such particle acceleration cannot occur without the excitation of some electromagnetic turbulence that also takes a significant fraction of the incoming energy:

$$W_{em} = \xi_B \gamma_{sh}^2 n_u mc^2. \tag{10}$$

How this turbulent component is generated will be discussed in Sect. 4. Of course a pre-existing mean magnetic field or background turbulence may be present in addition to self-generated turbulence.

In the next section we describe two limiting cases: that of a magnetized shock with a strong mean ambient field, and a weakly magnetized shock where the self-generated turbulence dominates the mean field. We then present a general phase diagram for the physics of these relativistic collisionless shock waves and discuss the consequences for Fermi acceleration.

[1] Of course shocks do not have to be ultra-relativistic and can have any supersonic speed $\lesssim c$. This is a critical aspect of gamma-ray burst afterglows where the jet shock slows from ultra-relativistic to trans-relativistic to non-relativistic speeds. Trans-relativistic shocks are difficult to describe analytically but can be treated directly with PIC or Monte Carlo techniques.

3.1 With a Mean Field

Early work on the theory and simulation of a relativistic shock with a transverse mean field in pair plasmas, with an obvious application to the termination shock of Pulsar Wind Nebulae, was worked out by J. Arons and collaborators in several steps (Alsop and Arons 1988; Hoshino et al. 1992; Gallant and Arons 1994) (see also Kennel and Coroniti 1984). In this 1-D simulation, the incoming pair plasma is partially reflected back at a magnetic barrier, which is formed as the transverse magnetic field is compressed in the slow-down of the flow. A fraction of the incoming plasma is transmitted whereas a significant fraction is reflected making two opposite loops in phase space that end up in the downstream flow, producing a synchrotron resonant electromagnetic field that propagates both forwards and backwards in the form of an extraordinary mode. The backward mode, because of its compressive nature, heats the downstream flow.

When the flow is composed of protons and electrons, the intense coherent forward wave produces, through a ponderomotive force, an electrostatic wake of large amplitude, that brings the incoming electrons to a relativistic temperature up to equipartition with the protons. This eventually quenches the formation of the electrostatic wake by the ponderomotive force, which is proportional to q^2/m, once the relativistic mass of the electrons reaches that of the proton. Once electrons have been heated to near equipartition with the protons, the formation of the shock becomes similar to the case of a pair plasma.

This scenario depends on a crucial parameter called "magnetization", which is defined as the ratio of the flux of magnetic energy across the shock to the flux of incoming matter energy:

$$\sigma \equiv \frac{B_\perp^2/4\pi}{\gamma_{\rm sh}^2 n_{\rm u} m c^2}. \tag{11}$$

Note that $B_\perp$ is defined in the shock frame, following our notation convention. The magnetization parameter is sometimes defined relative to the incoming kinetic energy $\gamma_{\rm sh}(\gamma_{\rm sh}-1)n_{\rm u}mc^2$, but in the ultra-relativistic limit $\gamma_{\rm sh} \gg 1$, these two definitions are equivalent.

Apart from the angular dependence, σ depends essentially on the magnetization of the ambient plasma, since $B_\perp = \gamma_{\rm sh} B_{\perp|\rm u}$. In the typical interstellar medium, $\sigma \sim 10^{-9}$, thus a ultra-relativistic shock in the interstellar medium corresponds to a very low magnetization.

3.2 Weakly Magnetized Shocks

In this subsection, we indicate the requirement for producing micro-turbulence in a weakly magnetized plasma. At very low magnetization, a self-consistent process develops such that a fraction of incoming particles is reflected back into the upstream region. These reflected particles trigger some form of streaming instability which generates precursor electromagnetic turbulence which eventually produces enough scattering to reflect particles back to the subshock.

Recall indeed from the previous section that particles that outrun the shock form a beam with opening angle $\simeq 1/\gamma_{\rm sh}$ in the rest frame of the background plasma. Such a configuration is prone to beam-plasma instabilities.

The shock thus behaves as a self-sustaining self-generating structure. Moreover this process can allow for the development of a superthermal tail, which expands the size of the precursor (which increases as some power of gyroradius $r_{\rm g}$, hence of particle energy). In this description, superthermal particles originate from shock-reflected particles; since both populations share a common mean energy per particle of $\gamma_{\rm sh}$, we will not distinguish one from the other in the following and we will use "returning particles" or "superthermal particles" interchangeably.

Consider a streaming instability, triggered by the beam of returning particles, with a growth rate $g_{\rm inst|u}$ defined in the rest frame of the background plasma. In the present situation, an instability is relevant only if its growth time is shorter than the time spent by the background plasma in the shock precursor. Assume also that, although weak, the background magnetic field nevertheless sets the size of the precursor: $\ell_{\rm prec|u} \sim \omega_{\rm L,0}^{-1}/\gamma_{\rm sh}^3$. Therefore a streaming instability emerges as a good candidate for generating turbulence in the precursor if its growth rate is such that

$$g_{\rm inst|u} > \gamma_{\rm sh}\omega_{\rm c}, \tag{12}$$

where $\omega_{\rm c} = eB_{|\rm u}/mc$ is the cyclotron frequency of the background plasma; the above uses $\omega_{\rm L,0} = \omega_{\rm c}/\gamma_{\rm sh}^2$ since $p_{|\rm u} \simeq \gamma_{\rm sh}^2 mc$ for a superthermal particle in the background plasma rest frame. This is a severe constraint at a relativistic shock; expressed for a proton plasma as a function of magnetization, this constraint becomes:

$$\sigma < \gamma_{\rm sh}^{-2}\frac{g_{\rm inst|u}^2}{\omega_{\rm pi}^2} \tag{13}$$

with $\omega_{\rm pi} \equiv (4\pi n_{\rm u}e^2/m_p)^{1/2}$. The process not only requires a very low magnetization, it also requires instabilities that work on micro-physical scales since the criterium (12) excludes slower MHD dynamics. This latter criterium is the essential reason for which the phenomenology of ultra-relativistic shocks displaying intense magnetic fields and superthermal tails, like the termination shocks of gamma-ray bursts and pulsar wind nebulae, unavoidably requires a micro-physical description.

3.3 Classification in Term of the Magnetization. Phase Diagram

In this subsection, we present the micro-instabilities that are possible candidates for the generation of the magnetic turbulence. The maximum growth rates of standard beam-plasma instabilities are (within a factor of order unity) $(\omega_{\rm pb}^2\omega_{\rm ep}/\gamma_{\rm b|u}^2)^{1/3}$ for the electrostatic two stream instability ($\gamma_{\rm b|u}$ represents the typical Lorentz factor of the beam of superthermal particles in the upstream plasma frame, see below), $(\omega_{\rm pb}^2\omega_{\rm ep})^{1/3}$ for the oblique two-stream instability, and $\omega_{\rm pb}$ for the Weibel instability, where $\omega_{\rm pb}$ is the (relativistic) plasma frequency of the beam of returning particles:

$$\omega_{\rm pb} = \left(\frac{4\pi n_{\rm cr|u}e^2}{\gamma_{\rm b}m}\right)^{1/2} \tag{14}$$

and $\omega_{\rm ep}$ represents the electron plasma frequency of the background plasma. Using Eq. (9) with $n_{\rm cr|u} = \gamma_{\rm sh}P_{\rm cr}/(\gamma_{\rm sh}mc^2)$, and $\gamma_{\rm b|u} \simeq \gamma_{\rm sh}^2$, one finds $\omega_{\rm pb} \simeq \sqrt{\xi_{\rm cr}}\omega_{\rm p}$, where the conversion parameter $\xi_{\rm cr}$ is defined by relation (9) and $\omega_{\rm p}$ is the electron plasma frequency $\omega_{\rm pe}$ in a pair plasma, or $\omega_{\rm pi}$ in a proton electron plasma.

Note that a growth time scale of the order of $\omega_{\rm pi}^{-1}$ can be compatible with condition (12) if the magnetization is low enough, but not the proton cyclotron time $\omega_{\rm ci}^{-1}$. A detailed analysis reveals that the only magnetized waves which could be excited are the electron whistler waves (Lemoine and Pelletier 2010).

The two stream instability and the oblique two stream instability grow through a Čerenkov resonance of the electrostatic (Langmuir) wave with the background electrons.

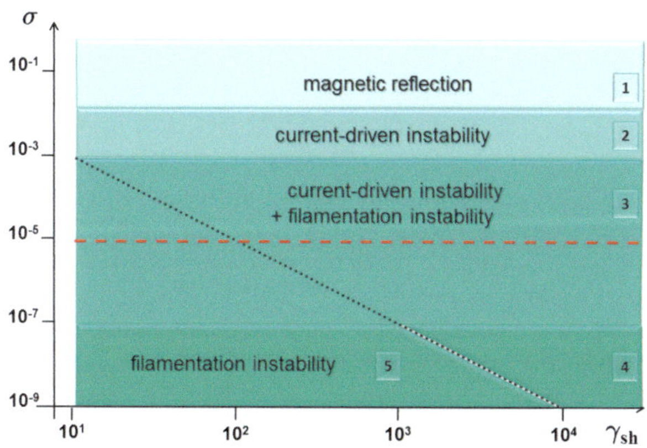

Fig. 2 Phase diagram of relativistic shocks ($\gamma_{\rm sh} \geq 10$) in the plane $(\gamma_{\rm sh}, \sigma)$. The *dotted diagonal line* delimits from above the region in which the filamentation instability can grow in the background plasma on a precursor crossing timescale, in the absence of plasma slow-down. In the presence of a background magnetic field, the background plasma is slowed down as it enters the precursor (see Sect. 4.1), hence the filamentation instability can grow in all of region 3. The *dashed horizontal line* delimits from above the region in which the scattering in the micro-turbulence is sufficiently strong to allow Fermi acceleration, see Sect. 3.4. In region 1, the shock forms through the reflection on the compressed background magnetic field; at lower magnetizations, the shock forms through the build-up of a microturbulent magnetic barrier. The dominant instabilities are: filamentation in region 5, filamentation and current-driven instability (see Sect. 4.1) in region 3, and the current driven instability in region 2

When the background electrons are heated to relativistic temperature, the electrostatic waves become superluminal and thus these instabilities are quenched by resonance suppression (Bret 2009; Lemoine and Pelletier 2011; Shaisultanov et al. 2012). Nevertheless, the oblique two stream instability certainly plays an important role at the beginning of the precursor by heating the background electrons.

A phase diagram involving the two parameters σ and $\gamma_{\rm sh}$ can be drawn to delimit the region in which micro-turbulence can be self-generated. For the most often discussed instability, namely the filamentation/Weibel instability, the criterion is simple: $\gamma_{\rm sh}^2\sigma < \xi_{\rm cr}$. This limit is a first order approximation to the development of Weibel-like turbulence in the precursor of relativistic shocks; in the following, we show that a refined description of the physics of the precursor introduces a "buffer effect" which modifies somewhat this limit at moderate magnetization.

3.4 Fermi Acceleration

Similarly to what has been proposed for non-relativistic shocks, relativistic shocks can develop a superthermal tail thanks to the Fermi process. Indeed the excitation of magnetic turbulence that extends in the precursor and is also transmitted to the downstream flow allows for fast scattering of superthermal particles that can cross the shock front back and forth and thus gain energy. However at relativistic shocks, the Fermi process has some specific features.

We assume that the magnetic disturbances that scatter particles are almost frozen in the plasma flow both upstream and downstream (this will turn out to be true with Weibel turbulence). The particle energy gain through the Fermi process depends on the relative velocity $\beta_{\rm rel}$ between the upstream and downstream flows. Let a particle, having a pitch angle cosine

μ_1 before scattering, cross the shock front, be scattered and come back with a pitch angle cosine μ_2. Its energy gain is

$$G = \frac{1-\beta_{\rm rel}\mu_1}{1-\beta_{\rm rel}\mu_2}. \tag{15}$$

As defined above, μ_1 and μ_2 are expressed in the upstream rest frame and $\beta_{\rm rel}$ represents the velocity of the downstream plasma relative to the upstream frame; changing the sign convention for $\beta_{\rm rel}$, one would obtain the same formula with μ_1 and μ_2 now both expressed in the downstream rest frame. In Eq. (15), and for the following discussion, it is implicitly assumed that the shock is unmodified by Fermi acceleration with a discontinuous jump between the upstream and downstream regions.

Choosing the downstream reference frame, $\beta_{\rm rel} \simeq 1 - 1/\gamma_{\rm sh}^2$ and the pitch angle cosine of a particle crossing the front towards downstream has $-1 < \mu_1 < \beta_{\rm sh}$; a particle coming back from downstream to upstream has $\beta_{\rm sh} < \mu_2 < 1$.

The average energy gain can be calculated easily in the case of non-relativistic shocks only, because the distribution functions are almost isotropic and μ_1 and μ_2 independent; one then finds:

$$\langle G\rangle = \left(1+\frac{2}{3}|\beta_{\rm rel}|\right)^2. \tag{16}$$

Moreover, because of the almost isotropic distributions, the probability of escape through advection in the downstream plasma can also be easily estimated: $P_{\rm esc} = 4|\beta_{\rm d}|$. The index of the power law distribution $\mathrm{d}N/\mathrm{d}p \propto p^{-s}$ is then obtained as Bell (1978):

$$s = 1 - \frac{\ln(1-P_{\rm esc})}{\ln\langle G\rangle} \simeq 1 + 3\left|\frac{\beta_{\rm d}}{\beta_{\rm rel}}\right|, \tag{17}$$

i.e. $s = 2$ for the strong adiabatic shock where $|\beta_{\rm rel}| = 3|\beta_{\rm d}|$.

For relativistic shocks, the strong anisotropy of the distribution functions makes these calculations more complicated, (e.g. Ellison et al. 1990a; Gallant and Achterberg 1999; Kirk et al. 2000; Achterberg et al. 2001) see however Vietri (2003), Lemoine and Pelletier (2003), or Keshet and Waxman (2005) for an alternative point of view. If one were to consider a random incident pitch angle cosine μ_1, then $\mu_2 \simeq 1 - \mathcal{O}(1/\gamma_{\rm sh}^2)$ would imply $\langle G\rangle \sim \gamma_{\rm sh}^2$. However, this maximum limit can only be achieved for particles crossing from downstream to upstream for the first time. Indeed, particles participating in a further Fermi cycle are caught up by the front upstream with $\mu_1 \simeq 1 - \mathcal{O}(1/\gamma_{\rm sh}^2)$ so that the gain falls to a few only; precisely it is always close to 2 after the first Fermi cycle (e.g. Achterberg et al. 2001; Lemoine and Pelletier 2003).

The index can be defined through the equation

$$\langle 1 - P_{\rm esc}(\mu_2)\rangle\langle G(\mu_1,\mu_2)^{s-1}\rangle = 1, \tag{18}$$

and because G remains close to 2 at each cycle, the index turns out to be close to

$$s \simeq 1 - \frac{\ln(1-P_{\rm esc})}{\ln\langle G\rangle}. \tag{19}$$

The numerical estimate is $1 - P_{\rm esc} \simeq 0.4$ leading to $s \simeq 2.3$. An illustration of the spectrum of accelerated particles for an unmagnetized, unmodified shock with $\gamma_{\rm sh} = 100$ is shown in Fig. 3.

As we describe in Sect. 6, basic requirements of energy and momentum conservation mean the test-particle power law shown in Fig. 3 will be modified to some extent if shocks accelerate cosmic rays efficiently. If the scattering mean free path is an increasing function

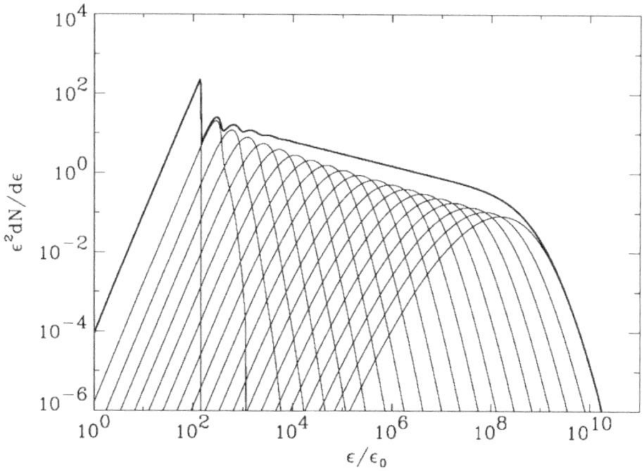

Fig. 3 Spectrum of superthermal particles as collected far downstream (*thick line*) for $\gamma_{sh} = 100$, obtained from a test-particle Monte Carlo simulation, assuming isotropic scattering both upstream and downstream of the shock, as in an unmagnetized shock (Lemoine and Pelletier 2003). The *thin lines* indicate the populations of particles that have experienced 0, 1, ... Fermi cycles through the shock: as the number of cycles increases, so does the mean energy of the population, but with a smaller number of particles due to the finite P_{esc}

of particle momentum as expected, the accelerated spectrum will obtain a concave shape being harder at high energies before a turnover produced by finite shock effects.

In addition, if the shock is sweeping up non-relativistic matter, such as the ISM, the injection of highly anisotropic thermal particles must be described consistently with CR production. Currently, the only techniques that can handle thermal particle injection and efficient non-linear CR production in relativistic shocks are PIC simulations and Monte Carlo techniques, described in greater detail in Sect. 6.

4 Physics of the Precursor

The main physical processes that develop in the precursor of a relativistic collisionless shock undergoing Fermi acceleration are the following:

(i) The incoming precursor plasma slows before the sharp transition at the viscous subshock due to transverse currents. This occurs with ordered or disordered magnetic fields for both pair and ionic plasmas and produces an important buffer effect in the precursor where a highly relativistic shock far upstream turns into a mildly relativistic one deep in the precursor, close to the shock front. If Fermi acceleration is efficient the backpressure of accelerated particles produces an additional shock modification forced by conservation of energy and momentum.

(ii) We argue that the Weibel/filamentation instability is the dominant generator of precursor magnetic turbulence; this instability does not quench when the electrons are heated to relativistic temperatures, but it saturates when the beam dispersion becomes too large.

(iii) Then we develop the important concept of the Weibel frame for both pair and ionic plasmas. This is the frame where the turbulence is purely magnetic and where particles undergo a simple pitch-angle scattering process.

(iv) We discuss the crucial heating process of background electrons and relate it to the drag force experience by the background particles.

(v) We discuss some aspects of the non-linear theory for the formation of the filaments.
(vi) We then return to the physics of particle scattering in Weibel turbulence and derive it from the the theory of Hamiltonian chaos (the details are given in Appendix A.1).

4.1 Dynamics in the Precursor with a Mean Field; Buffer Effect

The slow-down of the incoming plasma with respect to the shock front will be analyzed in a dedicated Section 4.5 for the case of an unmagnetized plasma. In this section we examine the simpler case of the slow-down due to a transverse mean field.

4.1.1 The Case of a Pair Plasma

Consider first the beam of returning particles. Because of the transverse mean magnetic field (aligned say along z), electrons and positrons are deflected in opposite directions (along $\boldsymbol{y}$) during their orbit around B and thus a current transverse to the shock normal and to the mean field is generated, with density $j_b = n_{cr}ec$ (measured here in the front frame). One can show that this current is compensated to an excellent accuracy and on a short timescale by a current carried by the incoming background plasma, once it enters the precursor. As the density of this background current density can be written $j_{bg} = \gamma_{sh} n_u e \beta_y c$ in the front frame, with $n_{cr} = \xi_{cr}\gamma_{sh}n_u$, one infers a transverse velocity $\beta_y \simeq \xi_{cr}$ (Lemoine et al. 2014a, 2014b).

The main slow-down effect on the incoming plasma is due to the deviation of background electrons and positrons due to the development of the compensating current. Indeed, one can show that the total Lorentz factor of the background plasma is conserved, while the development of a transverse velocity $\pm\xi_{cr}$ for electrons/positrons implies that the center-of-mass of the background plasma moves solely along $\boldsymbol{x}$ with a Lorentz factor:

$$\gamma_{cm} = \frac{\gamma_{sh}}{(1+\xi_{cr}^2\gamma_{sh}^2)^{1/2}}. \tag{20}$$

Thus any shock front of Lorentz factor larger than $1/\xi_{cr}$ produces a deceleration of the incoming plasma such that the center of mass moves with a Lorentz factor $1/\xi_{cr}$. This is a crucial buffer effect that transforms any ultra-relativistic shock into a mildly one in its precursor from the point of view of the growth of instabilities.

As discussed in Lemoine et al. (2014a, 2014b), this modification of the motion of the center-of-mass indeed modifies the criterion for the growth of instabilities in the precursor of a shock, due to time dilation effects. One can show in particular that the Weibel instability can now grow provided $\sigma < \xi_{cr}^3$, independently of the shock Lorentz factor, as illustrated in Fig. 2. Moreover, the above configuration of a compensated transverse current turns out to be unstable on a short timescale, leading to the growth of a current-driven filamentation instability with growth rate $\sim\omega_{pe}$, i.e. faster than the Weibel/filamentation rate. This latter instability can thus grow provided $\sigma < \xi_{cr}^2$ at any value of the shock Lorentz factor, thereby bridging the gap between weakly and moderately/strongly magnetized shock waves.

4.1.2 The Case of an Electron-Ion Shock

We consider first an electron-ion plasma where electrons have been rapidly heated to a relativistic temperature of order $m_p c^2$ (this heating will be explored in a next section). Their inertia is thus comparable to that of the protons, so that a significant fraction of the electrons is reflected back on the shock front along with protons, leading to a charge neutral beam of returning particles. In such a case, the situation is similar to the case of a pair plasma discussed above. If the electrons are relativistically hot, but below equipartition with the protons, i.e. $T_e = \xi_{\rm th} m_p c^2$ with $\xi_{\rm th} > m_e/m_p$, one can recover similar results, as will be presented elsewhere.

If the population of incoming electrons remains cold because the precursor is too short, then electrons are not reflected back and the beam of returning current carries a net electric charge $\rho_{\rm cr} = \xi_{\rm cr}\gamma_{\rm sh} n_{\rm u} e$ (front frame). As the background plasma compensates this electric charge, a force density develops along the y-direction, transverse to the shock normal and to the mean field B; in the relativistic MHD approximation, this force density reads:

$$f_y = -\rho_{\rm cr} E_y \sim \beta_{\rm sh}\xi_{\rm cr} n_{\rm u}\gamma_{\rm sh} e B. \tag{21}$$

The above equation uses a zero current density along x, which is guaranteed by the assumption of one-dimensional stationary dynamics in the shock front frame. The work of this force over the length of a Larmor gyration is $\xi_{\rm cr}\beta_{\rm sh}\gamma_{\rm sh}^2 n_{\rm u} m_p c^2$. Thus protons are deviated in the transverse direction with a velocity $\xi_{\rm cr}$ and the bulk motion is diminished as previously.

4.2 Weibel Turbulence

Because the mean magnetic field must be very weak and the growth of the instability faster than the cyclotron frequency of electrons in a pair plasma, or of protons in a protonic plasma, the expected turbulence must be excited by a fast micro-instability on short length scales. The typical scale must be at most the inertial scale of protons: $\delta_i \equiv c/\omega_{\rm pi}$; MHD develops at larger scales than this inertial scale.

The Weibel instability, which oversteps the other streaming instabilities and survives to electron heating, is mostly an electromagnetic instability with a very low phase velocity, and thus produces slowly propagating magnetic turbulence, quite suitable for particle scattering and thus for a Fermi process. The mechanism of the instability is simple: the beam of returning particles is supposed to be charge neutralized, although when a transverse magnetic wave has been produced, its quiver force separates electric charges; this generates an intense electric current, which in turn amplifies the primary magnetic wave; then it blows up (see e.g. Medvedev and Loeb 1999; Wiersma and Achterberg 2004; Lyubarsky and Eichler 2006; Milosavljević and Nakar 2006; Achterberg and Wiersma 2007; Achterberg et al. 2007). Its growth rate is fast, of the order of $\omega_{\rm pb}$, with respect to the background frame, and the instability is purely growing in this frame with a weak electrostatic component. A sketch of the mechanism of this instability in the precursor of a relativistic shock is presented in Fig. 4.

We already mentioned that relativistically hot electrons do not quench this Weibel instability (to be discussed shortly); however a large angular dispersion of the beam can stabilize the Weibel instability, as discussed e.g. in Rabinak et al. (2011), Lemoine and Pelletier (2011). A simple description of that saturation effect by the beam dispersion consists in treating the transverse dispersion of the beam as a temperature effect; this introduces an effective sound velocity $\beta_* c \sim c/\gamma_{\rm sh}$, see Eq. (69), which modifies the hydrodynamic response of the beam fluid to the quiver force that separates opposite charges. The instability then disappears (in this fluid description) when $\Im\omega \lesssim \beta_*\omega_{\rm p}$, i.e. when $\gamma_{\rm sh}$ becomes smaller than $1/\xi_{\rm cr}^{1/2}$. A more detailed analysis with some consequences of this saturation effect will be addressed in a forthcoming paper.

4.3 The Weibel Frame in a Pair Plasma

For the estimation of the efficiency of scattering and thus of the Fermi process, and also for estimating the radiation efficiency of an ultra-relativistic shock, it is very important to know what is the proper frame of Weibel turbulence.

Fig. 4 Development of the Weibel/filamentation instability in the precursor of a relativistic shock: as the background plasma enters the precursor defined by the region in which superthermal particles exist, the interpenetration of the two populations develops a filamentation instability: a transverse magnetic fluctuation of the background plasma leads to charge separation in this population, forming current filaments at the nodes of the fluctuation, which then feeds back positively on the fluctuation. See text for details

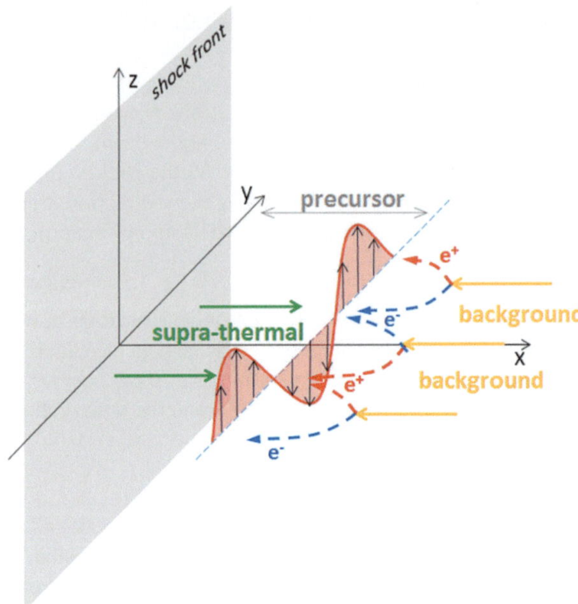

The Weibel frame is defined as the frame where the electrostatic potential vanishes. Its determination is detailed in Appendix A.1, where it is shown that it propagates at the speed ξ_{cr} with respect to the background plasma. This has two important consequences, namely that the high energy returning particles are scattered with respect to that frame and that the background plasma suffers a drag force which slow it down and heats the electrons. The Weibel frame is the suitable frame to calculate the growth rate of the instability, which is purely magnetic and purely growing, since the charge separation in the beam is exactly compensated by an opposite charge separation in the background plasma.

In the front frame, a strong electrostatic potential $\delta\Phi$ is set-up; its magnitude can be determined by the Lorentz transform from the Weibel frame to the front frame: $\delta\Phi = \gamma_w\beta_w\delta A_{x|w}$, defining β_w (resp. γ_w) as the velocity (resp. Lorentz factor) of the frame in which $\delta\Phi$ vanishes, relatively to the shock front frame. Similarly, the x-component of the magnetic potential reads $\delta A_x = \gamma_w\delta A_{x|w}$, so that $\delta\Phi = \beta_w\delta A_x \simeq -\beta_{sh}\delta A_x$. In the front frame, there must be a net total electric charge density fluctuation $\delta\rho$, also determined through a Lorentz transform: $\delta\rho = \gamma_w\beta_w\delta j_{x|w}/c$; the net current density fluctuation is $\delta j_x = \gamma_w\delta j_{x|w}$ so that $\delta\rho \simeq -\beta_{sh}\delta j_x/c$. It is assumed here that in the Weibel frame, the total charge density perturbation $\delta\rho_{|w}$ vanishes. The Maxwell equation that determines the electromagnetic field as a function of the current density, namely $\Delta\delta A_x = -4\pi j_x/c$, is then the same as the Poisson relation between the electrostatic potential and the electric charge.

To conclude this section on the Weibel instability in a pair shock, let us show that it is quite convenient to calculate the growth rate of the Weibel instability in the Weibel frame. Indeed in this frame $\delta\Phi_{|w} = 0$ and $\delta\rho_{tot|w} = 0$, which allows one to write two simple coupled equations to derive the instability growth rate in the cold approximation. Switching over to Fourier space, assuming $k_x = 0$ as before, we insert $\delta\rho_{b|w} = -\delta\rho_{bg|w}$ and $\beta_{bg|w} = -\xi_{cr}$ in the dynamical equation $\partial_\alpha\partial^\alpha\delta A^\mu = -4\pi\delta j^\mu$ and in the background dynamical equation, to obtain:

$$\left(-\omega^2 + \omega_p^2 + k_\perp^2 c^2\right)\delta A_{x|w} \simeq -4\pi\beta_{b|w}c^2\delta\rho_{bg|w},$$

$$\omega^2 \delta\rho_{\mathrm{bg|w}} = \xi_{\mathrm{cr}} \frac{\omega_{\mathrm{p}}^2}{4\pi} k_\perp^2 \delta A_{x|\mathrm{w}}. \tag{22}$$

The plasma frequency of the background plasma emerges in the first equation because of the response of the conduction current to the electromagnetic perturbation. In details:

$$\delta j_x = \sum_\alpha \rho_\alpha \delta\beta_{x\alpha} + \beta_\alpha \delta\rho_\alpha \tag{23}$$

and the conduction current density perturbation $\rho_\alpha \delta\beta_{x\alpha}$ is obtained through the dynamical equation for $\delta\beta_{x\alpha}$:

$$\rho_\alpha \delta\beta_{x\alpha} = -\frac{\omega_{p\alpha}^2}{4\pi\gamma_\alpha^2} \delta A_x. \tag{24}$$

The strong dependence on γ_α justifies the neglect of the conduction response of the beam of superthermal particles; furthermore, in the Weibel frame, $\gamma_{\mathrm{bg|w}} \simeq 1$.

One thus derives the dispersion relation to leading order in ω/kc:

$$\omega^2 = -\xi_{\mathrm{cr}} \beta_{\mathrm{b|w}} \omega_{\mathrm{p}}^2 \frac{k_\perp^2 c^2}{\omega_{\mathrm{p}}^2 + k_\perp^2 c^2}. \tag{25}$$

This is a purely growing magnetic mode. If we take into account the case $k_x \neq 0$, two modifications are introduced: a complete Laplacian operator in the Maxwell equation and a Doppler effect in the dynamical equation of the background plasma, with $\partial_t^2 \mapsto (\partial_t - \xi_{\mathrm{cr}} c \partial_x)^2$ which leads to a simple modification of the dispersion equation:

$$(\omega + \xi_{\mathrm{cr}} k_x c)^2 = -\xi_{\mathrm{cr}} \beta_{\mathrm{b|w}} \omega_{\mathrm{p}}^2 \frac{k_\perp^2 c^2}{\omega_{\mathrm{p}}^2 + k^2 c^2}. \tag{26}$$

The growth rate is invariant by parity; thus in the Weibel frame, forward and backward waves propagate with a frequency $\xi_{\mathrm{cr}} k_x c$ with the same growth rate. A strong Doppler effect is produced in the front frame, so that both forward and backward waves are caught up by the shock front.

4.4 The Weibel Instability in an Electron-Ion Shock

We now assume a plasma made of protons and hot electrons and extend the calculations to include the Landau effects in the electromagnetic response. The detailed calculations are presented in Appendix A.2.

As for the pair shock, one can derive the velocity of the Weibel frame by analyzing the generation of electric charge by the quiver force, taking into account the susceptibilities of the various plasma components. To circumvent the technical difficulty associated with the Lorentz transform of these susceptibilities from one frame to another, it is more convenient to rely on a Lorentz covariant formulation of linear response theory, as formulated by Melrose (1986) and Achterberg and Wiersma (2007), which involves the polarization tensor of species s $\alpha^{\mu}_{s,\nu}$ defined through $\delta j^\mu_s = \alpha^{\mu}_{s,\nu} \delta A^\nu$. The combination of Maxwell equations in the Lorentz gauge, namely $\partial_\mu \partial^\mu \delta A^\nu = -\sum_s \delta j^\nu_s$ and of the above response then leads to the definition of the Weibel frame as that in which

$$\sum_s \alpha^{01}_s = 0. \tag{27}$$

The calculations of α^{01}_s in a generic frame, for streaming hot electrons, are too lengthy to be reported here; these calculations will be presented elsewhere. Nevertheless, one obtains the same result as before, $\beta_{\mathrm{w|u}} \sim \xi_{\mathrm{cr}}$.

The concept of Weibel frame is essential in the prospect of particle acceleration and radiation, since this is the frame in which particles are purely scattered (at least almost, for there is an induction electric field significantly smaller than the magnetic field). The conclusion of this section is that particles are scattered by the magnetic field of Weibel turbulence with respect to a frame moving forward at sub-relativistic speed ($\sim\xi_{\rm cr}$) with respect to the background plasma.

4.5 Electron Heating and the Drag Force

Electrons of the background plasma can be heated through two possible electric fields generated by the Weibel instability: one is the strong motional electrostatic field measured in the front frame due to the Lorenz transform, oriented both perpendicular to the magnetic field fluctuation and to the shock normal; the other is the induction electric field, which derives from the growth of the magnetic field energy density, and which is significant during the stage of instability growth.

The electrostatic field $\delta E_\perp$ is associated with a potential fluctuation which has been obtained in the front frame through a Lorentz transform: $\delta\Phi = \beta_{\rm w}\delta A_x$. The heating due to this field is caused by the scattering of electrons in the Weibel frame, which leads to a kind of second order Fermi process in the front frame, because a variation of x-momentum in the Weibel frame produces a variation of energy in the front frame. This is the strongest electric field when measured in the front frame with $\delta E_\perp \sim \delta B$.

In the Weibel frame, the inductive electric field derives from:

$$\delta E_{x|\rm w} \equiv -\frac{1}{c}\partial_t \delta A_{x|\rm w} \sim \frac{g_{\rm inst|w}}{\omega_{\rm pi}} B_{|\rm w}. \tag{28}$$

Therefore $\delta E_x^2 \sim \xi_{\rm cr}\delta E_\perp^2$ in the front frame and $\xi_{\rm cr} < 1$; this inductive electric field nonetheless provides an efficient source of heating (Gedalin et al. 2012).

4.5.1 Scattering and Heating in the Electrostatic Field

In a first approximation, background electrons are scattered by a static magnetic field in the Weibel frame that moves at speed $\beta_{\rm w|u} \simeq \xi_{\rm cr}$ with respect to the background frame. Therefore, in the background frame, their temperature increases and they suffer a drag force. Because their energy does not change during scattering in the Weibel frame (as long as one neglects the induction field effect), they suffer a modification of their x-momentum $\Delta p_{x|\rm u}$ and energy $\Delta\epsilon_{|\rm u}$ in the background frame.

Together with $\Delta\epsilon_{|\rm w} = 0$, the Lorentz transform $\Delta\epsilon_{|\rm w} = \gamma_{\rm w|u}(\Delta\epsilon_{|\rm u} - \beta_w c\Delta p_{x|\rm u})$ provides an essential relation between energy gain and momentum variation:

$$\Delta\epsilon_{|\rm u} = \beta_{\rm w|u} c\Delta p_{x|\rm u}. \tag{29}$$

The stochastic force $\Delta p_{x|\rm u}/\Delta t_{|\rm u}$, after statistical averaging, leads to a macroscopic drag force $F_{\rm d|u}$ and the distribution of background electrons, isotropic in its rest-frame, gets heated at a rate (background frame):

$$k\dot{T} = \frac{1}{3}\beta_{\rm w|u} c F_{\rm d|u}. \tag{30}$$

The drag force is best evaluated in the Weibel frame through a Fokker-Planck term:

$$F_{\rm d|w} = \frac{1}{n}\frac{\partial(n\langle\Delta p_{x|\rm w}\rangle)}{\partial t} = \int {\rm d}^3 p_{|\rm w}\, p_{x|\rm w}\frac{\partial}{\partial p_{i|\rm w}} D_{ij}\frac{\partial}{\partial p_{j|\rm w}} f(\boldsymbol{p}_{|\rm w}) \tag{31}$$

with a momentum diffusion coefficient of the form

$$D_{ij} = e^2\langle \delta B_\perp^2 \rangle \tau_{c|w} h_{ij}^\perp \tag{32}$$

with $h_{ij}^\perp = \delta_{ij} - p_{i|w} p_{j|w} / p_{|w}^2$ the projector orthogonal to momentum. The timescale $\tau_{c|w}$ represents the coherence timescale of the random force associated to the Weibel turbulence. Using the Lorentz invariance of f and the fact that f depends only on $\epsilon_{|u}/T$ in the background frame, one can derive

$$F_{d|w} \sim \frac{\beta_{w|u} c}{kT} e^2 \langle \delta B_\perp^2 \rangle \tau_{c|w} \tag{33}$$

up to a prefactor of order unity.

Using the above, we can estimate the temperature that can be achieved and the slow-down of the background flow that ensues. As discussed further below—see Eq. (45) in particular—the precursor length is approximately δ_i / ξ_B, as measured in the background rest-frame, and therefore the background plasma experiences the drag force for a duration $\xi_B^{-1} \omega_{pi}^{-1}$. The temperature achieved is then roughly of order

$$\xi_{\rm th} \sim \left(\frac{\xi_{\rm cr}^2 e^2 \langle \delta B_\perp^2 \rangle \tau_{c|w} c^2}{m_p^2 c^4 \xi_B \omega_{\rm pi}} \right)^{1/2} \sim \xi_{\rm cr} (\tau_{c|w} \omega_{\rm pi})^{1/2}. \tag{34}$$

In the background frame, the dragging effect is seen as a force that pushes the plasma forward, with an intensity $F_{d|u} = F_{d|w}$. Over the precursor crossing timescale, the upstream plasma is displaced in energy by

$$\Delta\gamma \simeq \frac{F_d \delta_i}{m_p c \xi_B} \sim (\tau_{c|w} \omega_{\rm pi})^{1/2} \tag{35}$$

indicating that in the front frame, the variation in momentum is of order unity if $\tau_{c|w}\omega_{\rm pi} \sim 1$.

Therefore this scattering process acting on the background plasma is able to slow-down the whole incoming plasma, with possible consequences regarding the saturation of the Weibel instability in regards of the angular dispersion of the beam.

A more complete calculation should take into account the spatial profiles of the various quantities in a dynamical system; however the former estimates provide useful orders of magnitude regarding heating and slow down.

4.5.2 Heating in the Inductive Electric Field

The induction field acts during the phase of instability growth and its heating effect can be estimated with quasi-linear theory which leads to (in the Weibel frame)

$$n\dot{T} \simeq \int \mathrm{d}^3 p_{|w}\, \epsilon_{|w} \frac{\partial}{\partial p_{x|w}} \Gamma_{xx} \frac{\partial}{\partial p_{x|w}} f(\boldsymbol{p}_{x|w}) \tag{36}$$

with $\Gamma_{xx} \simeq (g_{\rm inst|w}\omega_{\rm pi})^2 e^2 \langle \delta B_\perp^2 \rangle \tau_{c|w}$. The heating rate is thus a factor $\omega_{\rm pi}^{-1} g_{\rm inst|w} / \xi_{\rm cr}$ times that in the stochastic electrostatic field; in the cold plasma limit, both are comparable because $g_{\rm inst|w} \sim \xi_{\rm cr}\omega_{\rm pi}$, but at larger temperatures, heating in the inductive field is slightly larger. However, this heating process in the inductive field only applies to the linear growth phase; it is much reduced in the non-linear phase. Overall, one thus expects to reach a scaling given by Eq. (34).

One can also estimate the electron temperature from the phenomenology of the relativistic regime of oscillations in very intense fields: $T_e \sim eB\ell_c$. The typical transverse scale of turbulence is the electron Debye length: $\ell_c = \lambda_D = \xi_{th}^{1/2}\delta_i$ in the hot electron regime. Then we get another estimate compatible with the previous one:

$$\xi_{th} \sim \xi_B \quad \text{or} \quad nT \sim \frac{B^2}{4\pi}. \tag{37}$$

This is a remarkable result that suggests that the thermal energy density follows the magnetic energy density and that confinement structures can occur in the background plasma.

4.5.3 Weibel Filaments

The above relation suggests that heating of the background plasma ensures that the thermal pressure inside the filaments counterbalance the magnetic stress. It is important here to note that the filaments are built out of the background plasma, not by filamentation of the beam of returning particles, as is often thought. Indeed, the transverse momentum of these particles in the background plasma frame is of the order of $\gamma_{sh}mc$, which is too large to allow them to be confined in structures of transverse size δ_i, since $e\delta B_{|u}\delta_i \sim \xi_B^{1/2}mc^2$.

The filaments observed in PIC simulations are thus real, but they form in the background plasma. This is tightly related to the fact that the background plasma moves at a velocity $-\beta_{w|u}$ with respect to the magnetic field. These filaments are some sort of coherent structures of a quasi bi-dimensional turbulence, where inverse cascade develops.

One can derive the general structure of these filaments in the transverse plane by considering the inhomogeneous equilibrium of the background plasma in the presence of a finite A_x component (background plasma frame). Using a non-relativistic description, appropriate to the transverse dynamics, one finds from the force balance in the transverse direction:

$$\nabla_\perp \left\{ \frac{1}{2}mv^2 + T \ln\left(\frac{n_\pm}{\bar{n}}\right) + q\xi_{cr}A_x \right\} = 0. \tag{38}$$

The possibly residual kinetic energy $E_K = mv^2/2$ is in the form of irrotational motions, the Lorenz driving being irrotational. One obtains Boltzmann type equilibria, assuming the same temperature for both populations of particles:

$$n = \bar{n}\exp\bigl[-(E_K + \xi_{cr}qA_x)/T\bigr]. \tag{39}$$

In a pair plasma, one gets a charge separation such that

$$(n_+ - n_-)e = -2n_0 e \exp(-E_K/T)\sinh(\xi_{cr}eA_x/T). \tag{40}$$

Because in motion of velocity $-\xi_{cr}$ with respect to the Weibel frame, its associated current generates a magnetic field such that:

$$\Delta_\perp A_x = -8\pi\bar{n}e\xi_{cr}\exp(-E_K/T)\sinh(\xi_{cr}eA_x/T). \tag{41}$$

Defining $a \equiv \xi_{cr}eA_x/T$, we obtain a sinh-Poisson equation, whose analytical solutions are known:

$$\Delta_\perp a = -\frac{\xi_{cr}^2}{\lambda_D^2}\exp(-E_K/T)\sinh a. \tag{42}$$

The solutions of that equation are known to be in the form of a double lattice of filaments of alternating polarity. The size of the filaments appears clearly as $\lambda = \lambda_D/\xi_{cr} = \sqrt{\xi_{th}}/\xi_{cr}\delta_i$. This estimate does not account for the kinetic energy contribution that increases the Debye length. However its contribution is expected to remain not larger than $\xi_{cr}eB\lambda \sim |a|T$, which is also the estimate of the achieved temperature. In a pair plasma that energy is sub-relativistic.

In the case of an electron-ion plasma, the solution are similar if the kinetic energy is negligible. Otherwise, it breaks the symmetry by giving more importance to the electron contribution. In that case instead of a sinh-Poisson equation, the filaments are described by a Poincaré equation, whose solutions are of Bennett equilibrium type with all the filaments having the same polarity and thus undergoing more merging than in the sinh-Poisson case:

$$\Delta_\perp a = -\frac{\xi_{\rm cr}^2}{\lambda_{\rm D}^2} e^a. \tag{43}$$

4.6 Scattering of Weibel Turbulence

The size of the precursor is determined by the residence time of returning particles. As mentioned earlier, and as seen from the background frame, they are outrun by the shock front when their pitch angle has opened up to $1/\gamma_s$ due to a scattering at a frequency ν_s such that:

$$\nu_{\rm s} \sim \frac{e^2 \langle \delta B_{|\rm u}^2 \rangle \tau_{\rm c|u}}{\gamma_{\rm b|u}^2 m^2 c^2}. \tag{44}$$

The Lorentz factor of beam particles in the background plasma rest frame $\gamma_{\rm b|u} \simeq \gamma_{\rm sh}^2$. The above estimate derives from a random phase approximation (Milosavljević and Nakar 2006; Pelletier et al. 2009) and turns out to be valid even with a more elaborated method based on Hamiltonian chaos, described in Appendix A.3.

Measured from the background frame one obtains:

$$\ell_{\rm prec|u} \sim \frac{c}{\gamma_{\rm sh}^2 \nu_{\rm s}} (1 - \beta_{\rm sh}) \sim \omega_{\rm pi}^{-1} \xi_B^{-1} \tag{45}$$

assuming that $\tau_{\rm c|u} \sim \omega_{\rm pi}^{-1}$. Thus the travel time of the background plasma across the precursor is about $(\xi_B \omega_{\rm pi})^{-1}$, which is longer than the growth time of the instability $(\sqrt{\xi_{\rm cr}}\omega_{\rm pi})^{-1}$, while the residence time of superthermal particles of Lorentz factor $\gamma_{|\rm u}$ in the precursor is of the order of

$$t_{\rm res|u} \sim 2\gamma_{\rm sh}^2 \ell_{\rm prec|u} \left(\frac{\gamma_{|\rm u}}{\gamma_{\rm b|u}} \right)^2 \tag{46}$$

This residence directly controls the acceleration timescale, which thus scales as the energy squared of the superthermal particle. This result has been confirmed by PIC simulations (Sironi et al. 2013).

The value of the coherence time for the superthermal particles is not a trivial issue, as mentioned in Achterberg et al. (2007). Indeed, in an idealized Weibel turbulence with $k_x = 0$, the conjugate momentum $p_x - eA_x/c$ is a conserved quantity of the particle trajectories. Since $p_x \sim \gamma_{\rm sh} mc$ while $eA_x/c \sim \xi_B^{1/2} \gamma_{\rm sh} mc$ in the front frame, the former largely dominates, hence superthermal particles would never be able to go back to the shock if the turbulence were exactly x-independent. Consequently, the finite k_x must play a key role in the transport of superthermal particles.

The general scaling of the residence time with energy ϵ, hence of the acceleration timescale $t_{\rm acc} \propto \epsilon^2$, has been obtained in the framework of a random phase approximation model in Pelletier et al. (2009). Nevertheless, one can also derive this scaling using a model of Hamiltonian chaos applied to the trajectories of superthermal particles. This leads to a validation of the scattering law in a more general situation than the random phase approximation. In Appendix A.3, we present a simple model of such chaotic dynamics; a more complete model will be presented in a forthcoming publication.

5 Phenomenological Consequences

5.1 Scattering Efficiency and Maximal Energy

5.1.1 Onset of Acceleration

As discussed in Sect. 2.2, an oblique background magnetic field in the upstream becomes essentially transverse in the downstream flow, leading to a super-luminal shock configuration. As discussed in Lemoine et al. (2006) and Pelletier et al. (2009), such a configuration inhibits Fermi acceleration unless the scattering frequency in the turbulence exceeds the gyrofrequency in the background field (which controls the rate at which particles are advected in the downstream plasma). In Lemoine et al. (2006), it was actually speculated that the development of Weibel micro-turbulence should lead to an efficient Fermi process and this point of view has been confirmed by PIC simulations.

For scattering in micro-turbulence in the downstream flow, the above condition for the development of Fermi cycles can be written:

$$\frac{e^2 \delta B_{|\mathrm{d}}^2 \tau_{\mathrm{c|d}}}{m^2 c^2 \gamma^2} > \frac{e B_{|\mathrm{d}}}{m c \gamma}, \tag{47}$$

with $\tau_{\mathrm{c|d}} \sim \omega_{\mathrm{pi}}^{-1}$. Here γ denotes the Lorentz factor of the accelerated particle in the downstream frame; for the first Fermi cycle, $\gamma \sim \gamma_{\mathrm{sh}}$. The above can be written in terms of the parameters σ and ξ_B as

$$\sqrt{\sigma} < \xi_B. \tag{48}$$

It thus indicates that only weakly magnetized shock waves should be able to accelerate particles efficiently. For typical values $\xi_B \sim 10^{-2}$ in the vicinity of the shock front, the above bound suggests that particle acceleration should take place in shocks with $\sigma \lesssim 10^{-4}$. This point of view, developed in Lemoine et al. (2006) and Pelletier et al. (2009) (see also Niemiec et al. (2006) for numerical simulations) has been beautifully confirmed by PIC simulations, e.g. Sironi et al. (2013).

5.2 Maximal Energies

The maximal energy that can be reached by the Fermi process in a shock of a given size and age depends on several factors, in particular the scattering rate [Eq. (44)], the energy loss or escape timescale, and the background magnetic field. The above discussion indicates that in the front frame, the upstream residence time typically dominates over the downstream residence time; this effect arises because the upstream fluctuations move at high velocity relative to the front, while the downstream perturbations are essentially static and isotropic in the shock front frame. However, as discussed in detail by Plotnikov et al. (2013), the inclusion of a small but finite background magnetic field changes the scale of the precursor.

To simplify the discussion, one can use the downstream residence time as a proxy for the acceleration timescale; one would obtain similar results for a more careful calculation taking into account the effect of a background magnetic field comparable to that of the interstellar medium and $\gamma_{\mathrm{sh}} \sim 100$, representative of gamma-ray bursts external shocks.

One can then show that the maximal proton energy is of order 10^{16} eV at the external shock of a gamma-ray bursts. In contrast, the maximum electron energy is limited by synchrotron energy losses to values of the order of 10 TeV, independent of the magnetic field and shock Lorentz factor (Kirk and Reville 2010; Plotnikov et al. 2013; Wang et al. 2013). This can give rise to GeV synchrotron photons at early times $\sim 10^2$–10^3 s, as observed by the Fermi satellite in some gamma-ray bursts.

5.3 Radiative Diagnostics

Weakly magnetized shocks, which are prone to Fermi acceleration, are those in which intense micro-turbulence can be excited in the shock precursor by streaming instabilities. The fate of this turbulence downstream of the shock, on timescales $\gg\omega_{\rm pi}^{-1}$ has been the subject of intense debate. The linear dissipation rate of microturbulence is $\gamma_k \simeq k^3c^3/\omega_{\rm pi}^2$ in a relativistic plasma (e.g. Chang et al. 2008), a result which extends into the mildly non-linear regime (Lemoine 2015a), suggesting that turbulence on scales $\sim\omega_{\rm pi}^{-1}$ is dissipated on $\omega_{\rm pi}^{-1}$ timescales. In contrast, the typical (comoving) size of the blast of a gamma-ray burst is of the order of $R/\gamma_{\rm sh} \sim 10^7 c/\omega_{\rm pi}$ (for an external density $n = 1\ \rm cm^{-3}$), orders of magnitude larger, and this length scale gives the order of magnitude over which most electrons radiate their energy through synchrotron radiation. An important question therefore is whether Weibel turbulence, self-generated in the shock precursor, can survive over the above time/length scales in order to provide the turbulence in which synchrotron photons are produced, see e.g. Medvedev and Loeb (1999), Gruzinov and Waxman (1999), Chang et al. (2008), Keshet et al. (2009), Lemoine (2013, 2015a, 2015b) for discussions of this issue.

While additional instabilities or inverse cascade might potentially help sustain or amplify this turbulence, one may also conceive a simpler scenario, in which the power spectrum of magnetic fluctuations at the shock contains long wavelength modes, with a long lifetime $\propto k^{-3}$. As discussed in Lemoine (2013), these modes could be excited by the highest energy particles accelerated at the shock front, which propagate over large distances in the upstream. The gradual dissipation of the micro-turbulence by phase mixing would then lead to a ξ_B which now depends on some power of the distance to the shock. The synchrotron spectrum produced by a population of particles interacting with such a dissipative turbulence bears some unique features, which could potentially be used as a diagnostic of Weibel turbulence (Lemoine 2013, 2015a). A generic prediction, in particular, is a significantly larger ratio of flux at high energy to flux at low energy, because high energy photons are produced by high energy electrons, which radiate on a shorter timescale, hence in a stronger turbulence. Such a signature may have been detected in some gamma-ray bursts, whose multi-wavelength light curve can be nicely explained by a synchrotron spectrum resulting from a dissipative turbulence with $\xi_B \propto (\omega_{\rm pi}x/c)^{-0.5}$.

6 Macroscopic Simulations of Relativistic Shocks

The physics of relativistic collisionless shocks together with its application to the phenomenology of powerful astrophysical sources, are complex non-linear multi-scale problems, which have been addressed with a variety of methods (see e.g. Bykov and Treumann 2011; Marcowith et al. 2016). Significant progress in our understanding of these phenomena has come through particle-in-cell (PIC) simulations, which probe the full non-linear relationship between the particles and the electromagnetic fields, see Sironi et al. (2015) and references therein. Such simulations are however computationally intensive and cannot yet explore the physics of shock waves on the long macroscopic time and space scales on which the source evolves, or on which radiation is produced. In this context, the analytical theory that we have proposed in the previous section is useful because it can be benchmarked against these PIC simulations on small physical scales and extrapolated to the macrophysical scales of interest. An alternative method is the numerical Monte Carlo technique which we describe in this section.

Monte Carlo methods offer a modeling technique that lies between PIC simulations and semi-analytic models. In contrast to semi-analytic descriptions which make a global diffusion approximation, particles pitch-angle scatter locally with the Monte Carlo technique (e.g., Ellison and Eichler 1984). The important difference is that for the diffusion approximation to be valid, particle distributions must be nearly isotropic in all frames and large flow gradients must be avoided. While these approximations are good for superthermal particles in non-relativistic shocks, they are never good for particles crossing relativistic shock front. In the Monte Carlo code, particles undergo a small pitch-angle scattering event after a time small compared to a gyroperiod. Thus the scattering is "local" and highly anisotropic distributions in sharp flow gradients can be treated directly. Of course, the properties of the scattering are not calculated from first principles, as done with PIC simulations. But scattering can be directly parameterized, as so far done for relativistic shocks (e.g., Ellison et al. 2013), or determined more fundamentally from CR-induced plasma instabilities, magnetic field amplification (MFA), and an analytic determination of scattering rates from the self-consistently determined magnetic turbulence power spectrum in non-relativistic shocks (e.g., Vladimirov et al. 2008; Bykov et al. 2014).

The analytical model described in the previous sections can be seen as an effective two-fluid theory, describing how self-generated electromagnetic turbulence affects and couples a fluid of superthermal particles with a background plasma fluid. As such, it can accurately model the back-reaction of the accelerated particles and of the background plasma on the turbulence. In contrast, the Monte Carlo model is a one-component theory. The pitch-angle scattering approximation treats all particles the same and Fermi accelerated particles arise smoothly from the background plasma which is a self-consistently described collection of particles rather than a separate fluid. The non-linear relationship between the Fermi accelerated population and the bulk plasma is determined by iteratively solving the energy-momentum tensor by forcing energy and momentum conservation across the shock.

Although the relativistic Monte Carlo simulations that follow do not make a distinction between the rest frame of the one-fluid plasma and the rest frame of the turbulence, in contrast to Sect. 4.2, they will be generalized in the near future to a more realistic setting in which the growth of turbulence and its rest frame are dictated by microphysical considerations. Monte Carlo simulations of non-relativistic shocks which incorporate these effects are described in the following as an illustration.

Most Monte Carlo models thus far have assumed plane, steady-state shocks with Fermi acceleration limited by free escape boundaries. Parallel (e.g., Ellison and Eichler 1984; Ellison et al. 2016), oblique (e.g., Baring et al. 1993; Summerlin and Baring 2012) and perpendicular (e.g., Takamoto and Kirk 2015) shock geometries have been modeled.

The main features of the Monte Carlo technique are: (1) a large dynamic range can be modeled with reasonable computing resources following particles consistently from thermal energies to the highest CR energies with no constraint on the anisotropy of the particle distribution; (2) the non-linear (NL) feedback on the shock structure from efficient Fermi acceleration can be determined consistently with injection and the production of ultra-high-energy CRs; (3) since particle anisotropies are treated consistently, no distinction in the particle transport need be made between non-relativistic, trans-relativistic, or ultra-relativistic shocks; (4) electrons and ions can be simultaneously injected and accelerated allowing effects depending on particle mass to be determined self-consistently; and (5) absolutely normalized radiation, from electrons and ions, can be calculated as a straightforward function of model and environmental parameters.

The main drawbacks of this method are that it is intrinsically steady state, generally done for plain-parallel shocks, and thus far too computationally slow to include in hydrodynamic evolutionary models such as those for young SNRs using semi-analytic descriptions of NL diffusive shock acceleration (e.g., Ellison et al. 2007, 2012; Lee et al. 2012; Zirakashvili and Ptuskin 2012).

We describe here some physical results obtained with the Monte Carlo technique which may help to understand the backreaction effects of the accelerated particles on the shock flow at large scales which can affect the particle spectra and radiation and which are difficult to simulate yet with the PIC simulations.

6.1 PIC vs Monte Carlo Simulations

At present, the self-consistent problem of relativistic collisionless shock structure, turbulence production, and particle acceleration can only be solved with full scale PIC simulations. A very important result of these simulations is the demonstration that the Weibel instability, as well as current-driven filamentation instabilities, produce short-wavelength turbulence which can scatter accelerated particles (see e.g. Spitkovsky 2008; Plotnikov et al. 2011; Sironi et al. 2013; Lemoine et al. 2014c). In the simulations performed thus far, the self-generated turbulence is seen to fill a layer of a few hundred ion skin depths close to the subshock. Another important result is that PIC simulations have shown that, regardless of the background magnetic field geometry, unmagnetized relativistic shocks can be efficient particle accelerators converting $\gtrsim 10\%$ of their ram pressure into accelerated particles (e.g. Sironi et al. 2013).

Despite these successes, the limitations on box size, run time, and dimensionality (see Jones et al. 1998; Vladimirov et al. 2008) inherent in PIC simulations force the consideration of less fundamental descriptions. This is especially true for non-relativistic and trans-relativistic shocks where the precursors may extend over the scales of hundreds or thousands of gyroradii of maximal energy CRs. By parameterizing the particle transport, the Monte Carlo technique we discuss can simulate the non-linear shock structure on astrophysically significant scales for arbitrary shock speeds (see Ellison et al. 1995; Summerlin and Baring 2012; Ellison et al. 2016). In the case of non-relativistic shocks, the Monte Carlo model can simultaneously include the non-linear shock structure, accelerated particle production, and magnetic turbulence generation accounting for both short scale and long-wavelength instabilities produced by the anisotropic CR particle distributions (e.g. Vladimirov et al. 2009; Bykov et al. 2014). It is even possible to include super-diffusive propagation (i.e. the so-called Lévy-walk models, e.g. Zimbardo and Perri 2013) consistently in the shock precursor (Bykov et al. 2017).

In principle, microphysical parameters determined with PIC simulations can be implemented in the Monte Carlo code to allow the analysis of large-scale CR modified shocks. While this will allow for a large increase in dynamic range, it remains uncertain whether or not self-generated CR-driven instabilities occur on large scales, particularly for fully relativistic shocks (see e.g. Sagi and Nakar 2012; Lemoine et al. 2014c). In the next section we present some Monte Carlo results showing the large-scale macroscopic structure of relativistic shocks. We also include a discussion of trans-relativistic and non-relativistic shocks to emphasize that the differences between shocks of varying speed is mainly one of analysis; mathematical techniques separate shocks more than the underlying physics which must be continuous through the trans-relativistic range.

7 Non-linear Shock Structure

It is reasonably certain from direct observations that non-relativistic shocks can inject and accelerate ions efficiently, while the electron acceleration efficiency, at least in interplanetary

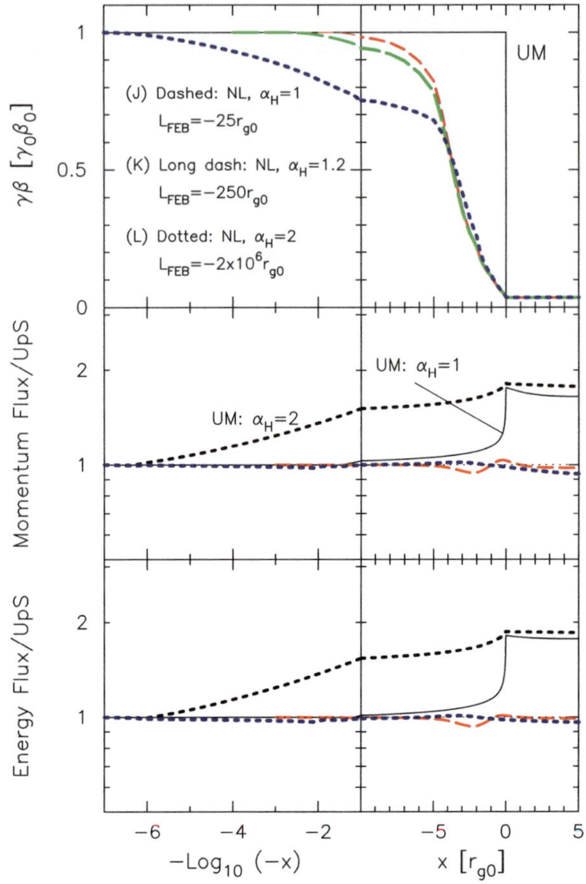

Fig. 5 The top panels show the non-linear (NL) shock structure in terms of the inverse density, $\gamma(x)\beta(x)/(\gamma_0\beta_0)$, for shocks with mean free paths having different momentum dependence illustrated in Fig. 6. This figure is adapted from Fig. 7 in Ellison et al. (2016) and that paper has full details. In all cases, the shock Lorentz factor is $\gamma_{sh} = 10$ and curves labeled UM are for unmodified shocks where the shock structure is discontinuous

shocks, is still uncertain (e.g., Dresing et al. 2016). While no direct observations exist for relativistic shocks, many models, particularly those for γ-ray bursts (GRBs), assume Fermi acceleration is efficient with $\gtrsim$10% of the shock ram kinetic energy placed in nonthermal particles. In many cases, this energy is assumed to be given to electrons since they produce readily observable synchrotron and inverse-Compton emission. To our knowledge, there are no self-consistent models of first-order Fermi shock acceleration that place more energy in electrons than ions. Unless shocks are accelerating a pair dominated plasma whatever energy fraction is assumed for electrons will be equaled or exceeded by ions even if they do not produce observable radiation.

If Fermi acceleration is efficient, the shock structure must be modified by the back pressure of CRs as they scatter in the shock precursor. This is illustrated in Fig. 5 which is adapted from Fig. 7 in Ellison et al. (2016). The shocks shown in Fig. 5 are relativistic with a Lorentz factor $\gamma_{sh} = 10$. For a given acceleration efficiency, non-relativistic shocks will have a more pronounced precursor structure because it is easier for CRs to scatter upstream in a non-relativistic flow. The Monte Carlo code determines the momentum and energy conserving shock structure by iteration, as described in Ellison et al. (1996, 2013).

7.1 Iteration of Shock Structure

If we assume that the shock is parallel and if we ignore the effects of escaping momentum and energy fluxes, the shock-frame, steady-state flux conservation relations are:

$$F_{\mathrm{n}}(x) = \gamma(x)n(x)\beta(x) = F_{\mathrm{n}}^{0}, \tag{49}$$

$$F_{\mathrm{px}}(x) = \gamma^{2}(x)\beta_{x}^{2}(x)\big[e(x) + P_{\mathrm{xx}}(x)\big] + P_{\mathrm{xx}}(x) = \Phi_{P0}, \tag{50}$$

$$F_{\mathrm{en}}(x) = \gamma^{2}(x)\beta_{x}(x)\big[e(x) + P_{\mathrm{xx}}(x)\big] = F_{\mathrm{en}}^{0}. \tag{51}$$

Here, F_{n}^{0}, Φ_{P0}, and F_{en}^{0} are the far upstream number, momentum, and energy fluxes, respectively, and P_{xx} is the xx-component of the pressure tensor, $\mathcal{P}$. Given the input quatities, F_{n}^{0}, Φ_{P0}, and F_{en}^{0}, these three equations have four unknowns: $n(x)$, $\gamma(x)$, $e(x)$, and $P_{\mathrm{xx}}(x)$, so an additional constraint is needed for solution. We use the adiabatic and gyrotropic equation of state:

$$e(x) = \frac{P(x)}{\Gamma(x) - 1} + \rho(x)c^{2}, \tag{52}$$

where $e(x)$ is the total energy density, $\rho(x)c(x)^{2}$ is the rest mass energy density, $\Gamma(x)$ is the adiabatic index, and $P(x) = P_{\mathrm{xx}}$ is the isotropic scalar pressure, i.e., $P = \mathrm{Tr}(\mathcal{P})/3$ (see Double et al. 2004, for more details). All of these quantities, along with the bulk flow Lorentz factor (also called the shock profile) $\gamma(x)$, vary with position x in the precursor.

The solution is accomplished in the Monte Carlo code by iteration. Starting with a shock profile, i.e., $\gamma(x)$, the simulation is run and $F_{\mathrm{n}}(x)$, $F_{\mathrm{px}}(x)$, and $F_{\mathrm{en}}(x)$ are measured at every x. If the measured fluxes don't match the known values F_{n}^{0}, Φ_{P0}, and F_{en}^{0} to within some error limits, $P(x)$, which is a local frame quantity, is determined using Eq. (50) or (51). Using this new $P(x)$, Eqs. (49)–(51) are used to predict the next shock profile, i.e., the next $\gamma(x)$. This process is iterated until the fluxes are conserved. An example from data presented in Ellison et al. (2016) is shown in Fig. 5. In contrast with semi-analytic techniques where multiple solutions can be obtained (see Drury 1983, for a review), unique solutions always result with the Monte Carlo approach because of the non-linear (NL) feedback between the thermal injection and the shock structure.

The above procedure becomes significantly more complicated with an oblique magnetic field. In this case, there are x-dependent parallel and perpendicular components of the magnetic field and Eqs. (49)–(51), are replaced by six equations which include magnetic pressure. This system has been derived by Double et al. (2004). Self-consistent solutions have been found for NL oblique non-relativistic shocks (e.g., Ellison et al. 1996), oblique relativistic unmodified shocks (e.g., Ellison and Double 2004), but we know of no solutions, other than from PIC simulations, for NL oblique relativistic shocks.

7.2 Particle Transport and Magnetic Turbulence

Pitch-angle particle scattering and convection are determined in one of two ways in the Monte Carlo code. In the simplest scheme the mean free path $\lambda(x, p)$ is a fully parameterized function of particle momentum p and position x relative to the subshock at $x = 0$. This method has been used extensively in previous non-relativistic applications (e.g., Berezhko and Ellison 1999) and is currently used for relativistic shocks.

In Fig. 6 we show examples of the scattering mfp for relativistic shocks ($\gamma_{\mathrm{sh}} = 10$) where the momentum dependence has been varied as follows (see Ellison et al. 2016, for full details):

Fig. 6 Mean free paths for the self-consistent shocks shown in Fig. 5 (see Ellison et al. 2016). The positions of the free escape boundaries (FEBs) are shown and it is noted that the momenta at which $\lambda_{mfp}(p_{FEB}) \simeq |L_{FEB}|$ (solid dots) are such that p_{FEB}/p_{max} is a strong increasing function of α_H

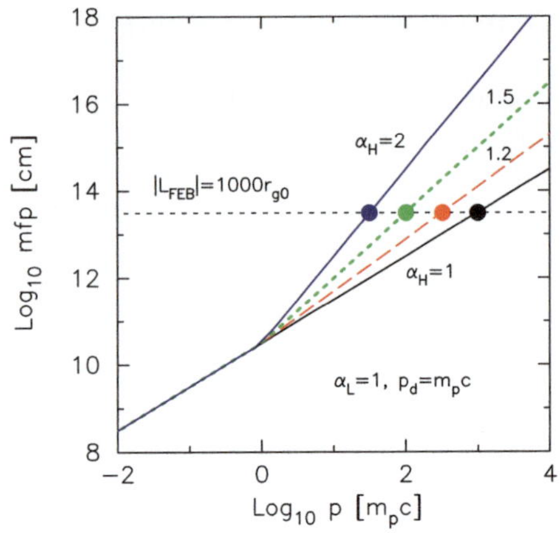

Fig. 7 Phase-space distributions with various indexes α_H and p_d, as indicated (see Ellison et al. 2016). Note the effect p_d has on the Fermi acceleration efficiency in the top two curves both with $\alpha_H = 2$. Model (J) is simulated with $\alpha_H = 1$ and the upstream FEB is at $-25 r_{g0}$. To have essentially the same p_{max}, model (L) with $\alpha_H = 2$ has a FEB at $-2 \times 10^6 r_{g0}$. All spectra are calculated at $x = 0$ in the rest frame of a shock with $\gamma_{sh} = 10$

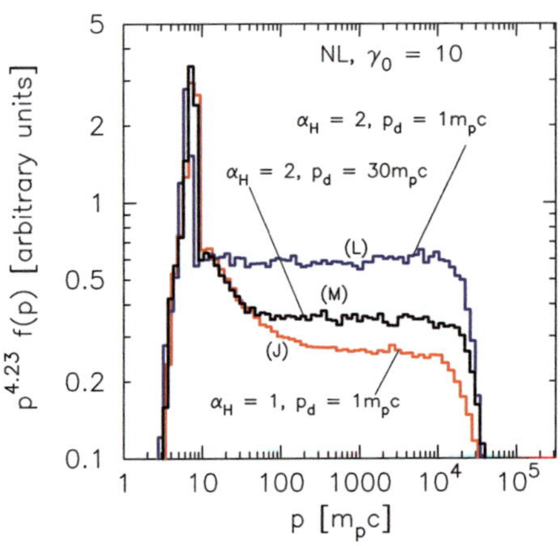

$$\lambda_{mfp}(p) = \eta_{mfp} \cdot r_g(p_d) \cdot (p/p_d)^{\alpha_L} \quad \text{for } p < p_d$$
$$= \eta_{mfp} \cdot r_g(p_d) \cdot (p/p_d)^{\alpha_H} \quad \text{for } p \geq p_d. \tag{53}$$

Here p is the particle momentum in the local frame, $\alpha_L \leq 1$, $\alpha_H \geq 1$, p_d is a dividing momentum between the low- and high-momentum ranges, $r_g(p_d) = p_d c/(eB_0)$ is the gyroradius for a proton with local-frame momentum p_d in the background field B_0, and $\eta_{mfp} \geq 1$ is a parameter that determines the strength of scattering. More complicated forms for $\lambda_{mfp}(p)$ can be used in the Monte Carlo simulation to model semi-analytic and/or PIC results where available. Note that $\lambda_{mfp}(p)$ was taken to be position independent in the results shown in Figs. 6 and 7.

The Bohm limit is described by $\eta_{mfp} = \alpha_L = \alpha_H = 1$, and the conditions $\alpha_L \leq 1$ and $\alpha_H \geq 1$ ensure that $\lambda_{mfp} \geq \eta_{mfp} r_g$ for all p.[2] In the simple geometry of a plane-parallel, steady-state shock, results scale with η_{mfp}. Large values of η_{mfp} imply weak scattering with long length and time scales.

[2]The case with $\eta_{mfp} = \alpha_L = \alpha_H = 1$ is often referred to as "Bohm diffusion" however particle trajectories are calculated in the Monte Carlo code without making a diffusion approximation.

In Fig. 6, $\alpha_H = 2$ is used as a typical value predicted by the analytical theory of Sect. 4.2 of particle transport in small-scale micro-turbulence.

Regardless of the form for $\lambda_{\rm mfp}$, pitch-angle scattering is modeled as follows: after a time $\delta t \ll \tau_g$ (τ_g is the gyro-period) a particle scatters isotropically and elastically in the local plasma frame through a small angle $\delta\theta \le \delta\theta_{\rm max}$. The maximum scattering angle in any scattering event is given by (see Ellison et al. 1990a)

$$\delta\theta_{\rm max} = \sqrt{6\delta t / t_c}, \tag{54}$$

where $\delta t = \tau_g / N_g$, $t_c = \lambda_{\rm mfp}/v$ is the collision time, v is the particle speed in the local frame, and N_g is a free parameter. Typically, N_g is chosen large enough to produce fine-scattering results that do not change substantially as N_g is increased further (see Summerlin and Baring 2012, for examples with values of N_g producing large-angle scattering).

All particles are injected far upstream with a thermal distribution and scatter and convect into and across the subshock into the downstream region. A critical approximation used to date in the Monte Carlo simulation is that the subshock is transparent, i.e., no attempt is made to describe the effects of a cross-shock potential, amplified magnetic turbulence, or other effects that may occur in the viscous subshock layer. This approximation has been relaxed in recent work with the parameterization of energy transfer between ions and electrons at the subshock and an injection threshold (Warren et al. 2015a).

Upon interacting with the downstream plasma, some particles will gain energy sufficient to allow them to scatter back across the subshock and be further accelerated. The fraction of particles that are injected, i.e., do manage to re-cross the subshock is determined stochastically and constitutes a "thermal leakage injection" model once the further assumption is made that the subshock is transparent. This injection model requires no additional parameters or assumptions once the scattering mean free path is defined in Eq. (53).

The additional parameter implemented by Warren et al. (2015a) to model radiation spectra from unmagnetized relativistic shocks is $f_{\rm ion}$ which is the fraction of far upstream ion ram kinetic energy transferred to electrons. The far upstream kinetic energy of an ion is $(\gamma_{\rm sh} - 1)m_i c^2$, where m_i is the ion mass. When an ion crosses the subshock from upstream to downstream for the first time, a part of this energy equal to $f_{\rm ion}(\gamma_{\rm sh} - 1)m_i c^2$ is removed from it. The total energy removed from all of the ions is divided equally among electrons and added to their energy as they cross the subshock into the downstream region for the first time. This parameterisation was implemented to account for the effect of the energy exchange between the ions and electrons in the subshock which is clearly seen in Fig. 11 from Sironi et al. (2013) where PIC simulations of a low magnetization ($\sigma < 10^{-5}$) ion-electron plasma shock were presented.

7.3 Particle Escape

While the majority of shocked particles convect downstream and don't re-cross the subshock to experience Fermi acceleration, a small fraction will continue to be accelerated until they run out of acceleration time or their diffusion length becomes comparable to the shock size. Both the acceleration rate and the effective shock size depend critically on self-generated turbulence which will always have a finite extent. The finite extent of self-generated turbulence guarantees that particle escape will play a role in Fermi acceleration.

In quasi-steady-state shocks, such as the Earth bow shock, or long-lived shocks, such as SNR blast waves, the effective shock size limits acceleration (except at earliest times for SNRs) and the escape of particles from the shock becomes an important part of the Fermi process. In fact, in high Mach number non-relativistic shocks, it isn't possible to conserve energy unless escape occurs and this has been modeled with a free escape boundary (FEB). The steeper spectrum produced by Fermi acceleration in relativistic shocks obviates the energy conservation argument, but the finite extent of self-generated turbulence means particle escape will occur here as well.

Because self-generated turbulence will decay in the downstream region, particle escape can occur from downstream as well as upstream from the precursor.[3] The Monte Carlo simulation limits acceleration with an upstream and/or downstream FEB, although lateral escape can also be modeled. Lateral escape may be important in objects as diverse as the Earth bow shock and GRB jets. How escape actually occurs will depend on the detailed geometry of the shock but all these scenarios produce a corresponding $p_{\max}$ with no differences large enough to be discerned from current observations. Even though the Monte Carlo model is steady-state, the acceleration time of individual particles is calculated and Fermi acceleration can also be limited by setting a maximum time (see Achterberg et al. 2001; Drury 2011, for a discussions of particle escape).

An upstream FEB has been used extensively in models of CR production in young SNRs (e.g., Ellison et al. 2007; Morlino et al. 2009; Caprioli et al. 2010) where there is typically a SNR age where a transition occurs between $p_{\max}$ being determined by the remnant age to being determined by the finite shock radius. We note that there is direct observational evidence for upstream escape at the quasi-parallel Earth bow shock (e.g., Trattner et al. 2013).

Using Eq. (53) to vary the momentum dependence of $\lambda_{\rm mfp}(p)$ we show some examples in Fig. 7. These are shock-frame, phase-space distributions for relativistic shocks with $\gamma_{\rm sh} = 10$. For these examples, the size of the precursor (i.e., the position of the upstream FEB) has been varied to produce approximately the same $p_{\max}$. The model parameters are described in detail in Ellison et al. (2016) but it is sufficient to say that the precursor size depends strongly on the momentum dependence of $\lambda_{\rm mfp}(p)$. For Model (J) with $\alpha_H = 1$, the upstream FEB is at $-25 r_{g0}$, while Model (L) with $\alpha_H = 2$ and essentially the same $p_{\max}$, has a FEB at $-2 \times 10^6 r_{g0}$. Here $r_{g0} = m_p u_0 c/(e B_0)$ where u_0 is the shock speed, B_0 is the far upstream magnetic field, and m_p, c, and e have their using meanings. A value of $\alpha_H > 1$ implies weak turbulence but this also means that energetic particles will be able to scatter well into the precursor opening the possibility of self generation of precursor turbulence.

7.4 Thermal Particle Injection

A collisionless shock moving through ambient material will heat all of the material and inject some fraction of that material into the Fermi acceleration mechanism. While it is clear that a background superthermal population will always be accelerated (e.g., Blandford and Ostriker 1978), there is compelling evidence that some fraction of the cold unshocked gas, upon being heated, will be Fermi accelerated as well (e.g., Blandford and Eichler 1987; Ellison et al. 1990b; Sironi et al. 2013). We characterize any process that turns cold, unshocked material into superthermal particles, thermal particle injection.

Thermal particle injection (TPI) has been considered a particularly difficult part of Fermi acceleration from both observational and theoretical aspects. By definition, thermal particles are always highly anisotropic at shocks of any speed, making injection difficult to describe with semi-analytic models based on the diffusion-advection equation. Injection can be

[3] In the often used approximation of an infinite, plane shock, particles do not "escape" downstream unless a boundary is imposed. Particles convect away from the shock but always have a finite, but increasingly small, probability of scattering back to the subshock.

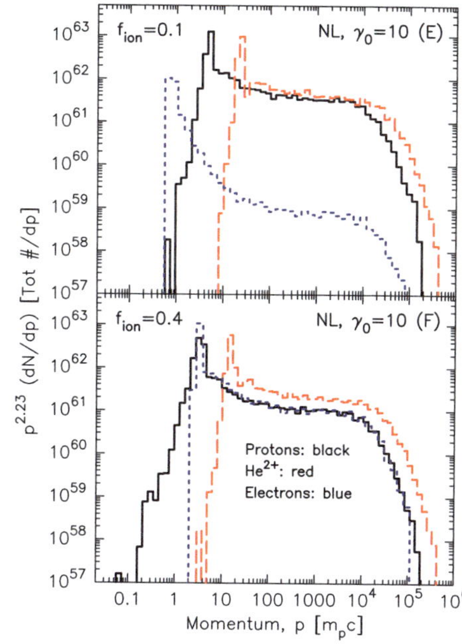

Fig. 8 Non-linear downstream spectra for models of relativistic ion-electron shocks of $\gamma_{sh} = 10$ with $f_{ion} = 0.1$ and $f_{ion} = 0.4$ simulated by Warren et al. (2015a). In *both panels* the *solid* (*black*) *curves* are protons, the *dashed* (*red*) *curves* are He^{2+}, and the *dotted* (*blue*) *curves* are electrons

treated more directly with the Monte Carlo code since there are no constraints on anisotropy. This makes it possible to describe injection in an internally self-consistent way within the assumptions made to describe particle transport with no additional assumptions.

However, as seen in PIC simulations, the subshock layer will be highly turbulent (e.g., Sironi et al. 2013; Caprioli et al. 2015) making the injection process fundamentally complex. While the simple approximations for particle scattering made by the Monte Carlo model cannot describe the full complexity of TPI, Monte Carlo results have been shown to be reasonably consistent with PIC simulations of relativistic shocks. Using similar parameters, the Monte Carlo simulation can reasonably reproduce the NL shock structure, particle spectral shape, and overall acceleration efficiency obtained by the PIC simulation (see Ellison et al. 2013, for details). We know of no semi-analytic method that can do this for relativistic shocks. In effect, the Monte Carlo injection approximations average over the complex microscopic details while ensuring that global momentum and energy conservation is maintained. These approximations allow parameterizations of PIC results that can then be applied on larger CR momentum, length, and time scales than are currently possible with PIC simulations.

7.5 Precursor Structure

In order for Fermi acceleration to work, some downstream particles must scatter back across the shock into the upstream region creating a precursor. While the length of this precursor is a strong inverse function of shock Lorentz factor, it must be present on some scale if CRs are accelerated beyond the initial boost they obtain in their first interaction with the shock.

We characterize the precursor scale with a penetration depth, $D_{pen}(p_{peak})$, where p_{peak} is the momentum where $p^{4.23} f(p)$ peaks. Examples are shown in Fig. 9 for an unmodified (UM) shock (top panel) and the NL shock (bottom panel) where momentum and energy are conserved and the shock is smoothed. While it is clear that the momentum of the peak, p_{peak},

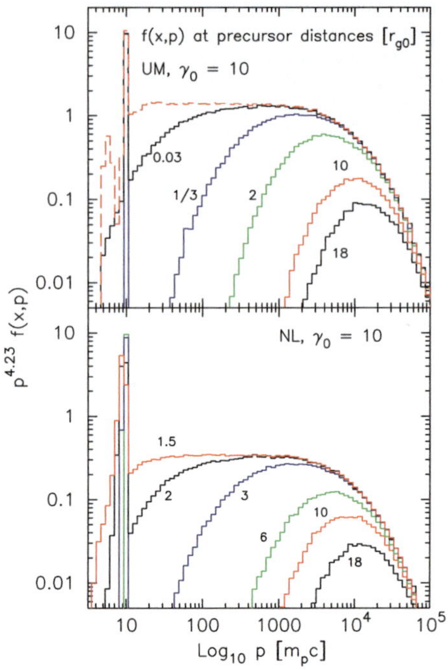

Fig. 9 The distribution function simulated with the Monte Carlo technique [represented as $p^{4.23} f(x, p)$] at the different distances in the shock precursor. The distance (which is measured in r_{g0}) is indicated by a label on the corresponding curve. The simulation results for an unmodified shock (UM) are shown in the *top panel* while $p^{4.23} f(x, p)$ for a simulated non-linear shock (NL) where momentum and energy are conserved is shown at the *bottom panel*

is not precise—the precursor distributions are not delta functions—there is a clear difference in $D_{\rm pen}(p_{\rm peak})$ between the two cases.

In Fig. 10 we plot $D_{\rm pen}$ vs. $p_{\rm peak}$ for shocks with varying Lorentz factors. In the UM shocks (top panel), $D_{\rm pen} \propto p_{\rm peak}^2$ in all cases. Note that the curves in the top panel are scaled to the $\gamma_{\rm sh} = 10$ result with a normalization factor $(\gamma_{\rm sh}/10)^2$ as shown. For a given p, the penetration depth scales as $1/\gamma_{\rm sh}^2$, as expected for an UM flow. The NL shocks show something very different. For $\gamma_{\rm sh} = 10$, 30, and 60, $D_{\rm pen} \propto p_{\rm peak}^{1/2}$. It is much easier for low momentum particles to diffuse upstream once the precursor develops. At the highest p shown in Fig. 10, the penetration depth becomes similar for the UM and NL cases. Note that the trans-relativistic UM $\gamma_{\rm sh} = 2$ shock obeys the $D_{\rm pen} \propto p_{\rm peak}^2$ scaling but doesn't conform to the $1/\gamma_{\rm sh}^2$ scaling.

Fermi acceleration in relativistic shocks has been suggested as a possible source of ultra-high-energy CRs (e.g., Keshet and Waxman 2005), and relativistic shocks may be responsible for non-thermal radiation observed from a number of astrophysical sources including γ-ray bursts (GRBs), type Ibc supernovae, and extra-galactic radio jets. In order to explain this CR production and radiation, electron and ion acceleration to ultra-relativistic energies must be modeled. Particle-in-cell simulations concentrate on microphysical aspects of relativistic shocks but lack the dynamic range needed to model nonlinear shock structure and Fermi acceleration from thermal injection energies to CR energies above $\sim 10^{15}$ eV. On the other hand, semi-analytic descriptions of magnetic field amplification and particle scattering can cover a large dynamic range but have not yet included non-linear effects from efficient Fermi acceleration. We believe the Monte Carlo technique offers a way to combine insights gained from semi-analytic studies with PIC results, such as the transfer of ion energy to electrons, in a NL, broadband model capable of producing radiation that can be compared with observations (e.g., Warren et al. 2015b, 2017).

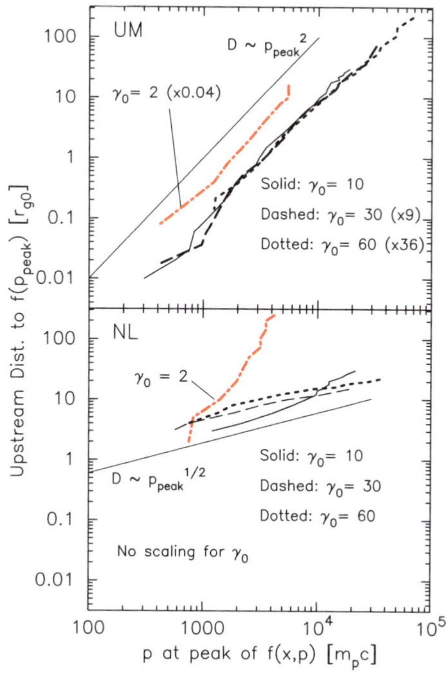

Fig. 10 The characteristic precursor scale derived as the distance $D_{pen}(p_{peak})$ from the shock to the precursor position where $p^{4.23} f(x, p)$ peaks. The momentum p_{peak} corresponds to the peak of $p^{4.23} f(x, p)$ in Fig. 9. We plot D_{pen} vs. p_{peak} for shocks with varying Lorentz factors. In the *top panel* the results for unmodified shocks (UM) are presented, while the *bottom panel* shows the results for non-linear shocks (NL) see Fig. 9

7.6 Lessons from Non-relativistic Shock models: Scattering Center Speed

Thus far, the calculation of self-generated turbulence in NL relativistic shocks has only been done with PIC simulations. Semi-analytic calculations have been performed in unmodified relativistic shocks, as discussed earlier. As a prelude to future relativistic shock Monte Carlo calculations, we show here some examples for NL non-relativistic shocks performed recently by Bykov et al. (2014). Despite the physical differences between relativistic and non-relativistic shocks discussed above, the non-relativistic results can serve as a useful guide.

The NL, non-relativistic Monte Carlo model includes magnetic field amplification from resonant instabilities, as well as short-wavelength (e.g., Bell 2004; Pelletier et al. 2006; Amato and Blasi 2009) and long-wavelength (e.g., Bykov et al. 2012a; Schure et al. 2012) non-resonant instabilities. In Fig. 11, taken from Bykov et al. (2014), we show downstream particle spectra (top panel) for examples with and without the long-wavelength, non-resonant instability. The examples shown were simulated with and without the turbulence cascade, as indicated. It is important to note that in these non-relativistic results, the scattering mean free path is self-consistently determined and has an arbitrary p and x dependence. The self-consistently determined diffusion coefficient is shown in the bottom panel in Fig. 11, again taken from Bykov et al. (2014). It is clear from Fig. 11 that the mfp does not have a simple Bohm-like momentum dependence. A very important distinctive feature of these non-relativistic results is the spectral hardness at high p. Such hard spectra are not expected in relativistic shocks, as shown in Fig. 8. The hard CR spectrum in NL, non-relativistic shocks provides the free energy needed for strong amplification of magnetic turbulence in the shock precursor. It remains to be seen how turbulence will generated in relativistic shocks.

Self-generated magnetic turbulence is essential for Fermi acceleration to work since particles must scatter nearly elastically off this turbulence to cross the subshock several times

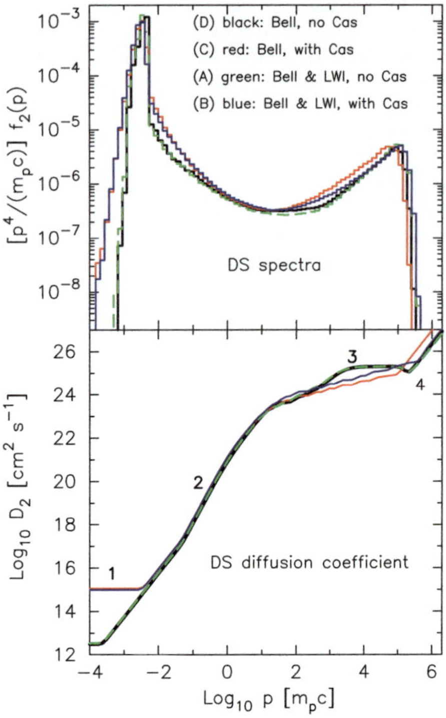

Fig. 11 Phase-space particle distributions, $[p^4/(m_pc)]f_2(p)$ (*top panel*), and diffusion coefficients (*bottom panel*) calculated downstream from the subshock in the nonlinear Monte Carlo model for shock velocity 5000 km s^{-1}, the far upstream density $n_0 = 0.3\ \mathrm{cm}^{-3}$, and magnetic field $B_0 = 3\ \mu\mathrm{G}$. The particle spectra and diffusion coefficients calculated consistently for different models of magnetic turbulence (from Bykov et al. 2014).

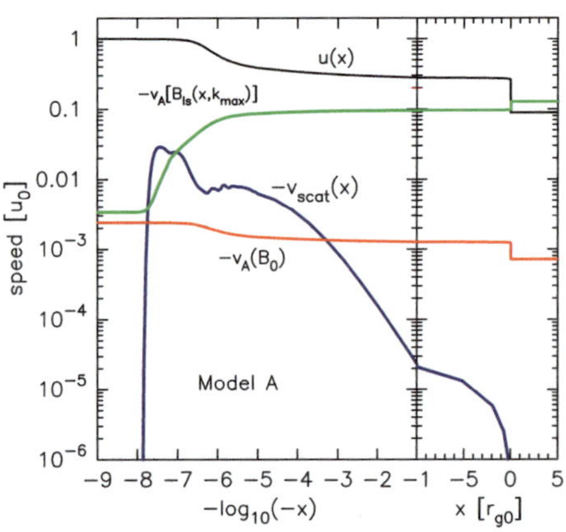

Fig. 12 The scattering center velocity profile (*solid line*) derived from the nonlinear Monte Carlo model for shock velocity 5000 km s^{-1}, the far upstream density $n_0 = 0.3\ \mathrm{cm}^{-3}$, and magnetic field $B_0 = 3\ \mu\mathrm{G}$. We also show the flow velocity profile (*dashed line*) and Alfvén velocities calculated with the amplified magnetic field $B_{ls}(x)$ (*dotted line*) and the initial field B_0 (*dot-dashed line*) (from Bykov et al. 2014)

and be accelerated above thermal energies. An important aspect of this is that the turbulence may move through the bulk plasma so the effective compression ratio for Fermi acceleration, $R_{\mathrm{eff}}(x) = v_{\mathrm{scat}}(x)/u_2$, may differ from the compression ratio of the bulk plasma, $R = u(x)/u_2$, where u_2 is the downstream flow speed in the shock rest frame.

Figure 12 shows an example from a non-relativistic shock calculation where v_{scat} has been determined self-consistently with the Monte Carlo simulation. While most semi-analytic calculations assume v_{scat} is the Alfvén speed, $v_a(x)$, calculated either with the ambient magnetic field or the amplified field, the self-consistent results obtained in Bykov et al.

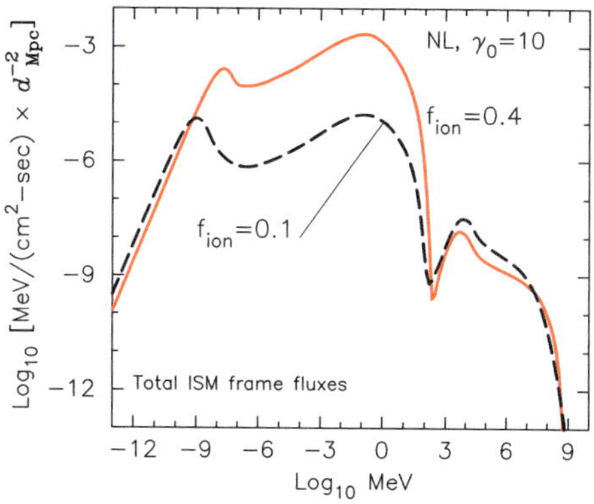

Fig. 13 Total observed photon energy flux for non-linear Monte Carlo models of relativistic ion-electron shocks of $\gamma_{sh} = 10$ with $f_{ion} = 0.1$ and $f_{ion} = 0.4$ simulated by Warren et al. (2015a)

(2014) show that it can differ substantially from $v_a(x)$ regardless of the magnetic field assumed.

The velocity of the scattering centers, i.e. the Weibel frame, has been determined analytically in the relativistic regime in Sect. 4.2. This motion is of course implicitly included self-consistently in PIC simulations, and it will be incorporated in future Monte Carlo simulations of relativistic shocks.

7.7 Radiation from Monte Carlo Shock Modeling

Within the assumptions of the TPI model, the Monte Carlo technique calculates the thermal and nonthermal plasma consistently. Therefore, the distribution functions, such as those shown in Fig. 7, contain the background plasma as well as the accelerated population. This means the ion and electron spectra are absolutely normalized for a given set of environmental and model parameters and the synchrotron, bremsstrahlung, and pion-decay emission is also absolutely normalized without requiring additional parameters. The current Monte Carlo code does not include synchrotron-self-Comptonization so an external photon field is required for inverse Compton. Additional parameters will also be required for the other radiation mechanisms if radiation is calculated from shock accelerated particles interacting with an external medium.

In Fig. 8, distributions from Warren et al. (2015a) are shown where the fraction of energy transferred from ions to electrons, f_{ion}, has been set to two different values as indicated. In both cases, the NL shock structure was calculated including electrons, protons, and He^{2+}. Fig. 13, also from Warren et al. (2015a), shows the broadband continuum radiation produced in these shocks. The transfer of ion energy to electrons radically influences the synchrotron emission. The emission at $\sim$1 MeV with $f_{ion} = 0.4$ is $\sim$100 times that for $f_{ion} = 0.1$. On the other hand, the pion-decay emission in the GeV-TeV region drops for $f_{ion} = 0.4$ since the ions have less energy. See Warren et al. (2015a) for a full discussion of these effects.

8 Summary

In this article, we have provided an analytical description of the main components of a relativistic collisionless shock, building on earlier work and borrowing from the results of particle-in-cell simulations. We have completed this discussion with a review of Monte Carlo techniques, which offer a complementary method to approach the mesoscale physics of these shock waves. As an executive summary, we review here some of the most salient points of the physics of relativistic collisionless shock waves:

(i) In unmagnetized shocks where the background magnetic field is dominated by self-generated turbulence, the shock forms thanks to a partial reflection of the incoming particles on a front produced by the enhancement of the electromagnetic energy density.
(ii) The electromagnetic precursor is generated by a streaming micro-instability. The Weibel instability emerges as the leading micro-instability in the unmagnetized regime. At weak but finite magnetization, a current driven instability is expected to play an important role as well.
(iii) At magnetization levels not far below unity (and higher), the shock is formed by the compression of the background magnetic field.
(iv) As we have argued, the Weibel turbulence which is excited in the precursor propagates forwards at a sub-relativistic speed (typically ξ_{cr} with respect to the background plasma, ξ_{cr} being the ratio of the cosmic ray pressure over the incoming energy density in the front frame).
(v) Electron heating in the precursor occurs through two processes: through the motion of the scattering centers with respect to the background, and through the induction electric field.
(vi) The motion of the Weibel frame with respect to the background, in addition to its heating effect, produces a slowing of the incoming plasma that might reduce its speed to a mildly relativistic one, thereby allowing a saturation of the Weibel instability by the dispersion effect of the beam.
(vii) Weibel turbulence gives rise to filaments in the background plasma, whereas the beam of superthermal particles does not suffer filamentation due to its high inertia.
(viii) The particle scattering rate decreases with the energy of the considered particle in proportion to γ^{-2}. This result first established with a random phase approximation has been confirmed by a Hamiltonian chaos investigation, and has also been confirmed by PIC simulations.
(ix) Because of this fast decrease of the scattering frequency, even a tiny mean field can stop the Fermi acceleration process; acceleration can take place in shock waves for which the magnetization $\sigma \lesssim \xi_B^2$, i.e., $\sigma \lesssim 10^{-4}$. In particular, if the Weibel instability dominates turbulence production on all scales into the non-linear regime, protons cannot be accelerated beyond 10^{16} eV in the termination shocks of gamma-ray bursts. Electrons can be efficiently accelerated up to 10 TeV or so in similar conditions and give rise to GeV synchrotron emission at early times.
(x) If relativistic shocks are as efficient in producing superthermal particles as suggested by PIC simulations, and often assumed in astrophysical models, non-linear effects from efficient CR production must be considered.
(xi) With efficient CR production, the shock structure, turbulence production, Fermi acceleration, and particle injection from the thermal background are coupled. By using phenomenological particle scattering, Monte Carlo methods can address this multi-scale coupling over large dynamic ranges in an internally consistent fashion.
(xii) The analysis of relativistic shocks is substantially more difficult than non-relativistic ones because particle distributions are not nearly isotropic in all frames and the diffusion approximation cannot be used directly. Nevertheless, the underlying physics of collisionless shock formation, particle injection, wave generation, and Fermi acceleration must be continuous from non-relativistic speeds through the trans-relativistic

regime. Monte Carlo techniques can address important non-linear aspects of this transition.

(xiii) A major direction for future work is to include magnetic field amplification for relativistic and trans-relativistic shocks. Additional work can be directed toward non-linear oblique shocks and using the Monte Carlo code in astrophysical applications.

Acknowledgements G. Pelletier and M. Lemoine were financially supported by the Programme National Hautes Énergies (PNHE) of the C.N.R.S. and by the ANR-14-CE33-0019 MACH project. A.M. Bykov was supported by RSF grant 16-12-10225.

Appendix A: Appendices of the First Part of the Paper

A.1 Determination of the Weibel Frame in a Pair Plasma

We consider the transverse perturbed motion in the linear approximation in the background plasma and in the relativistic beam of returning particles, assuming here a pair shock. For a species indexed by α, moving at velocity β_α in a frame to be specified later, with proper enthalpy density h_α and number density n_α, the perturbed dynamical and continuity equations can be written as

$$c^{-1}h_\alpha\gamma_\alpha\partial_t\boldsymbol{\delta u}_{\alpha,\perp}+\boldsymbol{\nabla}_\perp\delta p_\alpha=n_\alpha\gamma_\alpha q_\alpha\boldsymbol{\nabla}_\perp(\delta\Phi-\beta_\alpha\delta A_x), \tag{55}$$

$$c^{-1}\partial_t\delta(\gamma_\alpha n_\alpha)+\gamma_\alpha n_\alpha\boldsymbol{\nabla}_\perp\cdot\boldsymbol{\delta\beta}_{\alpha,\perp}=0. \tag{56}$$

It is understood that δp_α, $\boldsymbol{\delta u}_\alpha$, δn_α are first order perturbations while h_α, n_α etc. are zeroth-order uniform and constant in time variables. The above implicitly assumes that the electromagnetic perturbations are characterized by δA^0 and δA^x only, and that $\partial_x\cdot=0$ for all quantities.

For the sake of simplicity, we assume cold species and neglect the pressure perturbation. The more general case of warm plasmas is discussed further below. Then for each fluid (one fluid being composed of $+$ and $-$ species), meaning the superthermal fluid and the background plasma fluid, we derive from the above two equations an uncoupled equation for the evolution of its *apparent* charge perturbation:

$$\partial_t^2\delta\rho_\alpha=\frac{\omega_{p\alpha}^2}{4\pi}\Delta_\perp(\delta\Phi-\beta_\alpha\delta A_x), \tag{57}$$

where apparent charge density means $\rho_\alpha\equiv\gamma_\alpha e(n_{\alpha+}-n_{\alpha-})$; $\omega_{\mathrm{p},\alpha}$ represents the plasma frequency defined in the rest frame of the considered fluid, such that

$$\omega_{\mathrm{p},\alpha}^2\equiv\frac{4\pi n_\alpha e^2c^2}{h_\alpha/n_\alpha} \tag{58}$$

with the index α running over the superthermal (returning) particles and over the background plasma. For a plasma temperature $T\ll mc^2$ (cold non-relativistic plasma), $h_\alpha\simeq n_a mc^2$ and $\omega_{p\alpha}^2=4\pi n_\alpha e^2/m$. This is the case of the background plasma, for which n_α designs the density n_u of the ambient plasma. For a plasma of relativistic temperature $T>mc^2$, $h_\alpha\simeq 4p_a\simeq 4/3\Delta\gamma_\alpha n_\alpha mc^2$ and $\omega_{p\alpha}^2=3\pi n_\alpha e^2/(\Delta\gamma_\alpha m)$. This is the case of the plasma of returning particles for which: $n_\alpha\simeq\xi_{\mathrm{cr}}\gamma_{\mathrm{sh}}n_\mathrm{u}$, while the thermal Lorentz factor is $\Delta\gamma_\alpha\sim\gamma_{\mathrm{sh}}$ (front frame); thus its plasma frequency is such that $\omega_{\mathrm{pb}}\simeq\xi_{\mathrm{cr}}\omega_\mathrm{p}$, ω_p the plasma frequency of the background plasma.

We define the Weibel frame as that in which the instability becomes purely magnetic, i.e. where the electrostatic potential and the total electric charge density vanish. The evolution of the total charge density is governed by

$$4\pi\,\partial_t^2\delta\rho_{\rm tot} = \Delta_\perp\left\{\sum_\alpha \omega_{\rm p\alpha}^2(\delta\Phi - \beta_\alpha\delta A_x)\right\}. \tag{59}$$

In the Lorentz gauge, $\delta\Phi$ can be expressed directly in terms of the total charge perturbation; switching over to Fourier space, one infers

$$\left(\frac{\omega^4}{k^2} - \omega^2 + \sum_\alpha \omega_{\rm p\alpha}^2\right)\delta\Phi = \sum_\alpha \beta_\alpha\omega_{\rm p\alpha}^2\delta A_x. \tag{60}$$

Thus there is no source of electrostatic excitation when $\sum_\alpha \beta_\alpha\omega_{\rm p\alpha}^2 = 0$; this defines the Weibel frame, that always exists. For the case we are treating, the relation defining the Weibel frame is simply:

$$\xi_{\rm cr}\beta_{\rm b|w} + \beta_{\rm bg|w} = 0 \tag{61}$$

with $\beta_{\rm b|w}$ (resp. $\beta_{\rm bg|w}$) the velocity of the beam (resp. the background plasma) relative to this Weibel frame. The beam moves relativistically with respect to the background plasma, therefore either $\beta_{\rm b|w} \sim 1$ or $\beta_{\rm bg|w} \sim -1$; we will argue further on that the plasma filamentation is carried by the background, which motivates the former choice. Thus the Weibel frame moves forward (in the same direction as the shock front) relative to the background plasma, but at a low velocity

$$\beta_{\rm w|bg} = \frac{\xi_{\rm cr}}{1+\xi_{\rm cr}}. \tag{62}$$

One can also directly calculate the speed of the Weibel frame with respect to the front, written $\beta_{\rm w}$ (with other velocities are also written in the front frame):

$$\frac{\beta_{\rm w} - \beta_{\rm bg}}{1 - \beta_{\rm w}\beta_{\rm bg}} = \xi_{\rm cr}\frac{\beta_{\rm b} - \beta_{\rm w}}{1 - \beta_{\rm b}\beta_{\rm w}}, \tag{63}$$

which leads to

$$\beta_{\rm w} \simeq -\beta_{\rm sh} + \frac{\xi_{\rm cr}}{\gamma_{\rm sh}^2}. \tag{64}$$

It can be easily checked that the latter result is also obtained by a Lorentz transform of the result in the background frame.

A.2 Determination of the Weibel Frame in a Ionic Plasma with Hot Electrons

We assume now a plasma made of protons and hot electrons. When electrons are relativistically hot, they are in a quasi-Boltzmann equilibrium and also they are responsible for strong Landau effects in the electromagnetic response functions. The transverse and longitudinal response functions of background electrons are defined through

$$\delta j_{x,e} = \frac{\omega^2}{4\pi c}\chi_{{\rm T},e}\delta A_{x|{\rm u}}, \tag{65}$$

$$\delta\rho_e = -\frac{k^2}{4\pi}\chi_{{\rm L},e}\delta\Phi_{|{\rm u}} \tag{66}$$

with

$$\chi_{\mathrm{T},e} \simeq -\frac{\omega_{\mathrm{pi}}^2}{\xi_{\mathrm{th}} k^2 c^2}\left(1 - i\frac{\pi}{4}\frac{kc}{\omega}\right), \tag{67}$$

$$\chi_{\mathrm{L},e} \simeq \frac{\omega_{\mathrm{pi}}^2}{\xi_{\mathrm{th}} k^2 c^2}\left(1 + i\frac{\pi}{2}\frac{\omega}{kc}\right) \tag{68}$$

and let us recall the notation: $\xi_{\mathrm{th}} \equiv kT_e/(m_p c^2)$. Using a cold response for the background ions in their rest frame, i.e. $\delta j_{x,i} = -\omega_{\mathrm{pi}}^2/(4\pi)\delta A_{x|\mathrm{u}}$, inserting these responses in the equation for $\delta A_{x|\mathrm{u}}$, one obtains the dispersion relation:

$$\left(\omega^2 - k^2\beta_*^2 c^2\right)\left(1 + \frac{k^2c^2}{\omega_{\mathrm{pi}}^2} - i\frac{\pi}{4\xi_{th}}\frac{\omega}{kc}\right) = -\xi_{\mathrm{cr}}\beta_{\mathrm{b|u}}^2 k^2 c^2. \tag{69}$$

This dispersion relation neglects terms of order ω^2/k^2 in the second term on the l.h.s., and it neglects a contribution of the electrostatic potential on the r.h.s, which is of order $\beta_{\mathrm{w|u}} \sim \xi_{\mathrm{cr}}$ relative to the above term. The effective sound velocity $\beta_* c$ which enters the equation of motion for the charge perturbation of the beam is of order $1/\gamma_{\mathrm{b}}$, i.e. $\beta_*^2 c^2 = c_{\mathrm{s,b}}^2/\gamma_{\mathrm{sh}}^2$ with $c_{\mathrm{s,b}} = 1/\sqrt{3}$.

Let us write the complex frequency of the Weibel modes as follows: $\omega = iz\xi_{\mathrm{th}}kc$, where $z \in \mathbb{C}$. The modification of the transverse response by the electronic Landau effect is the main modification of the dispersion relation for Weibel modes, at least off the saturation regime, that reads (omitting terms of order $1/\gamma_{\mathrm{sh}}^2\xi_{\mathrm{cr}}^2$):

$$\xi_{\mathrm{th}}^2\frac{\pi}{4}z^3 + a\xi_{\mathrm{th}}^2 z^2 + \frac{\pi}{4}\beta_*^2 z + a\beta_*^2 - \xi_{\mathrm{cr}} = 0, \tag{70}$$

where $a \equiv 1 + k^2c^2/\omega_{\mathrm{pi}}^2$. The physical situation of interest corresponds to $\xi_{\mathrm{cr}} \gg \xi_{\mathrm{th}}$ where

$$z^3 \simeq \frac{4}{\pi}\frac{\xi_{\mathrm{cr}}}{\xi_{\mathrm{th}}^2}. \tag{71}$$

We thus find an instability with a growth rate:

$$g_{\mathrm{inst|u}} = \left(\frac{4\xi_{\mathrm{cr}}}{\pi\xi_{\mathrm{th}}^2}\right)^{1/3}\xi_{\mathrm{th}}kc \simeq (\xi_{\mathrm{th}}\xi_{\mathrm{cr}})^{1/3}kc. \tag{72}$$

This growth rate is not significantly different from the cold one for $kc = \omega_{\mathrm{pi}}$ and $\xi_{\mathrm{th}} \sim \xi_{\mathrm{cr}}$. The threshold due to beam dispersion is found by keeping all the terms except the third degree term and assuming $\xi_{\mathrm{cr}} - a\beta_*^2$ positive but arbitrary small. Then we find the unstable root:

$$z \simeq \frac{4}{\pi\beta_*^2}\left(\xi_{\mathrm{cr}} - a\beta_*^2\right), \tag{73}$$

which leads to the saturation condition: $\gamma_{\mathrm{bg}} = \xi_{\mathrm{cr}}^{-1/2}$ in the front frame.

A.3 The Scattering Law Derived from the Hamiltonian Chaos

We thus consider returning cosmic rays in the Weibel frame with a very high Lorentz factor γ, such that their motion in the x-direction is almost constant ($v_x = c$ with a very good accuracy); then, only their non-relativistic transverse motion matters with $v_\perp \sim c/\gamma_{\mathrm{sh}}$. We assume that the magnetic modes carry an x-wave number k_0 which is small compared to the transverse ones $k_0 \ll k_\perp \sim \delta_i^{-1}$, as it appears to be in numerical simulations. Thus the cosmic rays can suffer resonances with Weibel modes: $k_\perp v_\perp \pm k_0 c = 0$.

The most simple description consists in considering the transverse dynamics only, the contribution of the x-coordinate amounting to a time dependent driving with $k_0 x(t) = k_0 ct$, with a frequency $\omega_0 \equiv k_0 c \ll \omega_{pi}$. In a 2D description, with y denoting the transverse coordinate, this simple description involves a Lagrangian of the form:

$$\begin{aligned}\mathcal{L} &= \frac{\dot{y}^2}{2} + \varepsilon \cos(\omega_0 t) \sum_n a_n \cos(k_n y) \\ &= \frac{\dot{y}^2}{2} + \frac{\varepsilon}{2} \sum_n a_n \big[\cos(k_n y - \omega_0 t) + \cos(k_n y + \omega_0 t)\big],\end{aligned} \tag{74}$$

where the small amplitude parameter is defined as

$$\varepsilon \equiv \frac{q \delta A_{x|\mathrm{w}}}{\gamma m c^2}. \tag{75}$$

The corresponding Hamiltonian describes a Chirikov standard map, with the small ϵ parameter characterizing the perturbation which will provide the transition to chaotic dynamics.

Let us define a local coordinate of phase space $\theta_n \equiv k_n y(t) \mp \omega_0 t$. A resonance is defined by $\dot{\theta}_n = 0$. Around such a resonance, the Lagrangian can be approximated as the time average:

$$\bar{\mathcal{L}}_n = \frac{1}{2k_n^2} (\dot{\theta}_n \pm \omega_0)^2 + \frac{\varepsilon}{2} a_n \cos\theta_n. \tag{76}$$

The conjugate momentum of θ_n is

$$J_n \equiv \frac{\partial \bar{\mathcal{L}}_n}{\partial \dot{\theta}_n} = \frac{\dot{\theta}_n \pm \omega_0}{k_n^2}. \tag{77}$$

and the corresponding Hamiltonian is:

$$\bar{H}_n = \frac{k_n^2}{2} J_n^2 \mp \omega_0 J_n - \frac{\varepsilon}{2} a_n \cos\theta_n. \tag{78}$$

This time independent Hamiltonian describes the energy invariant around each resonance. The n-resonance occurs for $J_n = J_n^{(r)} = \pm\omega_0/k_n^2$ (which is nothing but $k_n v_n = \pm\omega_0$). Assuming for the time being that the resonances are sufficiently isolated, then setting $J_n = J_n^{(r)} + \delta J_n$ around such a resonance, one finds a pendulum like Hamiltonian, with oscillations described by

$$\delta \dot{J}_n = \frac{\ddot{\theta}_n}{k_n^2} = -\frac{\varepsilon}{2} a_n \sin\theta_n. \tag{79}$$

The oscillations in θ_n have a pulsation Ω_n in the bottom of the well such that:

$$\Omega_n^2 \equiv \frac{\varepsilon}{2} k_n^2 a_n. \tag{80}$$

The separatrix being defined as the region which separates oscillations from rotations, one derives the maximum amplitude of the separatrix around the n-resonance in the J_n direction of phase space as $\Delta J_n = 2\sqrt{2}\Omega_n/k_n^2$. In contrast, the distance between two neighboring resonances is $\omega_0(k_n^{-2} - k_{n-1}^{-2}) \simeq 2\omega_0 \delta k_n / k_n^3$. The simple heuristic threshold for chaos, proposed by Chirikov (1979), is that the separatrix amplitude be larger than the spacing between resonances, giving: $\Omega_n > \omega_0 \delta k_n / k_n^2$, which is always easily satisfied in our case, the transverse spectrum of Weibel instability being a continuum.

Therefore, above a very low threshold, the resonances overlap and the transverse motions of cosmic rays randomly jump from one resonance to another one. Using Eq. (79), one infers the evolution of J_n after i jumps:

$$\langle \Delta J_n^2 \rangle \sim \frac{\varepsilon^2}{8} a_n^2 \Delta t^2 i, \tag{81}$$

with $\Delta t \simeq \omega_0^{-1}$ the time taken to change resonance, which plays the role of coherence time. The above assumes a random θ_n in resonance n at step i. Now, the variation in J_n is related to the variation of the pitch angle α of the particle, since $\Delta J_n = \Delta v_\perp / k_n \simeq \Delta\alpha / k_n$. As $t = i\,\Delta t \to +\infty$, one thus infers a scattering frequency

$$\nu_s = \frac{\langle \Delta\alpha^2 \rangle}{2t} \propto \varepsilon^2 k_n^2 \propto \xi_B \omega_{\rm pi} \frac{\gamma_{\rm sh}}{\gamma^2}. \tag{82}$$

References

A. Achterberg, J. Wiersma, The Weibel instability in relativistic plasmas, I: linear theory. Astron. Astrophys. **475**, 1–18 (2007). doi:10.1051/0004-6361:20065365

A. Achterberg, Y.A. Gallant, J.G. Kirk, A.W. Guthmann, Particle acceleration by ultrarelativistic shocks: theory and simulations. Mon. Not. R. Astron. Soc. **328**, 393–408 (2001). doi:10.1046/j.1365-8711.2001.04851.x

A. Achterberg, J. Wiersma, C.A. Norman, The Weibel instability in relativistic plasmas, II: nonlinear theory and stabilization mechanism. Astron. Astrophys. **475**, 19–36 (2007). doi:10.1051/0004-6361:20065366

D. Alsop, J. Arons, Relativistic magnetosonic solitons with reflected particles in electron-positron plasmas. Phys. Fluids **31**, 839–847 (1988). doi:10.1063/1.866765

E. Amato, P. Blasi, A kinetic approach to cosmic-ray-induced streaming instability at supernova shocks. Mon. Not. R. Astron. Soc. **392**, 1591–1600 (2009). doi:10.1111/j.1365-2966.2008.14200.x

M.G. Baring, D.C. Ellison, F.C. Jones, The injection and acceleration of particles in oblique shocks—a unified Monte Carlo description. Astrophys. J. **409**, 327–332 (1993). doi:10.1086/172666

M.C. Begelman, J.G. Kirk, Shock-drift particle acceleration in superluminal shocks—a model for hot spots in extragalactic radio sources. Astrophys. J. **353**, 66–80 (1990)

A.R. Bell, The acceleration of cosmic rays in shock fronts, I. Mon. Not. R. Astron. Soc. **182**, 147–156 (1978)

A.R. Bell, Turbulent amplification of magnetic field and diffusive shock acceleration of cosmic rays. Mon. Not. R. Astron. Soc. **353**, 550–558 (2004)

E.G. Berezhko, D.C. Ellison, A simple model of nonlinear diffusive shock acceleration. Astrophys. J. **526**, 385 (1999)

R. Blandford, D. Eichler, Particle acceleration at astrophysical shocks: a theory of cosmic ray origin. Phys. Rep. **154**, 1–75 (1987)

R.D. Blandford, C.F. McKee, Fluid dynamics of relativistic blast waves. Phys. Fluids **19**, 1130–1138 (1976)

R.D. Blandford, J.P. Ostriker, Particle acceleration by astrophysical shocks. Astrophys. J. **221**, 29–32 (1978)

A. Bret, Weibel, two-stream, filamentation, oblique, Bell, Buneman... Which one grows faster? Astrophys. J. **699**, 990–1003 (2009). doi:10.1088/0004-637X/699/2/990

A.M. Bykov, R.A. Treumann, Fundamentals of collisionless shocks for astrophysical application, 2: relativistic shocks. Astron. Astrophys. Rev. **19**, 42 (2011). doi:10.1007/s00159-011-0042-8

A.M. Bykov, D.C. Ellison, M. Renaud, Magnetic fields in cosmic particle acceleration sources. Space Sci. Rev. **166**, 71–95 (2012a). doi:10.1007/s11214-011-9761-4

A. Bykov, N. Gehrels, H. Krawczynski, M. Lemoine, G. Pelletier, M. Pohl, Particle acceleration in relativistic outflows. Space Sci. Rev. **173**, 309–339 (2012b)

A.M. Bykov, D.C. Ellison, S.M. Osipov, A.E. Vladimirov, Magnetic field amplification in nonlinear diffusive shock acceleration including resonant and non-resonant cosmic-ray driven instabilities. Astrophys. J. **789**, 137 (2014). doi:10.1088/0004-637X/789/2/137

A.M. Bykov, D.C. Ellison, S.M. Osipov, A nonlinear Monte Carlo model of super-diffusive shock acceleration with magnetic field amplification. Phys. Rev. E **95**, 033207 (2017). doi:10.1103/PhysRevE.95.033207

D. Caprioli, E. Amato, P. Blasi, Non-linear diffusive shock acceleration with free-escape boundary. Astropart. Phys. **33**, 307–311 (2010)

D. Caprioli, A.R. Pop, A. Spitkovsky, Simulations and theory of ion injection at non-relativistic collisionless shocks. Astrophys. J. Lett. **798**, 28 (2015). doi:10.1088/2041-8205/798/2/L28

P. Chang, A. Spitkovsky, J. Arons, Long-term evolution of magnetic turbulence in relativistic collisionless shocks: electron-positron plasmas. Astrophys. J. **674**, 378–387 (2008). doi:10.1086/524764

B.V. Chirikov, A universal instability of many-dimensional oscillator systems. Phys. Rep. **52**, 263–379 (1979). doi:10.1016/0370-1573(79)90023-1

F. de Hoffmann, E. Teller, Magneto-hydrodynamic shocks. Phys. Rev. **80**, 692–703 (1950). doi:10.1103/PhysRev.80.692

G.P. Double, M.G. Baring, F.C. Jones, D.C. Ellison, Magnetohydrodynamic jump conditions for oblique relativistic shocks with gyrotropic pressure. Astrophys. J. **600**, 485–500 (2004)

N. Dresing, S. Theesen, A. Klassen, B. Heber, Efficiency of particle acceleration at interplanetary shocks: statistical study of STEREO observations. Astron. Astrophys. **588**, 17 (2016). doi:10.1051/0004-6361/201527853

L.O. Drury, An introduction to the theory of diffusive shock acceleration of energetic particles in tenuous plasmas. Rep. Prog. Phys. **46**, 973–1027 (1983)

L.O. Drury, Escaping the accelerator: how, when and in what numbers do cosmic rays get out of supernova remnants? Mon. Not. R. Astron. Soc. **415**, 1807–1814 (2011). doi:10.1111/j.1365-2966.2011.18824.x

D.C. Ellison, G.P. Double, Diffusive shock acceleration in unmodified relativistic, oblique shocks. Astropart. Phys. **22**, 323–338 (2004)

D.C. Ellison, D. Eichler, Monte Carlo shock-like solutions to the Boltzmann equation with collective scattering. Astrophys. J. **286**, 691–701 (1984)

D.C. Ellison, F.C. Jones, S.P. Reynolds, First-order Fermi particle acceleration by relativistic shocks. Astrophys. J. **360**, 702–714 (1990a). doi:10.1086/169156

D.C. Ellison, E. Moebius, G. Paschmann, Particle injection and acceleration at earth's bow shock—comparison of upstream and downstream events. Astrophys. J. **352**, 376–394 (1990b)

D.C. Ellison, M.G. Baring, F.C. Jones, Acceleration rates and injection efficiencies in oblique shocks. Astrophys. J. **453**, 873 (1995)

D.C. Ellison, M.G. Baring, F.C. Jones, Nonlinear particle acceleration in oblique shocks. Astrophys. J. **473**, 1029 (1996)

D.C. Ellison, D.J. Patnaude, P. Slane, P. Blasi, S. Gabici, Particle acceleration in supernova remnants and the production of thermal and nonthermal radiation. Astrophys. J. **661**, 879–891 (2007)

D.C. Ellison, P. Slane, D.J. Patnaude, A.M. Bykov, Core-collapse model of broadband emission from SNR RX J1713.7-3946 with thermal X-rays and gamma rays from escaping cosmic rays. Astrophys. J. **744**, 39 (2012)

D.C. Ellison, D.C. Warren, A.M. Bykov, Monte Carlo simulations of nonlinear particle acceleration in parallel trans-relativistic shocks. Astrophys. J. **776**, 46 (2013). doi:10.1088/0004-637X/776/1/46

D.C. Ellison, D.C. Warren, A.M. Bykov, Particle spectra and efficiency in nonlinear relativistic shock acceleration—survey of scattering models. Mon. Not. R. Astron. Soc. **456**, 3090–3099 (2016)

Y.A. Gallant, A. Achterberg, Ultra-high-energy cosmic ray acceleration by relativistic blast waves. Mon. Not. R. Astron. Soc. **305**, 6–10 (1999). doi:10.1046/j.1365-8711.1999.02566.x

Y.A. Gallant, J. Arons, Structure of relativistic shocks in pulsar winds: a model of the wisps in the Crab Nebula. Astrophys. J. **435**, 230–260 (1994). doi:10.1086/174810

M. Gedalin, E. Smolik, A. Spitkovsky, M. Balikhin, Electron heating by filamentary instability. Europhys. Lett. **97**, 35002 (2012). doi:10.1209/0295-5075/97/35002

A. Gruzinov, E. Waxman, Gamma-ray burst afterglow: polarization and analytic light curves. Astrophys. J. **511**, 852–861 (1999). doi:10.1086/306720

M. Hoshino, J. Arons, Y.A. Gallant, A.B. Langdon, Relativistic magnetosonic shock waves in synchrotron sources—shock structure and nonthermal acceleration of positrons. Astrophys. J. **390**, 454–479 (1992)

F.C. Jones, J.R. Jokipii, M.G. Baring, Charged-particle motion in electromagnetic fields having at least one ignorable spatial coordinate. Astrophys. J. **509**, 238–243 (1998). doi:10.1086/306480

C.F. Kennel, F.V. Coroniti, Magnetohydrodynamic model of Crab nebula radiation. Astrophys. J. **283**, 710–730 (1984). doi:10.1086/162357

U. Keshet, E. Waxman, Energy spectrum of particles accelerated in relativistic collisionless shocks. Phys. Rev. Lett. **94**(11), 111102 (2005)

U. Keshet, B. Katz, A. Spitkovsky, E. Waxman, Magnetic field evolution in relativistic unmagnetized collisionless shocks. Astrophys. J. Lett. **693**, 127–130 (2009). doi:10.1088/0004-637X/693/2/L127

J.G. Kirk, P. Duffy, Topical review: particle acceleration and relativistic shocks. J. Phys. G, Nucl. Part. Phys. **25**, 163–194 (1999). doi:10.1088/0954-3899/25/8/201

J.G. Kirk, B. Reville, Radiative signatures of relativistic shocks. Astrophys. J. Lett. **710**, 16–20 (2010). doi:10.1088/2041-8205/710/1/L16

J.G. Kirk, A.W. Guthmann, Y.A. Gallant, A. Achterberg, Particle acceleration at ultrarelativistic shocks: an eigenfunction method. Astrophys. J. **542**, 235–242 (2000). doi:10.1086/309533
S.H. Lee, D.C. Ellison, S. Nagataki, A generalized model of nonlinear diffusive shock acceleration coupled to an evolving supernova remnant. Astrophys. J. **750**, 156 (2012)
M. Lemoine, Synchrotron signature of a relativistic blast wave with decaying microturbulence. Mon. Not. R. Astron. Soc. **428**, 845–866 (2013). doi:10.1093/mnras/sts081
M. Lemoine, Nonlinear collisionless damping of Weibel turbulence in relativistic blast waves. J. Plasma Phys. **81**(1), 455810101 (2015a). doi:10.1017/S0022377814000920
M. Lemoine, The synchrotron self-Compton spectrum of relativistic blast waves at large Y. Mon. Not. R. Astron. Soc. **453**, 3772–3784 (2015b). doi:10.1093/mnras/stv1800
M. Lemoine, G. Pelletier, Particle transport in tangled magnetic fields and Fermi acceleration at relativistic shocks. Astrophys. J. Lett. **589**, 73–76 (2003)
M. Lemoine, G. Pelletier, On electromagnetic instabilities at ultra-relativistic shock waves. Mon. Not. R. Astron. Soc. **402**, 321–334 (2010)
M. Lemoine, G. Pelletier, Dispersion and thermal effects on electromagnetic instabilities in the precursor of relativistic shocks. Mon. Not. R. Astron. Soc. **417**, 1148–1161 (2011). doi:10.1111/j.1365-2966.2011.19331.x
M. Lemoine, G. Pelletier, B. Revenu, On the efficiency of Fermi acceleration at relativistic shocks. Astrophys. J. Lett. **645**, 129–132 (2006). doi:10.1086/506322
M. Lemoine, G. Pelletier, L. Gremillet, I. Plotnikov, A fast current-driven instability in relativistic collisionless shocks. Europhys. Lett. **106**, 55001 (2014a). doi:10.1209/0295-5075/106/55001
M. Lemoine, G. Pelletier, L. Gremillet, I. Plotnikov, Current-driven filamentation upstream of magnetized relativistic collisionless shocks. Mon. Not. R. Astron. Soc. **440**, 1365–1378 (2014b). doi:10.1093/mnras/stu213
M. Lemoine, G. Pelletier, L. Gremillet, I. Plotnikov, Current-driven filamentation upstream of magnetized relativistic collisionless shocks. Mon. Not. R. Astron. Soc. **440**, 1365–1378 (2014c). doi:10.1093/mnras/stu213
Y. Lyubarsky, D. Eichler, Are gamma-ray burst shocks mediated by the Weibel instability? Astrophys. J. **647**, 1250–1254 (2006). doi:10.1086/505523
A. Marcowith, A. Bret, A. Bykov, M.E. Dieckman, L. O'C Drury, B. Lembège, M. Lemoine, G. Morlino, G. Murphy, G. Pelletier, I. Plotnikov, B. Reville, M. Riquelme, L. Sironi, A. Stockem Novo, The microphysics of collisionless shock waves. Rep. Prog. Phys. **79**(4), 046901 (2016). doi:10.1088/0034-4885/79/4/046901
M.V. Medvedev, A. Loeb, Generation of magnetic fields in the relativistic shock of gamma-ray burst sources. Astrophys. J. **526**, 697–706 (1999). doi:10.1086/308038
D.B. Melrose, *Instabilities in Space and Laboratory Plasmas* (1986), p. 288
M. Milosavljević, E. Nakar, The cosmic-ray precursor of relativistic collisionless shocks: a missing link in gamma-ray burst afterglows. Astrophys. J. **651**, 979–984 (2006)
G. Morlino, E. Amato, P. Blasi, Gamma-ray emission from SNR RX J1713.7-3946 and the origin of galactic cosmic rays. Mon. Not. R. Astron. Soc. **392**, 240–250 (2009)
J. Niemiec, M. Ostrowski, M. Pohl, Cosmic-ray acceleration at ultrarelativistic shock waves: effects of downstream short-wave turbulence. Astrophys. J. **650**, 1020–1027 (2006). doi:10.1086/506901
G. Pelletier, M. Lemoine, A. Marcowith, Turbulence and particle acceleration in collisionless supernovae remnant shocks, I: anisotropic spectra solutions. Astron. Astrophys. **453**, 181–191 (2006)
G. Pelletier, M. Lemoine, A. Marcowith, On Fermi acceleration and magnetohydrodynamic instabilities at ultra-relativistic magnetized shock waves. Mon. Not. R. Astron. Soc. **393**, 587–597 (2009)
I. Plotnikov, G. Pelletier, M. Lemoine, Particle transport in intense small-scale magnetic turbulence with a mean field. Astron. Astrophys. **532**, 68 (2011). doi:10.1051/0004-6361/201117182
I. Plotnikov, G. Pelletier, M. Lemoine, Particle transport and heating in the microturbulent precursor of relativistic shocks. Mon. Not. R. Astron. Soc. **430**, 1280–1293 (2013)
I. Rabinak, B. Katz, E. Waxman, Long-wavelength unstable modes in the far upstream of relativistic collisionless shocks. Astrophys. J. **736**, 157 (2011). doi:10.1088/0004-637X/736/2/157
E. Sagi, E. Nakar, On particle acceleration rate in gamma-ray burst afterglows. Astrophys. J. **749**, 80 (2012). doi:10.1088/0004-637X/749/1/80
K.M. Schure, A.R. Bell, L. O'C Drury, A.M. Bykov, Diffusive shock acceleration and magnetic field amplification. Space Sci. Rev. **173**, 491–519 (2012)
R. Shaisultanov, Y. Lyubarsky, D. Eichler, Stream instabilities in relativistically hot plasma. Astrophys. J. **744**, 182 (2012). doi:10.1088/0004-637X/744/2/182
L. Sironi, A. Spitkovsky, J. Arons, The maximum energy of accelerated particles in relativistic collisionless shocks. Astrophys. J. **771**, 54 (2013)

L. Sironi, U. Keshet, M. Lemoine, Relativistic shocks: particle acceleration and magnetization. Space Sci. Rev. **191**, 519–544 (2015). doi:10.1007/s11214-015-0181-8

A. Spitkovsky, On the structure of relativistic collisionless shocks in electron-ion plasmas. Astrophys. J. Lett. **673**, 39–42 (2008)

E.J. Summerlin, M.G. Baring, diffusive acceleration of particles at oblique, relativistic, magnetohydrodynamic shocks. Astrophys. J. **745**, 63 (2012)

M. Takamoto, J.G. Kirk, Rapid cosmic-ray acceleration at perpendicular shocks in supernova remnants. Astrophys. J. **809**, 29 (2015). doi:10.1088/0004-637X/809/1/29

K.J. Trattner, F. Allegrini, M.A. Dayeh, H.O. Funsten, S.A. Fuselier, D. Heirtzler, P. Janzen, H. Kucharek, D.J. McComas, E. Möbius, T.E. Moore, S.M. Petrinec, D.B. Reisenfeld, N.A. Schwadron, P. Wurz, The free escape continuum of diffuse ions upstream of the Earth's quasi-parallel bow shock. J. Geophys. Res. Space Phys. **118**, 4425–4434 (2013). doi:10.1002/jgra.50447

M. Vietri, On particle acceleration around shocks, I. Astrophys. J. **591**, 954–961 (2003). doi:10.1086/375534

A.E. Vladimirov, A.M. Bykov, D.C. Ellison, Turbulence dissipation and particle injection in nonlinear diffusive shock acceleration with magnetic field amplification. Astrophys. J. **688**, 1084–1101 (2008)

A.E. Vladimirov, A.M. Bykov, D.C. Ellison, Spectra of magnetic fluctuations and relativistic particles produced by a nonresonant wave instability in supernova remnant shocks. Astrophys. J. Lett. **703**, 29–32 (2009)

X.Y. Wang, R.Y. Liu, M. Lemoine, On the origin of $>$10 GeV photons in gamma-ray burst afterglows. Astrophys. J. Lett. **771**, 33 (2013). doi:10.1088/2041-8205/771/2/L33

D.C. Warren, D.C. Ellison, A.M. Bykov, S.H. Lee, Electron and ion acceleration in relativistic shocks with applications to GRB afterglows. Mon. Not. R. Astron. Soc. **452**, 431–443 (2015a). doi:10.1093/mnras/stv1304

D.C. Warren, D.C. Ellison, A.M. Bykov, S.H. Lee, Electron and ion acceleration in relativistic shocks with applications to GRB afterglows. Mon. Not. R. Astron. Soc. **452**, 431–443 (2015b). doi:10.1093/mnras/stv1304

D.C. Warren, D.C. Ellison, M.V. Barkov, S. Nagataki, Nonlinear particle acceleration and thermal particles in grb afterglows. Astrophys. J. **835**, 248 (2017). doi:10.3847/1538-4357/aa56c3

J. Wiersma, A. Achterberg, Magnetic field generation in relativistic shocks. An early end of the exponential Weibel instability in electron-proton plasmas. Astron. Astrophys. **428**, 365–371 (2004). doi:10.1051/0004-6361:20041882

G. Zimbardo, S. Perri, From Lévy walks to superdiffusive shock acceleration. Astrophys. J. **778**, 35 (2013). doi:10.1088/0004-637X/778/1/35

V.N. Zirakashvili, V.S. Ptuskin, Numerical simulations of diffusive shock acceleration in SNRs. Astropart. Phys. **39**, 12–21 (2012). doi:10.1016/j.astropartphys.2011.09.003

MIX
Papier aus verantwortungsvollen Quellen
Paper from responsible sources
FSC® C105338

If you have any concerns about our products,
you can contact us on
ProductSafety@springernature.com

In case Publisher is established outside the EU,
the EU authorized representative is:
Springer Nature Customer Service Center GmbH
Europaplatz 3, 69115 Heidelberg, Germany

Printed by Libri Plureos GmbH
in Hamburg, Germany